Vermutlich hat der Leser bereits bemerkt, daß der vorliegende Band von neuen Herausgebern redigiert ist. Ich habe in Anbetracht meines Alters den Springer-Verlag ersucht, mich nach dem Erscheinen des Bandes 27 von meinen redaktionellen Pflichten zu entbinden. Dies war nach 32jähriger Arbeit an diesem Werk kein leichter Entschluß. Ich werde den wissenschaftlichen Kontakt mit manchen Autoren vermissen, ebenso mit jenen Lesern, die mir im Zusammenhang mit diesem oder jenem der fast 200 Beiträge freundlichst geschrieben haben.

Was die Zukunft unserer „Fortschritte" betrifft, so bin ich überzeugt, daß sie in sehr guten und verläßlichen Händen liegt, und ich wünsche meinen Kollegen, den Herren W. HERZ, H. GRISEBACH und A. I. SCOTT, zu ihrer Arbeit guten Erfolg. Weiters vertraue ich darauf, daß die künftigen Autoren wie auch unser stets wachsender Leserkreis den neuen Herausgebern ebensolches Vertrauen schenken werden wie dem Unterzeichneten.

The readers of this Volume have certainly noticed the change in Editors. After the publication of Volume 27, I have asked the Springer-Verlag to relieve me from my duties, because of my advanced age, although this was not an easy decision after 32 years of editorial work. I shall sorely miss the scientific contact with a number of contributing authors and with those readers who have written to me concerning one or the other of the almost 200 review articles published in this Series.

I strongly feel that the future of the "Fortschritte" is in excellent hands and I wish much success to my colleagues, Drs. W. HERZ, H. GRISE-BACH and A. I. SCOTT. They will certainly enjoy, as did the undersigned, the support of the authors and of the ever-increasing circle of our readers.

Der vorliegende Band der „Fortschritte" ist der erste, der nicht die bisher übliche Zeile „herausgegeben von L. ZECHMEISTER" auf der Titelseite trägt. Professor ZECHMEISTERS persönlicher Einsatz und der hohe Standard, den er als Herausgeber setzte, haben diese Zeile zu einem Gütezeichen für alle Wissenschaftler werden lassen, die auf dem Gebiete der organischen Naturstoffe arbeiten. Wir möchten deshalb unserer Hoffnung Ausdruck geben, daß wir, seinem Beispiel folgend, das hohe Niveau der Serie aufrechterhalten können.

The present volume of "Fortschritte" is the first which does not carry the familiar line "Edited by L. ZECHMEISTER". Professor ZECHMEISTER's devotion to the success of this series and his exacting editorial standards have made this line a by-word for all scientists interested in the chemistry of natural products. We hope to emulate the example he has set and to maintain the high level of the Series in the future.

W. HERZ, Tallahassee
H. GRISEBACH, Freiburg
A. I. SCOTT, New Haven

FORTSCHRITTE DER CHEMIE ORGANISCHER NATURSTOFFE

PROGRESS IN THE CHEMISTRY OF ORGANIC NATURAL PRODUCTS

BEGRÜNDET VON · FOUNDED BY

L. ZECHMEISTER

HERAUSGEGEBEN VON · EDITED BY

W. HERZ   **H. GRISEBACH**   **A. I. SCOTT**
TALLAHASSEE, FLA.   FREIBURG i. BR.   NEW HAVEN, CONN.

ACHTUNDZWANZIGSTER BAND
TWENTY-EIGHTH VOLUME

VERFASSER · AUTHORS
W. BROMER · H. EGGE · K. EITER
R. EYJÓLFSSON · D. GROSS · H. HIKINO · Y. HIKINO · B. G. JACKSON
R. B. MORIN · J. E. PIKE · E. W. WARNHOFF
H. WIEGANDT · E. WONG

MIT 14 ABBILDUNGEN · WITH 14 FIGURES

1970

WIEN · SPRINGER-VERLAG · NEW YORK

ISBN-13: 978-3-7091-7125-7   e-ISBN-13: 978-3-7091-7123-3
DOI: 10.1007/978-3-7091-7123-3

Inhaltsverzeichnis

Contents

Oligosaccharide der Frauenmilch. Von H. WIEGANDT und H. EGGE, Institut für Physiologische Chemie, Universität Marburg, Marburg/Lahn, BRD 404

Glucagon: Chemistry and Action. By W. BROMER, The Lilly Research Laboratories, Eli Lilly and Company, Indianapolis, Indiana, USA ... 429

Errata

p. 271, line 2 read "ecdysterols exhibit negative and positive", for "ecdysterols exhibit positive and negative".

p. 292, line 5 read "since the methyl", for "since no methyl".

p. 297, Table 3 read "Osmundaceae" and "*Osmunda asiatica*", for "Osumundaceae" and "*Osumunda asiatica*".

Structural and Biogenetic Relationships
of Isoflavonoids

By E. WONG, Palmerston North, New Zealand

Contents   Page

I. Introduction

The isoflavonoids are natural phenolic compounds having in common the 1,2-diphenylpropane skeleton (1). Structurally and biogenetically they are closely related to the flavonoid compounds which are derivatives of 1,3-diphenylpropane (2).

(1)   (2)

Within the isoflavonoids are included many classes of natural products. Isoflavones, isoflavanones, rotenoids, pterocarpans and coumestans are well established members of this group but others such as isoflavans, 3-aryl-4-hydroxycoumarins, coumaronochromones, and hydroxy- and dehydro-variants of pterocarpans and rotenoids have only recently been established as natural products (143). The skeletal structures for the various classes of isoflavonoid compounds, arranged in order of their oxidation level, are given in *Chart 1*.

Isoflavones have been the subject of several previous surveys (191, 187, 36). The first review which treats the isoflavonoid compounds together as a group is that of OLLIS in 1962 (140). The book by DEAN (1963) (52) also covers very thoroughly the chemistry of many of the individual classes. The number of isoflavonoid compounds (aglycones) listed in these two comprehensive works (covering the period up to about 1960) is 51. Since that date the number of naturally-occurring isoflavonoid compounds known has more than doubled, the present figures being: isoflavones 58, other isoflavonoids 77. It can thus be seen that the isoflavonoids constitute a large and structurally varied group of natural products.

Isoflavonoids, in contrast to the flavonoids, are of very limited taxonomic distribution in the higher plants, being confined very largely to the Papilionatae sub-family of the Leguminosae. The systematic distribution and taxonomic significance of these compounds within this plant family has recently been discussed by HARBORNE (98). It has generally been considered that flavonoids and isoflavonoids are not synthesised by bacteria and fungi (*98*). The very recent report (*106*) of the production of isoflavones by the bacterium *Mycobacterium phlei* has been traced to small amounts of genisteinglucoside in the growth medium (*17 a*).

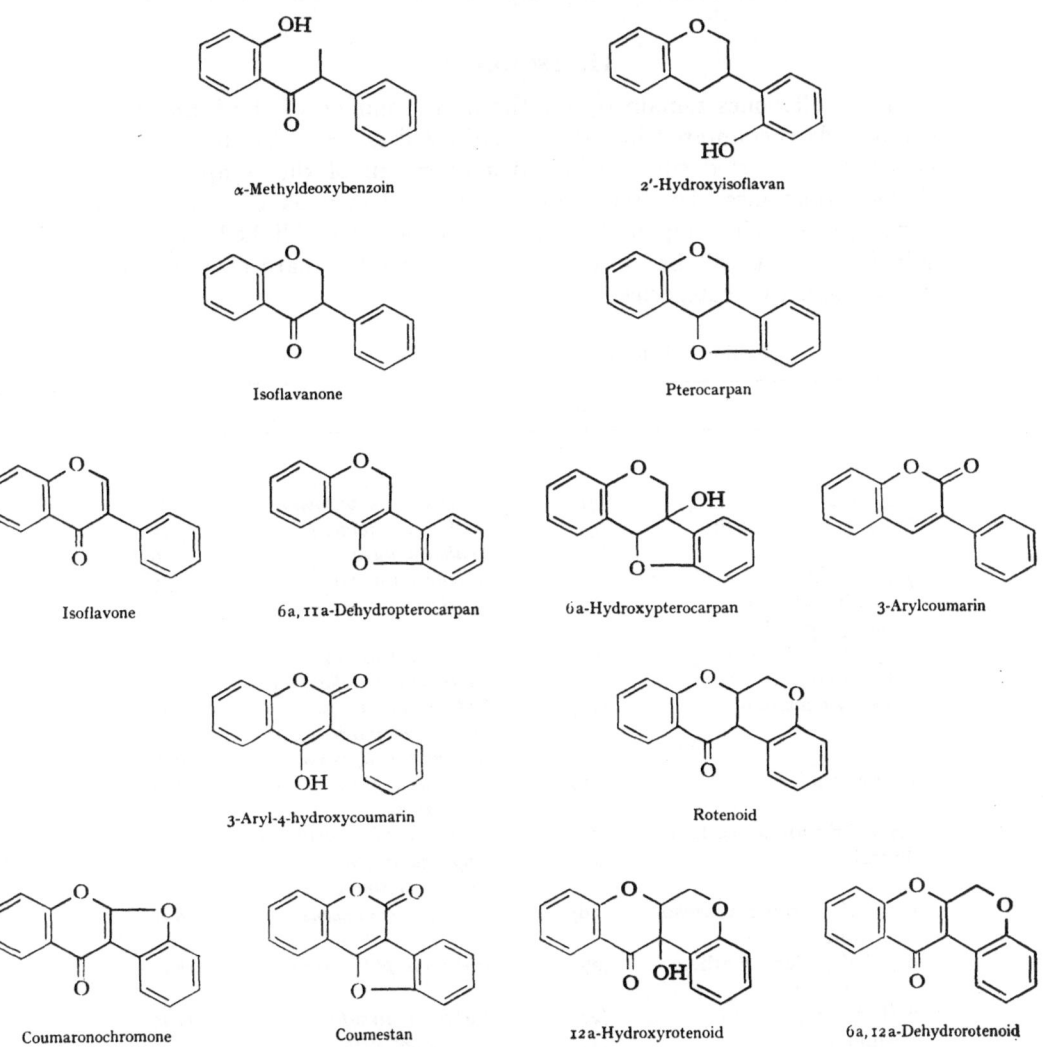

α-Methyldeoxybenzoin   2'-Hydroxyisoflavan

Isoflavanone   Pterocarpan

Isoflavone   6a, 11a-Dehydropterocarpan   6a-Hydroxypterocarpan   3-Arylcoumarin

3-Aryl-4-hydroxycoumarin   Rotenoid

Coumaronochromone   Coumestan   12a-Hydroxyrotenoid   6a, 12a-Dehydrorotenoid

Chart 1. Skeletal Structures for the Isoflavonoid Classes

In this review the structure and occurrence of all known naturally-occurring isoflavonoid compounds is surveyed. Recent advances in their phytochemistry will be emphasised and developments in their synthesis and stereochemistry since 1960 noted. Apart from providing a broad view of the structural variability within this group of natural products, this survey aims to provide a comparative structural basis for deducing the biogenetic relationships among the various classes of iso-flavonoids.

II. Isoflavones

The isoflavones remain by far the most common of the isoflavonoid compounds encountered in nature. *Table 1* lists the naturally-occurring isoflavones reported since 1960. Just over half of the compounds are simple isoflavones. The remainder may be classed as complex isofla-vones since each compound contains one or more C-linked isoprenoid substituents. It will be convenient to discuss the natural isoflavones under these two categories.

Table 1. *Natural Isoflavones Reported Since 1960*

Name	Formula	Plant Source	References
Simple Isoflavones			
Isoformononetin	(3)	*Machaerium villosum*	(139)
Di-O-methyldaidzein	(4)	*Dalbergia violacea*	(142)
Durlettone	(5)	*Millettia dura*	(146)
7,3',4'-Trihydroxyisoflavone	(6)	*Machaerium villosum*	(139)
7,4'-Dihydroxy-3'-methoxyisoflavone	(7)	*Machaerium villosum*	(139)
Calycosin (3'-Hydroxy-formonoetin)	(8)	*Baptisia lecontei*	(133)
		Pterocarpus dalbergioides	(152)
5-Methylgenistein	(17)	*Cytisus laburnum*	(38)
		Genista hispanica	(151)
		Laburnum alpinum	(65)
Pratensein	(20)	*Trifolium pratense, T. sub-terraneum*	(194)
6,7,4'-Trihydroxyisoflavone	(23)	*Soya hispida* (fermented)	(95)
Texasin	(25)	*Baptisia australis*	(135)
		Platymiscium praecox	(84)
6,7,3',4'-Tetramethoxyiso-flavone	(26)	*Pterodon pubescens*	(84)
6,7,2',3',4'-Pentamethoxy-isoflavone	(27)	*Pterodon pubescens*	(84)
6-Hydroxygenistein	(29)	*Baptisia hirsuta*	(133)
Irisolidone	(32)	*Iris nepalensis*	(162)
Irisolone	(35)	*Iris nepalensis*	(83)
7-Methyltectorigenin	(33)	*Dalbergia sissoo*	(11)
		Pterocarpus angolensis	(136)
7,4'-Dimethyltectorigenin	(34)	*Dalbergia sissoo*	(9)
Derrustone	(21)	*Derris robusta*	(63)
Milldurone	(28)	*Millettia dura*	(146)
		Pterodon pubescens	(84)

Name	Formula	Plant Source	References
Complex Isoflavones			
Derrubone	(43)	*Derris robusta*	*(63)*
Piscerythrone	(47)	*Piscidia erythrina*	*(67)*
Piscidone	(48)	*Piscidia erythrina*	*(67)*
Dehydroneotenone	(51)	*Pachyrrhizus erosus*	*(51)*
		Neorautanenia edulis,	*(34)*
		N. amboensis	
Robustone	(44)	*Derris robusta*	*(63)*
Robustone methyl ether	(45)	*Derris robusta*	*(63)*
Durmillone	(50)	*Millettia dura*	*(146)*
		Millettia ferruginea	*(105)*
Ichthynone	(46)	*Piscidia erythina*	*(170)*
Ferrugone	(49)	*Millettia ferruginea*	*(105)*
Toxicarol isoflavone	(56)	*Derris malaccensis*	*(104)*
Chandalone	(55)	*Derris scandens*	*(66)*
Scandenone (Warangalone)	(53)	*Derris scandens*	*(66, 156)*
Scandinone	(54)	*Derris scandens*	*(156)*
Auriculatin	(52)	*Millettia auriculata*	*(172)*
Munetone	(57)	*Mundulea sericea*	*(60, 59)*

1. Simple Isoflavones

All the 39 known simple isoflavones can be grouped into 16 oxygenation patterns, as can be seen in *Chart 2*. The general chemistry of isoflavones has been well covered previously (*140, 52*). In the following paragraphs some of the more interesting features of the recently studied compounds will be briefly mentioned.

7,3′,4′-Trihydroxyisoflavone (6), previously known only as a synthetic product, has recently been isolated from the heart wood of the Brazilian jacaranda tree *(Machaerium villosum)*, together with its 3′-methyl ether (7) and isoformononetin (3) as new natural products (*139*). The discovery of this trihydroxy-compound completes the list of parent isoflavones having the simpler, more commonly encountered A- and B-ring hydroxylation patterns in C_6-C_3-C_6 compounds.

Baptigenin, the aglycone of baptisin, an isoflavone glucoside found in the roots of *Baptisia tinctoria* together with pseudobaptisin, probably has the 7,3′,4′,5′-tetrahydroxy structure (12). FARKAS and co-workers (*70*) have synthesised this compound and found its properties to be in good agreement with those reported for the natural product.

6,7,4′-Trihydroxyisoflavone (23), parent phenol of the previously known 6-methoxy-derivative, afromosin (24), was recently found in fermented soybean and shown to have antioxidant properties (*95*). It is

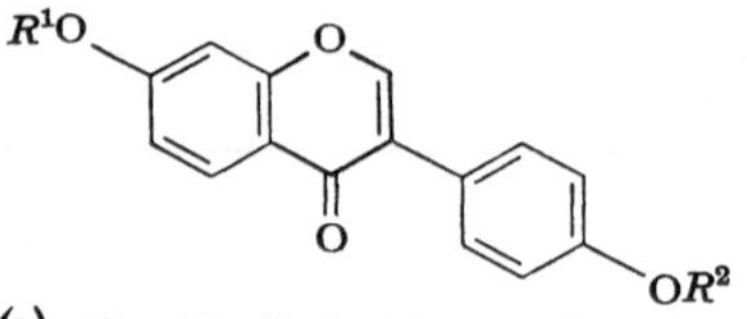

(1) $R^1 = R^2 = H$. Daidzein

(2) $R^1 = H$, $R^2 = Me$. Formononetin

(3) $R^1 = Me$, $R^2 = H$. Isoformononetin

(4) $R^1 = R^2 = Me$

(5) $R^1 = Me$, $R^2 = -CH_2CH{=}CMe_2$. Durlettone

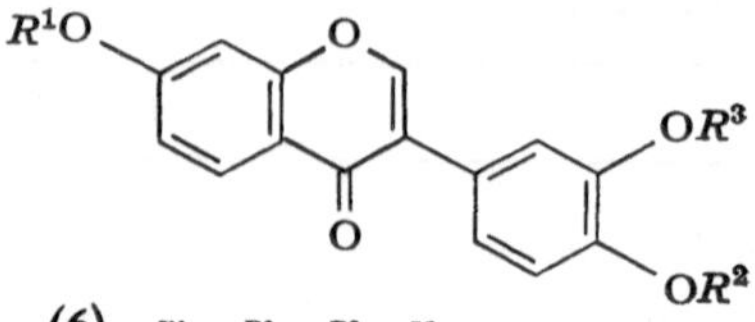

(6) $R^1 = R^2 = R^3 = H$

(7) $R^1 = R^2 = H$, $R^3 = Me$

(8) $R^1 = R^3 = H$, $R^2 = Me$ Calycosin

(9) $R^1 = R^2 = R^3 = Me$ Cabreuvin

(10) $R^1 = H$, $R^2 + R^3 = CH_2$ ψ-Baptigenin

(11) $R^1 = -CH_2CH{=}CHMe_2$, $R^2 + R^3 = CH_2$ Maximin

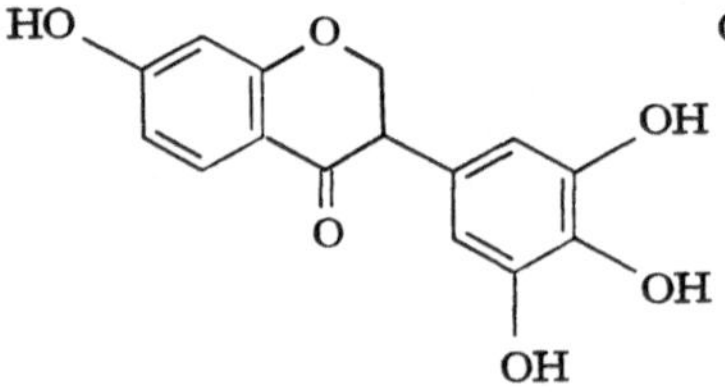

(12) Baptigenin

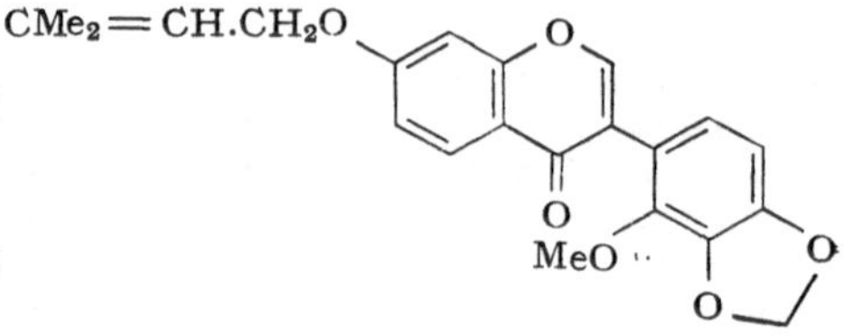

(13) Maxima substance C

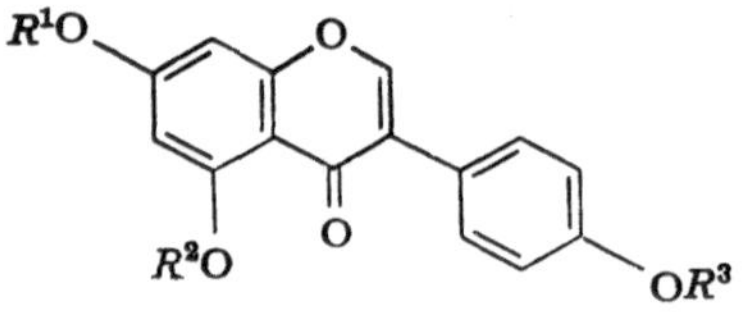

(14) $R^1 = R^2 = R^3 = H$. Genistein

(15) $R^1 = R^2 = H$, $R^3 = Me$. Biochanin A

(16) $R^1 = Me$, $R^2 = R^3 = H$. Prunetin

(17) $R^1 = R^3 = H$, $R^2 = Me$. 5-Methylgenistein

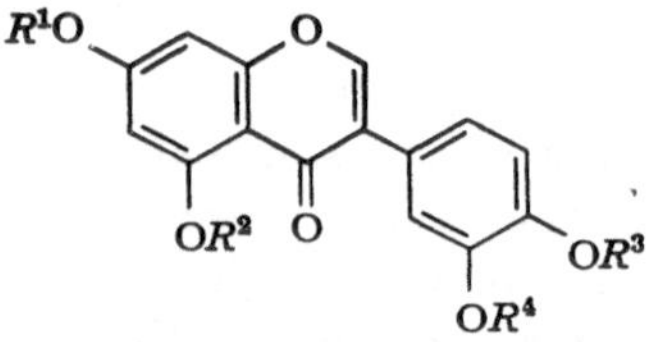

(18) $R^1 = R^2 = R^3 = R^4 = H$. Orobol

(19) $R^1 = Me$, $R^2 = R^3 = R^4 = H$. Santal

(20) $R^1 = R^2 = R^4 = H$, $R^3 = Me$. Pratensein

(21) $R^1 = R^2 = Me$, $R^3 + R^4 = CH_2$ Derrustone

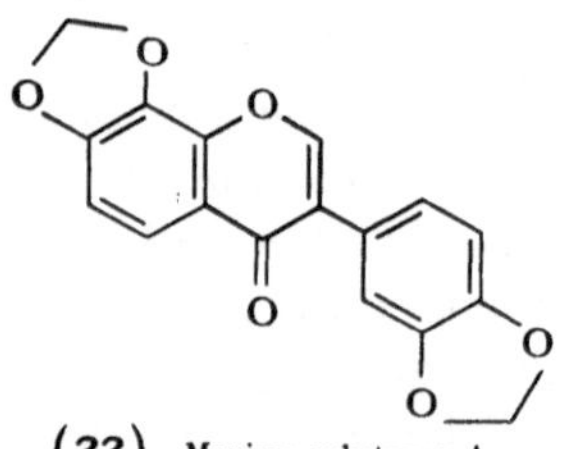

(22) Maxima substance A

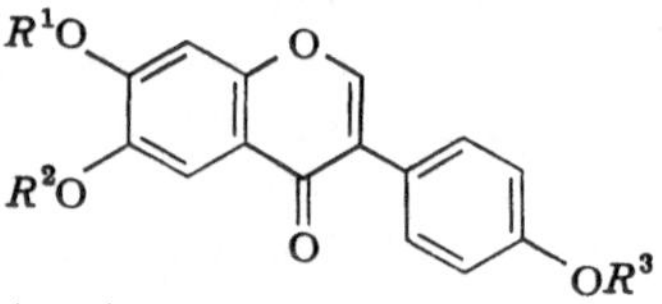

(23) $R^1 = R^2 = R^3 = H$

(24) $R^1 = H$, $R^2 = R^3 = Me$. Afromosin

(25) $R^1 = R^2 = H$, $R^3 = Me$. Texasin

Chart 2. Constitutions of Simple Isoflavones

References, pp. 63—73

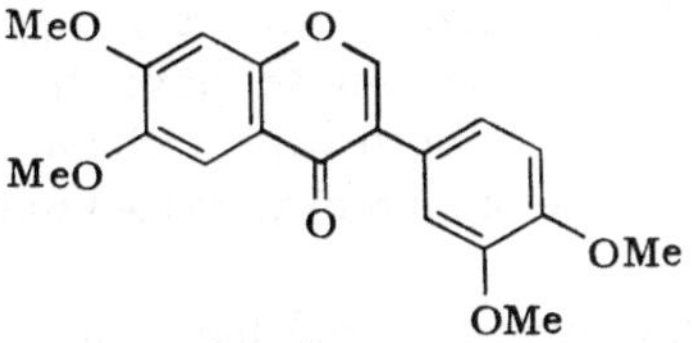

(26)

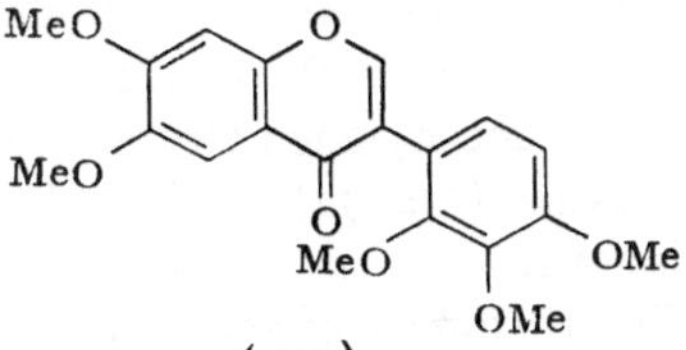

(27)

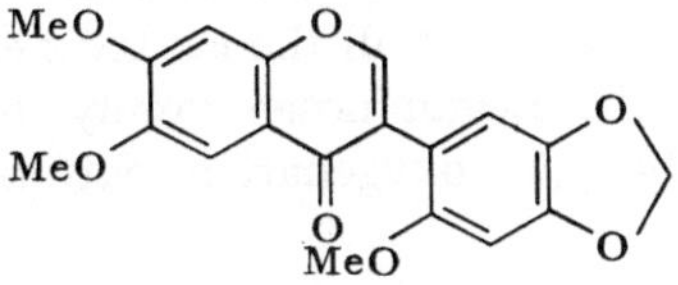

(28) Milldurone

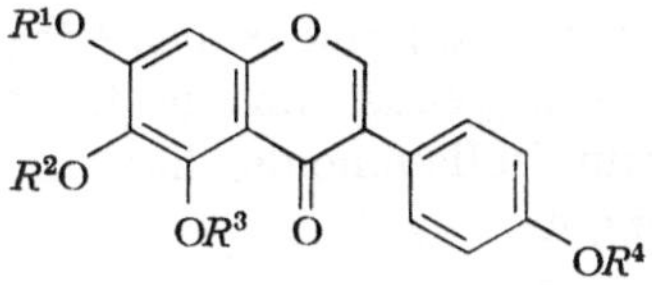

(29) $R^1 = R^2 = R^3 = R^4 = $ H. 6-Hydroxygenistein

(30) $R^1 = R^3 = R^4 = $ H, $R^2 = $ Me. Tectorigenin

(31) $R^1 = R^3 = $ Me, $R^2 = R^4 = $ H. Muningin

(32) $R^1 = R^3 = $ H, $R^2 = R^4 = $ Me. Irisolidone

(33) $R^1 = R^2 = $ Me, $R^3 = R^4 = $ H. 7-Methyltectorigenin

(34) $R^1 = R^2 = R^4 = $ Me, $R^3 = $ H. 7,4'-Dimethyltectorigenin

(35) $R^1 + R^2 = CH_2$, $R^3 = $ Me, $R^4 = $ H. Irisolone

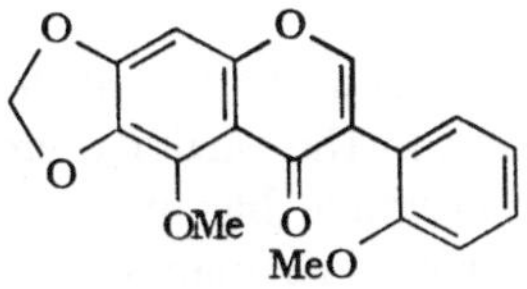

(36) Tlatlancuayin

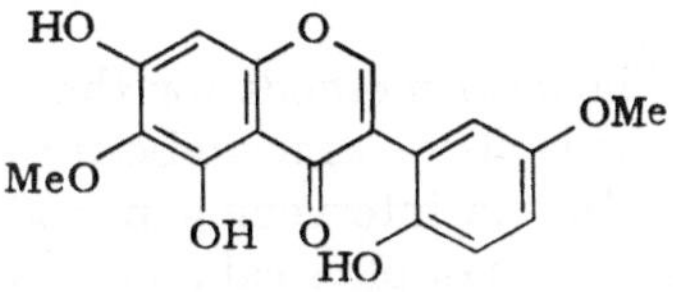

(37) Podospicatin

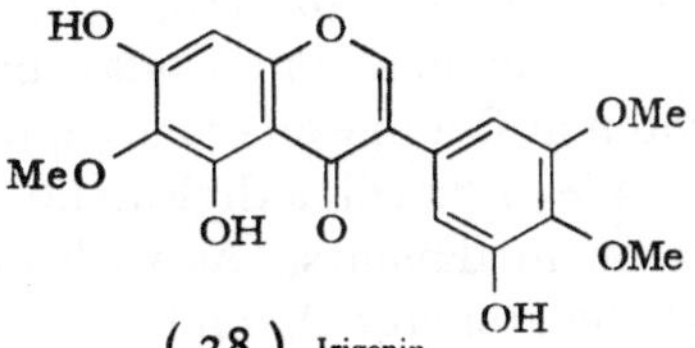

(38) Irigenin

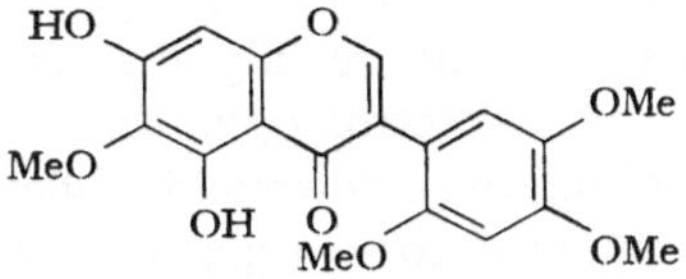

(39) Caviunin

Chart 2 (continued)

possible that the same compound is present in soybean as a glycoside since hydrolysis of extracts of the unprocessed material yielded the same antioxidant activity. Several more new examples of 6,7-substituted isoflavones (25)–(27) have recently been found in Brazilian plants (84).

Maxima substance A previously (*163*) assigned the 6,7:3′,4′-bis-methylenedioxyisoflavone structure, has been shown to have the isomeric 7,8-methylenedioxy structure (*22*) (*122*). Both isomers have been synthesised (*122, 71*). This is the first example of an 8-oxygenated natural isoflavone.

The rhizomes of some *Iris* species (family Iridaceae) have long been known to contain tectorigenin (30) and irigenin (38) as 7-glucosides (*140*). Recent studies on *I. nepalensis* (*83, 162*) have resulted in the isolation of *irisolidone* (32) and *irisolone* (35). It is of interest that all the isoflavones from this genus, one of the few outside the Leguminosae family to contain isoflavonoids, have in common the 5,6,7-oxygenation pattern in ring A.

Many of the earlier known isoflavones (*140, 52*) have in recent years been found in additional plant sources. Systematic survey of *Baptisia* species (*132, 125*) has revealed the presence of daidzein (**1**), formononetin (**2**), genistein (**14**), ψ-Baptigenin (**10**), and orobol (**18**). ψ-Baptigenin has also been found in *Pterocarpus erinaceus* (*19*) and *Maackia amurensis* (*183*); genistein in *Laburnum alpinum* (*65*); and formononetin in *Pterocarpus vidalianus* (*148*), *Machaerium villosum* (*139*), *Pericopsis* spp. (*107*) and *Castanospermum australe* (*62*). Afromosin (**24**), first isolated from the heartwood of *Afromosia elata*, appears to have a wide distribution. It has recently been found in *Baptisia australis* (*125*), *Pericopsis* spp. (*107*), *Myrocarpus fastigiatus* and *Myroxylon balsamum* (cabreuva wood) (*99*), *Amphimas pterocarpoides* (*19*) and *Castanospermum australe* (*62*). Biochanin A (**15**) further occurs in *Dalbergia sissoo* (*11*).

Chemical methods for the characterisation and structural determination of isoflavones have been well summarised in the earlier reviews (*140, 52*). In the intervening period the major advance has been in the increased use of physical methods, in particular nuclear magnetic resonance spectroscopy, as an aide in structural elucidation.

The n. m. r. characteristics of flavonoid compounds, including isoflavones, have been surveyd by various workers (*110, 128, 134, 16*). Isoflavones are readily distinguished from flavones and isoflavanones by n. m. r. spectroscopy. The olefinic proton at C-2 in isoflavones appears as a chacteristic downfield singlet at $\sim\tau$ 2.2 ($\sim$ 1.7 in dimethylsulphoxide) as compared to $\sim\tau$ 3.3 for the C-3 proton in flavones. As with other flavonoids, the pattern of substitution of the aromatic A- and B-rings in isoflavones can usually be readily determined from the n. m. r. spectrum. More recent work on benzene-induced solvent effects on methoxyl resonances appear useful for determining relative orientations of methoxyl groups in highly substituted flavonoid compounds (*193*). Thus aromatic methoxyl groups between *ortho* substituents possess the same chemical shift in benzene and deuterochloroform solution, while the shift of a methoxyl adjacent to at least one proton lies at o.3—o.5 ppm higher field in benzene solution than in deuterochloroform.

The mass spectral properties of a few isoflavones have been reported. The spectra of these compounds are charcterised by strong molecular ion peaks, reflecting the stability of the isoflavone ring system. Fragmentation via a retro Diels-Alder process, involving the two bonds β to ring A *(Chart 3)*, can usually be recognised. The spectrum of di-methyldaidzein (40) for example, shows two peaks corresponding to fragments (41) and (42), derived respectively from rings A and B (*12*) This breakdown pattern however is generally found also in flavones and other flavonoid compounds.

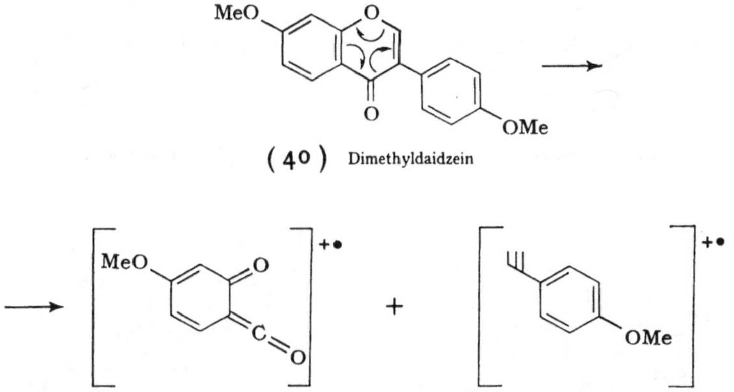

(4 0)   Dimethyldaidzein

(41)   (4 2)

Chart 3. Fragmentation of Isoflavones

2. Complex Isoflavones

The complex isoflavones recently reported (Table 1) have nearly all come from rotenoid-bearing plants belonging to the *Millettia, Piscidia, Neorautanenia* and *Derris* genera. Their structures are given in formulas (43)–(57) *(Chart 4)*. In complexity these compounds resemble jamaicin (58) and mundulone (59), earlier-known examples of isoflavones from rotenoid-bearing plants. Osajin (60) and pomiferin (61) complete the list of known complex isoflavones.

The most significant aspect of recent chemical studies on these compounds has been the invaluable assistance of n. m. r. in structure elucidation. The presence and nature of the isoprenoid substituent in these isoflavones is readily revealed; the 2,2-dimethylchromene and C- or O-linked 3,3-dimethylallyl groupings exhibit distinguishing features in the n. m. r. spectrum (*110, 146*). The contribution of mass spectrometry, in the other hand, has been limited. The highly substituted iso-

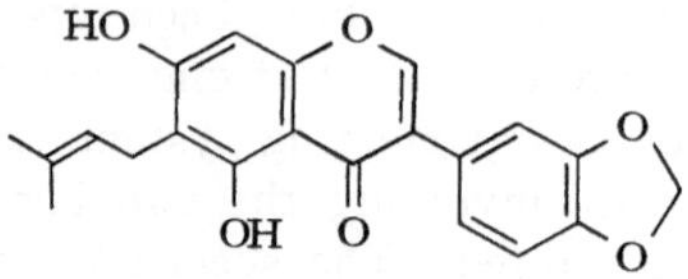

(43) Derrubone

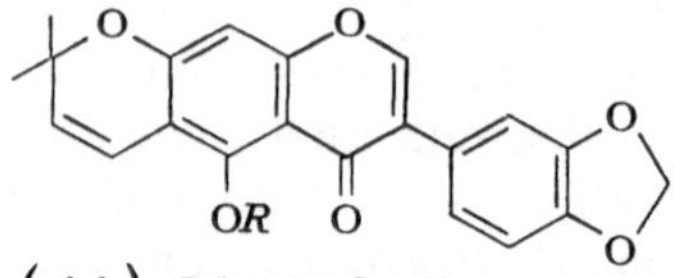

(44) Robustone $R = H$
(45) Robustone methyl ether $R = Me$

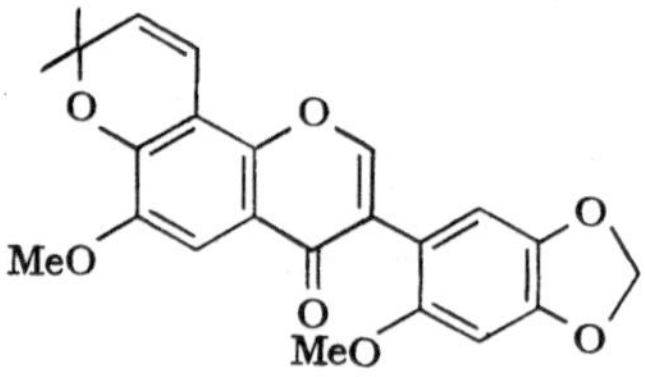

(46) Ichthynone

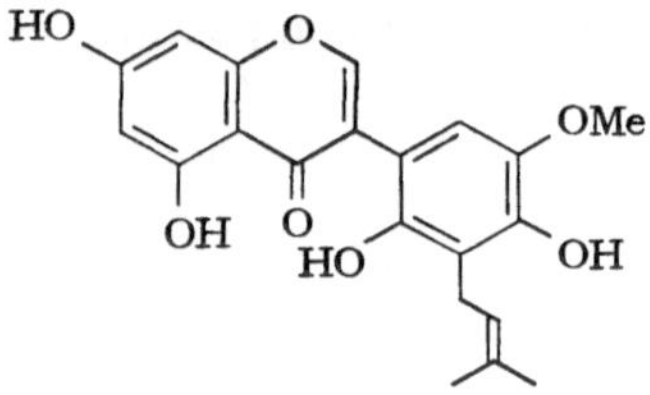

(47) Piscerythrone

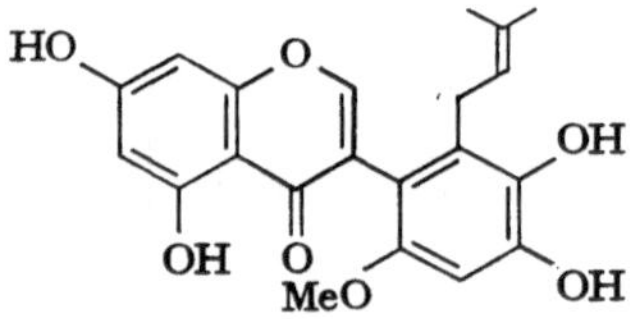

(48) Piscodone

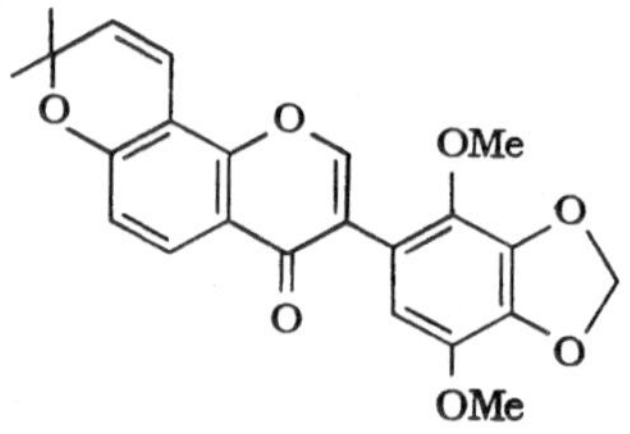

(49) Ferrugone

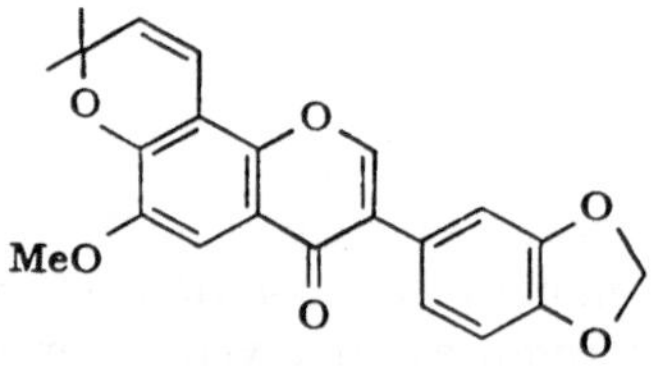

(50) Durmillone

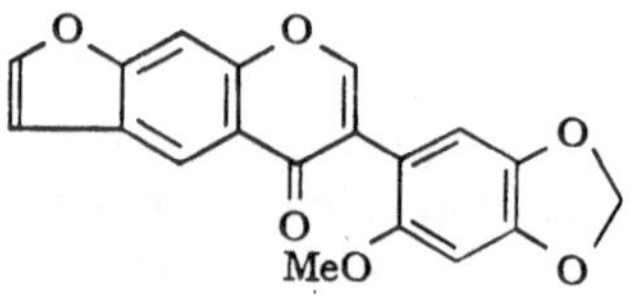

(51) Dehydroneotenone

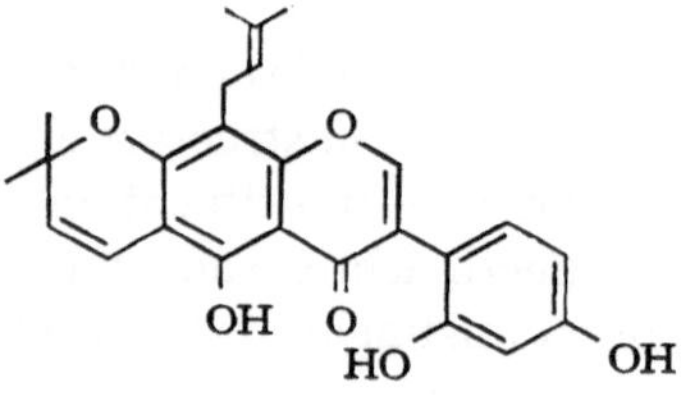

(52) Auriculatin

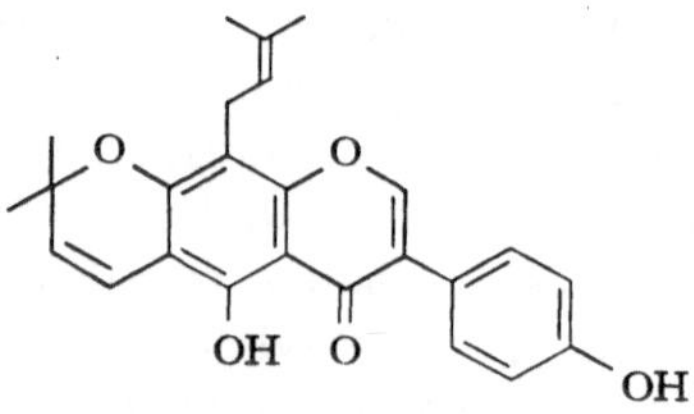

(53) Scandenone

Chart 4. Naturally-Occurring Complex Isoflavones

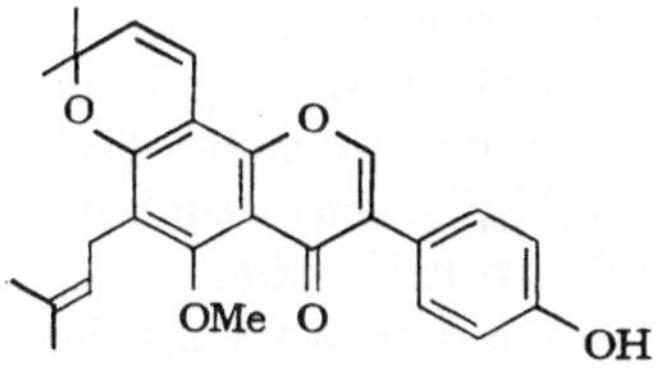

(54) Scandinone

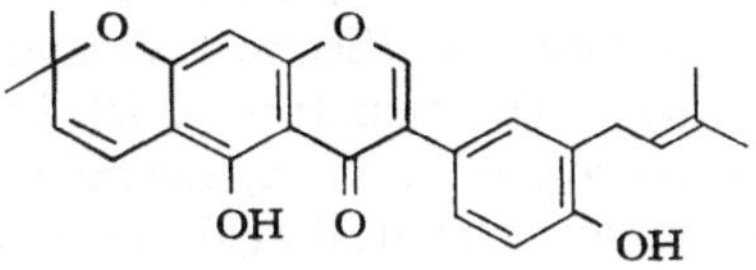

(55) Chandalone

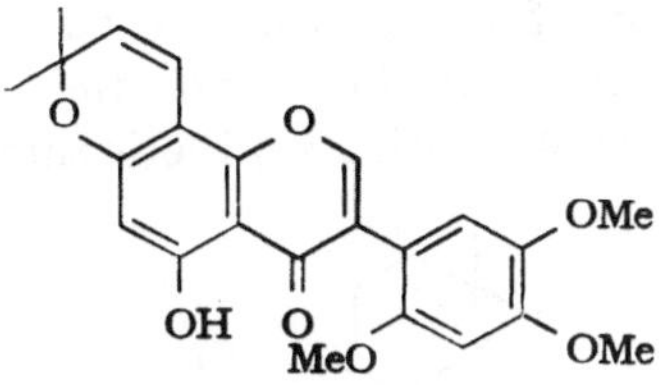

(56) Toxicarol Isoflavone

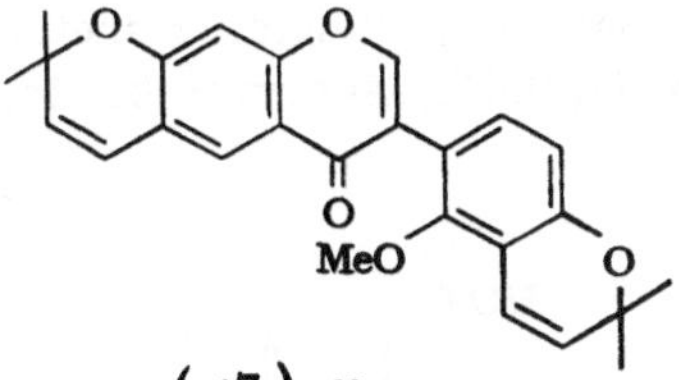

(57) Munetone

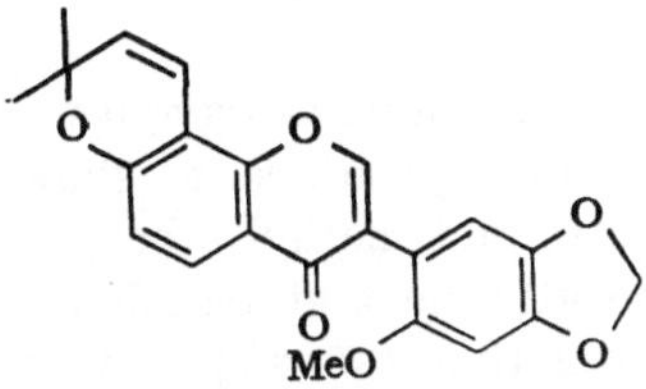

(58) Jamaicin

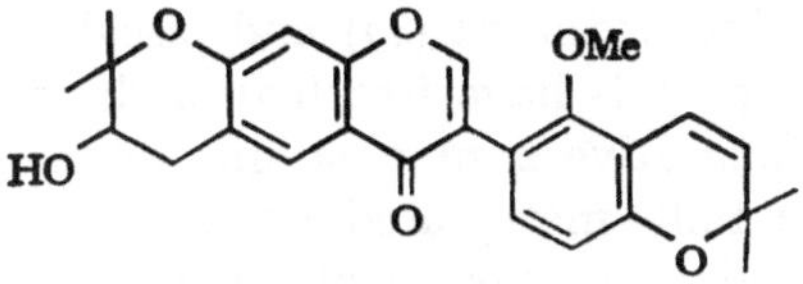

(59) Mundulone

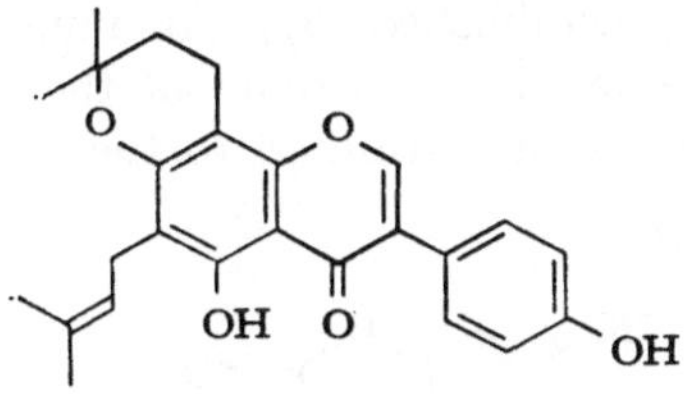

(60) Osajin

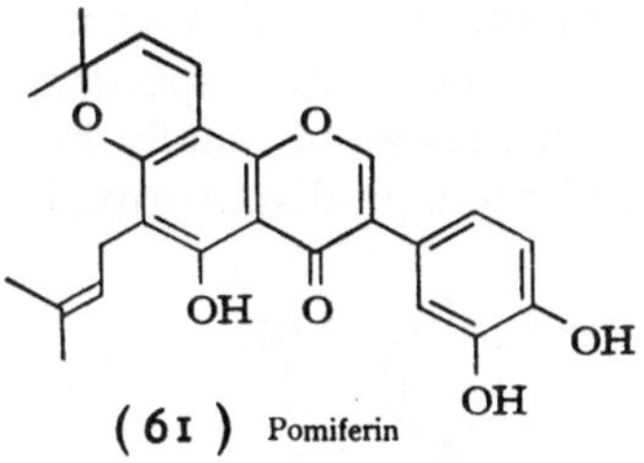

(61) Pomiferin

Chart 4 (continued)

flavones are even more stable to fragmentation than the simpler compounds, leading to mass spectra in which the molecular ion is by far the
largest peak. Minor breakdown processes however can occur which are
peculiar to the substitution pattern in the particular compound (*157*).

Ichthynone (46), *piscerythrone* (47), and *piscidone* (48) occur in the Jamaican Dogwood *(Piscidia erythrina)* together with jamaicin (58), several rotenoids (Section IV) and the coumaronochromone, lisetin (65) (*170, 67*). The structure of ichthynone, $C_{21}H_{14}O_5$, was deduced from a combination of n. m. r., degradative, and synthetic studies (*61, 170*). Spectral and chemical evidence revealed the presence of the 2,2-dimethyl-chromene, methoxyl and methylenedioxyl substituents. Chemical shift data and the coupling pattern of aromatic protons indicated that the 5-, 2'-, and 5'-positions are unsubstituted. The exact location of the individual substituents was deduced from the structure of two key degradation products: 6-methoxypiperylic acid (62) and the acid $C_{12}H_{13}O_4$ (OMe), shown to be (63) by comparison with a synthetic sample.

(62)   (63)

Piscerythrone (47) and piscidone (48) are isomeric isoflavones. Both compounds have free hydroxyls at the 5- and 7-positions, as shown by comparative n. m. r. studies of their acetates and methyl ethers. Since the fully methylated derivatives of piscerythrone and piscidone were found to be non-identical, it was suspected that the major structural difference between the two compounds involves the position of the di-methylallyl side chain. The structure assigned to piscerythrone (64) has been confirmed by oxidation to the coumaronochromone, lisetin (65), the structure of which was independently established (*67*). An unusual feature of the n. m. r. spectrum of piscidone, the occurrence of the C-2 proton signals at unusually high field, has been attributed to the steric effect of the 2',6'-disubstituted B ring (*67*).

(64) Piscerythrone   (65) Lisetin

Ferrugone (49) occurs with *durmillone* (50) in *Millettia ferruginea*, the Abyssinian Berebera tree, another rotenoid-bearing plant (*105*). N. m. r.

data readily revealed the nature of the substituent groups and the pattern of substitution of the rings. Comparison of the chemical shift of the 6′-proton in ring B in the two solvents deuterochloroform and benzene further revealed that this proton is next to a methoxyl. A choice between the alternative arrangements (66) and (67) for ring B substituents was made on the basis of finding apiolic acid (68) as a product of alkaline hydrogen peroxide oxidation of ferrugone.

(66)   (67)   (68) Apiolic acid

Structure (50) for durmillone was similarly deduced from n. m. r. data, in this instance without recourse to any chemical degradation reactions. Durmillone has also been isolated from *M. dura* and the same structure (50) was assigned independently by OLLIS and co-workers (*146*).

Scandenone (53), *scandinone* (54), and *chandalone* (55) from *Derris scandens* (*156, 66*) represent complex isoflavones containing both di-methylchromene and dimethylallyl residues. Osajin (60), which repre-sents with pomiferin (61) an earlier example of this type, was also isolated from this tree (*156*). While no rotenoids have been reported from this species, three 3-aryl-4-hydroxycoumarins, closely related structurally to scandenone and scandinone, were found. These will be considered in Section VII. The structures of these isoflavones were established again mainly by n. m. r. studies. Scandenone is an isomer of osajin and scandi-none methyl ether was found to be identical with osajin dimethyl ether, thus establishing the different arrangements of the isoprenoid residues in the two compounds.

Dehydroneoteonone (51) from *Neorautanenia pseudopachyrrhiza* and related species (Tribe Phaseoleae) represents the only known isoflavone with a furano substituent. However the identical substitution pattern is present in five other isoflavonoid compounds found in these species (*34, 51*). The structure of dehydroneotenone was originally established as an oxidation product of the isoflavanone neotenone (69). Linear fusion of the furano ring was established by the identification of the compound (70) with the acid obtained from hydrogen peroxide oxidation of dehydroneotenone (*51*).

Toxicarol isoflavone was the first complex isoflavone isolated from rotenoid-bearing plants (*100*). Structure (56) was tentatively proposed for this isoflavone on the basis of its close structural similarity to the rotenoid toxicarol (*71*). This structure has recently been confirmed by spectroscopic studies (*104*).

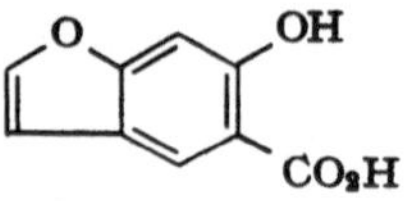

(70)

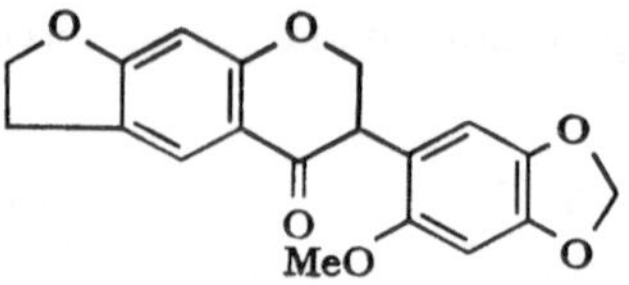

(69) Neotenone

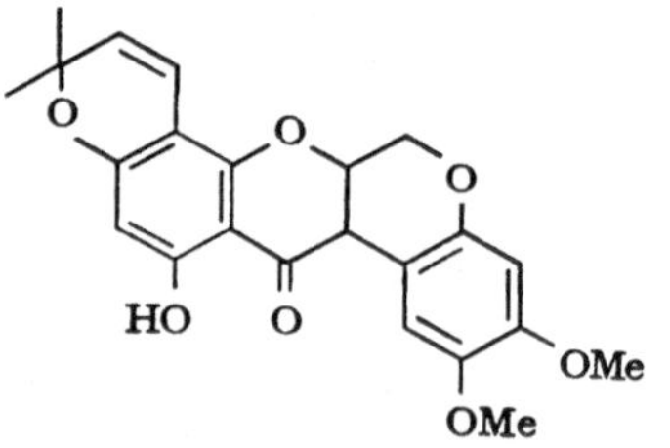

(71) Toxicarol

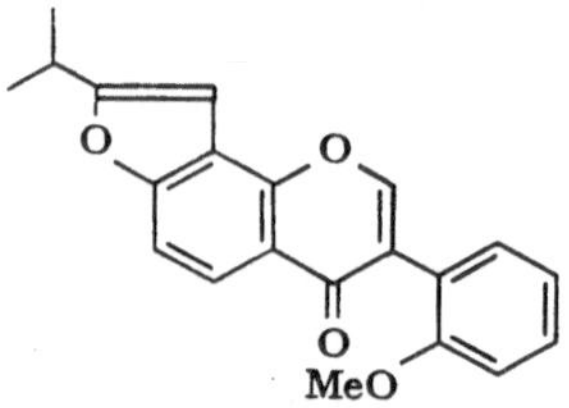

(72)

Munetone was originally (57) given the usual constitution (72). Recent n. m. r. studies (59) have shown that its correct structure is (57). This has been conformed by its partial synthesis from mundulone (59) (60).

3. Isoflavone Glycosides

All naturally occurring isoflavone glycosides currently known are listed in *Table 2*. Some of these glycosides have been known since a very early date (*140, 52*) but the majority have been reported only in recent years, mainly as a result of more systematic analysis of plant extractives. Thus the chromatographic survey of *Baptisia* species has led to the recognitition of a host of new isoflavone glycosides (*133, 135, 132, 125*). In some cases, e. g. in subterranean clover (*Trifolium subterranum*), may of the constituents previously isolated as free isoflavones have been shown to exist *in vivo* predominantly as glycosides (*17*).

General methods of isolation and characterisation of phenolic glycosides, applicable to flavonoid compounds and isoflavones have been summarised (97). N. m. r. spectroscopy has also been used for the study of glycosidic linkages (*129, 166, 129*). Since the earlier studies of FARKAS and his colleagues (*68*) little work on the synthesis of isoflavone glycosides has been reported in recent years.

References, pp. 63—73

Table 2. *Natural Isoflavone Glycosides*

Name	Constitution	Plant Source	References
Puerarin	Daidzein-8-C-glucoside (**75**)	*Pueraria thunbergina*	(*138*)
	Puerarin-xyloside	*Pueraria thunbergina*	(*138*)
	Puerarin-4',6''-diacetate (**76**)	*Pueraria tuberosa*	(*23*)
Daidzein	Daidzein-7-glucoside	*Trifolium praetense*	(*184*)
		Baptisia lecontei	(*132*)
	Daidzein-7-rhamnosyl-glucoside	*Baptisia lecontei*	(*132*)
Ononin	Formononetin-7-glucoside	*Baptisia australis*	(*125*)
		Trifolium pratense	(*184, 169*)
ψ-Baptisin	ψ-Baptigenin-7-rhamno-sylglucoside	*Baptisia lecontei*	(*132*)
	Calycosin-7-glucoside	*Baptisia lecontei*	(*132*)
	Calycosin-7-rhamnosyl-glucoside	*Baptisia lecontei*	(*132*)
	Texasin-7-glucoside	*Baptisia australis*	(*125*)
Wistin	Afromosin-7-glucoside	*Wistaria spp.*	(*173*)
		Baptisia australis	(*125*)
Genisten	Genistein-7-glucoside	*Ulex nanus, Adeno-carpus complicatus*	(*151*)
		Genista raetum	(*150*)
Sophoricoside	Genistein-4'-glucoside	*Sophora japonica*	(*140*)
Sophorabioside	Genistein-7-neohesperido-side	*Sophora japonica*	(*69*)
	Genistein-7-rhamnosyl-glucoside	*Baptisia lecontei*	(*132*)
Sphaerobioside	Genistein-7-rutinoside	*Baptisia sphaerocarpa*	(*167*)
Sissotrin	Biochanin-A-7-glucoside	*Cicer arietinum*	(*199*)
		Trifolium pratense	(*184, 169*)
		Dalbergia sissoo	(*10*)
	Biochanin-A-7-glucoside-5-malonate (**74**)	*Trifolium pratense*	(*184*)
Lanceolarin	Biochanin-A-7-apiosyl-glucoside (**73**)	*Dalbergia lanceolaria*	(*130*)
Prunetrin	Prunetin-4'-glucoside	*Prunus species*	(*140*)
	6-Hydroxygenistein-7-rhamosylglucoside	*Baptisia hirsuta*	(*133*)
	Orobol-7-rhamnosyl-glucoside	*Baptisia lecontei*	(*132*)
Tectoridin	Tectorigenin-7-glucoside	*Iris tectorum*	(*140*)
	7-Methyltectorigenin-4'-rhamnosylglucoside	*Dalbergia sissoo*	(*2*)
Iridin	Irigenin-7-glucoside	*Iris species*	(*140*)

Of the glycosides in *Table 2* the majority are 7-glucosides or 7-rhamosyl-glucosides. Less frequently encountered are 4'-glucosides and 4'-rhamosyl-glucosides. In the case of some of the biosides the nature and positions of the glycosidic linkages have not been fully elucidated. The structures of three of the glycosides listed in *Table 2* are worthy of special mention.

Lanceolarin from the root bark of *Dalbergia lanceolaria* has been shown to be the 7-β-apiosylglucoside of biochanin A, with the position of linkages between apiose and glucoside being 1 → 2 (73) (*130*). This is a further example of the occurrence of the uncommon branched-chain sugar apiose in glycosidic combination.

(73) Lanceolarin

(74) Biochanin A-7-glucoside-5-malonate

R = —CH₂CO₂H

(75) R = H Puerarin

(76) R = CH₃CO— Puerarin-4',6''-diaceta

The novel acylated derivative, *biochanin A-7-glucoside-5-malonate* (74) was recently isolated from red clover (*Trifolium pratense*) by Tamura et al. (*184*). The glycoside on mild acid hydrolysis yield biochanin A-7-glucoside; together with carbon dioxide and acetic acid as degradation products of the malonyl substituent. Assignment of the acyl group to the phenolic hydroxyl rather than to the sugar moiety was based on the presence in the i. r. spectrum of a band at 1678 cm⁻¹ which was assigned to a carbonyl group in a —CO—O-Aryl linkage.

The roots of *Pueraria* species (*P. thunbergiana*, *P. pseudohirsuta*, and *P. thomsoni*) are used in traditional Chinese medicine. Murakami et al. (*138*) isolated from these roots puerarin (75) and puerarin-xyloside which were the first isoflavone-C-glycosides reported to occur in nature. From the roots of *P. tuberosa* Bhutani et al. (*23*) recently isolated a diacetyl derivative of puerarin and established its constitution as (76). This isoflavone thus has the distinction of combining in the same molecule the three rare structure features: C-glycosylation, esterification of sugar hydroxyl by simple aliphatic acid, and esterification of a phenolic hydroxyl.

4. Synthesis of Isoflavones

Two methods have become well established for the synthesis of isoflavones. The ethyl orthoformate method of VENKATARAMAN and the ethoxalyl chloride method of BAKER and OLLIS both use deoxybenzoins as intermediates (*140*). Isoflavones synthesised by one or other of these routes in recent years include afromosin (*5*), caviunin (*58*), cabreuvin (*76*), tlatlancuayin (*6*) maxima substance A (*122*), maximin (*75*, *121*), maxima substance C (*4*), 7-methyltectorigenin (*81*), irisolone (*72*) and podospicatin (*33*).

The WESSELY-MOSER rearrangement of a 5,7,8-substituted isoflavone to the 5,6,7-substituted isomer under alkaline conditions has been exploited by FARKAS and his colleagues for the synthesis of many isoflavones of the latter type (*68*), including tectorigenin (*186*), irigenin (*68*) and caviunin (*186*).

Several interesting new routes to isoflavones, not having recourse to deoxybenzoins as intermediates, have been described recently. Based on an earlier postulated pathway for isoflavone biosynthesis, a useful synthetic route to isoflavones *via* boron trifluoride-catalysed rearrangement of chalcone epoxides has been developed by SESHADRI and co-workers (*94*, *21*, *22*, *113*). The method is illustrated in *Chart 5* for the synthesis of afromosin (*113*).

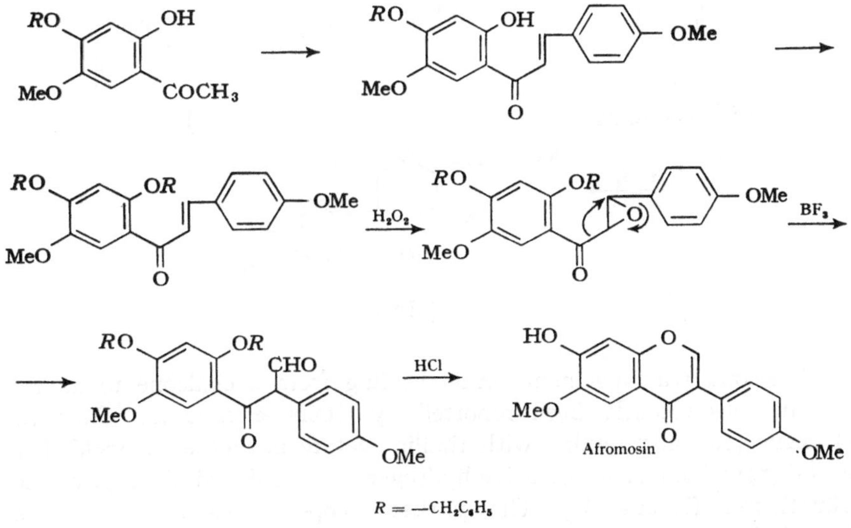

$R = —CH_2C_6H_5$

Chart 5. Synthesis of Afromosin via Chalcone Epoxide Rearrangement

A formyldeoxybenzoin is formed as the product of the rearrangement step and this, on treatment with acid, readily cyclises to an isoflavone.

The protection of all free hydroxyl groups in the chalcone molecule, including that at the 2'-position is necessary.

The ease of the rearrangement reaction is affected by the position of alkoxyl substituents in the aromatic rings. A 4-alkoxyl group in the B ring (chalcone numbering) is apparently necessary for rearrangement to take place (*21, 22*). The electron donating effect of the 4-alkoxyl group presumably operates to favour the initial formation of the carbonium intermediate (77), leading to the subsequent migration of the benzoyl moiety. A methoxy substituent at the 2-position of ring B however prevents reaction (*21*). Chalcone epoxides having resorcinol, pyrogallol or hydroxyquinol patterns in ring A undergo reaction as illustrated by the synthesis of formononetin (*94*), ψ-baptigenin (*94*), 7,8,3',4'-tetramethoxyisoflavone (*22*), cabreuvin (*22*) and afromosin (*113*) by this route. Isoflavones having a phloroglucinol grouping in ring A on the other hand could not be obtained by this method (*113*).

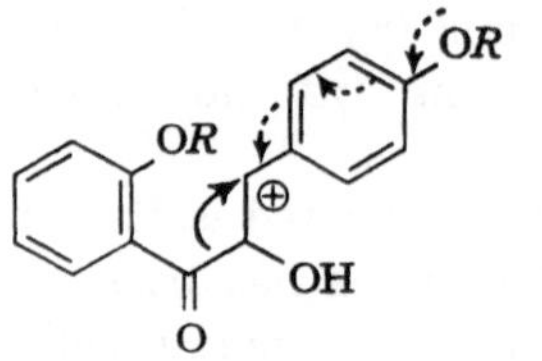

(77)

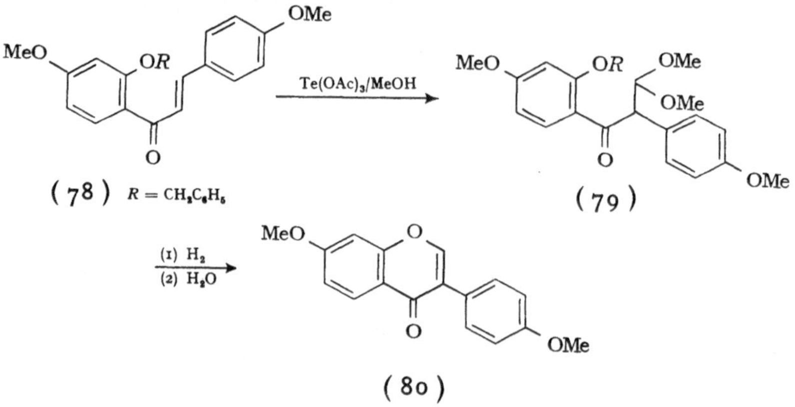

(78)   R = CH₂C₆H₅

(79)

(1) H₂
(2) H₂O

(80)

A second rearrangement route, leading from a chalcone to an isoflavone, has recently been reported by OLLIS et al. (*144*). Thus the chalcone (78) on refluxing with thallic acetate in methanol, yielded an acetal (79) which by consecutive hydrogenolysis and hydrolysis gave 7,4'-dimethoxyisoflavone (80). The synthetic scope of this reaction has not been defined but the rearrangement step has been shown by tracer studies to involve the B ring (*144*), in line with known biosynthetic results (*86*). This reaction thus offers an interesting laboratory parallel to isoflavone biosynthesis (*196*).

References, pp. 63—73

A novel route to isoflavones involving the coupling of an salicylic acid derivative to an arylacetaldehyde fragment in the form of an enamine is shown in *Chart 6 (185)*. 2'-Hydroxy isoflavones can be synthesised by this method.

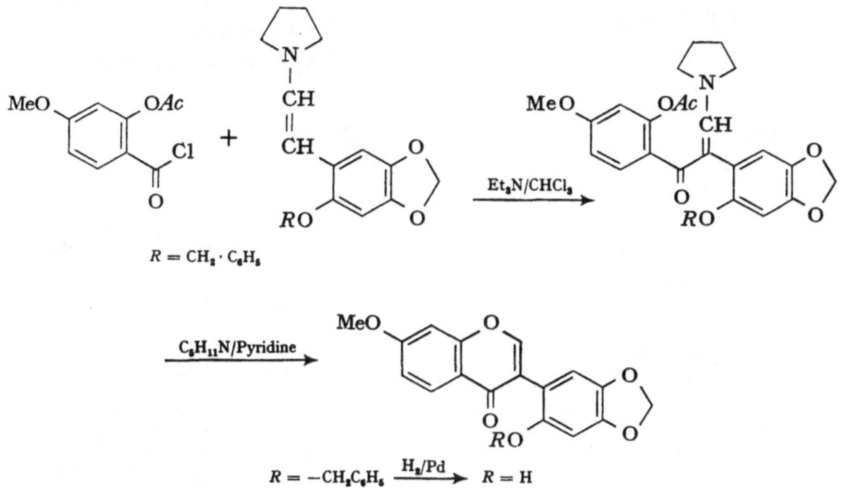

$R = CH_2 \cdot C_6H_5$

$R = -CH_2C_6H_5 \xrightarrow{\ H_2/Pd\ } R = H$

Chart 6. Synthesis of Isoflavones via Salicylic Acid and Arylacetaldehyde Derivatives

III. Isoflavanones

In contrast to the large number of isoflavones encountered as natural products, only nine isoflavanones are at present known *(Chart 7)*. Of these the following have been reported since the last reviews *(140, 52)*.

Dalbergioidin (85) and *ougenin* (86) were isolated by SESHADRI and co-workers from the heartwood of an Indian timber, *Ougeinia dalbergioides (8)*. Dalbergioidin formed a tetramethyl ether which readily gave a 2,4-dinitrophenylhydrazone and an oxime and was identified as 5,7,2',4'-tetramethoxyisoflavanone. Dalbergioidin is therefore the parent phenol of the previously known isoflavanones, ferreirin (82) and homo-ferreirin (83). The last compound was also isolated from *O. dalbergioides* in this work; and has been shown to occur also in the roots of *Cicer arietinum (93)*.

Ougenin (86) is currently the only C-methylated isoflavonoid known. Methylation of homoferreirin under conditions where C-methylation also takes place gave a product identical with ougenin dimethyl ether (90). C-methylation has also been carried out with the isoflavone (91). (±) Ougenin has recently been synthesized *(111)*. It is interesting to note that these heartwood isoflavanones of *O. dalbergioides* occur also in the leaves of the tree *(3)*.

2*

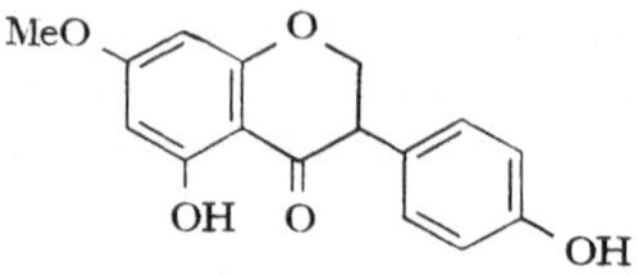

(81) Padmakastein

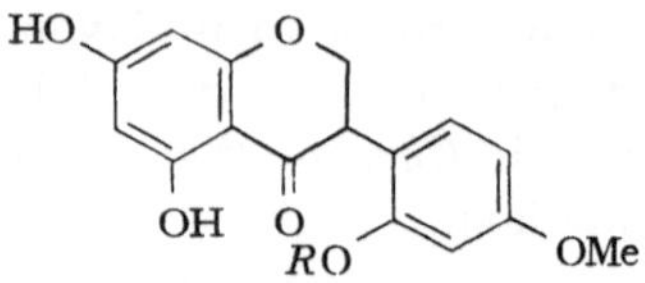

(82) Ferreirin R = H

(83) Homoferreirin R = Me

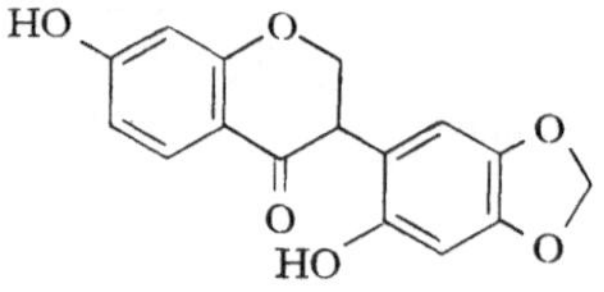

(84) Sophorol

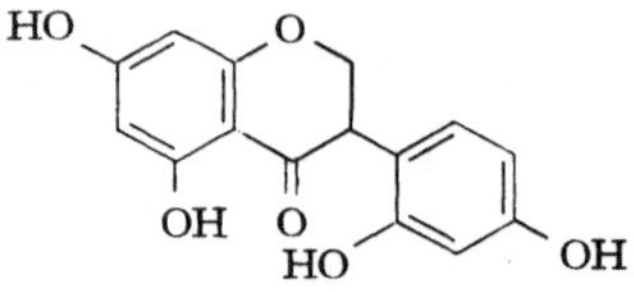

(85) Dalbergioidin

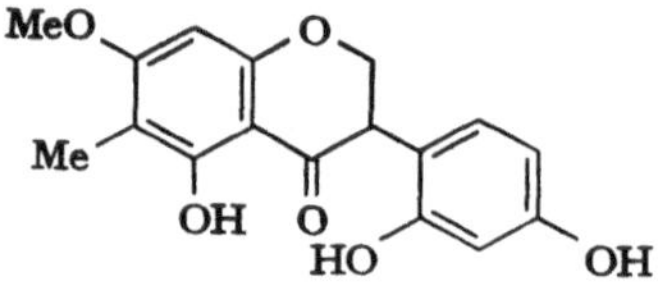

(86) Ougenin

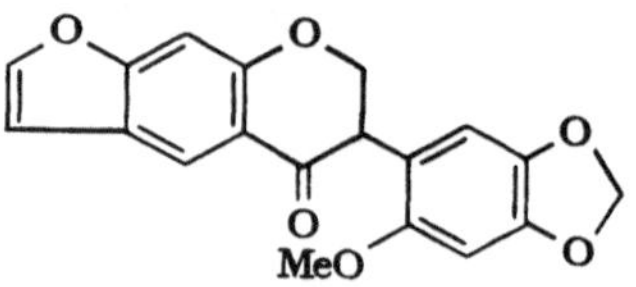

(87) Neotenone

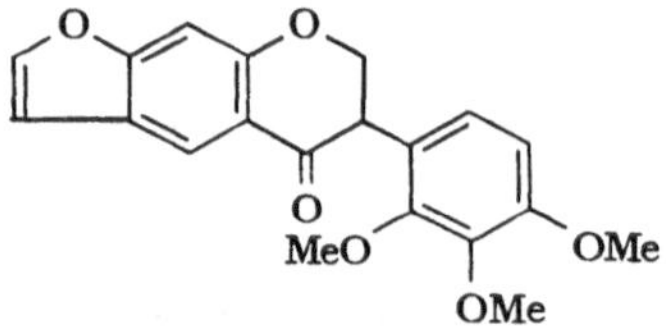

(88) Nepseudin

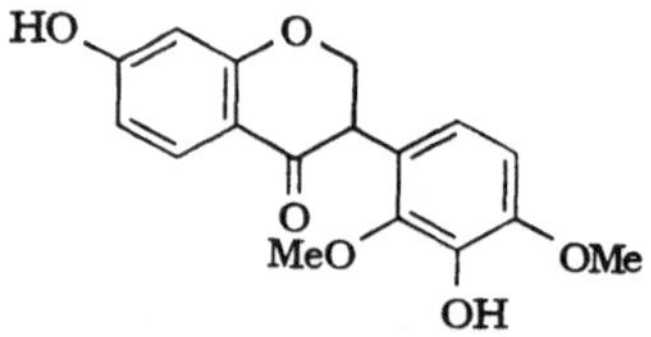

(89) Violanone

Chart 7. Natural Isoflavanones

Investigation of the rotenoid-bearing plant, *Neorautanenia pseudo-pachyrrhiza*, by CROMBIE and WHITING (*51*) has revealed the presence of two furano-isoflavanones. *Neotenone*, $C_{19}H_{14}O_6$, has the structure (87). Despite similar u. v. and i. r. spectral characteristics and rotenoid-like colour reactions, neotenone was found not to be a rotenoid. Oxidation with manganese dioxide gave a dehydro-compound which turned out to be an isoflavone, dehydroneotenone, of structure (92). This was proved by hydrolysis to the corresponding deoxybenzoin (93) which on treatment with ethyl orthoformate-pyridine piperidine regenerated the isoflavone. Dehydroneotenone on reduction with potassium borohydride gave the isoflavanol (94) which on oxidation with chromium trioxide yielded neotenone. Neotenone has subsequently been found to be present also in the related species *N. edulis*, *N. amboensis* (*34*) and *Pachyrrhizus erosus* (*51*). The natural occurrence of the isoflavone dehydroneotenone in these species (*34*, *51*) has already been mentioned.

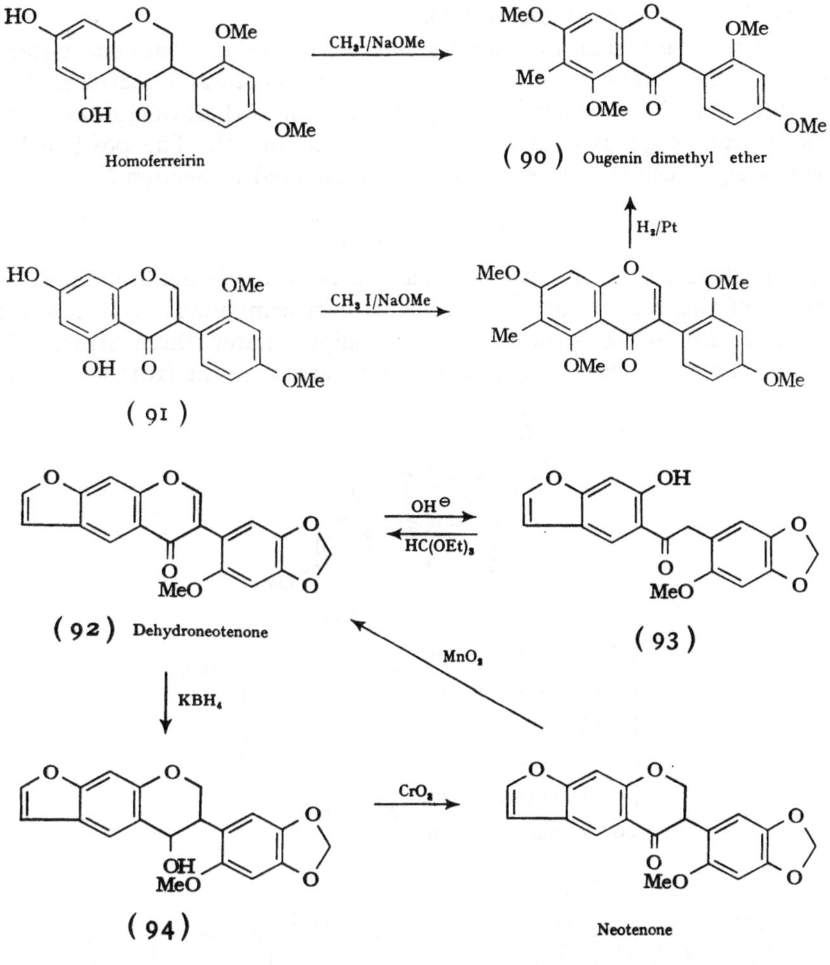

Homoferreirin

(90) Ougenin dimethyl ether

(91)

(92) Dehydroneotenone

(93)

(94)

Neotenone

Nepseudin ($C_{20}H_{18}O_6$) (88), the second isoflavanone from *N. pseudo-pachyrrhiza*, is similar to neotenone but has the rare 2′3′4′-oxygenated ring B replacing the more usual 2′4′5′-oxygenation. As with neotenone, its structure was proved by degradative studies of the isoflavone obtained by oxidation (*51*). Nepseudin has been synthesised (*74*).

Violanone (89) has been stated (*143*) to be a constituent of *Dalbergia violacea*, a source of isoflavans (Section VI).

In line with previous experience, the isoflavanones described above were isolated only in the optically inactive form, even though in the case of the *N. pseudopachyrrhiza* compounds, nothing more alkaline than Woelm "neutral" alumina was used in the isolation (*51*). Sophorol (84) (*179*) ([α]$_D$ + 9.5 acetone, — 13.6 ethanol) thus remains the only optically active isoflavanone so far isolated.

It can be seen that although the number of isoflavanones encountered is small, they show an interesting range of structural variations. It is notable that with the exception of padmakastein (81) all the isoflavanones known are oxygenated at the 2′-position in ring B. The possible biogenetic significance of this fact will be considered in Section X.

IV. Rotenoids

The chemistry of the rotenoids has been covered in detail in an earlier review of this series (*46*). The earlier known compounds (95)–(100) (*96*) revealed interesting structural relationships. Later three structurally related rotenoids with a linear C/D/E ring arrangement (101)–(103) were

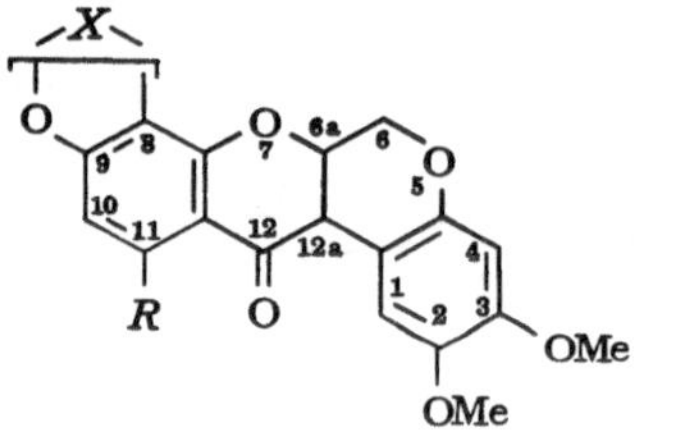

(95) Rotenone $R = H$
(96) Sumatrol $R = OH$
X = CH_2=CMe–CH–CH$_2$

(97) Deguelin $R = H$
(98) Toxicarol $R = OH$
X = Me$_2$C–CH=CH

(99) Elliptone $R = H$
(100) Malaccol $R = OH$
X = CH=CH

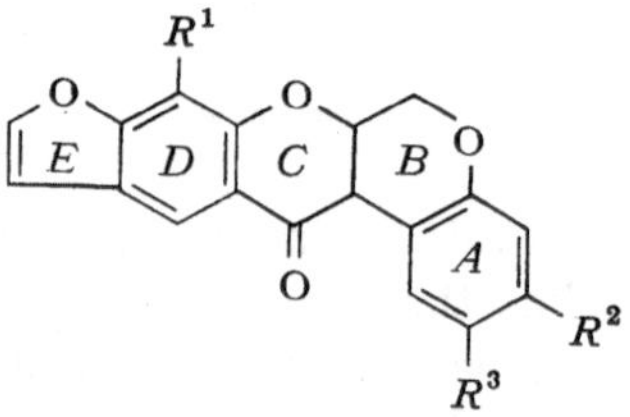

(101) Pachyrrhizone $R^1 = OMe$, $R^2 + R^3 = -OCH_2O-$
(102) Dolineone $R^1 = H$, $R^2 + R^3 = -OCH_2O-$
(103) Erosone $R^1 = H$, $R^2 = R^3 = OMe$

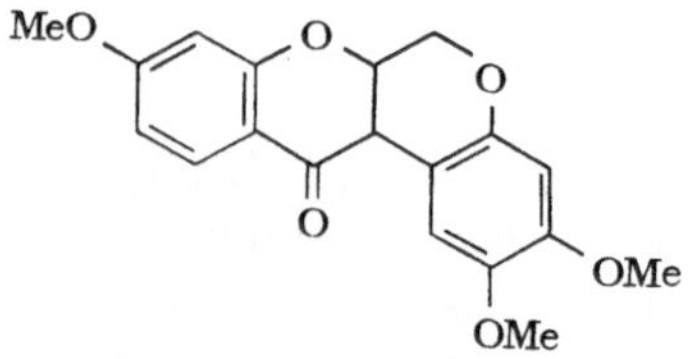

(104) Munduserone

isolated from *Pachyrrhizus erosus* and the related *Neorautanenia* species
(*46*). Munduserone (**104**) from *Mundulea sericea*, is the simplest rotenoid
yet known (*141*).

The absolute stereochemistry of rotenone is shown in (**105**) (*35*).
Comparison of the optical rotatory dispersion (o. r. d.) characteristics
of most of the other rotenoids has shown that they all have the same
absolute configuration at C-6a and C-12a (*141*) (*56*).

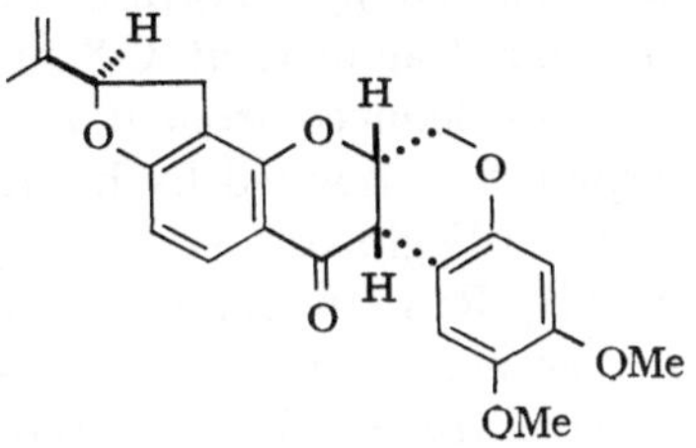

(105) (6aS, 12aS) Rotenone

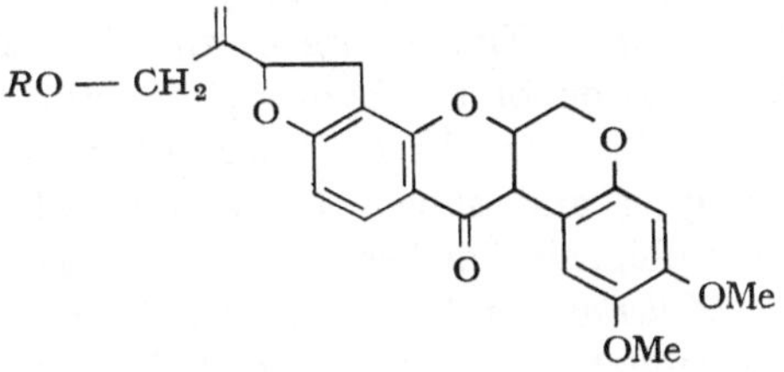

(106) Amorphin R = vicianosyl

(107) Amorphigenin R = H

Since the 1963 survey (*46*), three new rotenoids (**106**), (**108**), (**109**) have
been described. The existence in nature of 12a-hydroxyrotenoids and
probably also of 6a,12a-dehydrorotenoids, exemplified by (**110**) and (**111**)
respectively, has also been revealed (*67*, *146*).

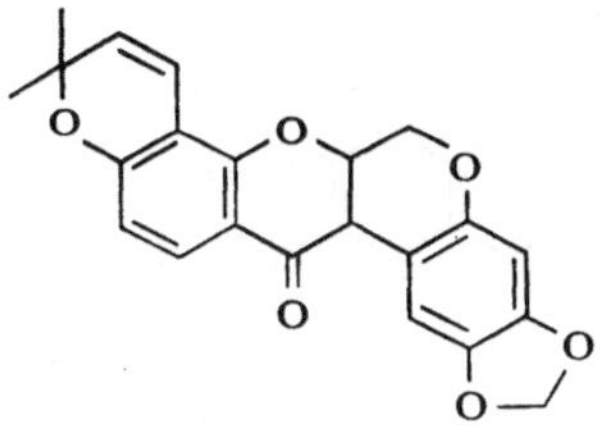

(108) Millettone

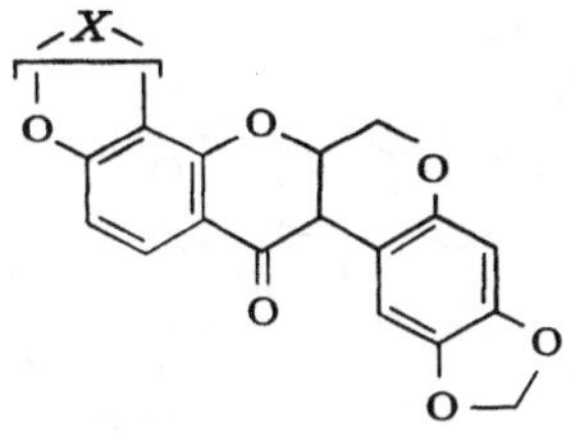

(109) Isomillettone X = $CH_2=CMe-CH-CH_2$

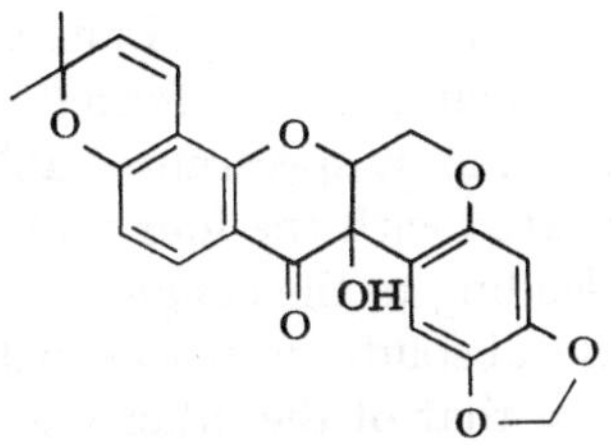

(110) Millettosin

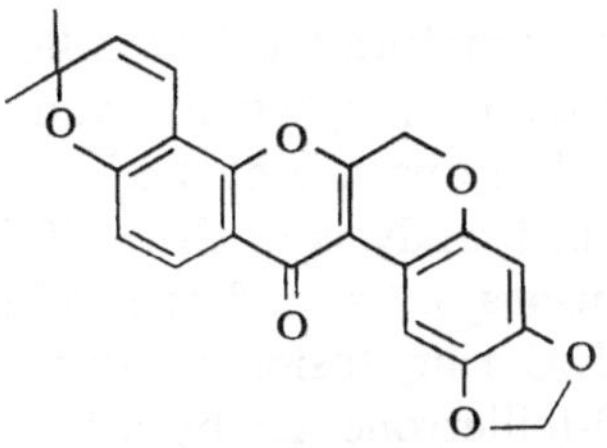

(111) Dehydromillettone

Amorphin (106), the first rotenoid-glycoside, was isolated in 1965 by CROMBIE and collaborators from the seeds of *Amorpha fruticosa* (*39*). On acidic or enzymic hydrolysis amorphin yielded glucose, arabinose, and the aglycone *amorphigenin*, a new rotenoid characterised as 8'-hydroxyrotenone (107).

The presence in amorphigenin of the same A/B/C/D fragment as in rotenone (95) was demonstrated by standard rotenoid degradation reactions supported by spectroscopic evidence. The n. m. r. characteristics of rotenone have previously been described by CROMBIE and LOWN (*49*). The nature of the five-carbon-atom fragment at C-8 and the location of the hydroxyl function were also deduced from n. m. r. data. The disaccharide fragment from amorphin was isolated by hydrogenolysis and shown to be 6-O-α-L-arabinosyl-D-glucose (vicianose). Amorphigenin has been studied independently by Russian workers (*120*).

Millettone (108) and *isomillettone* (109), together with rotenone (95) and sumatrol (96), were isolated as rotenoid constituents from the roots of *Piscidia erythrina* by OLLIS and coworkers (*67*). *Dehydromillettone* (111) was also isolated in very small amount from the same source. Millettone, together with the two previously-known rotenoid-type compounds tephrosin (112) and dehydrodeguelin (113), was also isolated by OLLIS *et al.* (*146*) from extracts of *Millettia dura*. *Millettosin* (110) a new 12a-hydroxyrotenoid, related to millettone as tephrosin (112) is to deguelin (97), also occurs in this plant. The products millettone, millettosin and tephrosin isolated from *M. dura* were all optically active (laevorotatory).

(112) Tephrosin   (113) Dehydrodeguelin

The structure (108) of millettone was derived from a detailed analysis of its n. m. r. spectrum and confirmed by chemical transformations (*146*). The absence of methoxyl signals in the n. m. r. spectrum enabled a detailed analysis of the ABCD system associated with the four protons in positions 6, 6a, and 12a to be carried out, leading to the assignment of a *cis* B/C ring fusion in this compound. The absolute stereochemistry of (−)-millettone has been found to conform to that of the other known rotenoids [see (105)].

Isomillettone (109) was isolated as an inseparable mixture with millettone and was detected and characterised by a careful analysis of the n. m. r. spectrum of the mixture (67). Complete coincidence of the signals due to ten protons indicated that both components of this mixture had the same rotenoid skeleton and the reduced signals due to the 2,2-dimethylchromene residue in millettone (108) was exactly compensated by signals of the appropriate intensity to be associated with a residue of type X in the formula (109).

Millettosin (110) was isolated in small quantity only and its constitution was proposed on the basis of mass spectral data (146). The fragmentation patterns of rotenoids in mass spectrometry have been reported (164). Nearly all the compounds studied showed intense peaks corresponding to the following type of fragmentation (Chart 8). The fragmentation of millettosin followed this scheme closely.

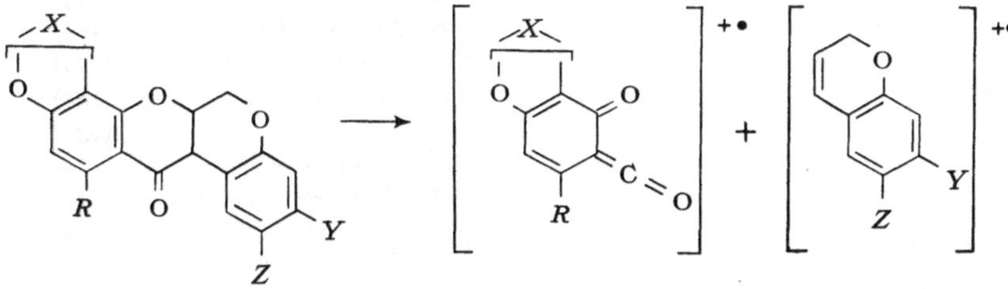

Chart 8. Fragmentation of Rotenoids

The presence of 12a-hydroxyrotenoids such as (±)-tephrosin (112) in plant extracts has been considered to be due to artefact formation (46), probably *via* aerial oxidation (47). The isolation of (—)-milletosin and (—)-tephrosin from *M. dura*, however, suggests that they are likely to be natural products (146). In view of the ease of dehydration of 12a-hydroxyrotenoids (47), the existence as natural products of 6a,12a-dehydrorotenoids, such as dehydrodeguelin in *M. dura* and dehydromillettone in *P. erythrina*, cannot be regarded as established.

V. Pterocarpans

The name pterocarpan has been given (101) to the coumaranochroman ring system (114) (systematic name: 6a,11a-dihydro-6-H-benzofuro-[3,2-c][1]benzopyran). The fully systematic (Ring Index) numbering in use with this nomenclature (102) is also shown.

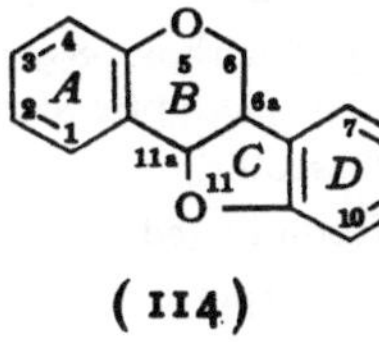

(114)

1. Natural Pterocarpans

The first two members of this series, pterocarpin (115) and homopterocarpin (116) were characterised in 1940 (140). Since 1960 13 additional naturally occurring pterocarpans have been described (Table 3); together with the three closely related derivatives *pisatin* (117), *variablin* (118) and *anhydrovariablin* (119). Two pterocarpan glucosides are also known.

Table 3. *Natural Pterocarpans*

Name	Formula	Plant Source	References
Demethylhomopterocarpin	(120)	*Swartzia madagascariensis*	(103)
		Andira inermis	(44)
		Dalbergia spp.	(143)
		Machaerium spp.	(143)
Homopterocarpin	(116)	*Pterocarpus indicus*	(20)
		Swartzia madagascariensis	(103)
		Pericopsis angolensis	(103)
		Machaerium villosum	(143)
Maackiain	(121)	*Maackia amurensis*	(180)
		Andira inermis	(43)
		Sophora japonica	(174)
		Pterocarpus dalbergioides	(152)
		Swartzia madagascariensis	(103)
		Dalbergia spruceana	(143)
Pterocarpin	(115)	*Pterocarpus indicus*	(20, 45)
		Swartzia madagascariensis	(103)
		Sophora subprostrata	(174)
Philenopteran	(132)	*Lonchocarpus laxiflorus*	(155)
9-O-Methylphilenopteran	(133)	*Lonchocarpus laxiflorus*	(155)
3-Hydroxy-4,9-dimethoxy-pterocarpan	(124)	*Swartzia madagascariensis*	(103)
3,4,9-Trimethoxypterocarpan	(125)	*Swartzia madagascariensis*	(103)
3,4-Dihydroxy-8,9-methylenedioxypterocarpan	(130)	*Dalbergia spruceana*	(143)
4-Hydroxy-3-methoxy-8,9-methylenedioxypterocarpan	(131)	*D. spruceana*	(143)
3-Hydroxy-4-methoxy-8,9-methylenedioxypterocarpan	(126)	*D. spruceana*	(143)
		Swartzia madagascariensis	(103)
3,4-Dimethoxy-8,9-methylenedioxypterocarpan	(127)	*S. madagascariensis*	(103)
		Neorautanenia ficifolia	(28)
Phaseollin	(137)	*Phaseolus vulgaris*	(159)
Neodulin (edulin)	(136)	*Neorautanenia edulus*	(34)
Leiocarpin	(138)	*Apuleia leiocarpa*	(84)

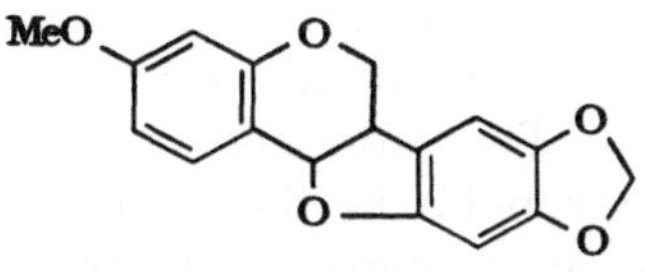

(115) Pterocarpin

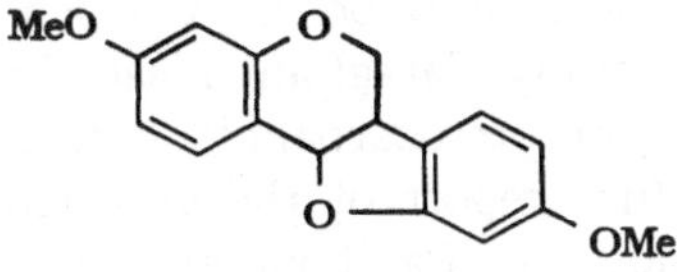

(116) Homopterocarpin

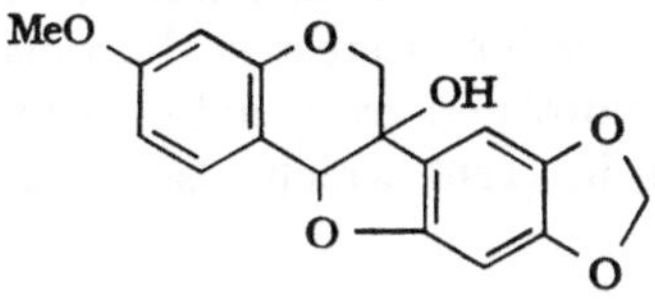

(117) Pisatin

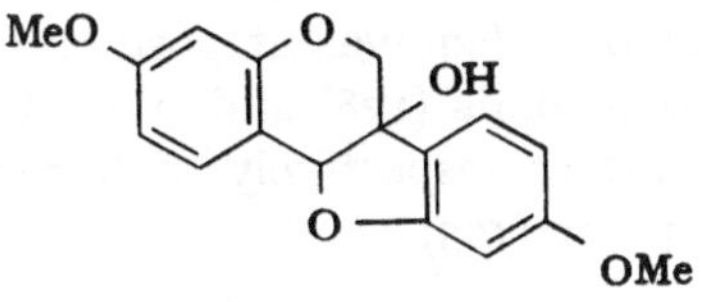

(118) Variablin

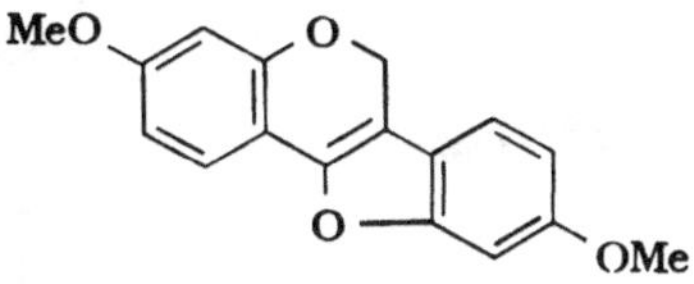

(119) Anhydrovariablin

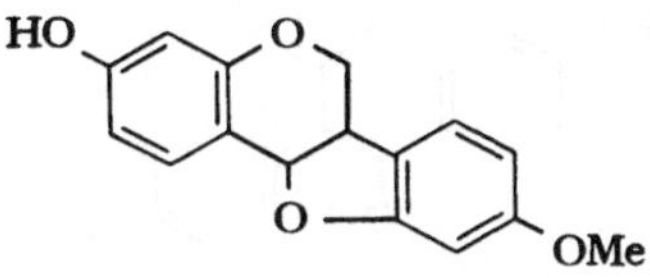

(120) Demethylhomopterocarpin

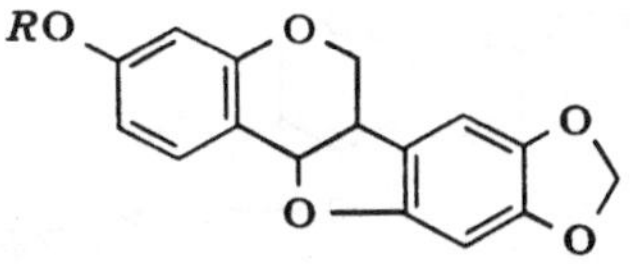

(121) Maackiain R = H

(122) Trifolirhizin ((−)-maackiainglucoside)

R = glucosyl

(123) Sophorajaponicin ((+)-maackiainglucoside)

R = glucosyl

The simpler pterocarpans, pterocarpin (115), homopterocarpin (116), demethylhomopterocarpin (120) and maackiain (demethylpterocarpin) (121) are of common occurrence, usually as heartwood constituents. (−)-Maackiain was first described as its β-glucoside, trifolirhizin (122) (31, 32). The aglycone could not be obtained from acid hydrolysis of the glucoside but was isolated pure after enzymic treatment (31). The enantiomorphic (+)-maackiain-β-glucoside (123) (sophorajaponicin) has since been found by SHIBATA and NISHIKAWA (174) to occur in Sophora japonica. Free maackiain also exists in this plant very probably in the (±) form. From two other related Sophora species these workers isolated (−)-maackiain-β-glucoside (174). Pterocarpans can exist therefore in both antipodal forms in nature.

The heartwood of the Rhodesian tree, *Swartzia madagascariensis* (sub-family Caesalpinioideae) has been shown to contain a remarkable array of pterocarpanoid compounds (*103*). This work also constitutes the first report of the occurrence of isoflavonoids in the Leguminosae outside the Papilionatae sub-family. The new pterocarpan structures characterised are shown in formulas (**124**)–(**127**). The special feature in these compounds is the oxygenation of C-4 which is uncommon in the isoflavones (equivalent position C-8). Of greater interest is the isolation of compounds (**128**) and (**129**), being dehydro and oxygenated dehydro-derivatives respectively of homopterocarpin (**116**) which also occurs in this plant (*103*).

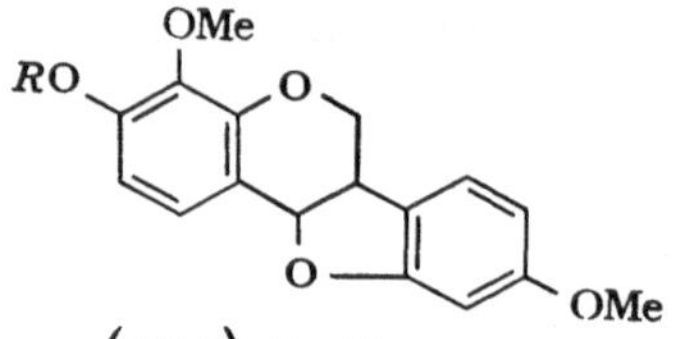

(**124**) R = H
(**125**) R = Me

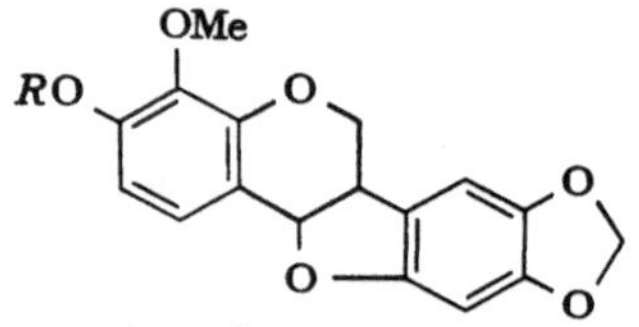

(**126**) R = H
(**127**) R = Me

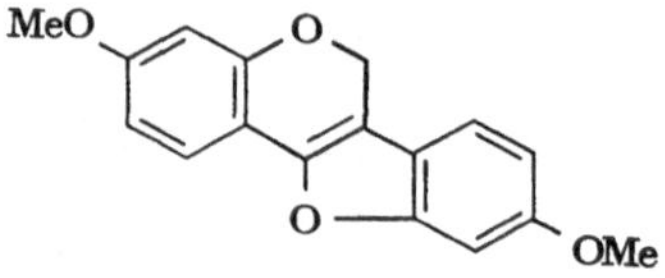

(**128**)

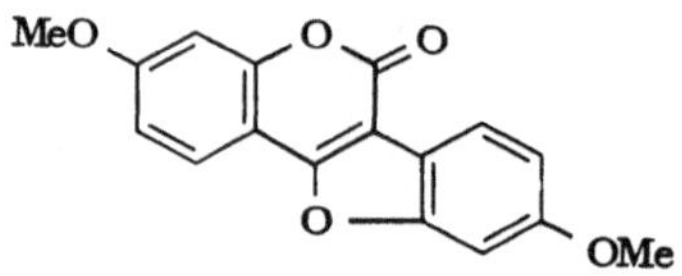

(**129**)

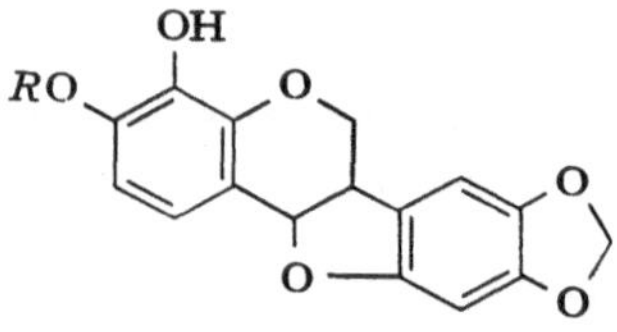

(**130**) R = H
(**131**) R = Me

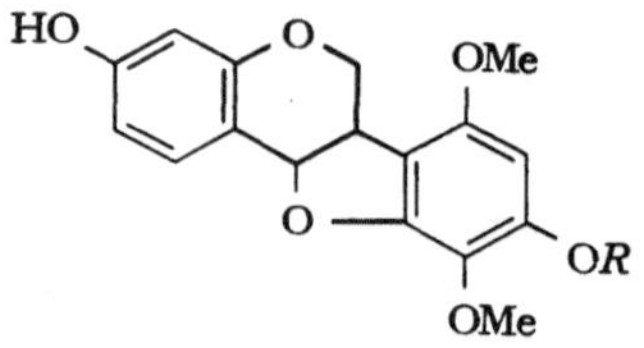

(**132**) Phileopteran R = H
(**133**) 9-O-Methylphilenopteran R = Me

The 6a-hydroxy-pterocarpan derivative, pisatin (**117**) is formed in pods of pea *(Pisum sativum)* after fungal infection (*160*). A similar 6a-hydroxy compound, variablin (**118**) has been stated to occur in *Dalbergia spruceana* (*143*). Anhydrovariablin, which corresponds to the dehydro compound (**128**) from *S. madagascariensis*, is also present, together with compounds (**130**) and (**131**) (*143*), closely related structurally to the *S. madagascariensis* constituents (124)–(127). The occurrence of (+)-pterocarpanoids in *D. spruceana* and *Machaerium* spp. has also been reported (*143*).

Philenopteran (132) and *9-O-methylphilenopteran* (133) occur with two structurally closely related isoflavans in *Lonchocarpus laxiflorus* (*155*). Determination of the position of substituents in these highly substituted compounds was accomplished by detailed n. m. r. spectral studies, including benzene-induced methoxyl proton shifts. PELTER and AMENECHI pointed out in this work (*155*) that in the pterocarpan series, unlike in other isoflavonoids, the mass spectra cannot be used to assign the various substituent groups to rings A and B because every fragment can be reasonably formulated as arising from either ring. In the spectrum of di-O-methylphilenopteran, for example, the ion at m/e 161 may be written as (134) or (135), arising from ring A or ring B respectively.

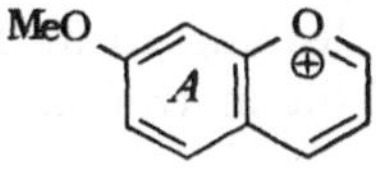

(134)

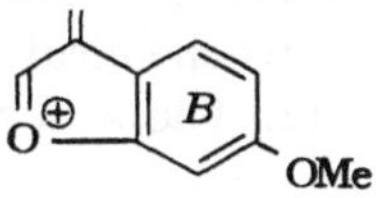

(135)

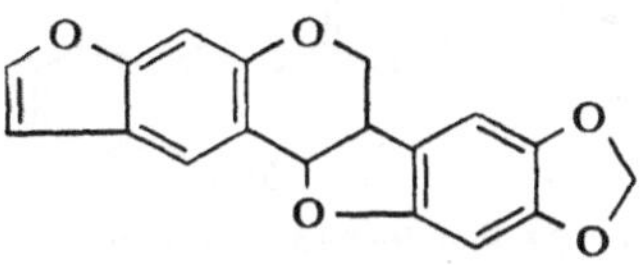

(136) Neodulin

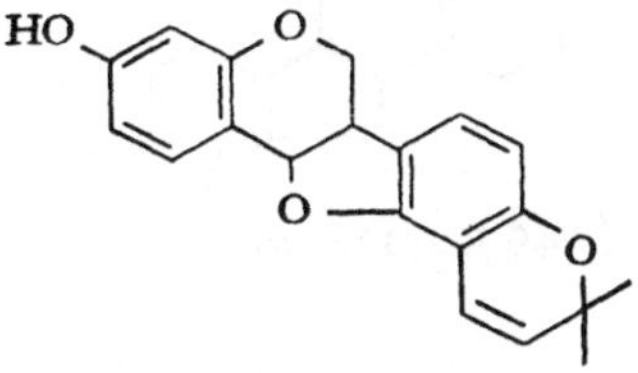

(137) Phaseollin

Neodulin (136) and *phaseollin* (137) are examples of pterocarpans with extra isoprenoid-derived rings and both are from plants of the tribe Phaseoleae (*34, 159*). Like pisatin (117) phaseollin is produced in the plant in response to fungal infection. These compounds are known to be inhibitory to a wide range of phytopathogenic fungi. In a recent study (*161*), the antifungal activity of some naturally occurring pterocarpans towards *Monilinia fructicola* was compared. Compounds found to be as active as (—)-phaseollin and (+)-pisatin were: (±)-pisatin, (—)-maackiain, (±)-maackiain, (—)-homopterocarpin and (±)-variablin (homopisatin).

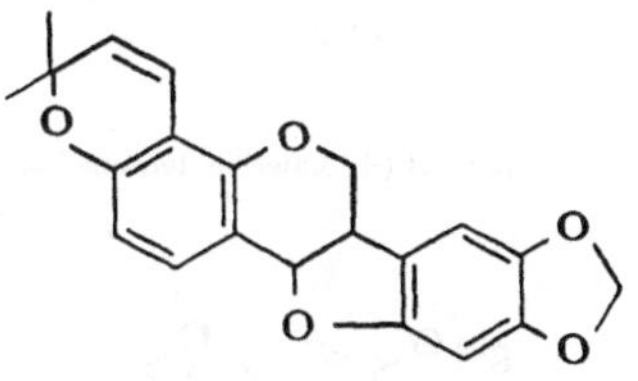

(138) Leiocarpin

Leiocarpin (138), occurring in the Brazilian tree *Apuleia leiocarpa*, is an interesting recent addition to the pterocarpanoid class (*84*).

2. Stereochemistry of Pterocarpans

With the interesting exception of (+)-pisatin and the (+)-ptero-carpans referred to (*143, 174*), the majority of known natural ptero-carpans have negative $[\alpha]_D$ values. These compounds have been taken to have the same absolute stereochemistry and results from o. r. d. com-parisons are consistent with this view (*181, 155, 143*). Their absolute configuration has been established as 6aR, 11aR, as shown in type formula (**139**). The assign-ment of this absolute configuration is based on evi-dence from the following independent series of investi-gations. (1) (—)-7,2'-Dimethoxy-4,'5'-methylene-dioxyisoflavan (**141**) obtained from reduction of (—)-pterocarpin (**140**) is enantiomeric by o. r. d. and c. d. (circular dichroism) down to 222 nm with the 3S-isoflavan analogue (**142**) (*41, 188*). The absolute configuration of (**142**) is known (*41, 192*) from the stereochemical course of the reaction leading to its formation (*Chart 9*). (2) The absolute configuration of C-6a in trofolirhizin (**143**)

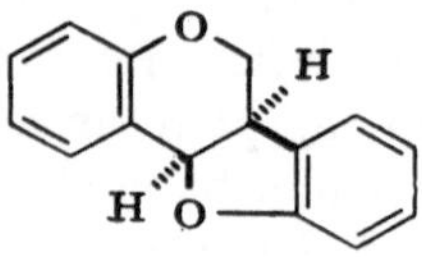

(**139**)

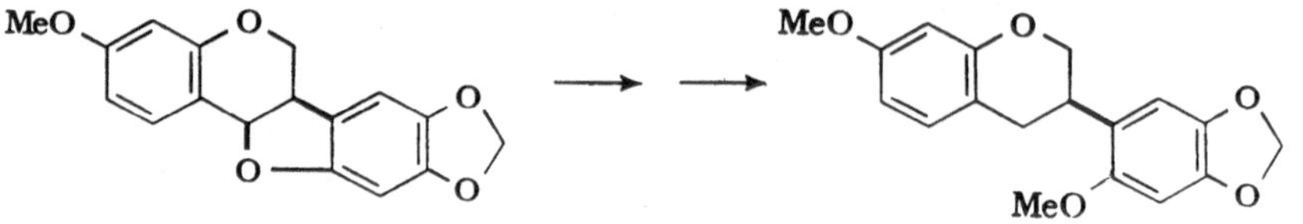

(**140**) (—)-Pterocarpin

(**141**) (—)-7,2'-Dimethoxy-4,'5'-methylenedioxyisoflavan.

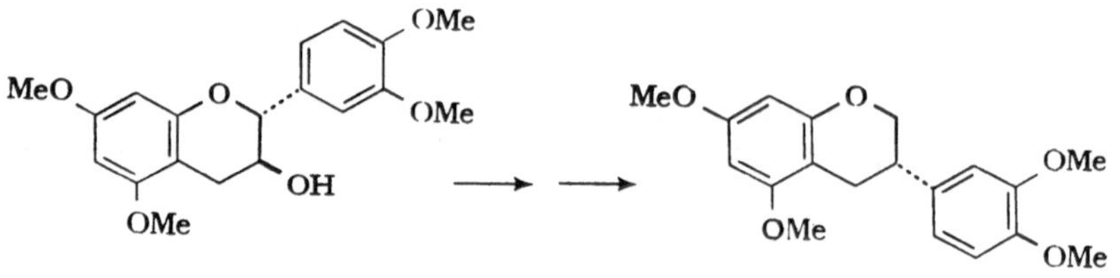

(2 S,3 S)-(+)-Catechin tetramethyl ether

(**142**) (S)-(—)-5,7,3',4'-Tetramethoxyisoflavan.

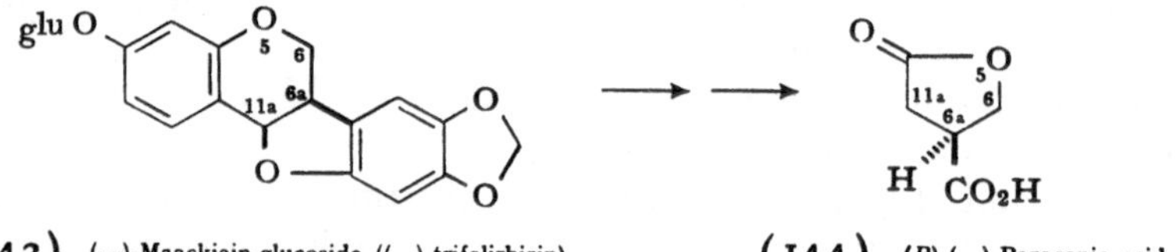

(**143**) (—)-Maackiain-glucoside ((—)-trifolirhizin)

(**144**) (R)-(—)-Paraconic acid

Chart 9. Stereochemical Correlations of (—)-Pterocarpans

References, pp. 63—73

has been established as R by degradation to (R)-$(-)$-paraconic acid (**144**) (*109*). (3) The relative configuration of the C-6a, C-11a, asymmetric centres for homopterocarpin is *cis* on the basis of n. m. r. evidence (*182*). This being accepted, the absolute configuration for the $(-)$-pterocarpans follows necessarily as 6aR, 11aR, (**139**) and that of the $(+)$-enantiomers as 6aS, 11aS.

In the n. m. r. spectrum of pterocarpans, the heterocyclic ring protons give rise to a highly complex four-spin system, with long range coupling preventing first oder analysis (*32*, *159*). A recent detailed analysis by PACHLER and UNDERWOOD (*149*) however has allowed accurate assignments of the coupling constants between these protons to be made. The *cis* arrangement of the 6a- and 11a-protons was confirmed; furthermore, the conclusion was reached that the normal preferred conformation of the (6aR, 11aR)-pterocarpan ring system is (**145**) rather than the alternative (**146**). In this conformation, rings A, C and D of the two enantiomorphic forms are superimposable. This fact has been used by PERRIN and CRUICKSHANK (*161*) to explain their interesting finding that both $(+)$- and $(-)$-forms of some pterocarpans possess comparable antifungal activity.

(**145**)   (**146**)

3. Reactions and Synthesis of Pterocarpans

Reductive ring opening of pterocarpans to 2'-hydroxyisoflavans constitutes an important degradative reaction of these compounds *(Chart 10)* (*34, 28, 155*). Acid treatment of pterocarpans yields resinous products (*31, 174*) presumably from intermediate formation of isoflavenes. These can be obtained from pterocarpans by controlled action of acid (*18*). Reaction of pterocarpans with potassium amide in liquid ammonia leads to benzofuran derivatives of type (**147**) (*18*). Oxidation of the isoflavenes (**148**) and (**149**) with osmium tetroxide followed by the action of alkali, leads to the synthesis of $(\pm)$ pisatin (**150**) and $(\pm)$-variablin (**151**) (*18*).

The ring cyclisation of 4-isoflavanols possessing a free 2'-hydroxyl group is a general method for the synthesis of pterocarpans. The intermediate isoflavans can be readily obtained from sodium borohydride reduction of 2′-hydroxyisoflavones *(Chart 11)*. Natural pterocarpans thus synthesised (in racemic form) include: pterocarpin (*185*), homopterocarpin (*182, 1*), demethylhomopterocarpin (*44*), 3,4,9-trimethoxypterocarpan and 3-hydroxy-4,9-dimethoxypterocarpan (*119*).

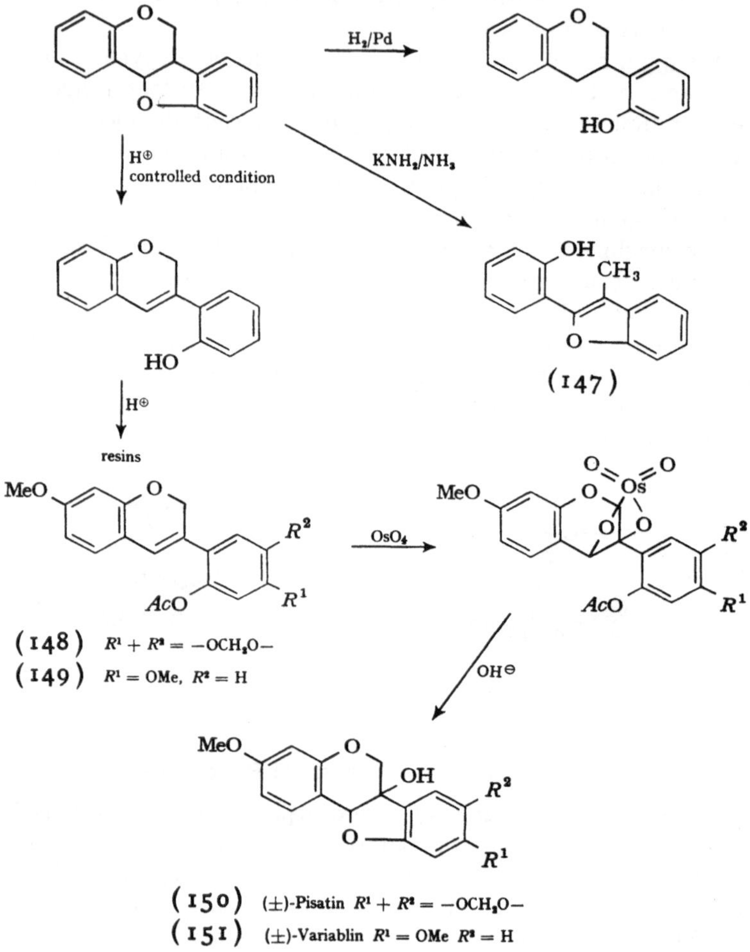
H₂/Pd
HO
H⊕
controlled condition
KNH₂/NH₃
OH
CH₃
HO
(147)
H⊕
resins
MeO
R²
OsO₄
AcO
R¹
(148) R¹ + R² = –OCH₂O–
(149) R¹ = OMe, R² = H
OH⊖
OH
(150) (±)-Pisatin R¹ + R² = –OCH₂O–
(151) (±)-Variablin R¹ = OMe R² = H

Chart 10. Some Reactions of Pterocarpans

A second synthetic route to pterocarpans, involving coumestans (Section VIII) and dehydropterocarpans as intermediate products, has been developed by Fukui and co-workers (*73*). The general scheme is shown also in *Chart 11*. Compounds synthesised by this route include dehydrohomopterocarpin (*29*), pterocarpin (*73*), maackiain (*77*) and 4-methoxypterocarpin (*28, 78*).

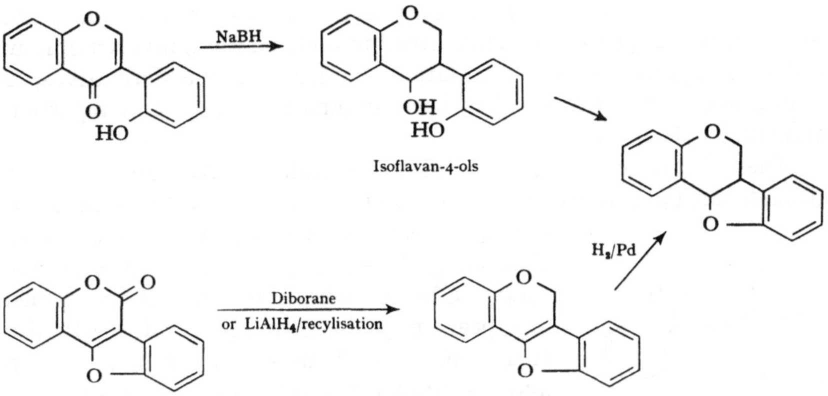

Chart 11. General Synthetic Routes to Pterocarpans

VI. Isoflavans

Isoflavans represent the most reduced of the isoflavonoid modifications. For many years this class was exemplified in nature by only one compound, equol (152). Since equol has been isolated only from the urine of animals, doubt has been cast on its classification as a natural product (*143*). Recent studies have shown that it is a metabolic product of isoflavones in poultry (*37*) and sheep (*30*). The occurrence of isoflavans in plants however has very recently been established by three independent groups of workers (*124, 174a, 155*).

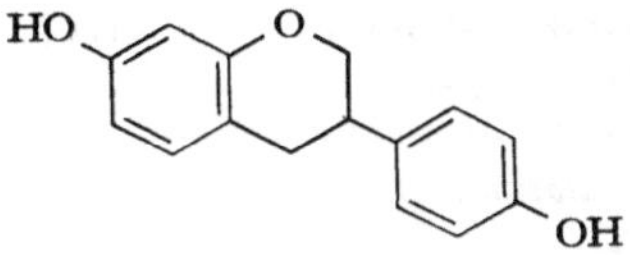

(152) Equol

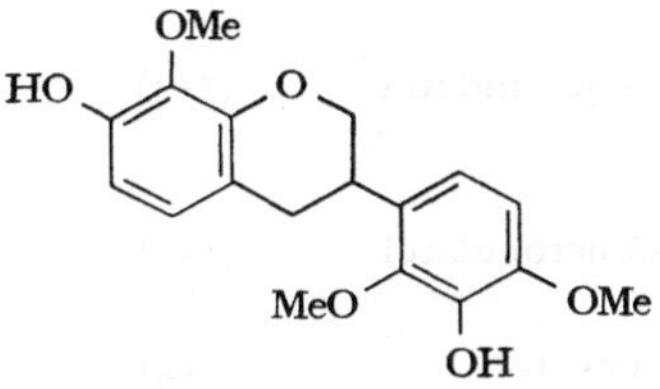

(153) Duartin

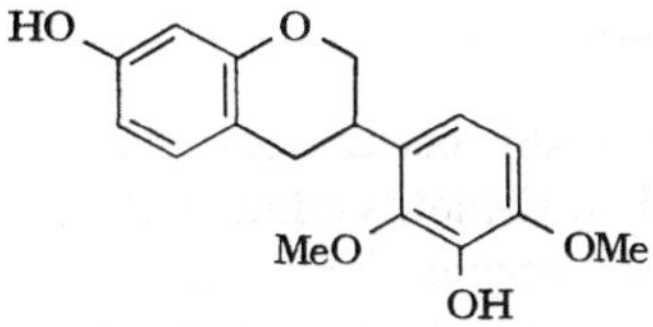

(154) Mucronulatol

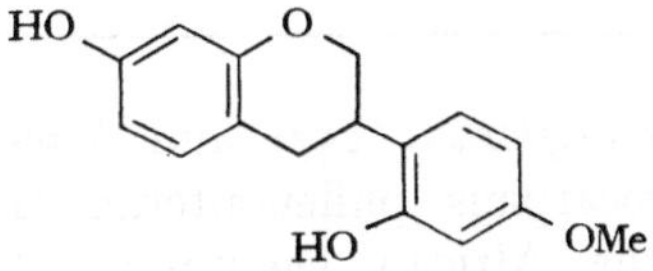

(155) Vestitol

Duratin (153), *mucronulatol* (154), and *vestitol* (155) occur in various Brazilian woods (Table 4). Their structural elucidation, involving n. m. r. and mass spectrometry, degradation via isoflavanone derivatives, and confirmation by synthesis, has been summarised (*143*) and reported in preliminary form (*124*).

These isoflavans were isolated in optically active forms and their absolute configurations have been studied (*123*) by comparison of their o. r. d. curves in the region 200–300 nm with that of (3 S)-(—)-5,7,3′,4′-tetramethoxyisoflavan [(142), *Chart 9*] whose absolute stereochemistry has previously been unequivocally established (*141, 188, 123*). Thus (—)-vestitol, (—)-duartin and (—)-mucronulatol were all shown (*123*) to have the 3 S-configuration, as represented in type formula (156).

(156) (3S)-Isoflavans

O. r. d. comparison confined to the long wavelength region only has been shown to be unsuitable for deciding configurational relationships in isoflavans (*123, 155, 188*). The configuration of (—)-equol, previously assigned 3 R, has now been shown to be 3 S on the basis of o. r. d. comparison at shorter wavelengths (*123, 188*).

Table 4. *Natural Isoflavans*

Name	Formula	Source	References
(—)-Duartin	(153)	*Machaerium acutifolium* *M. mucronulatum, M. opacum, M. villosum*	(*124*)
(—)-Mucronulatol	(154)	*M. acutifolium, M. mucro- nulatum, M. opacum, M. vestitum, M. villosum*	(*124*)
(±)-Mucronulatol	(154)	*M. mucronulatum, Dalbergia variabilis*	(*124*)
(+) Vestitol	(155)	*D. variabilis, Machaerium vestitum*	(*124*)
(+)-Laxifloran	(157)	*Lonchocarpus laxiflorus*	(*155*)
(+)-Lonchocarpan	(158)	*Lonchocarpus laxiflorus*	(*155*)
(+)-Licoricidin	(166a)	*Glycyrrhiza glabra*	(*174a*)

Laxifloran (157) and *lonchocarpan* (158) occur together with the pterocarpans philenopteran (132) and 9-O-methylphilenopteran (133) in the African leguminous tree *Lonchocarpus laxiflorus* (*155*). The assignment of the substituent groups in the highly substituted B-ring in these compounds presented problems which were solved by application of a combination of physical methods. On mass spectrometric evidence lonchocarpan was found to have one hydroxyl and three methoxyl groups in ring B. Comparison of the n. m. r. spectra of lonchocarpan dimethyl ether in deutero-chloroform and benzene indicated that two methoxyl groups are adjacent to the aromatic protons. Of the three possible substitution patterns (159)–(161) for ring B, (159) was thus excluded and (161) was favoured over (160) on biogenetic grounds. This structure was proved by synthesis, thus establishing the oxygenation pattern of lonchocarpan. The exact location of the B-ring hydroxyl group in lonchocarpan itself was deduced from a combined study on methoxyl-proton shifts and i. r. stretching frequencies.

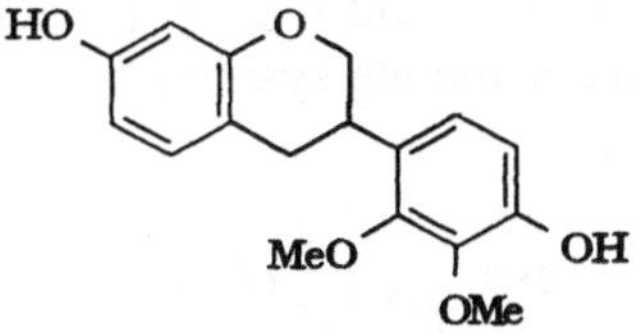

(**157**) Laxifloran

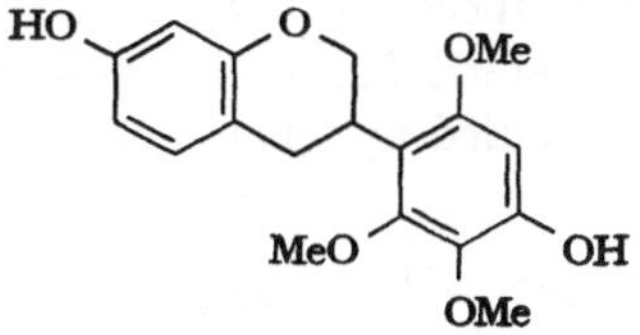

(**158**) Lochocarpan

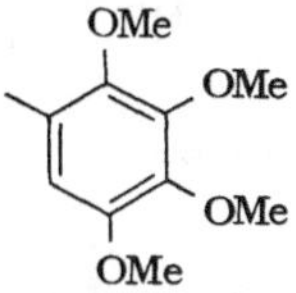

(**159**) 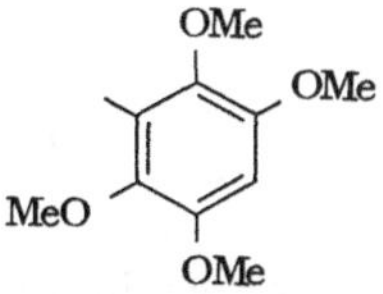   (**160**) 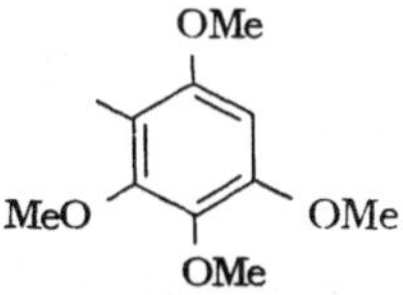   (**161**)

(+)-Laxifloran dimethyl ether and (+)-lochocarpan dimethyl ether have been shown to have the R configuration at C-3 (162), (163) (*155*) in contrast to the enantiomeric 3S-configuration assigned to the isoflavans (153), (155) from *Machaerium* spp. (*123*) described earlier.

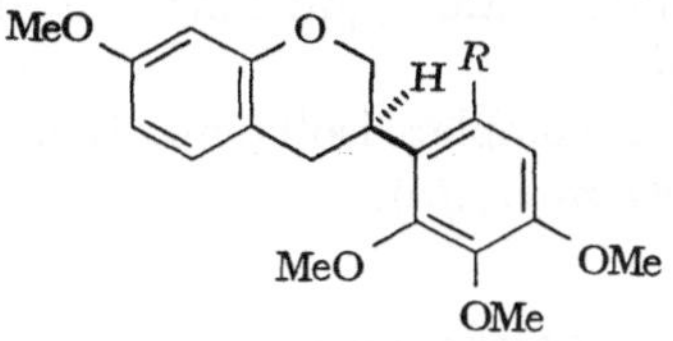

(**162**) (+)-Laxifloran dimethyl ether $R = H$

(**163**) (+)-Lonchocarpan dimethyl ether
$R = OMe$

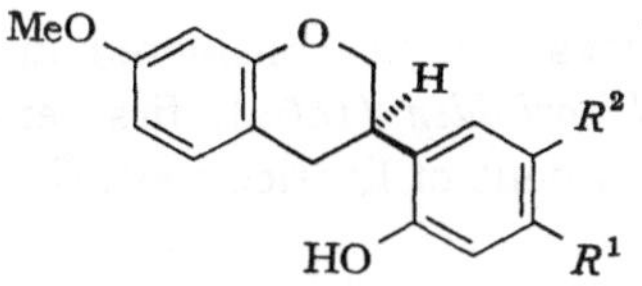

(**164**) (3 R)-(−)- 7-Methoxy-2′-hydroxy-
4′,5′-methylenedioxyisoflavan
$R^1 + R^2 = -OCH_2O-$

(**165**) (3 R)-(−)-7,4′-Dimethoxy-2′-hydroxyisoflavan
$R^1 = OMe, \ R^2 = H$

3*

The absolute configuration for (162) and (163) is related to that of (6aR,11aR)-pterocarpans (section V-2). (+)-Laxifloran dimethyl ether (162) had o. r. d. characteristics similar to those of the derived isoflavans (3R)-(−)-7-Methoxy-2'-hydroxy-4',5'-methylenedioxyisoflavan (164) and (3R)-(−)-7,4'-Dimethoxy-2'-hydroxy-isoflavan (165) in the region 300–220 nm. (+)-Lonchocarpan domethyl ether (163) however, gave a dissimular o. r. d. curve suggesting an enantiomeric configuration to the compounds (162), (164) and (165). That this was not so was proved by the preparation of this compound from (−)-philopteran (166). The atypical o. r. d. curve for (163) was attributed to the different conformation adopted by the heterocyclic ring, due to steric effects of the 2',6'-substituted ring B. This conformational effect was also reflected in the n. m. r. spectrum (155). Similar effects on o. r. d. and n. m. r. properties due to steric factors in these conformationally mobile systems have also been studied by OLLIS and co-workers (123).

(166) (−)-Philenopteran   (166a) Licoricidin

It is noteworthy therefore that as in pterocarpans, isoflavans can exist in antipodal forms in nature. Furthermore, in cases where both classes of compounds are found in the same plant, the stereochemistry with respect to the equivalent assymetric carbon centre (C-3 and C-6a) in the two classes is identical. Thus in *L. laxiflorus*, a common R configurational relationship is found (155), whilst in *Machaerium vestitum* and *Dalbergia variablis* (143) (+)-demethylhomopterocarpin, belonging to the S series, co-occurs with the (3S)-isoflavans (−)-mucronulatol and (+)-vestitol (124, 123). The possible biogenetic significance of these findings will be considered in Section X.

Licorioidin (166a), first example of a complex isoflavan, is a new constituent of licorice root, *Glycyrrhiza glabra* (174a).

VII. 3-Aryl-4-hydroxycoumarins

The recognition of this class of isoflavonoid compounds as natural products dates only from 1964, with the report by PELTER and associates (115) of the structurel elucidation of *lonchocarpic acid* (167) and *scandenin* (168) from *Derris scandens*. These workers later characterised *robustic acid* (169) and its methyl ether (170) from the related *D. robusta* species (*114*). OLLIS and associates, working independently, isolated in addition *lonchocarpin* (171) (from *D. scandens*) (*66*) and *robustin* (172), *robustin methyl ether* (173) and *derrusnin* (174) (from *D. robusta*) (*63*). These compounds thus form a remarkable series of closely related new plant constituents. The co-occurrence of similarly substituted complex isoflavones in these plants (*66, 63, 156*) has already been mentioned.

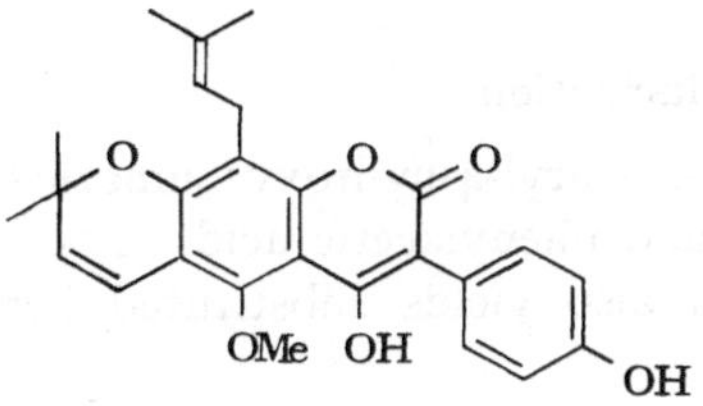

(167)   Lonchocarpic acid

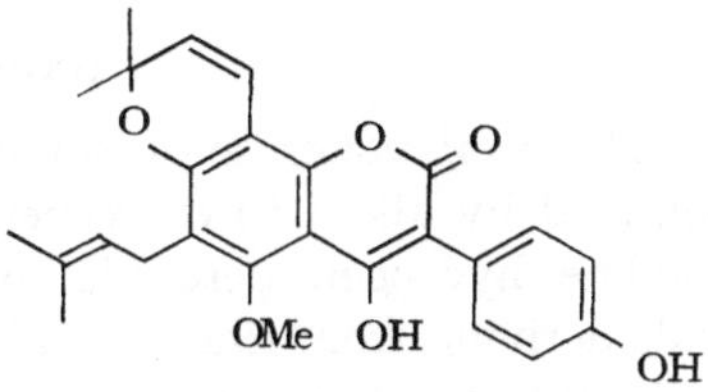

(168)   Scandenin

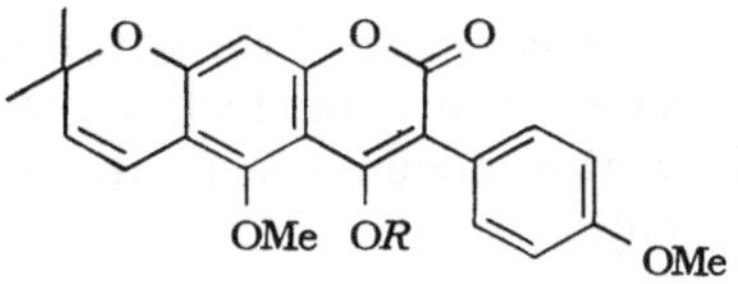

(169)   Robustic acid R = H
(170)   Robustic acid methyl ether R = Me

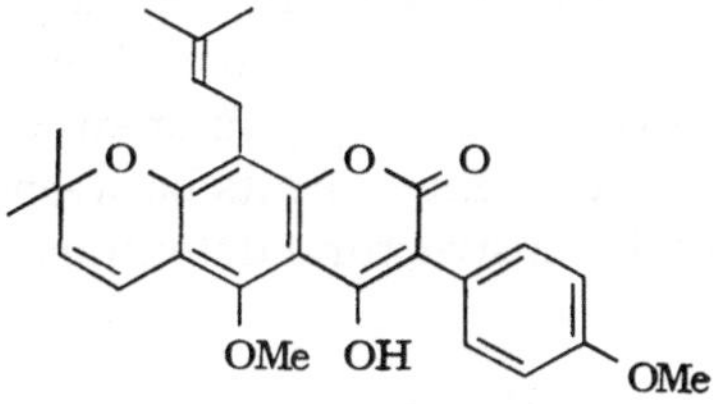

(171)   Lonchocarpin

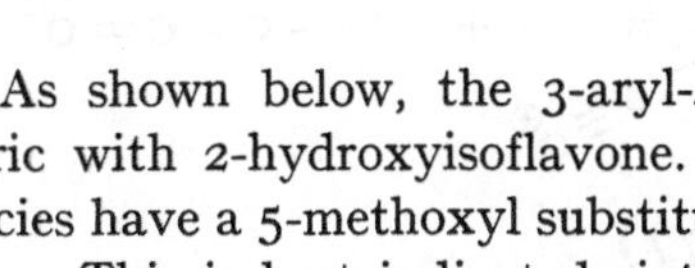

(172)   Robustin R = H
(173)   Robustin methyl ether R = Me

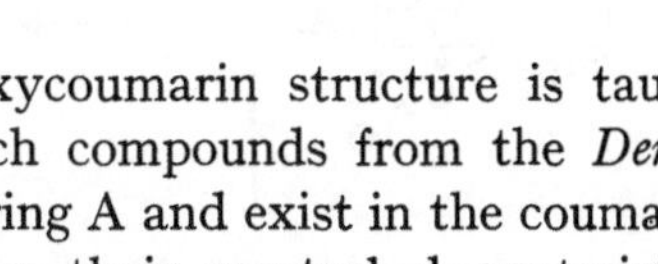

(174)   Derrusnin

As shown below, the 3-aryl-4-hydroxycoumarin structure is tautomeric with 2-hydroxyisoflavone. All such compounds from the *Derris* species have a 5-methoxyl substituent in ring A and exist in the coumarin form. This is best indicated, *inter alia*, by their spectral characteristics (u. v. λ max $\sim$ 230, $\sim$ 270, and $\sim$ 350 nm; i. r. $\nu_{co} \sim$ 1710 cm^{-1}) which resemble those of coumarins rather than isoflavones (*63, 66*). When a free 5-hydroxyl group is present in the molecule however, the 2-hydroxyisoflavone tautomer exists in large proportion in equilibrium (*114*). This is due presumably to the stabilising effect of hydrogen bonding between the peri 5-hydroxyl and 4-carbonyl groups. 3-Aryl-4-hydroxycoumarins can therefore be regarded as isoflavonoid counterparts (isoflavonols) of the well known flavonols in the flavonoid series.

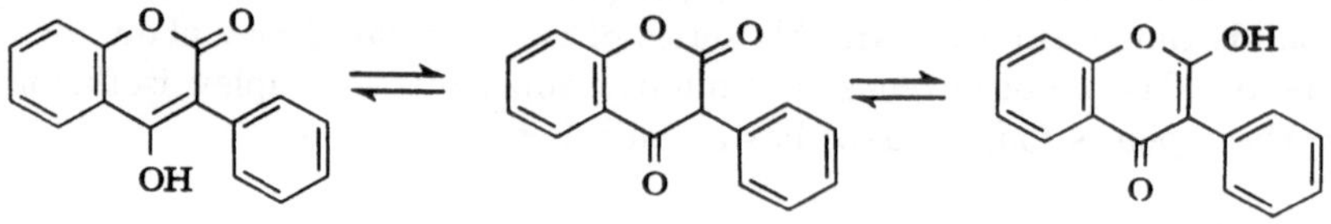

1. Structure Elucidation

Like isoflavones and isoflavanones, 3-aryl-4-hydroxycoumarins are degraded by alkali to deoxybenzoins and phenylacetic acids (*115*, *114*). Alkaline hydrogen peroxide oxidation also yields substituted benzoic acids deriving from rings A and B (*63*).

In the structure elucidation of the complex 3-aryl-4-hydroxycoumarins from *Derris* spps., physical methods, particularly n. m. r. and mass spectrometry were indispensable. OLLIS (*143*), in reviewing the work on the structural assignment of these compounds, emphasised the advantages of investigating a group of structurally-related natural products together; correlation experiments between the different compounds facilitated their individual structural determinations (*63*, *66*).

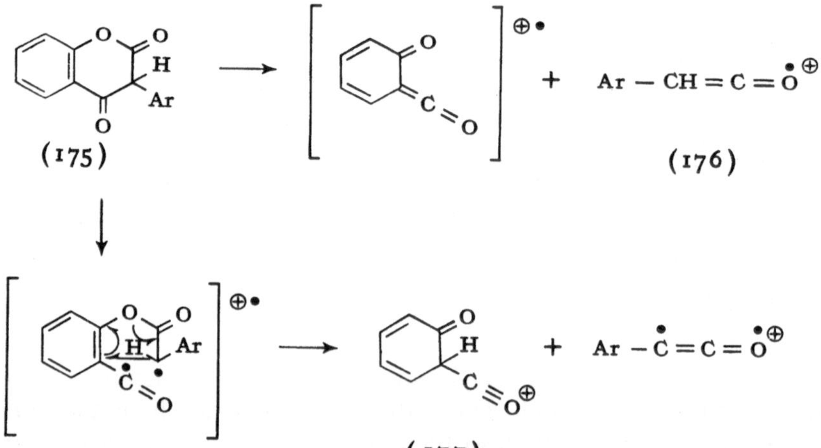

Chart 12. Fragmentation Patterns of 3-Aryl-4-hydroxycoumarins

Mass spectrometry has been particularly useful for the structure elucidation of 3-aryl-4-hydroxycoumarins (*115*, *114*). These compounds apparently fragment in the form (175) by the two routes shown in *Chart 12*

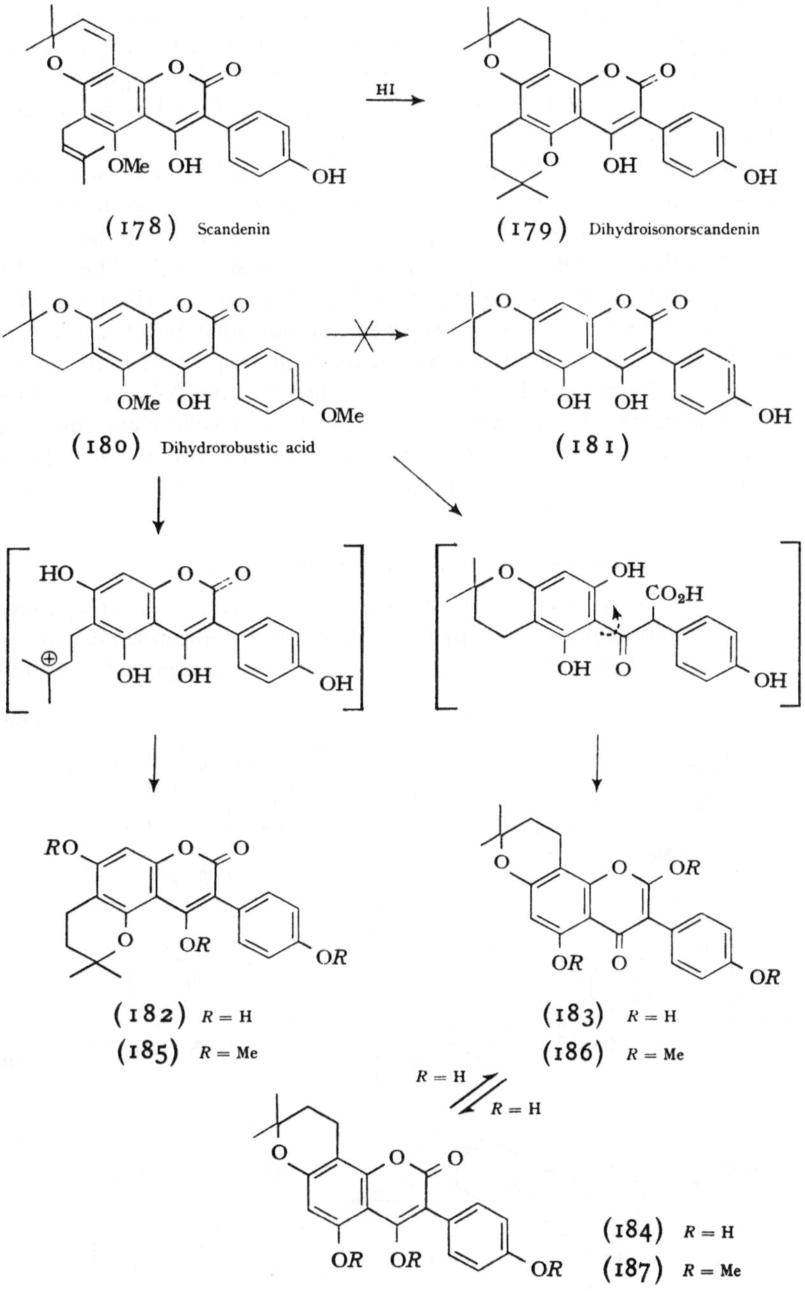

(178) Scandenin

(179) Dihydroisonorscandenin

(180) Dihydrorobustic acid

(181)

(182) R = H
(185) R = Me

(183) R = H
(186) R = Me

R = H
R = H

(184) R = H
(187) R = Me

Chart 13. Rearrangement Reactions of Scandenin and Dihydrorobustic Acid on Demethylation

(*115*, *158*). All four fragments are in fact capable of stabilising the positive charge. According to the scheme shown, the original hydrogen atom of the heterocyclic ring becomes attached to both the fragments, (176) containing ring B, and (177) containing ring A. This has been proved by deuterium-exchange studies (*158*).

Demethylation studies in this series of complex structures led to the observation of various rearrangement reactions. Reaction of scandenin (178) with hydriodic acid gave dihydroisonorscandenin (179), the constitution of which was proved by synthesis (*115*). The products of demethylation of dihydrorobustic acid (180) are summarised in *Chart 13* (*114*). The expected product (181) was not obtained but the three produced, (182)–(184), were characterised as their respective methyl ethers (185)–(187). These products clearly indicate that WESSELEY-MOSER type rearrangements involving both of the heterocyclic rings must have occurred. The isolation of (186) as a major product also showed that the 2-hydroxyisoflavone tautomer is important in 5-hydroxy-compounds of this type (*114*).

2. Synthesis of 3-Aryl-4-hydroxycoumarins

The 3-aryl-4-hydroxycoumarins are easily synthesised from condensation of a 2-hydroxydeoxybenzoin with methyl chloroformate followed by hydrolysis (*82*). Two novel routes to these compounds involving

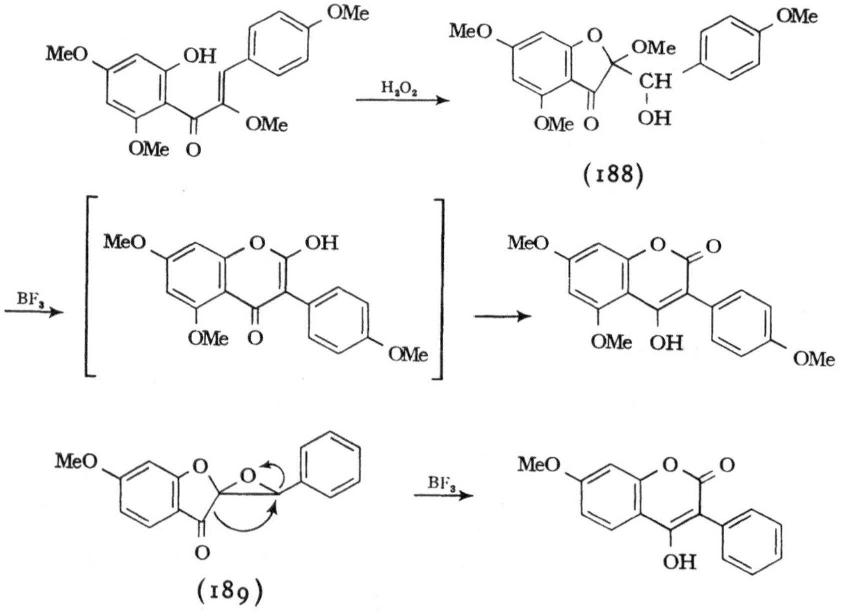

Chart 14. Novel Synthetic Routes to 3-Aryl-4-hydroxycoumarins

boron trifluoride-etherate catalysed rearrangements of either a 2-α-hydroxybenzyl-2-methoxycoumaranone (188) (*112*) or a related aurone epoxide (189) (*80*) are illustrated in *Chart 14*. It should be noted that in these reactions, it is the benzoyl group of ring A that undergoes migration. A similar type of reaction leading to 2-substituted isoflavones has previously been described by DEAN and PODIMUANG (*53*).

VIII. Coumestans

The trivial name coumestan has been proposed (*54*) for the coumaronocoumarin structure (167) which represents the highest possible level of oxidation of the isoflavonoid skeleton. The structural similarity of coumestans to both 3-aryl-4-hydroxycoumarins and 6a,11a-dehydro-pterocarpans (168) is obvious.

(167) Coumestan   (168)

1. Natural Coumestans

The first example of a natural coumestan, wedelolactone (169), was reported by GOVINDACHARI *et al.* in 1957 (*85*). Since then 16 additional compounds having this basic ring system have been isolated from various leguminous plants, the earlier ones (*171*) being coumestrol (171) from *Trifolium repens* and *Medicago sativa*, erosnin (172) from *Pachyrrhizus erosus*, psoralidin (173) from *Psoralea corylifolia* and norwedelolactone (170) from *Wedelia calendulacea* and *Eclipta alba*. BICKOFF and associates more recently isolated from alfalfa *(Medicago sativa)* trifoliol (174) (*126*), medicagol (175) (*127*), lucernol (176) and sativol (177) (*177*), and the three compounds represented by formulas (178)–(180) (*25, 178, 24*).

(169) Wedelolactone *R* = Me
(170) Norwedelolactone *R* = H

(171) Coumestrol

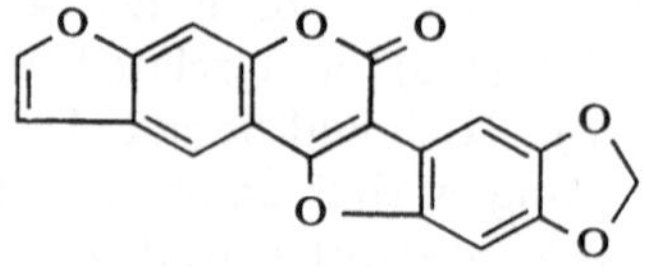

(172) Erosnin

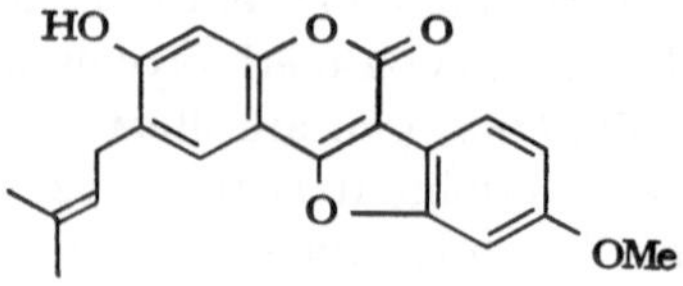

(173) Psoralidin

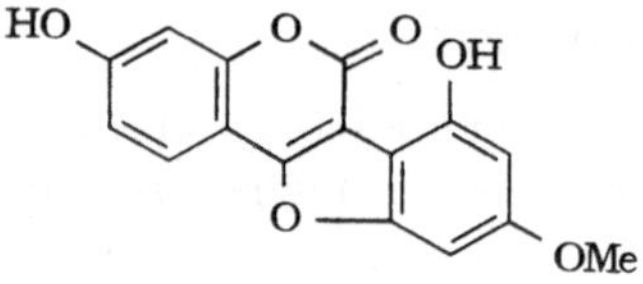

(174) Trifoliol

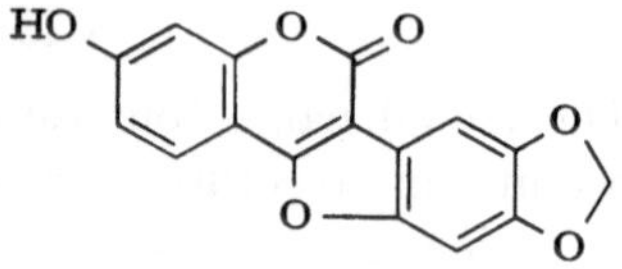

(175) Medicagol

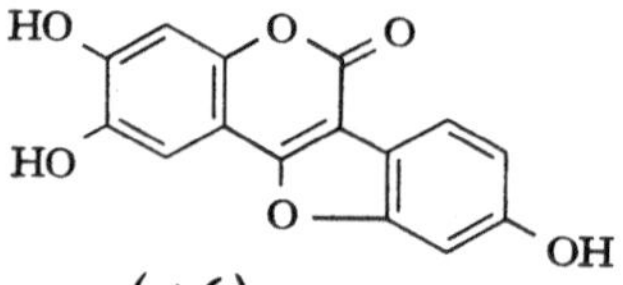

(176) Lucernol

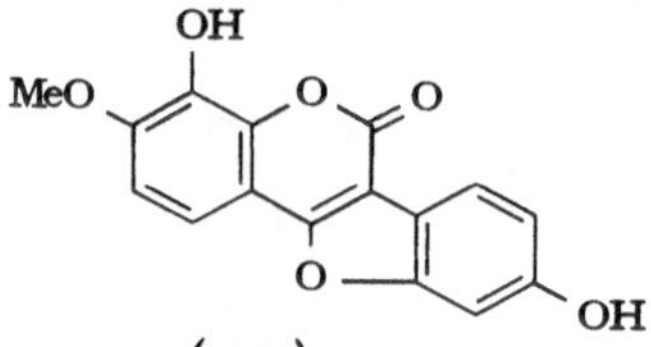

(177) Sativol

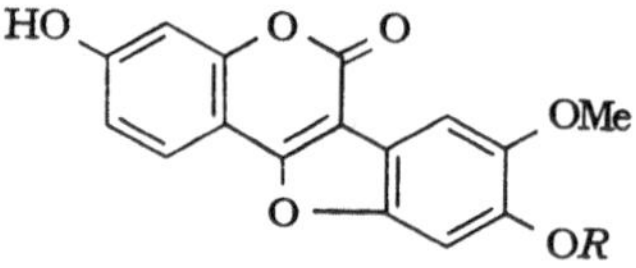

(178) R = H
(179) R = Me

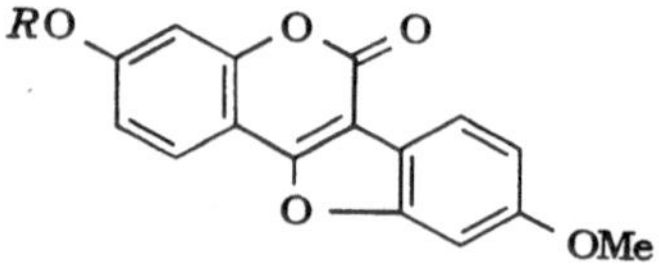

(180) R = H
(181) R = Me

These authors, while advocating (*126, 25*) the coumestan nomenclature, with the numbering system shown in (**182**), used indiscriminately at the same time the nomenclature and numbering system shown in (**183**). Thus (**178**) was named 3'-methoxycoumestrol (*25*) but (**179**) became 7-hydroxy-11,12-dimethoxycoumestan (**178**) and (**180**) was variously called 4'-O-methylcoumestrol (*24*) and (incorrectly) 4'-methoxycoumestrol (*25*). To add to the confusion, compound (**181**), isolated from *Swartzia madagascariensis*, has been named 3,9-dimethyl-6-oxopterocarp-6a-en by HARPER *et al.* (*103*) (reflecting its undoubted relationship with the many ptero-carpans co-occurring in that plant). The name coumestan seems to have become established and should be used as a generic mane for these compounds. Unfortunately the numbering system used with this nomenclature (**182**) is rather arbitrary and lacks the advantages of one based on either the fully systematic Ring Index system (**184**) or that of an isoflavone derivative (**185**). Ideally, from the broad point of view of these compounds as isoflavonoids, the retention of the number system of the isoflavones as in (**185**) is to be preferred. This comment applies equally to the other isoflavonoids with additional cyclic systems, viz. pterocarpans and rote-noids.

References, pp. 63—73

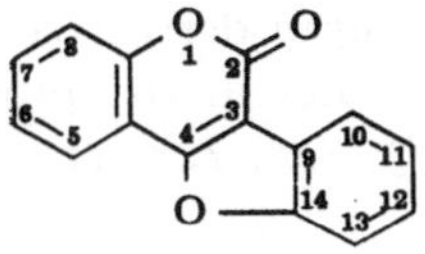

(182)

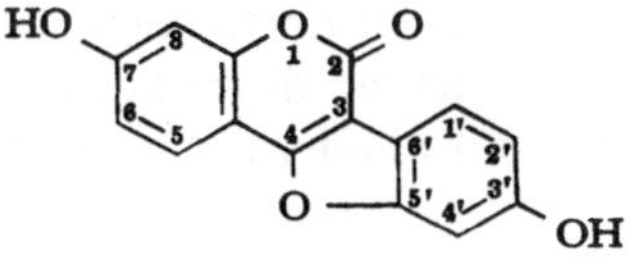

(183) Coumestrol

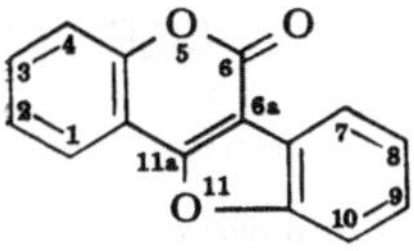

(184)

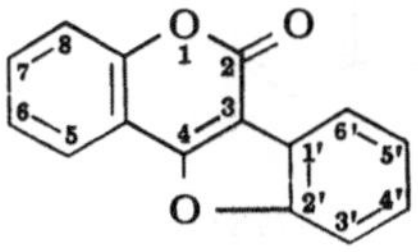

(185)

Sojagol (186) occurs in the root of mung bean (Phaseolus aureus) (201). The presence of the rare 2,2-dimethylchroman ring system is an interesting feature in this compound. The most recent additions to the list of natural coumestans are isoglycyrol (187), glycyrol (188) and 5-O-methylglycyrol (189), three new constituents of the roots of licorice (Glycyrrhiza spp.) (168).

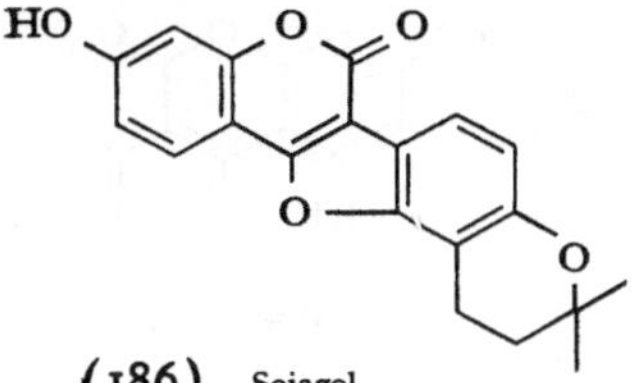

(186) Sojagol

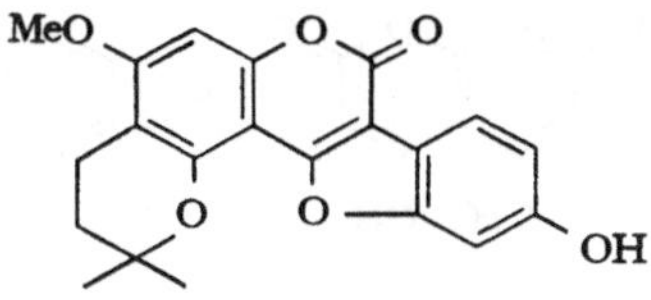

(187) Isoglycyrol

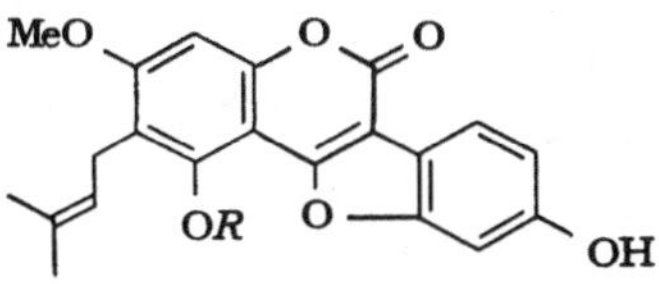

(188) Glycyrol R = H
(189) 5-O-Methylglycyrol R = Me

2. Structure Elucidation and Synthesis

The u. v. spectra of coumestans are similar to those of flavones and flavonols (126, 177, 25). Their i. r. spectra however, differ from those of the γ-chromones in having a band $\nu_{co} \sim 1710$ cm^{-1} due to the lactone carbonyl group (126).

Coumestans in general can be systematically degraded by methylative ring opening ((190), $R = CO_2Me$) hydrolysis ((190), $R = CO_2H$), and decarboxylation to their more soluble benzofuran derivatives ((190), $R = H$). Further degradation *via* ozonolysis results in the production of acidic and aldehydic fragments *(Chart 15)* (*85, 126, 127, 177, 25*).

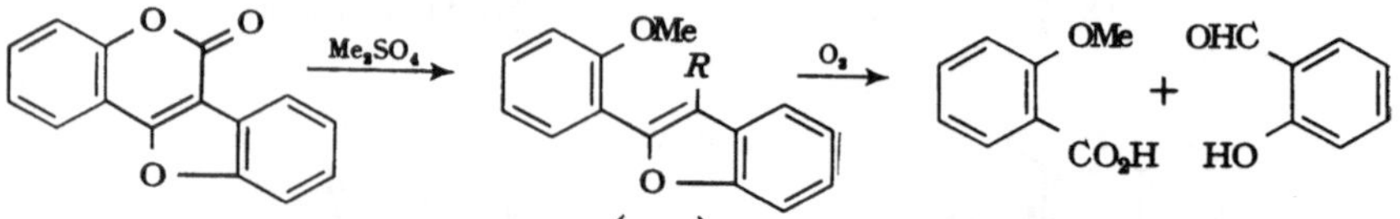

Chart 15. Systematic Degradation of Coumestans

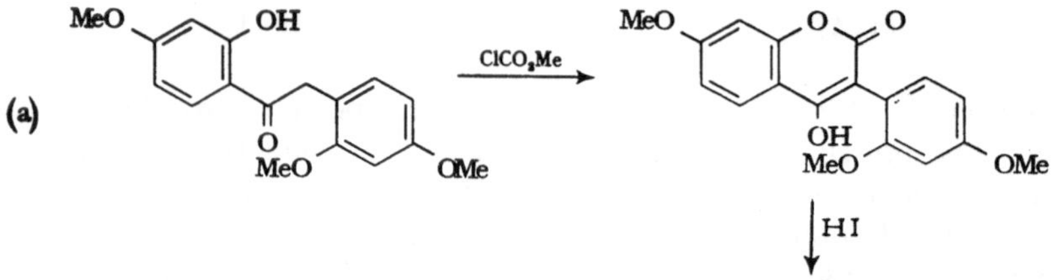

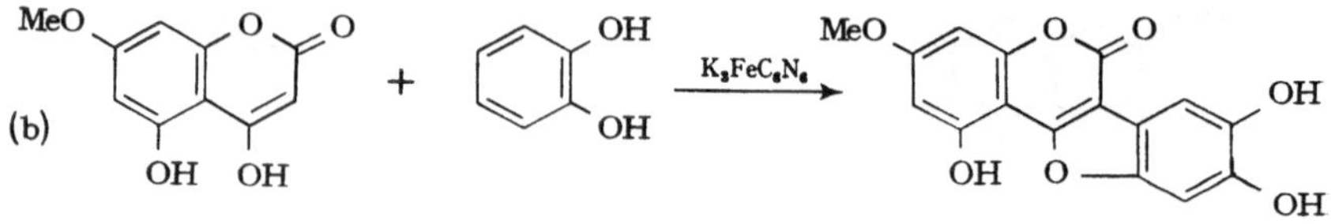

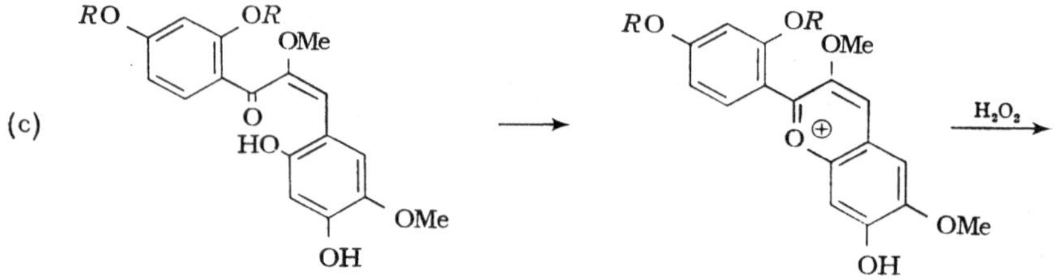

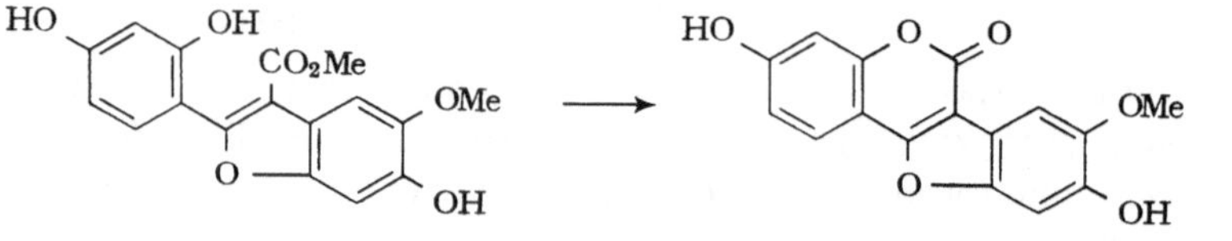

Chart 16. Synthetic Routes to Coumestans

References, pp. 63—73

Several approaches have been employed for the synthesis of coumestans. The most direct route involves the cyclization of a 3-aryl-4-
hydroxycoumarin possessing a 2'-hydroxyl group in ring B. The synthesis of coumestrol by this method (64) is illustrated in Chart 16.
Lucernol has also been synthesised by this route (118). 4-Hydroxy-
coumarins have successfully been converted to coumestans by condensation with catechol under dehydrogenating conditions. WANZLICK et al.
(190) first introduced this method for the synthesis of wedelolactone
(scheme b, Chart 16). Medicagol (127) and 7-hydroxy-11,12-dimethoxy-
coumestan (131) have similarly been synthesised via this route.

A novel general route to coumestans, due to JURD (116), is illustrated
in scheme c (Chart 16) for the synthesis of 7,12-dihydroxy-11-methoxy-
coumestan (192) (25). The key step in this process involves the hydrogen
peroxide oxidation of an appropriately substituted 2'-hydroxy-3-methoxy-
flavylium salt of type (191). Other coumestans synthesised by this method
include trifoliol (126), medicagol (117) and coumestrol (116).

Of biosynthetic interest is the reported formation of the coumestan
structure from 6a,11a-dehydropterocarpans (type formula (168)). The
methylene group in these compounds is easily oxidised to C=O by
chromium trioxide (29), or heating in air (55). The natural co-occurrence
of these two types of isoflavonoids in S. madagascariensis (103) has been
mentioned.

IX. Other Types of Isoflavonoids

Within this heading are included three compounds each of which is
the sole natural example thus far of a new isoflavonoid type. The
α-methyldeoxybenzoin, angolensin (193), has been known since 1952.
It occurs in the heartwood of Pterocarpus species, usually in company of
other isoflavonoids. It has also been isolated recently from teak wood
(Pericopsis spp.) (107). The absolute configuration of (−)-angolensin
has been determined as R, as shown in (193) (145, 42).

(193) (−)-Angolensin   (194) Pachyrrhizin

Pachyrrhizin from Pachyrrhizus erosus, was first characterised as
the 3-arylcoumarin (194) in 1957 by SIMONITSCH et al. (176). Its occurrence in the related species Neorautanenia pseudopachyrrhiza, N. edulus
and N. amboensis has since been reported (34, 51). Pachyrrhizin, to-

gether with neotenone (81), dehydroneotenone (51), neodulin (136),. erosnin (172) and dolineone (102) constitute a remarkable series of constituents within these species, all having the same substitution pattern but each representing a different class of isoflavonoid.

Among the isoflavonoid constituents of *Piscidia erythrina*, *lisetin* has recently been found to be a coumaronochromone, with the constitution shown in (195) (67). The oxidation level of lisetin, $C_{15}H_3O_3(OH)_3(OCH_3)$ (C_5H_9), excluded its formulation as an isoflavone but indicated a coumestan or a coumaronochromone type structure. The i. r. data for lisetin $(\nu_{co} = 1693$ cm$^{-1})$ indicated however that lisetin was not a compound of the former type $(\nu_{co} \sim 1700{-}1740$ cm$^{-1})$. Lisetin trimethyl ether (196), on hydrolysis by dilute alkali, yielded the 3-aryl-4-hydroxycoumarin (197). This provided confirmatory evidence that lisetin is a coumaronochromone and not a coumestan, since coumestans give hydroxy-acids under these conditions. The alkaline hydrolysis of coumaronochromones to 3-aryl-4-hydroxycoumarins has been suggested as a useful diagnostic reaction for this new type of natural product (67).

(195) Lisetin R = H
(196) R = Me

(197)

Final proof of the structure of lisetin was provided by its partial synthesis from the isoflavone piscerythrone (64), by means of oxidation with potassium ferricyanide.

X. Biogenesis of Isoflavonoids

The biosynthetic pathway for flavonoid compounds is now known in broad outline. The scheme shown in *Chart 17*, based on results from tracer studies to date (88, 196), summarises the probable biogenetic relationships among the different classes of flavonoids.

Isoflavonoids, on structural grounds, have long been considered to be related biogenetically to the flavonoids (79, 165, 92). Experimental support for this view came initially from tracer studies involving isoflavones. More recently these have been supplemented by studies involving other classes of isoflavonoids.

Chemicogenetic evidence for the close biosynthetic relationship between isoflavones and flavonoid compounds has recently been presented
(*197*). Chemically induced isoflavone deficient mutants of subterranean
clover *(Trifolium subterraneum)* also show modified patterns of the other
flavonoid constituents. One mutant, for example, had a reduced level of
all flavonoids concomitant with reduced isoflavones, whilst in another
all flavonoid constituents other than the isoflavones were greatly increased. These and other effects were explained in terms of genetic
blockages at different points of a common biosynthetic pathway for
isoflavones and the other flavonoids (*197*).

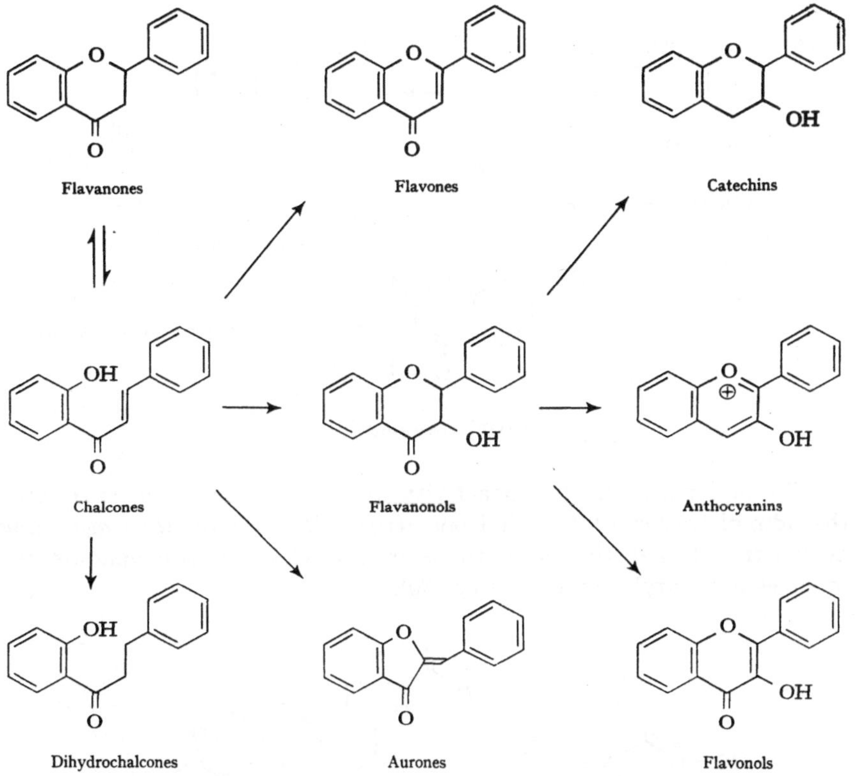

Flavanones   Flavones   Catechins

Chalcones   Flavanonols   Anthocyanins

Dihydrochalcones   Aurones   Flavonols

Chart 17. Probable Biogenetic Relationships among Flavonoids

In the following sections, isoflavonoid biogenesis will be examined
from two main points of view: (1) the biogenetic relationship of the group
as a whole to the flavonoids, and (2) the biogenetic relationship among
the isoflavonoid classes themselves.

1. 1,2-Aryl Migration as a Common Feature

The possession of a common 1,2-diphenylpropane skeleton (1) for the isoflavonoids implies that a rearrangement step is the distinguishing feature of their derivation from intermediates of the flavonoid series. The early work of Grisebach and his associates has demonstrated conclusively that formation of the isoflavone skeleton from precursors of the flavonoid pathway involves a 1,2-phenyl shift of the B-ring (86). Furthermore, results from feeding experiments (90, 91) using specifically labelled C_{15}-precursors of type (198) have clearly shown that the rearrangement step takes place at the chalcone stage or later. These earlier studies on isoflavone biosynthesis have been summarised (87).

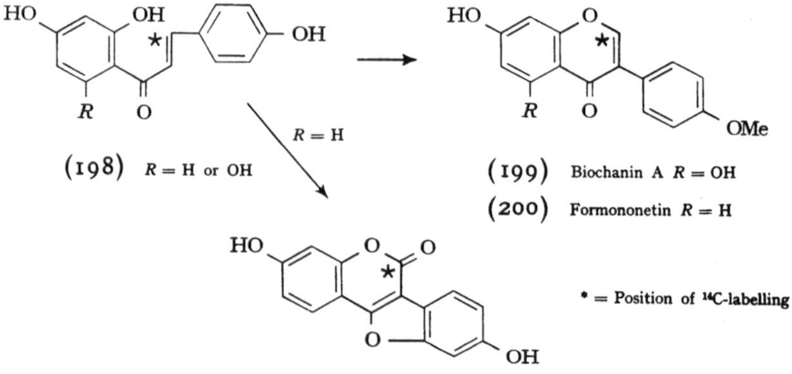

(198)   R = H or OH

(199)   Biochanin A   R = OH
(200)   Formononetin   R = H

(201)   Coumestrol

* = Position of ^{14}C-labelling

The incorporation of radioactivity at C-2 of coumestrol, (201) after the administration of the chalcone ((198), R = H) to *Medicago sativa* demonstrated that the biosynthesis of this class of isoflavonoids also involves a 1,2-aryl migration step (89).

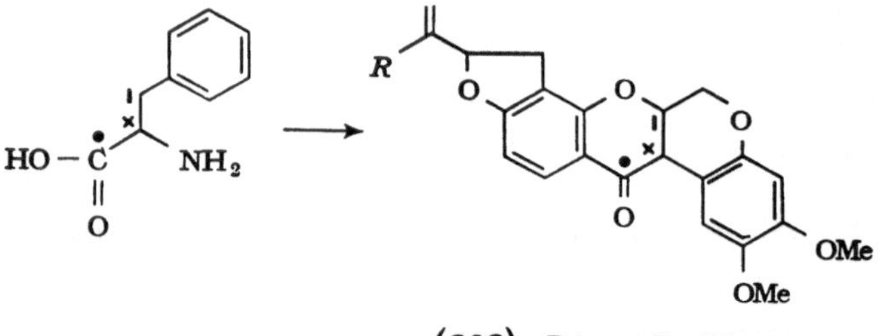

(202)   Rotenone   R = CH₃
(203)   Amorphigenin   R = CH₂OH

Chart 18. Labelling of Rotenoids from ^{14}C-Phenylalanine

Similarly, ·1,2-migration in the course of the elaboration of the rotenoid skeleton has recently been demonstrated by CROMBIE and THOMAS (50, 48). Feeding of ^{14}C-phenylalanine labelled variously at the 1-, 2- or 3-position to *Derris elliptica* resulted in the labelling of rotenone (202) in the 12-, 12a- or 6a-position respectively *(Chart 18)*. Similar results have been obtained for the biosynthesis of amorphigenin (203) in *Amorpha fruticosa* seeds (50).

While the occurrence of such a rearrangement step in the elaboration of the other classes of isoflavonoids has not been experimentally verified, there is little doubt that a similar step involving the 1,2-migration of the B ring is a characteristic feature in the biosynthesis of all isoflavonoids.

2. Branching Point from the Flavonoid Pathway

Given the proposition that all the isoflavonoids are derived from the flavonoid pathway, the question arises as to whether the different classes originate from the normal flavonoid pathway by parallel routes or whether they are formed sequentially one from another. The former supposition requires that several rearrangement-type reactions can take place, leading independently to different classes of isoflavonoid products, whilst in the latter scheme, a unique rearrangement step from some point on the flavonoid pathway leads to a primary isoflavonoid class which is the common progenitor of all the other classes of isoflavonoids.

Possession by plants of the capability to rearrange a flavonoid to an isoflavonoid structure is considered to be an isolated characteristic (98), since it is limited essentially to one group of phylogenetically related plants. In view of this rarity, the most plausible assumption is that only one such rearrangement step was evolved, leading from the flavonoid pathway to a unique isoflavonoid prototype. Once this unique enzymic step was acquired, the elaboration of the profusion of other isoflavonoid classes represented only a secondary phenomenon, requiring in essence only oxidation and reduction processes which need not necessaily be specific to isoflavonoid biosynthesis.

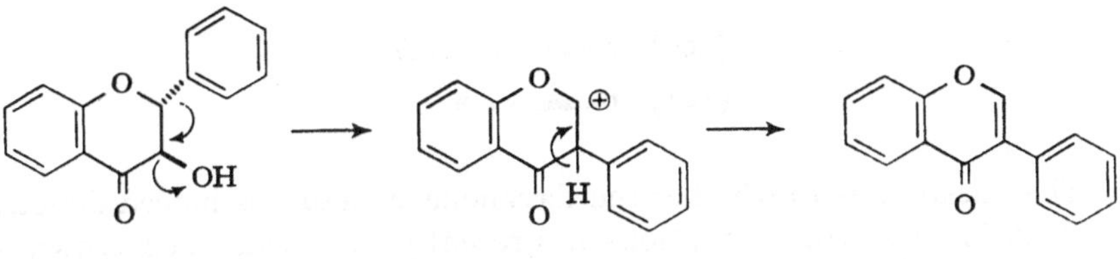

Chart 19. Previously Postulated Scheme for Isoflavone Biosynthesis

The important question now becomes — at what stage of the flavonoid pathway does this key rearrangement step take place? Earlier speculations (*165, 86*) have favoured the flavanonol structure as a suitable intermediate for conversion to an isoflavone, with the rearrangement step taking place in concert with elimination of the 3-hydroxyl group (*87*). This scheme *(Chart 19)*, although attractive on both mechanistic and stereochemical grounds, has been found to be untenable on the basis of tracer studies (*195, 199, 13, 15*).

Another possibility which has been considered was for the rearrangement to take place at the flavanone stage, after heterocyclic ring formation (*26*). Flavanones are isomeric with chalcones and an enzyme catalysing their conversion has recently been isolated and purified (*200, 137*). The flavanone naringenin (**204**) has been found to be a good precursor for the isoflavone biochanin A (199) (*153*), as well as other 5-oxygenated flavonoids (*88*), and the direct involvement of the flavanone structure in these transformation found apparent experimental verification when Grisebach and his co-workers (*154*) showed that the natural (—)-enantiomer of naringenin (**204**) was incorporated to a much greater extent than the (+)-enantiomer into biochanin A and other flavonoids.

(**204**)   Naringenin

(**205**)   Isoliquiritigenin   (**206**)   (—)-Liquiritigenin

(**207**)   Formononetin R = Me
(**208**)   Daidzein R = H

The question as to whether the flavanone structure is indeed directly involved in these transformations has recently been examined critically by Wong (*196*), who found that both *in vitro* with the purified enzyme

and *in vivo*, the chalcone isoliquiritigenin (**205**) and the flavanone liquiritigenin (**206**) are interconvertible. Parallel competitive feeding experiments in which either (a) ^{14}C-isoliquiritigenin diluted with an equal
amount of (—)-liquiritigenin, or (b) [14]C-(—)-liquiritigenin diluted similarly with isoliquiritigenin were fed as precursors to clover seeds showed
that the isoflavone products formononetin (**207**) and daidzein (**208**) had
higher specific activity in experiment (a) than in experiment (b) indicating that the isoflavone compounds were derived more directly from
the chalcone than from the flavanone. These conclusions have recently
been confirmed by double labelling studies (*198*). These results therefore
require that chalcones are converted to isoflavones without prior ring
closure to flavanones. The recent report by OLLIS *et al.* (*144*) of the
oxidative rearrangement of chalcones to isoflavonoid compounds by
thallic acetate provides a chemical demonstration of the feasibility of
this mode of biosynthesis.

3. Nature of the Primary Isoflavonoid Product

If a chalcone rearranges oxidatively to an isoflavonoid structure then
the primary isoflavonoid product could be an isoflavone. If no change
in the oxidation level is involved, then an isoflavanone would logically
be the primary product. In either case, the biogenetic relationship between isoflavone and isoflavanone may be assumed to be very close
(Chart 20).

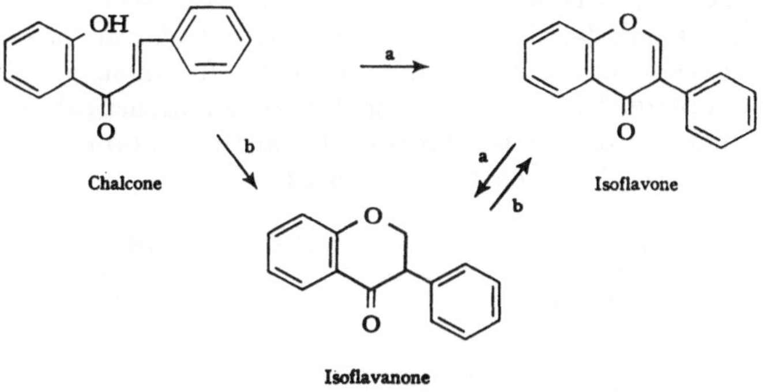

Chalcone

Isoflavone

Isoflavanone

Chart 20. Possible Relationships among Chalcone, Isoflavone and Isoflavanone

The nature of this biogenetic relationship has recently been studied
by GRISEBACH and ZILG (*93*). Feeding of tritium labelled dihydrodaidzein
(**209**) to mung bean seedlings *(Phaseolus aureus)* resulted in the labelling
of daidzein (**208**) but the extent of incorporation (.04–.09%) was only
about the same as that of the ^{14}C-liquiritigenin (**210**) fed simultaneously.

4*

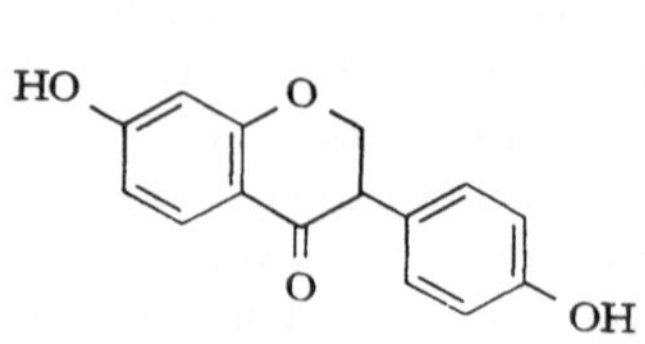

(209) Dihydrodaidzein 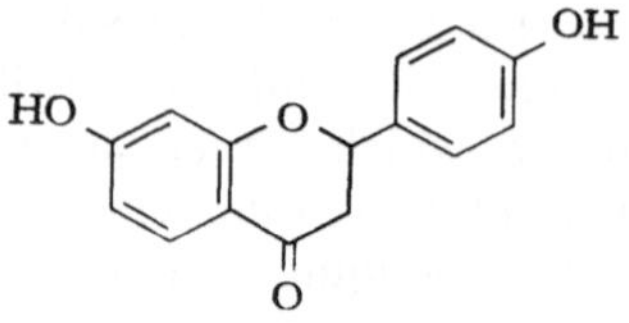   (210) Liquiritigenin

(In view of the discussion in the previous section concerning the role of chalcones and flavanones, incorporation of flavanones such as (210) in experiments of this type presumably proceeds *via* initial ring opening to chalcones.) In the same work (93), it was shown that the incorporation of the doubly-labelled flavanone (211) into formononetin in *Cicer arietinum* seedlings resulted in a 99% loss of the tritium activity (*Chart 21*).

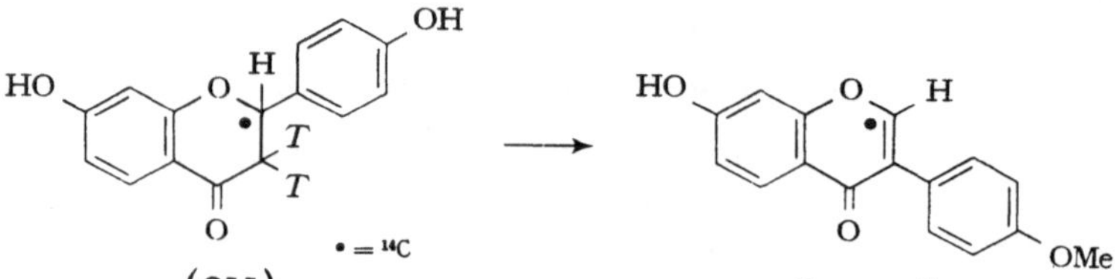

(211)   • = ^{14}C Formononetin

Chart 21. Incorporation of Flavanone into Isoflavone with loss of Tritium Activity

These results, although not unequivocal, would seem on balance to favour the sequence (a) chalcone → isoflavone → isoflavanone rather than the alternative sequence (b) which implicates isoflavanone as the primary isoflavonoid product (*Chart 20*). It may also be biogenetically significant that, in contrast to the isoflavones, the known natural isoflavanones nearly all have 2'-oxygenation in ring B.

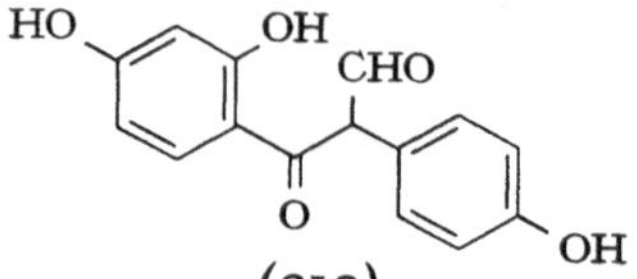

(212) 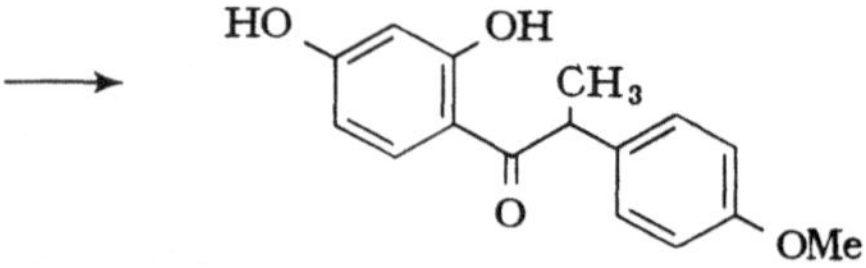   (213) Angolensin

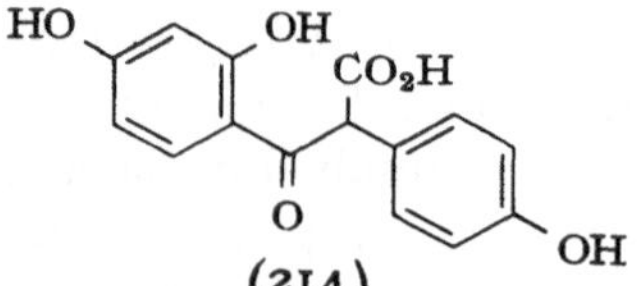

(214) 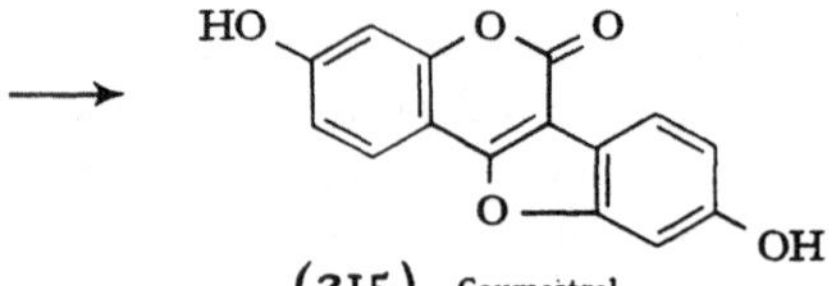   (215) Coumestrol

The existence of angolensin (213) as a natural product would seem at first to indicate that an open-chain intermediate, such as exemplified by (212), could constitute the primary isoflavonoid product from chalcone (86). Oxidation-reduction modifications of this intermediate could then lead to other isoflavanoids as illustrated by the hypothetical sequence (212) → (214) → (215) for coumestrol biosynthesis (86). In contrast to this view it is more likely that angolensin represents a metabolic product of other isoflavonoids. Recent studies (175) of isoflavone metabolism in sheep have shown that a compound of the angolensin type is a metabolic product of formononetin (216). This isoflavone co-occurs with angolensin in nearly all known sources of the latter. *Pterocarpus spicatus*, for example, contains in addition to these two constituents the branched C_9 compound (—)-p-hydroxyhydratropic acid (217) (45). These compounds could well be related metabolically as in the following sequence.

(216)   (217)

4. Biogenetic Relationships among the Isoflavonoid Classes

a) Direct Experimental Evidence

Direct experimental evidence for the interconversion *in vivo* of different classes of isoflavonoids is available only in a few cases *(Chart 22)*. BARZ and GRISEBACH (14) have shown in tracer experiments that the isoflavone diadzein (219) is converted into the corresponding coumestan, coumestrol (220). Tritium-labelled dihydrodaidzein (221) is similarly incorporated into coumestrol in *Phaseolus aureus*, with the extent of incorporation (.002–.02%, dilution factor 410–460) being about the same as that of ^{14}C-4',7-dihydroxyflavanone (218) fed simultaneously as a control (201). The conversion of dihydrodaidzein to daidzein in this plant (93) has already been mentioned. Unfortunately, results to date from tracer studies of this type have not allowed the precise nature of the biosynthetic relationship between these classes of compounds to be established.

Assuming the existence of a sequential biosynthetic relationship among the different classes of isoflavonoids, and in the absence of unequivocal direct evidence, it is worthwhile to see whether a study of the structures, distribution and co-occurrence patterns of all the natural isoflavonoids described in the foregoing sections would lead to deductions of plausible biosynthetic relationships among these compounds. Some conclusions drawn from such a survey are presented in the remaining sections of this review.

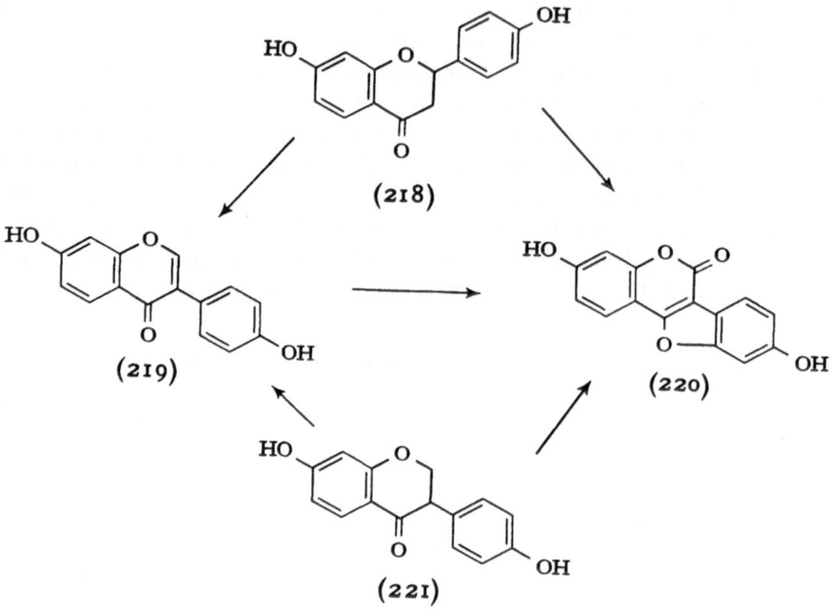

Chart 22. Interconversion of Compounds of Different Classes Shown by Tracer Studies

b) Co-occurrence of Classes of Isoflavonoids

The most striking fact regarding the isoflavonoids as a group is that the isoflavones are much more abundant than the other types. Furthermore, isoflavones are the compounds most frequently found co-occurring with the other types of isoflavonoids. Many examples of the co-occurrence of different classes of isoflavonoids in the same or related plants have already been cited. Table 5 summarises some of the more outstanding examples of this phenomenon. These facts are in excellent accord with the assumption that isoflavones are the primary isoflavonoid products from the flavonoid pathway, thus serving as progenitors of the other types of isoflavonoids. As can be seen in *Chart 1*, the isoflavones occupy a central intermediate position with respect to the oxidation level of the nucleus. They are capable of thus being converted to any of the other types by a minimum of oxidation-reduction steps.

Table 5. *Examples of Co-occurrence of Different Isoflavonoid Types*

	Isoflavones	Rotenoids	3-Aryl-4-hydroxy coumarins	Isoflavanones	Pterocarpans	Isoflavans	Coumestans	Other types
Millettia spp.	+	+						
Piscidia erythrina	+	+						+
Derris malaccensis	+	+						
Derris scandens	+		+					
Derris robusta	+		+					
Maackia amurensis	+			+	+			
Pterocarpus spp.	+				+			+
Dalbergia spp.	+			+	+	+		
Machaerium spp.	+				+	+		
Lonchocarpus laxiflorus					+	+		
Swartzia madagascariensis					+		+	
Pachyrrizus erosus	+	+		+			+	+
Neorautanenia spp.	+	+		+	+			+

c) Isoflavanones–Dehydropterocarpans–Pterocarpans–Isoflavans–Coumestans

A second very striking fact is that all known natural pterocarpans, dehydropterocarpans and 6a-hydroxypterocarpans, without exception belong to the 5-deoxy series of compounds. Most, but not all, of the isoflavans and coumestans also conform to this pattern. The possession by only these classes of isoflavonoids of this characteristic can be taken as an indication of some special biosynthetic relationship existing between them. *A priori*, this is a reasonable expectation on grounds of their common possession of the extra heterocyclic ring. The co-occurrence in *Swartzia madagascarinesis* (*103*) of the three compounds: homopterocarpin (**116**), dehydrohomopterocarpin (**128**) and 3,9-dimethyl-6-oxo-pterocarp-6a-en (**129**), all having the same pattern of substitution in the aromatic rings, but each representing a different class of these tetracyclic compounds, provides good circumstantial evidence for this belief.

The biogenetic relationship of isoflavans to pterocarpans is not at first sight obvious. Chemically, however, 2′-hydroxyisoflavans are readily derived from pterocarpans by hydrogenolysis *(Chart 10)*. It is accordingly satisfying to note that all of the natural isoflavans known are 2′-oxygenated. Stereochemical evidence may also be cited to support the belief that isoflavans are related to pterocarpans. Both types of compound can exist in nature in either of two enantiomorphic forms, but in the cases known (*155, 143*) where both types co-occur, the stereochemistry of the compounds has been found to coincide.

Assuming then that a close biosynthetic relationship for all of these types of isoflavonoids exists, a plausible sequence of biosynthesis can be postulated as shown in *Chart 23*. This scheme assigns a key position to the dehydropterocarpan structure (**222**). Chemical analogies for the transformations (**222**) → (**223**), (**223**) → (**224**), (**222**) → (**225**) have already been mentioned (*77, 78, 34, 28, 29, 55*).

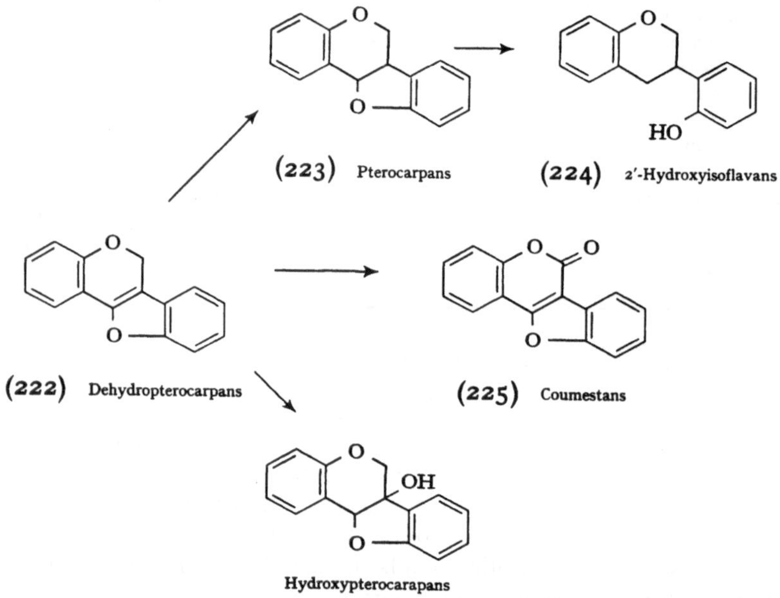

(223) Pterocarpans   **(224)** 2'-Hydroxyisoflavans

(222) Dehydropterocarpans   **(225)** Coumestans

Hydroxypterocarapans

Chart 23. Possible Biogenetic Relationships among Some Isoflavonoid Classes

If this scheme is valid then the generation of a dehydropterocarpan structure would become a key step in the pathway. A possible way to accomplish this, involving the enol form of an isoflavanone (**226**), is shown in *Chart 24* below. This scheme has two merits. By invoking an oxidative cyclisation step involving the oxygen at C-4 one in effect introduces a 2'-oxygenation into ring B solely as a consequence of the proximity of ring B to the C-4 oxygen function. This provides an explanation for a major characteristic difference between flavonoid and iso-flavonoid compounds.

References, pp. 63—73

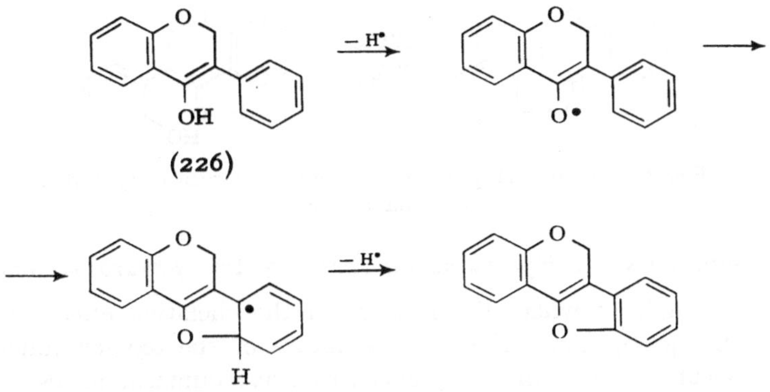

(226)

Chart 24. Possible Biosynthetic Route to Dehydropterocarpans

A second advantage of the scheme shown is that it could explain the predominance of 5-deoxy compounds in these types of isoflavonoids. Since the enol form is the active intermediate, the presence of a 5-hydroxyl group in an isoflavanone would be expected to reduce activity. The keto form is expected to be stabilised at the expense of the enol form, on account of the opportunity for hydrogen bond formation between the hydroxyl and carbonyl groups. Keto-enol tautomerism also offers an explanation for the lack of optical activity encountered in isoflavanones (*40*).

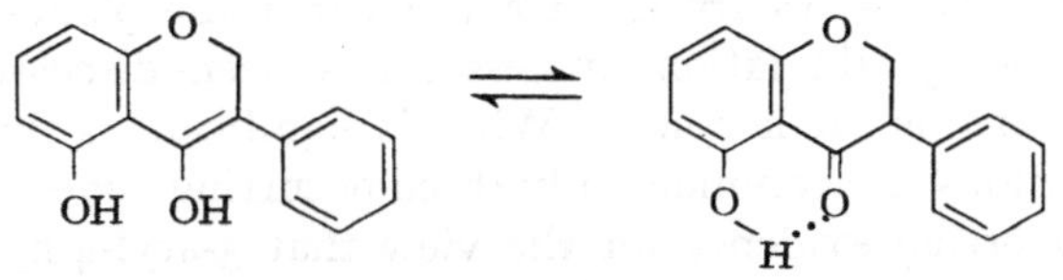

Hydrolysis of a dehydropterocarpan would yield a 2′-hydroxyisoflavanone (*Chart 25*). The prevalence of 2′-oxygenation in natural isoflavanones has already been mentioned. It should be pointed out that the reverse reaction (*Chart 25*) could represent an alternative scheme for the biosynthesis of the dehydropterocarpan structure. This scheme is acceptable if an alternative explanation for the prevalence of 2′-oxygenation in isoflavanones is also forthcoming. The co-occurrence of sophorol (**84**) and maackiain (**122**) in *Maackia amurensis* (*183*) can be used as supporting evidence for one or the other of these views.

In a very recent study on coumestrol (**215**) biosynthesis, DEWICK *et al.* (*55a*) showed by dilution analysis that 2′-hydroxydaidzein was formed in mung bean seedlings after administration of daidzein (**208**). These workers suggested that 2′-hydroxylation of an isoflavone or isoflavanone is a probable step in the biosynthesis of coumestans.

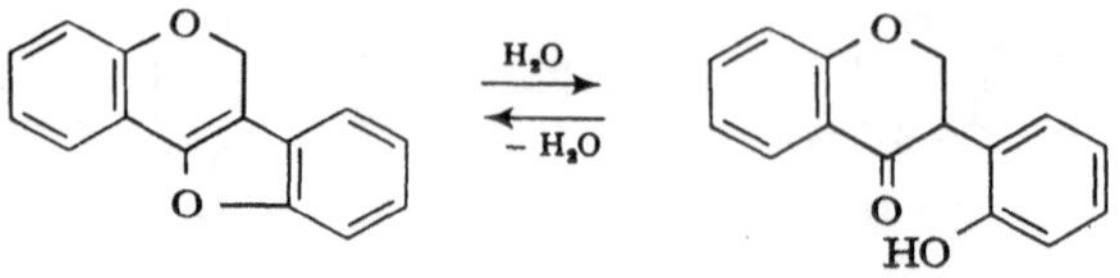

Chart 25. Possible Relationship between Dehydropterocarpans and 2'-Hydroxy-isoflavanones

d) Isoflavones – 2-Hydroxyisoflavones – 3-Aryl-4-hydroxycoumarins

An example of where the absence of the chelating effect of a peri 5-hydroxyl group can influence the nature of the oxygen function at C-4 is clearly shown in the 3-aryl-4-hydroxycoumarin series. As explained earlier, a 2-hydroxyisoflavone structure is tautomeric with the 3-aryl-4-hydroxycoumarin skeleton and the former tautomeric modification becomes important when a 5-hydroxyl group is present in the molecule (114). An examination of the known natural 3-aryl-4-hydroxy-coumarins shows in fact that none of them posseses a free 5-hydroxyl group. Instead, these compounds all have a methoxy-substituent at C-5, which elsewhere is of infrequent occurrence in the isoflavonoids.

From the foregoing, it is clear that biogenetically, 3-aryl-4-hydroxy-coumarins can be regarded as equivalent to 2-hydroxyisoflavones (156, 66). A second notable feature of the natural 3-aryl-4-hydroxycoumarin structures is that they are further substituted with isoprenoid residues, as are the isoflavones occurring with them in these plants. The nature of these secondary alkylation processes has been discussed (147) and need not concern us further here. What is significant is that these complex modifications are common to both co-occurring types, which can be taken as supporting evidence for the view that 3-aryl-4-hydroxycouma-rins are derived directly from isoflavones. A probably biosynthetic sequence accommodating all the points considered is given in Chart 26.

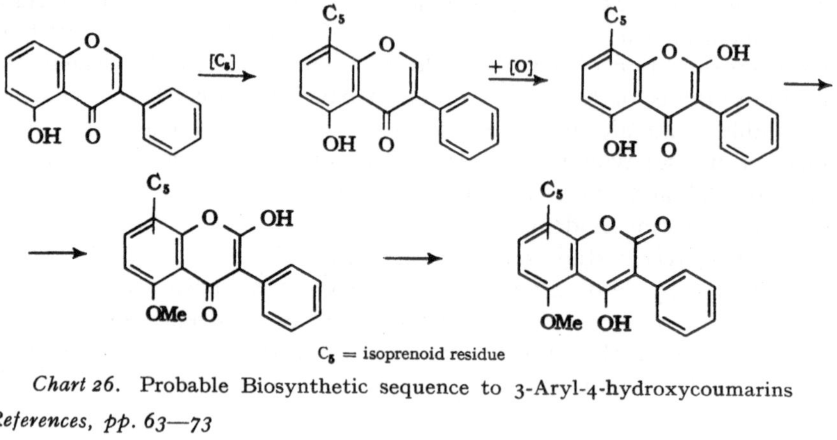

C_5 = isoprenoid residue

Chart 26. Probable Biosynthetic sequence to 3-Aryl-4-hydroxycoumarins

As a corollary to the scheme presented in *Chart 26*, it should be possible to find in nature 2-hydroxyisoflavones possessing a free 5-hydroxyl group. Free redical-type oxidation of compounds of this class in a manner analogous to that postulated for the cyclisation of isoflav-3-en-4-ols *(Chart 24)* could in theory lead to coumestans *(66)* and coumaronochromones *(Chart 27)*. However, very recent results *(55a)* have shown that 3-aryl-4-hydroxycoumarins are not precursors of coumestans, and a more probable route to coumaronochromones is considered in the next section.

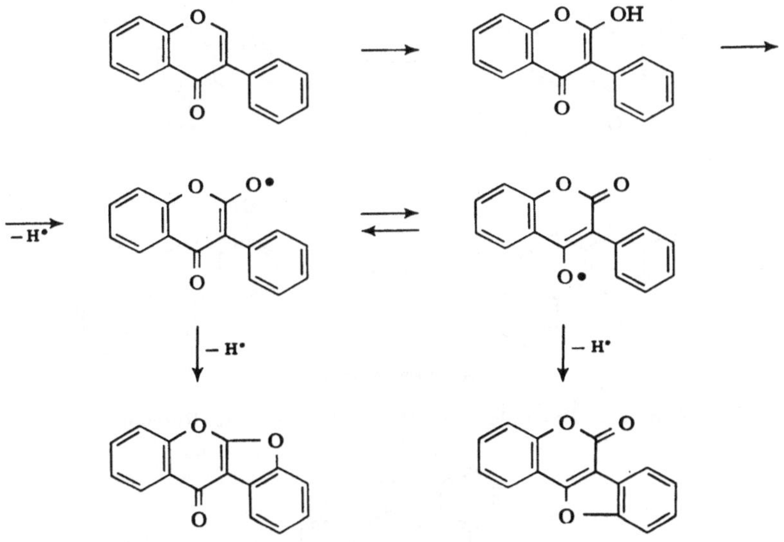

Chart 27. Hypothetical Scheme to Coumestans and Coumaronochromones

e) Isoflavones – Rotenoids – Coumaronochromones

The biosynthetic relationship of rotenoids to isoflavones has been the subject of previous speculations *(92, 46, 141)*. The major characteristic structural features in the rotenoids are the presence of the extra methylene at C-6 and extra C_5-alkylation in ring A. All available evidence is consistent with the view that the extra methylene comes from a 2'-methoxyl group of isoflavones. Tracer studies by CROMBIE and co-workers *(48)* have established that methionine is a relatively efficient source of this extra carbon. Rotenone *(202)* and amorphigenin *(203)* incorporated (Me-^{14}C)-methionine into C-6 to the extent of about $^1/_2$–$^1/_3$ times the figure for a ring B methoxyl group.

Again when one examines the structures of the isoflavones co-occurring with rotenoids, including: piscerythrone (47), piscidone (48), dehydroneotenone (51), ichthynone (46), ferrugone (56), toxicarol isoflavone
(49), auriculatin (52) and munetone (57), the two characteristic rotenoid
features, *viz* C_5-alkylation and 2'-oxygenation, are again present. The
predominance of 2'-O-methyl ether formation in these isoflavones is in
further accord with expectations. Thus it can be strongly argued that
the sequence of steps shown in *Chart 28* is very probably involved in the
biosynthesis of rotenoids.

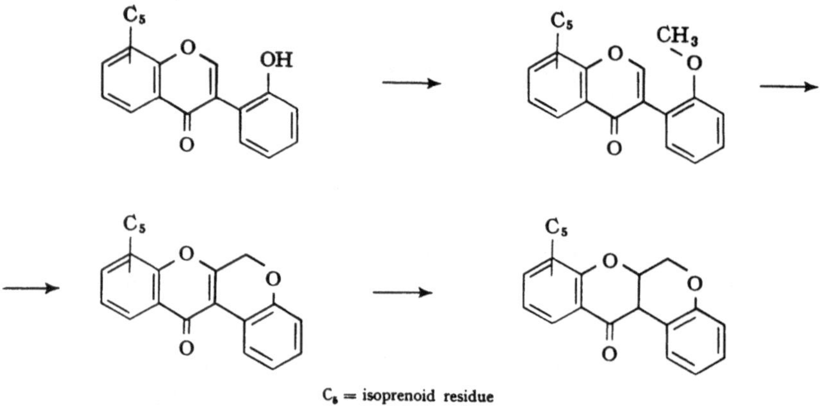

C_8 = isoprenoid residue

Chart 28. Probable Biosynthetic Sequence to Rotenoids

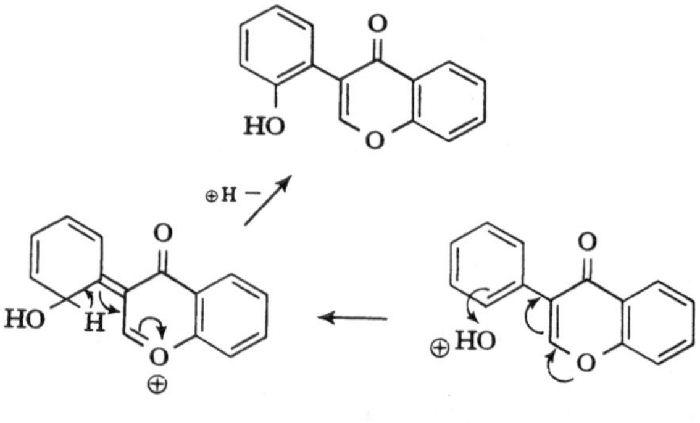

Chart 29. Mechanism for 2'-Hydroxylation of Isoflavones (*143*)

The key step in the scheme above is the introduction of the 2'-hy-
droxyl group into ring B of an isoflavone. A mechanistic rationalisation
for the greater likelihood of this reaction in isoflavones has been proposed
by OLLIS (*143*) and is reproduced in *Chart 29* above. The co-occurrence
of the coumaronochromone lisetin (65) with the isoflavone piscerythrone

(64) in the rotenoid bearing plant *Piscidia erythrina* readily fits into the scheme involving 2'-hydroxyisoflavones as key intermediates. Thus oxidation of (64) by free-radical processes (67) would lead to (65) as has already been demonstrated *in vitro* by ferricyanide oxidation (67).

In connection with the view that a 2'-methoxyl group is the key structural element for the extra ring formation in rotenoids, it is interesting to note that photo-oxidation of the flavonol methyl ether (227) has led to the formation of the compound (228) (189). The structural similarity of the chromanol derivative peltogynol (229) to (228) is suggestive (189).

(227)   (228)

(229) Peltogynol

A more recent example of a $(C_{15} + C)$-type compound may well be eucomin (231) which has been described as a homo-isoflavone (27). Biosynthetically it would seem more plausible to regard it as a homo-aurone, with its derivation occurring *via* the 2-methoxychalcone (230).

(230)   (231) Eucomin

f) Summary of Biogenetic Interrelationships

To summarise the views presented in the foregoing sections concerning the biogenesis of the individual classes of compounds, a biosynthetic scheme relating all the isoflavonoid classes is given in *Chart 30*. It is

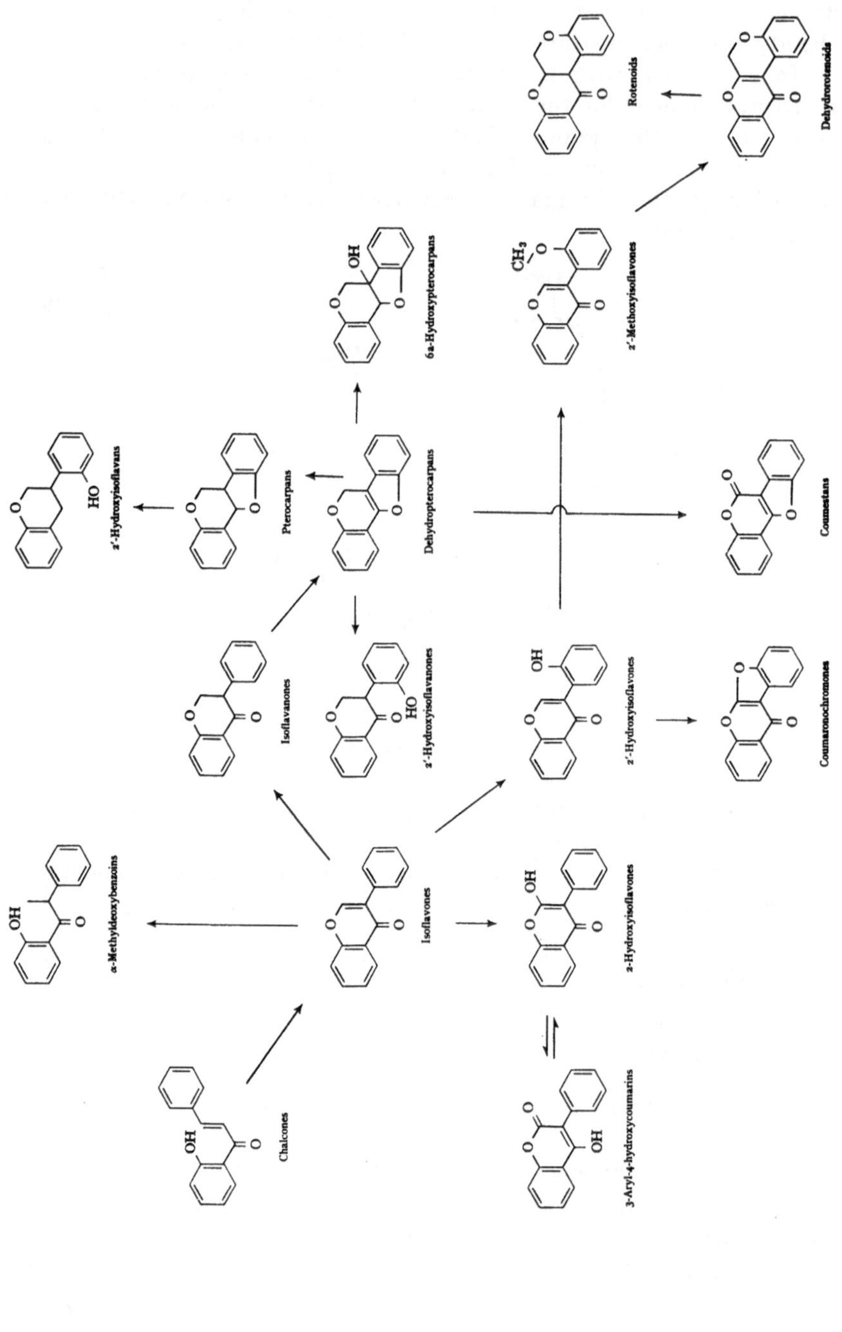

Chart 30. Possible Biogenetic Relationships among the Different Classes of Isoflavonoids

not meant to imply that this scheme will be correct in detail, but the broad relationships indicated seem highly plausible and should serve as useful starting points for more detailed biosynthetic studies. These, to be most meaningful, should ideally be at the enzymatic level. Useful enzymic studies in related areas of biosynthesis have already been made (*137, 105a, 186a*); it is hoped that further progress in this field of natural products biochemistry will not be long in forthcoming.

References

1. AGHORAMURTHY, K., A. S. KUKLA and T. R. SESHADRI: Synthesis of a Racemate of Homopterocarpin. Current Sci. (India) **30**, 218 (1961).

2. AHLUWALIA, V. K., G. P. SACHDEV and T. R. SESHADRI: Chemical Components of Immature Green Pods of *Dalbergia sissoo*. Indian J. Chem. **3**, 474 (1965).

3. AHLUWALIA. V. K., G. P. SACHDEV and T. R. SESHADRI: Chemical Investigation of the Leaves of *Ougeinia dalbergioides*. Indian J. Chem. **4**, 250 (1966).

4. ANIRUDHAN, C. A. and W. B. WHALLEY: The Synthesis of "Maxima Substance C". J. Chem. Soc. (London) **1963**, 6049.

5. ARORA, S. K., A. C. JAIN and T. R. SESHADRI: A New Synthesis of Afromosin. J. Indian Chem. Soc. **38**, 61 (1961).

6. ARORA, S. K., A. C. JAIN and T. R. SESHADRI: Synthesis of Tlatlancuayin. Tetrahedron **18**, 559 (1962).

7. AUDIER, H.: Études des Composés Flavoniques par Spectrométrie de masse. Bull. Soc. Chim. France 1966, 2892.

8. BALAKRISHNA, S., J. D. RAMANATHAN, T. R. SESHADRI and B. VENKATARAMANI: Special Chemical Components of the Heartwood of *Ougeinia dalbergioides* Linn. Proc. Roy. Soc. (London) **A** 268, 1 (1962).

9. BANERJI, A., V. V. S. MURTI and T. R. SESHADRI: Occurrence of 7,4'-Dimethyltectorigenin in the Flowers of *Dalbergia sissoo*. Current Sci. (India) **34**, 431 (1965).

10. BANERJI, A., V. V. S. MURTI and T. R. SESHADRI: Isolation of Sissotrin. A New Isoflavone Glycoside from the Leaves of *Dalbergia sissoo*. Indian J. Chem. **4**, 70 (1966).

11. BANERJI, A., V. V. S. MURTI, T. R. SESHADRI and R. S. THAKUR: Chemical Components of the Flowers of *Dalbergia sissoo*. Isolation of 7-Methyltectorigenin, a New Isoflavone. Indian J. Chem. **1**, 25 (1963).

12. BARNES, C. S. and J. L. OCCOLOWITZ: The Mass Spectra of Some Naturally Occurring Oxygen Heterocycles and Related Compounds. Austral. J. Chem. **17**, 975 (1964).

13. BARZ, W. und H. GRISEBACH: Über die Bedeutung von 3,5,7,4'-Tetrahydroxyflavanon (Dihydrokaempferol) für die Biosynthese von Isoflavonen. Z. Naturforsch. **21b**, 47 (1966).

14. BARZ, W. und H. GRISEBACH: Über die Umwandlung von Daidzein in die Isoflavonoide der Luzerne. Z. Naturforsch. **21b**, 1113 (1966).

15. BARZ, W., L. PATSCHKE and H. GRISEBACH: The Role of Dihydroflavonols in Flavonoid Biosynthesis. Chem. Commun. **1965**, 400.

16. BATTERHAM, T. J. and R. J. HIGHET: Nuclear Magnetic Resonance Spectra of Flavonoids. Austral. J. Chem. **17**, 428 (1964).

17. BECK, A. B.: The Oestrogenic Isoflavones of Subterranean Clover. Austral. J. Chem. **15**, 223 (1964).

17a. BENTLEY, R.: private communication to H. GRISEBACH.

18. BEVAN, C. W. L., A. J. BIRCH, B. MOORE, S. K. MUKERJEE: A Partial Synthesis of ($\pm$)-Pisatin. Some Remarks on the Structure and Reactions of Pterocarpin. J. Chem. Soc. (London) **1964**, 5991.

19. BEVAN, C. W. L., D. E. U. EKONG, M. E. OBASI and J. W. POWELL: West African Timbers XIII. Extracts from the Heartwood of *Amphimas pterocarpoides* and *Pterocarpus erinaceus.* J. Chem. Soc. (London) **C 1966**, 509.

20. BHRARA, S. C., A. C. JAIN and T. R. SESHADRI: A New Examination of the Special Components of *Pterocarpus indicus* Heartwood. Current Sci. (India) **33**, 303 (1964).

21. BHRARA, S. C., A. C. JAIN and T. R. SESHADRI: Some Reactions of Substituted 2'-Benzyloxychalkone Epoxides. Tetrahedron **20**, 1141 (1964).

22. BHRARA, S. C., A. C. JAIN and T. R. SESHADRI: Scope of Isoflavone Synthesis Using 2'-Benzyloxychalkone Epoxides. Tetrahedron **21**, 963 (1965).

23. BHUTANI, S. P., S. S. CHIBBER and T. R. SESHADRI: Components of the Roots of *Pueraria tuberosa*: Isolation of a New Isoflavone-C-glycoside (Di-O-acetyl-puerarin). Indian J. Chem. **7**, 210 (1969).

24. BICKOFF, E. M., A. L. LIVINGSTON, S. C. WITT, R. E. LUNDIN and R. R. SPENCER: Isolation of 4'-O-Methylcoumestrol from Alfalfa. J. Agric. Food Chem. **13**, 597 (1965).

25. BICKOFF, E. M., R. R. SPENCER, B. E. KNUCKLES and R. E. LUNDIN: 3'-Methoxycoumestrol from Alfalfa. Isolation and Characterisation. J. Agric. Food Chem. **14**, 444 (1966).

26. BIRCH, A. J.: Biosynthetic Pathways. In: T. SWAIN, Chemical Plant Taxonomy, p. 141. London: Academic Press. 1963.

27. BOEHLER, P. and CH. TAMM: The Homo-isoflavones, A New Class of Natural Product. Isolation and Structure of Eucomin and Eucomol. Tetrahedron Letters **1967**, 3479.

28. BOUWER, D. C. v. d. M. BRINK, J. P. ENGLEBRECHT and G. J. H. RALL: Neorautanenia Isoflavonoids Part III. 4-Methoxypterocarpin. A New Pterocarpan from *Neorautanenia ficifolia.* J. S. African Chem. Inst. **21**, 159 (1968).

29. BOWYER, W. J., J. N. CHATTERJEA, S. P. DHOUBHADEL, B. O. HANDFORD and W. B. WHALLEY: The Chemistry of the "Insoluble Red Woods". Part IX Homopterocarpin and Pterocarpin. J. Chem. Soc. (London) **1964**, 4212.

30. BRADEN, A. W. H., N. K. HART and J. A. LAMBERTON: The Oestrogenic Activity and Metabolism of Certain Isoflavones in Sheep. Austral. J. Agric. Research **18**, 335 (1967).

31. BREDENBERG, J. B-Son and P. K. HIETALA: Investigation of the Structure of Trifolirhzin, an Antifungal Compound from *Trifolium pratense* L. Acta Chem. Scand. **15**, 696 (1961).

32. BREDENBERG, J. B-Son and J. N. SHOOLERY: A Revised Structure for Pterocarpin. Tetrahedron Letters **1961**, 285.

33. BRIGGS, L. H. and T. CEBALO: Podocarpaceae X. Synthesis of Podospicatin and its Trimethyl Ether. Tetrahedron **19**, 2301 (1963).

34. BRINK, C. v. d. M., J. J. DEKKER, E. C. HANEKOM, D. H. MEIRING and G. J. H. RALL: Neorautanenia Isoflavonoids I. Interconversion of Neodulin and Dehydroneotenone. J. S. African Chem. Inst. **18**, 21 (1965).

35. BÜCHI, G., L. CROMBIE, P. J. GODIN, J. S. KALTENBRONN, K. S. SIDDALINGAIAH and D. A. WHITING: The Absolute Configuration of Rotenone. J. Chem. Soc. (London) **1961**, 2843.

36. CAMPBELL, N. In: E. H. RODD, Chemistry of Carbon Compounds, Vol. IV B, p. 921. Amsterdam: Elsevier. 1959.

37. CAYEN, M. N., G. TANG and R. H. COMMON: Urinary Conversion Products of Biochanin A and Formononetin in the Fowl. Biochim. Biophys. Acta 111, 349 (1965).
38. CHOPIN, J., M.-L. BOUILLANT et P. LEBRETON: Sur la méthyl-5 génistéine, nouvelle isoflavone naturelle extraite du Cytise (*Cytisus laburnum* L.). C. R. hebd. Séances Acad. Sci. 251, 736 (1960).
39. CLAISSE, J., L. CROMBIE and R. PEARCE: Structure and Stereochemistry of the Vicianoside Amorphin, the First Rotenoid Glycoside. J. Chem. Soc. (London) 1964, 6023.
40. CLARK-LEWIS, J. W.: Configurations of Optically Active Flavonoid Compounds. Rev. Pure Appl. Chem. 12, 96 (1962).
41. CLARK-LEWIS, J. W., I. DAINIS and G. C. RAMSAY: Flavan Derivatives XIV. The Absolute Configurations of Some 1,2-Diarylpropane Derivatives and of Some Isoflavans. Austral. J. Chem. 18, 1035 (1965).
42. CLARK-LEWIS, J. W., and R. W. JEMISON: The Absolute Configurations of (—)-Angolensin and Some Related 1,2-Diarylpropanes. Austral. J. Chem. 18, 1791 (1965).
43. COCKER, W., T. DAHL, C. DEMPSEY and T. B. H. McMURRAY: Extractives from Woods I. Extractives from *Andira inermis*. J. Chem. Soc. (London) 1962, 4906.
44. COCKER, W., T. B. H. McMURRAY and P. A. STANILAND: A Synthesis of Demethylhomopterocarpin. J. Chem. Soc. (London) 1965, 1034.
45. COOKE, R. G. and I. D. RAE: Isoflavonoids. I. Some New Constituents of *Pterocarpus indicus* Heartwood. Austral. J. Chem. 17, 379 (1964).
46. CROMBIE, L.: Chemistry of the Natural Rotenoids. Fortschr. Chem. organ. Naturstoffe 21, 275 (1963).
47. CROMBIE, L. and P. J. GODIN: Structure and Stereochemistry of the Rotenolones, Rotenolols, Isorotenolones and Isorotenolols. J. Chem. Soc. (London) 1961, 2861.
48. CROMBIE, L., C. L. GREEN and D. A. WHITING: Biosynthesis of Rotenoids. The Origin of C-6 and C-6a. J. Chem. Soc. (London) C 1968, 3029.
49. CROMBIE, L. and J. W. LOWN: Proton Magnetic Studies of Rotenone and Related Compounds. J. Chem. Soc. (London) 1962, 775.
50. CROMBIE, L. and M. B. THOMAS: Biosynthesis of Rotenoids. Evidence for 1,2-Aryl Migration and the Isoflavonoid Construction of Rings A, C and D in Rotenone and Amorphigenin. J. Chem. Soc. (London) C 1967, 1796.
51. CROMBIE, L. and D. A. WHITING: The Extractives of *Neorautanenia pseudopachyrrhiza*. The Isolation and Structure of a New Rotenoid and Two New Isoflavanones. J. Chem. Soc. (London) 1963, 1569.
52. DEAN, F. M.: Naturally Occurring Oxygen Ring Compounds. London: Butterworth. 1963.
53. DEAN, F. M. and V. PODIMUANG: The Course of the Algar-Flynn-Oyamada (A. F. O.) Reaction. J. Chem. Soc. (London) 1965, 3978.
54. DESCHAMPS-VALLET, C. et C. MENTZER: Sur une nouvelle voie d'accès à la série du «coumestane». C. R. hebd. Séances Acad. Sci. 251, 736 (1960).
55. DEWICK, P. M., W. BARZ and H. GRISEBACH: A Possible Chemical Analogy for Coumestan Biosynthesis. Chem. Commun. 1969, 466.
55a. DEWICK, P. M., W. BARZ and H. GRISEBACH: Biosynthesis of Coumestrol in Phaseolus Auveus Roxb. Phytochem. 9, 775 (1970).
56. DJERASSI, C., W. D. OLLIS and R. C. RUSSELL: The Relative Stereochemistry of the Rotenoids. J. Chem. Soc. (London) 1961, 1448.

57. DUTTA, N. L.: Chemical Investigations of *Mundulea suberose II*. Constitution of Munetone, the Principal Crystalline Product of the Root Bark. J. Indian Chem. Soc. **36**, 165 (1959).

58. DYKE, S. F., W. D. OLLIS and M. SAINSBURY: Synthesis of Isoflavones III. Caviunin. J. Org. Chem. (USA) **26**, 2453 (1961).

59. DYKE, S. F., W. D. OLLIS and M. SAINSBURY: The Constitution of Munetone. Proc. Chem. Soc. (London) **1963**, 179.

60. DYKE, S. F., W. D. OLLIS and M: SAINSBURY: Synthetis of Isoflavones IV. Munetone J. Chem. Soc. (London)C **1966**, 749.

61. DYKE, S. F., W. D. OLLIS, M. SAINSBURY and J. S. P. SCHWARZ: The Extractives of *Piscidia erythrina* L — II. Synthetical Evidence Concerning the Structure of Ichthynone. Tetrahedron **20**, 1331 (1964).

62. EADE, R. A., H. HINTERBERGER and J. J. H. SIMES: Extractives of Australian Timbers III Afromosin (Castanin, 6,4'-Dimethoxy-7-hydroxyisoflavone) from *Castanospermum australe*. Austral. J. Chem. **16**, 188 (1963).

63. EAST, A. J., W. D. OLLIS and R. E. WHEELER: Natural Occurrence of 3-Aryl-4-hydroxycoumarins. Part 1. Phytochemical Examination of *Derris robusta* (Roxb.) Benth. J. Chem. Soc. (London), C **1969**, 365.

64. EMERSON, O. H. and E. M. BICKOFF: Synthesis of Coumestrol, 3,9-Dihydroxy-6H-benzofuro[3,2-c][1]benzopyran-6-one. J. Amer. Chem. Soc. **80**, 4381 (1958).

65. ERDTMAN, H. and T. NORIN: Heartwood Constituents of *Laburnum alpinum* Bercht. and Presl. Acta Chem. Scand. **17**, 1781 (1963).

66. FALSHAW, C. P., R. A. HARMER, W. D. OLLIS, R. E. WHEELER, V. R. LALITHA and N. V. SUBBA RAO: Natural Occurrence of 3-Aryl-4-hydroxycoumarins. Part II. Phytochemical Examination of *Derris scandens* (Roxb.) Benth. J. Chem. Soc. (London) C **1969**, 374.

67. FALSHAW, C. P., W. D. OLLIS, J. A. MOORE and K. MAGNUS: The Extractives of *Piscidia erythrina* L — III. The Constitutions of Lisetin, Piscidone and Piscerythrone. Tetrahedron, Supplement No. 7, 333 (1966).

68. FARKAS, L.: Natural Isoflavones and their Glycosides. In: T. S. GORE, B. S. JOSHI, S. V. SUNTHANKER and B. D. TILAK, Recent Progress in the Chemistry of Natural and Synthetic Colouring Matters and Related Fields, p. 279. London: Academic Press. 1962.

69. FARKAS, L. and M. NOGRADI: The Structure of Sophorabiose. Tetrahedron Letters **1964**, 3919.

70. FARKAS, L., J. VARADY und A. GOTTSEGEN: Untersuchung der Glykoside der *Baptisia tinctoria* R. Br., II. Synthese des 7,3',4',5'-Tetrahydroxyisoflavons (Baptigenin). Chem. Ber. **96**, 1865 (1963).

71. FUKUI, K. and T. MATSUMOTO: Synthetic Studies of Sesamol Derivatives VI. The Synthesis of 6,7,3',4'-bis(methylenedioxy)isoflavone. An Isomer of the Maxima Substance A, and Related Compounds. Bull. Chem. Soc. Japan **38**, 612 (1965).

72. FUKUI, K. and T. MATSUMOTO: The Synthesis of Trisolone: Bull. Chem. Soc Japan **38**, 887 (1965)

73. FUKUI, K. and M. NAKATAMA: Total Synthesis of (±)-Pterocarpin and (±)-Pisatin. Tetrahedron Letters **1966**, 1805.

74. FUKUI, K. and M. NAKAYAMA: Total Synthesis of Nepseudin. Bull. Chem. Soc. Japan **41**, 1385 (1968).

75. FUKUI, K., M. NAKAYAMA and M. HATANAKA: Synthesis of Maxima Substance B. Bull. Chem. Soc. Japan **35**, 1929 (1962).

76. FUKUI, K., M. NAKAYAMA and M. HATANAKA: Synthesis of Cabreuvin. Nippon Kagaku Zasshi 84, 189 (1963). [Chem. Abstr. 59, 13926 (1963)].

77. FUKUI, K., M. NAKAYAMA, H. TSUGE and K. TSUZUKI: The Synthesis of (±)-Maackiain. Experientia 24, 536 (1968).

78. FUKUI, K., M. NAKAYAMA and K. TSUZUKI: The Synthesis of (±)-4-Methoxypterocarpin. Experientia 25, 122 (1969).

79. GEISSMAN, T. A. and E. HINREINER: Theories of the Biogenesis of Flavonoid Compounds. I and II. Botan. Rev. 18, 77, 165 (1952).

80. GEOGHEGAN, M., W. I. O'SULLIVAN and E. M. PHILBIN: Flavonoid Epoxides II. A New Synthesis of 4-Hydroxy-3-phenylcoumarins. Tetrahedron 22, 3209 (1966).

81. GHANIM, A., A. ZAMAN and A. R. KIDWAI: Synthesis of 7-Methyltectorigenin. Tetrahedron Letters 1964, 185.

82. GILBERT, A. H., A. McGOOKIN and A. ROBERTSON: Isoshekkangenin and the Synthesis of 4-hydroxycoumarins. J. Chem. Soc. 1957, 3740.

83. GOPINATH, K. W., A. R. KIDWAI and L. PRAKASH: The Chemical Examination of Iris Nepalensis — I Structure of Irisolone. Tetrahedron 16, 201 (1961).

84. GOTTLIEB, O. R.: Private Communication.

85. GOVINDACHARI, T. R., K. NAGARAJAN, B. R. PAI and P. C. PARTHASARTHY: Chemical Investigations of Wedelia calendulacea. II. The Position of the Methoxyl group in Wedelolactone. J. Chem. Soc. (London) 1957, 545.

86. GRISEBACH, H.: The Biosynthesis of Isoflavones. In: W. D. OLLIS, Chemistry of Natural Phenolic Compounds, p. 59. Oxford: Pergamon Press. 1961.

87. GRISEBACH, H.: Biosynthesis of Flavonoids. In: T. W. GOODWIN, Chemistry and Biochemistry of Plant Pigments, p. 279. London: Academic Press. 1965.

88. GRISEBACH, H.: Recent Investigations on the Biosynthesis of Flavonoids. In: T. J. MABRY, R. E. ALSTON and V. C. RUNECKLES, Recent Advances in Phytochemistry, Vol. I., p. 379. New York: Appleton-Century-Crofts. 1968.

89. GRISEBACH, H. und W. BARZ: Zur Biogenese der Isoflavone VIII. Mitt.: 4,2',4'-Trihydroxy-chalkon-4'-glucosid als Vorstufe für Coumoestrol, Formononetin und Daidzein in der Luzerne (Medicago sativa L.). Z. Naturforsch. 19b, 569 (1964).

90. GRISEBACH, H. und G. BRANDNER: Über die Biogenese von Biochanin A und Formononetin in der Kichererbse. Z. Naturforsch. 16b, 2 (1961).

91. GRISEBACH, H., und G. BRANDNER: Einbau des 2',4,4',6'-Tetrahydroxychalkon-2'-glucosid-[β-¹⁴C] in Isoflavone. Experientia 18, 400 (1962).

92. GRISEBACH, H. and W. D. OLLIS: Biogenetic Relationships between Coumarins, Flavonoids, Isoflavonoids, and Rotenoids. Experientia 11, 4 (1961).

93. GRISEBACH, H. und H. ZILG: Über die Bedeutung der Isoflavanone bei der Isoflavonbiosynthese. Z. Naturforsch. 23b, 494 (1968).

94. GROVER, S. K., A. C. JAIN and T. R. SESHADRI: A Convenient Synthesis of Isoflavones and Di-O-Methylangolensin Using Aryl Migration. Indian J. Chem. 1, 517 (1963).

95. GYORGY, P., K. MURATA and H. IKEHATA: Antioxidants Isolated from fermented Soybeans (Tempeh). Nature 203, 870 (1964).

96. HALLER, H. L., L. D. GOODHUE and H. A. JONES: The Constituents of Derris and Other Rotenone-bearing Plants. Chem. Rev. 30, 33 (1942).

97. HARBORNE, J. B.: Phenolic Glycosides and their Natural Distribution. In: J. B. HARBORNE, Biochemistry of Phenolic Compounds, p. 129. London: Academic Press. 1964.

98. HARBORNE, J. B.: Comparative Biochemistry of the Flavonoids. London: Academic Press. 1967.

99. HARBORNE, J. B., O. R. GOTTLIEB and M. MAGALHAES: Occurrence of the Isoflavone Afromosin in Cabreuva Wood. J. Org. Chem. (USA) **28**, 881 (1963).

100. HARPER, S. H.: The Active Principles of Leguminous Fish-Poison Plants. Part V. *Derris malaccensis* and *Tephrosia toxicaria*. J. Chem. Soc. (London) **1940**, 1178.

101. HARPER, S. H., A. D. KEMP and W. G. E. UNDERWOOD: Heartwood Constituents of *Swartzia madagascariensis*. Chem. and Ind. **1965**, 562.

102. HARPER, S. H., A. D. KEMP and W. G. E. UNDERWOOD: Heartwood Constituents of *Swartzia madagascariensis*. Chem. Commun. **1965**, 309.

103. HARPER, S. H., A. D. KEMP, W. G. E. UNDERWOOD and R. V. M. CAMPBELL: Pterocarpanoid Constituents of the Heartwoods of *Pericopsis angolensis* and *Swartzia madagascariensis*. J. Chem. Soc. (London) C **1969**, 1109.

104. HARPER, S. H. and W. G. E. UNDERWOOD: The Active Principles of Leguminous Fish Poison Plants. Part X. Toxicarol Isoflavone. J. Chem. Soc. (London) **1965**, 4203.

105. HIGHET, R. J. and P. F. HIGHET: The Structure of Two Isoflavones from the Abyssinian Berebera Tree. J. Org. Chem. (USA) **32**, 1055 (1967).

105a. HILLIS, W. E. and N. ISHIKURA: An Enzyme from *Eucalyptus* which Converts Cinnamoyl Triacetic Acid into Pinosylvin. Phytochem. **8**, 1079 (1969).

106. HUDSON, A. T. and R. BENTLEY: The Isolation of Isoflavonoids from Bacteria. Chem. Commun. **1969**, 830.

107. IMAMURA, H., Y. TANNO, T. TAKAHASHI: Isolation and Identification of Four Isoflavone Derivatives from Eurasian teak *Pericopsis*: Mikuzai Gakkaishi **14**, 295 (1968) [Chem. Abstr. **70**, 44835 (1969)].

108. ITAHAKI, Y., T. KUROKAWA, S. SASAKI, C. T. CHANG and F. C. CHEN: The Mass Spectra of Chalcones, Flavones and Isoflavones. Bull. Chem. Soc. Japan **39**, 538 (1966).

109. ITO, S., Y. FUJISE and A. MORI: Absolute Configuration of Pterocarpinoids. Chem. Commun. **1965**, 595.

110. JACKMAN, L. M.: Some Applications of Nuclear Magnetic Resonance Spectroscopy in Natural Product Chemistry. Fortsch. Chem. org. Naturstoffe **23**, 315 (1965).

111. JAIN, A. C., P. LAL and T. R. SESHADRI: Synthesis of ($\pm$)-Ougenin. Indian J. Chem. **7**, 61 (1969).

112. JAIN, A. C., V. K. ROHATGI and T. R. SESHADRI: A Novel Synthesis of 3-Phenyl-4-hydroxycoumarins. Tetrahedron **23**, 2499 (1967).

113. JAIN, A. C., P. D. SARPAL and T. R. SESHADRI: Synthesis of Afromosin Using the Chalcone Epoxide Method. Indian J. Chem. **3**, 369 (1965).

114. JOHNSON, A. P. and A. PELTER: The Structure of Robustic Acid, A New 4-Hydroxy-3-phenylcoumarin. J. Chem. Soc. (London) **1966**, 606.

115. JOHNSON, A. P., A. PELTER and P. STAINTON: Extractives from *Derris scandens*. Part I. The Structures of Scandenin and Lonchocarpic Acid. J. Chem. Soc. (London) **1966**, 192.

116. JURD, L.: Anthocyanins and Related Compounds IV. The Synthesis of Coumestrol and related Coumarinobenzofurans from Flavylium Salts. J. Org. Chem. (USA) **29**, 3036 (1964).

117. JURD, L.: Synthesis of 7-Hydroxy-5',6'-methylenedioxybenzofurano(3',2' : 3,4)-coumarin (Medicagol). J. Pharm. Sci. **54**, 1221 (1965).

118. KALRA, V. K., KUKLA and T. R. SESHADRI: Synthesis of Lucernol and Sativol Dimethyl Ether. Tetrahedron Letters **1967**, 2153.

119. KALRA, V. K., A. S. KUKLA and T. R. SESHADRI: Synthesis of ($\pm$)-7,8,4'-Trimethoxypterocarpan and ($\pm$)-7-Hydroxy-8,4'-Dimethoxypterocarpan. Tetrahedron **23**, 3221 (1967).

120. KONDRATENKO, E. S., A. U. KASYMOV and N. K. ABUBAKIROV: The Structure of Amorphygenine. Khim. Prir. Soedin **3**, 307 (1967). [Chem. Abstr. **68**, 21865 (1968)].

121. KUKLA, A. S. and T. R. SESHADRI: Synthesis of Maxima Substance B. J. Sci. Ind. Res. (India) **21 B**, 97 (1962).

122. KUKLA, A. S. and T. R. SESHADRI: Constitution and Synthesis of Maxima Isoflavones — A and — B. Tetrahedron **18**, 1443 (1962).

123. KUROSAWA, K., W. D. OLLIS, B. T. REDMAN, I. O. SUTHERLAND, O. R. GOTTLIEB and H. M. ALVES: The Absolute Configurations of the Animal Metabolite. Equol, Three Narurally Occurring Isoflavans, and One Natural Isoflavanquinone. Chem. Commun. **1968**, 1265.

124. KUROSAWA, K., W. D. OLLIS, B. T. REDMAN, I. O. SUTHERLAND, A. B. DE OLIVEIRA, O. R. GOTTLIEB and H. M. ALVES: The Natural Occurrence of Isoflavans and an Isoflavanquinone. Chem. Commun. **1968**, 1263.

125. LEBRETON, P., K. R. MARKHAM, W. T. SWIFT, OUNG-BORAN and T. J. MABRY: Flavonoids of *Baptisia australis* (Leguminosae). Phytochem. **6**, 1675 (1967).

126. LIVINGSTON, A. L., E. M. BICKOFF, R. E. LUNDIN and L. JURD: Trifoliol, a New Coumestan from Ladino Clover. Tetrahedron **20**, 1963 (1964).

127. LIVINGSTON, A. L., S. C. WITT, R. E. LUNDIN and E. M. BICKOFF: Medicagol, a New Coumestan from Alfalfa. J. Org. Chem. (USA) **30**, 2353 (1965).

128. MABRY, T. J., J. KAGAN and H. ROESLER: Nuclear Magnetic Resonance Analysis of Flavonoids. The University of Texas Publication No 6418. Austin. 1964.

129. MABRY, T. J., J. KAGAN and H. ROESLER: N. M. R. Spectra of the Trimethylsilyl Ethers of Flavonoid Glycosides. Phytochem. **4**, 177 (1965).

130. MALHOTRA, A., V. V. S. MURTI and T. R. SESHADRI: Lanceolarin, A New Isoflavone Glycoside of *Dalbergia Lanceolaria*. Tetrahedron **23**, 405 (1967).

131. MALLESWAR, D., G. SRIMANNARAYANA, V. SUNDARAMURTY and N. V. SUBBA RAO: Synthesis of 7-Hydroxy-11,12-dimethoxycoumestan. A Component of Alfalfa. Current Sci. (India) **36**, 346 (1967).

132. MARKHAM, K. R. and T. J. MABRY: The Identification of Twenty-three 5-Deoxy- and Ten 5-Hydroxyflavonoids from *Baptisia lecontei* (Leguminosae): Phytochem. **7**, 791 (1968).

133. MARKHAM, K. R., T. J. MABRY and T. W. SWIFT: New Isoflavones from the Genus *Baptisia* (Leguminosae). Phytochem. **7**, 803 (1968).

134. MARKHAM, K. R., W. RAHMAN, S. JEHAN and T. J. MABRY: NMR Analysis and Synthesis of 6- and 8-C-Methylisoflavones. J. Heterocyclic Chem. **4**, 61 (1967).

135. MARKHAM, K. R., W. T. SWIFT and T. J. MABRY: A New Isoflavone Glycoside from *Baptisia australis*. J. Org. Chem. (USA) **33**, 462 (1968).

136. MORGAN, J. W. W. and R. J. ORSLER: Isolation of 7-Methyltectorigenin from the Heartwood of Muninga *(Pterocarpus angolensis)*. Chem. and Ind. **1967**, 1173.

137. MOUSTAFA, E. and E. WONG: Purification and Properties of Chalcone-Flavanone Isomerase from Soya Bean Seed. Phytochem. **6**, 625 (1967).

138. MURAKAMI, T., Y. NISHIKAWA and T. ANDO: Constituents of Japanese and Chinese Crude Drugs IV. Constituents of Pueraria root. Chem. Pharm. Bull. (Tokyo) **8**, 688 (1960).

139. Oliveira, A. B., O. R. Gottlieb and W. D. Ollis: Chemsitry of Brazilian Leguminosae XVII — Constituents of *Machaerium villosum*. An Acad. Brasil. Cienc. **40**, 147 (1968). [Chem. Abstr. **70**, 44797 (1969)].

140. Ollis, W. D.: The Isoflavonoids. In: T. A. Geissman, The Chemistry of Flavonoid Compounds, p. 353. Oxford: Pergamon Press. 1962.

141. Ollis, W. D.: Structural Relationships Involving the Rotenoids. In: H. R. Arthur, Symposium on Phytochemistry, Hong Kong 1961, p. 128. Hong Kong University Press. 1964.

142. Ollis, W. D.: The Neoflavanoids, a New Class of Natural Products. Experientia **22**, 777 (1966).

143. Ollis, W. D.: New Structural Variants Among the Isoflavonoid and Neoflavonoid Classes. In: T. J. Mabry, R. E. Alston and V. C. Runeckles, Recent Advances in Phytochemistry ,Vol. I, p. 329. New York: Appleton-Century-Crofts. 1968.

144. Ollis, W. D., K. L. Ormand and I. O. Sutherland: The Oxidative Rearrangement of Chalcones by Thallic Acetate. A Chemical Analogy for Isoflavone Biosynthesis. Chem. Commun. **1968**, 1237.

145. Ollis, W. D., M. V. J. Ramsay and I. O. Sutherland: The Absolute Configuration of (—)-Angolensein. Austral. J. Chem. **18**, 1787 (1965).

146. Ollis, W. D., C. A. Rhodes and I. O. Sutherland: Extractives of *Millettia dura* (Dunn). Constitutions of Durlettone, Durmillone, Milldurone, Millettone and Millettosin. Tetrahedron **23**, 4741 (1967).

147. Ollis, W. D. and I. O. Sutherland: Isoprenoid units in Natural Phenolic Compounds. In: W. D. Ollis, Recent Developments in the Chemistry of Natural Phenolic Compounds, p. 74. Oxford: Pergamon Press. 1961.

148. Orth, H. and P. Forschner: Formononetin and Extraneous Substances in *Pterocarpus vidalianus*. Holzforschung **19**, 111 (1965). [Chem. Abst. **64**, 2325 (1966).]

149. Pachler, K. G. R. and W. G. E. Underwood: A Proton Magnetic Resonance Study of Some Pterocarpan Derivatives. The Conformation of the 6a,11a-Dihydro-6H-benzofuro[3,2-c]-[1]benzopyran Ring System. Tetrahedron **23**, 1817 (1967).

150. Paris, R.-R. et G. Faugeras: Sur les flavonoides des fleurs du *Retama raetam* Webb. et Berth. Isolement d'une isoflavone identifiée au génistoside. C. R. hebd. Séances Acad. Sci. **257**, 1728 (1963).

151. Paris, R.-R., et G. Faugeras: Sur les isoflavones des Papilionacées-Génistées. Isolement du genistoside à partir de l'*Ulex nanus* Forst. et de l'*Adenocarpus complicatus* Gay et du méthyl-génistéol à partir du *Genista hispanica* L. C. R. hebd. Séances Acad. Sci. **261**, 1761 (1965).

152. Parthasarathy, M. R., R. N. Puri and T. R. Seshadri: New Components of *Pterocarpus dalbergioides* Heartwood. Indian J. Chem. **7**, 118 (1969).

153. Patschke, L., W. Barz und H. Grisebach: Über den Einbau von 5,7,4'-Trihydroxyflavanon-[2,6,8,10^{14}-C$_4$] in Cyanidin und die Isoflavone Biochanin-A und Formononetin. Z. Naturforsch. **19b**, 1110 (1964).

154. Patschke, L., W. Barz und H. Grisebach: Stereospezifischer Einbau von (—)5,7,4'-Trihydroxyflavanon in Flavonoide und Isoflavone. Z. Naturforsch. **21b**, 201 (1966).

155. Pelter, A. and P. I. Amenechi: Isoflavonoid and Pterocarpinoid Extractives of *Lonchocarpus laxiflorus*. J. Chem. Soc. (London) **C 1969**, 887.

156. Pelter, A. and P. Stainton: Extractives from *Derris scandens* II. Isolation of Osajin and Two Isoflavones, Scandenone and Scandinone. J. Chem. Soc. (London) **C 1966**, 701.

157. PELTER, A., P. STAINTON and M. BARBER: The Mass Spectra of Oxygen Heterocycles II. The Mass Spectra of Some Flavonoids. J. Heterocyclic Chem. **2**, 262 (1965).

158. PELTER, A., P. STAINTON, A. P. JOHNSON and M. BARBER: The Mass Spectra of Oxygen Heterocycles I. The 4-Hydroxy-3-phenylcoumarins (Isoflavonols) J. Heterocyclic Chem. **2**, 271 (1965).

159. PERRIN, D. R.: The Structure of Phaseollin. Tetrahedron Letters **1964**, 29.

160. PERRIN, D. R. and W. BOTTOMLEY: Studies on Phytoalexins. V. The structure of Pisatin from *Pisum sativum* L. J. Amer. Chem. Soc. **84**, 1919 (1962).

161. PERRIN, D. R., and I. A. M. CRUICKSHANK: The Antifungal Activity of Pterocarpans towards *Monilinia fructicola*. Phytochem. **8**, 971 (1969).

162. PRAKASH, L., A. ZAMAN and A. R. KIDWAI: The Chemical Examination of *Iris nepalensis* III. Isolation and Structure of Irisolidone. J. Org. Chem. (USA) **30**, 3561 (1965).

163. RANGASWAMI, S. and B. V. R. SASTRY: Constitution of Maxima Substance A. Proc. Indian Acad. Sci. **44A**, 279 (1956).

164. REED, R. I. and J. M. WILSON: Electron Impact and Molecular Dissociation. Part XII. The Cracking Patters of Some Rotenoids and Flavones. J. Chem. Soc. (London) **1963**, 5949.

165. ROBINSON, R.: The Structural Relations of Natural Products. Oxford: Clarendon Press. 1955.

166. ROESLER, H., T. J. MABRY, M. F. CRANMER and J. KAGAN: The Nuclear Magnetic Resonance Analysis of the Disaccharide in Flavonoid Rhamoglucosides. J. Org. Chem. (USA) **30**, 4346 (1965).

167. ROESLER, H., T. J. MABRY und J. KAGAN: Sphaerobiosid, ein Isoflavonglykosid aus *Baptisia sphaerocarpa* Chem. Ber. **98**, 2193 (1965).

168. SAITOH, T. and S. SHIBATA: Chemical Studies on the Oriental Plant Drugs XXII. Some New Constituents of Licorice Root. 2. Glycyrol, 5-O-Methylglycyrol and Isoglycyrol. Chem. Pharm. Bull. (Tokyo) **17**, 729 (1969).

169. SCHULTZ, G.: Isoflavonglucoside Formononetin-7-glucosid und Biochanin A-7-glucosid in *Trifolium pratense* L. Naturwiss. **52**, 517 (1965).

170. SCHWARZ, J. S. P., A. I. COHEN, W. D. OLLIS, E. A. KALZKA and L. M. JACKMAN: The Extractives of *Piscidia erythrina* I. The Constitution of Ichthynone. Tetrahedron **20**, 1317 (1964).

171. SESHADRI, T. R.: Advances in the Phytochemistry of Isoflavonoids. In: H. R. ARTHUR, Symposium on Phytochemistry, Hong Kong 1961, p. 145. Hong Kong University Press. 1964.

172. SHABBIR, M., A. ZAMAN, L. CROMBIE, B. TUCK and D. A. WHITING: Structure of Auriculatin. Extractive of *Millettia auriculata*. J. Chem. Soc. (London) C **1968**, 1899.

173. SHIBATA, S., T. MURATA and M. FUJITA: Studies on the Constituents of Japanese and Chinese Crude Drugs. X. Wistin, A New Isoflavone Glucoside of *Wistaria* spp. Chem. Pharm. Bull. (Tokyo) **11**, 382 (1963).

174. SHIBATA, S. and Y. HISHIKAWA: Studies on the Constituents of Japanese and Chinese Crude Drugs VII. On the Constituents of the Roots of *Sophora subprostrata* Chun et T. Chen and *Sophora japonica* L. Chem. Pharm. Bull. (Tokyo) **11**, 167 (1963).

174a. SHIBATA, S. and T. SAITOH: The Chemical Studies on the Oriental Plant Drugs XIX. Some New Constituents of Licorice Root. The Structure of Licoricidin. Chem. Pharm. Bull. (Tokyo) **16**, 1932 (1968).

175. Shutt, D. A. and A. W. H. Braden: The Significance of Equol in Relation to the Oestrogenic Responses in Sheep Ingesting Clover with a High Formononetin Content. Austral. J. Agric. Res. **19**, 545 (1968).

176. Simonitsch, E., H. Frei und H. Schmid: Die Konstitution des Pachyrrhizins. Monatsh. Chem. **88**, 541 (1957).

177. Spencer, R. R., E. M. Bickoff, R. E. Lundin and B. E. Knuckles: Lucernol and Sativol, Two New Coumestans from Alfalfa *(Medicago sativa)*. J. Agric. Food Chem. **14**, 162 (1966).

178. Spencer, R. R., B. E. Knuckles and E. M. Bickoff: 7-Hydroxy-11,12-dimethoxycoumestan. Characterisation and Synthesis. J. Org. Chem. (USA) **31**, 988 (1966).

179. Suginome, H.: Oxygen Heterocycles. A New Isoflavanone from *Sophora japonica* L. J. Org. Chem. (USA) **24**, 1655 (1959).

180. Suginome, H.: Maackiain. Bull. Chem. Soc. Japan **39**, 1529 (1966).

181. Suginome, H.: The Absolute and Relative Stereochemical Correlation of Isoflavanone Sophorol and Naturally-occurring Isoflavan Derivatives. Bull. Chem. Soc. Japan **39**, 1544 (1966).

182. Suginome, H. and T. Iwadare: Sauerstoff-Heteroringe. Die Konfiguration und Synthese des d,l-Homopterocarpins. Experientia **18**, 163 (1962).

183. Suginome, H. and T. Kio: The Co-occurrence of Isoflavonoids at Different Oxidation Levels. Bull. Chem. Soc. Japan **39**, 1541 (1966).

184. Tamura, S., C. F. Chang, A. Suzuki and S. Kumai: Chemical Studies on "Clover Sickness" Part I. Isolation and Structural Elucidation of Two New Isoflavonoids in Red Clover. Agr. Biol. Chem. **33**, 391 (1969).

185. Uchiyama, M. and M. Matsui: A New Approach to the Synthesis of Isoflavones, 2'-Hydroxyisoflavones and an Alternative Synthesis of ($\pm$)-pterocarpin. Agric. Biol. Chem. **31**, 1490 (1967).

186. Varady, J.: A New Method for the Ring Isomerisation of Isoflavones. Direct Synthesis of Tectorigenin, 4-Methyltectorigenin, Caviunin and Other Isoflavones. Tetrahedron Letters **1965**, 4273.

186a. Vaughan, P. F. T., V. S. Butt, H. Grisebach and L. Schill: Hydroxylation of Flavonoids by a Phenolase Preparation from Leaves of Spinach Beet. Phytochem. **8**, 1373 (1969).

187. Venkataraman, K.: Flavones and Isoflavones. Fortschr. Chem. organ. Naturstoffe **17**, 1 (1959).

188. Verbit, L. and J. W. Clark-Lewis: Optically Active Aromatic Chromophores — VIII. Studies in the Isoflavonoid and Rotenoid Series. Tetrahedron. **24**, 5519 (1968).

189. Waiss, A. C. and J. Corse: Photooxidative Cyclization of Quercetin Pentamethyl Ether. J. Amer. Chem. Soc. **87**, 2068 (1965).

190. Wanzlick, H. W., R. Gritzky und H. Heiderpriem: Die Synthese des Wedelolactons. Chem. Ber. **96**, 305 (1963).

191. Warburton, W. K.: The Isoflavones. Quart. Rev. (Chem. Soc. London) **8**, 67 (1954).

192. Weinges, K.: The Neighbouring Group Effect in the Conversion of Tetramethyl-(+)-catechin into 2-Chlorotetramethyl-(—)-isocatechin. Proc. Chem. Soc. (London) **1964**, 138.

193. Wilson, R. G., J. H. Bowie and D. H. Williams: Solvent Effects in NMR Spectroscopy. Solvent shifts of Methoxyl Resonances in Flavones Induced by Benzene; An Aid to Structure Elucidation. Tetrahedron **24**, 1407 (1968).

194. Wong, E.: Pratensein. 5,7,3'-Trihydroxy-4'-methoxyisoflavone. J. Org. Chem. (USA) **28**, 2336 (1963).

195. WONG, E.: Flavonoid Biosynthesis in *Cicer arietinum*. Biochim. Biophys. Acta **111**, 358 (1965).

196. WONG, E.: The Role of Chalcones and Flavanones in Flavonoid Biosynthesis. Phytochem. **7**, 1751 (1968).

197. WONG, E. and C. M. FRANCIS: Flavonoids in Genotypes of *Trifolium subterraneum* — II. Mutants of the Geraldton Variety. Phytochem. **7,** 2131 (1968).

198. WONG, E. and H. GRISEBACH: Further Studies on the Role of Chalcone and Flavanone in Biosynthesis of Flavonoids. Phytochem. **8**, 1419 (1969).

199. WONG, E., P. I. MORTIMER and T. A. GEISSMAN: Flavonoid Constituents of *Cicer arietinum*. Phytochem. **4**, 89 (1965).

200. WONG, E. and E. MOUSTAFA: Flavanone Biosynthesis. Tetrahedron Letters **1966**, 3021.

201. ZILG, H. and H. GRISEBACH: Biosynthesis of Isoflavones — XVII. Identification and Biosynthesis of Coumestanes in *Soja hispida*. Phytochem. **7**, 1765 (1968).

(Received, December 1, 1969)

Recent Advances in the Chemistry of Cyanogenic Glycosides

By R. EYJÓLFSSON, Copenhagen

With 1 Figure

Contents

Acknowledgements. I am very grateful to Professor ANDERS KJÆR, The Technical University of Denmark, for encouraging the preparation of this review as well as for his perusal of the manuscript which resulted in numerous corrections and changes. I am also much obliged to the head of this laboratory, Professor HELMER KOFOD, for excellent working facilities and for his interest in my studies.

References, pp. 103—108

I. Introduction

The cyanogenic glycosides, here defined as glycosidic derivatives of α-hydroxynitriles, represent a rather limited class of natural products, which are widely distributed in the plant kingdom and, to a small extent, even in animals. A characteristic feature of these glycosides is their ability to release hydrocyanic acid on treatment with dilute acids or appropriate enzymes. The term "cyanogenic" is used to designate this property, regardless of whether pure substances, plants, or animals, are serving as the source. In the latter cases the term "cyanophoric" is occasionally employed synonymously.

Cyanogenesis in plants was probably first discovered by SCHRADER in 1803 (*103*) working with bitter almonds. In 1830, ROBIQUET and BOUTRON-CHARLARD (*100*) succeeded in isolating the parent glycoside, namely amygdalin. Over the years, a total of 18 cyanogenic glycosides have been isolated and characterized more or less completely (*Table 1*, p. 76). It will be noted that the majority of these compounds has been isolated in the era of classical organic chemistry and that progress in discovering new compounds, not to mention new structural types, has been surprisingly slow. It is worth remembering here that the mechanism of cyanogenesis has been established only in the minority of known cyanogenic species.

The cyanogenic glycosides have last been reviewed in 1958 by DILLEMANN (*36*). Since then, no complete reviews in this field have appeared. It is the purpose of the present article to survey the more recent advances and, hopefully, to stimulate continued interest in these interesting compounds.

II. General Chemical Considerations
Structural Relationships

The cyanogenic glycosides may conveniently be divided into four groups according to the chemical structure of the aglycone:

1) Glycosides derived from 2-hydroxy-2-phenylacetonitrile or derivatives thereof (*i. e.* hydroxy- or alkyl-substituted in the aromatic ring) *(amygdalin type)* (1), (3), (5), (6), (7), (9), (10), (11), (14), (15), (16).

2) Glycosides derived from saturated, aliphatic aglycones *(linamarin type)* (2), (13).

3) Glycosides with an aglycone containing a double bond positioned α,β to the nitrile group *(acacipetalin type)* (8), (18).

4) Glycosides with an unsaturated, alicyclic aglycone *(gynocardin type)* (4), (17).

Most cyanogenic glycosides are simple glucosides, the only exceptions being amygdalin (1), lucumin (15), and vicianin (6), which contain disaccharide residues.

Table 1. *The Known Cyanogenic Glycosides*[a]

Compound	Date of discovery	References[b]

(1) Amygdalin — Robiquet and Boutron-Charlard (1830) — (*100, 29*)

(2) Linamarin (Manihotoxin, Phaseolunatin) — Jorissen and Hairs (1891) — (*72, 32*)

(3) Dhurrin — Dunstan and Henry (1902) — (*38, 82*)

(4) Gynocardin — Power and Gornall (1904) — (*92, 33, 41, 44*)

(5) Sambunigrin — Bourquelot and Danjou (1905) — (*20, 53*)

[a] Compounds are arranged in chronological order according to the approximate date of their discovery.

[b] If two or more references are given, the first refers to the date of discovery.

References, pp. 103—108

(Table 1, continued)

Compound	Date of discovery	References
(6) Vicianin	BERTRAND (1906)	(*13, 76*)
(7) Prunasin	HÉRISSEY (1907)	(*69, 74*)
(8) Acacipetalin	STEYN and RIMINGTON (1935)	(*109, 98*)
(9)ᶜ Nandina-glucoside (Nandina-cyanogen)	FINNEMORE and LARGE (1936) (?), ABROL, CONN and STOKER (1966)	(*50, 5*)

ᶜ In the structures presented in this table, missing drawn lines to chiral centres indicate that the stereochemistry is unknown.

(Table 1, continued)

Compound	Date of discovery	References

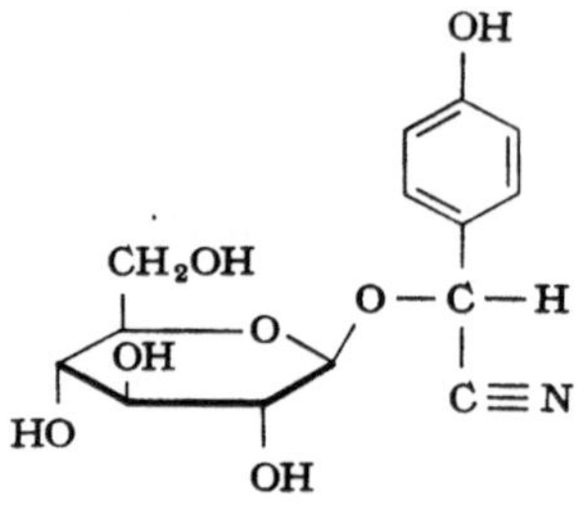

(10)   Taxiphyllin (Phyllanthin)

FINNEMORE, REICHARD and
  LARGE (1936) (?),
TOWERS, MCINNES and   (*51*, *115*)
  NEISH (1964)

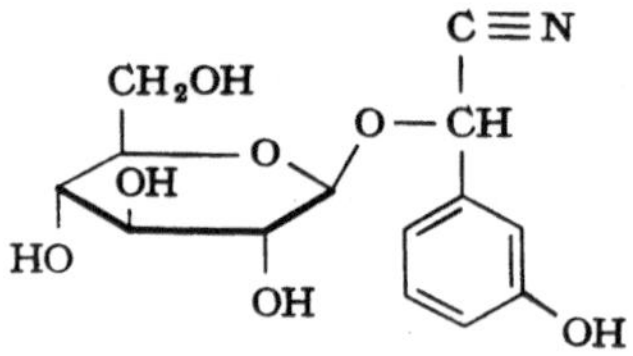

(11)   Zierin

FINNEMORE and COOPER   (*48*)
  (1936)

$$C_{14}H_{20(22)}O_{10}N_2 ?$$
$$(12)$$
Acalyphin

This compound is reported to be   RIMINGTON and ROETS (1937) (*99*)
  highly unstable, yielding HCN,
  glucose, and coloured sub-
  stance(s) on hydrolysis. No
  further reports have been presented on the structure of this
  substance

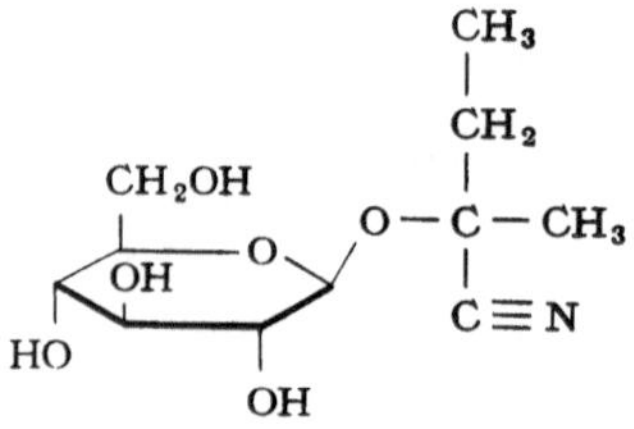

(13)   Lotaustralin[d]

FINNEMORE and COOPER
  (1938) (?),
BISSETT, CLAPP, COBURN,   (*49*, *15*)
  ETTLINGER and LONG Jr.
  (1969)

[d] The two enantiomeric β-D-glucopyranosides are collectively called methyllinamarins (*15*).

References, pp. 103—108

(*Table 1, continued*)

Compound	Date of discovery	References

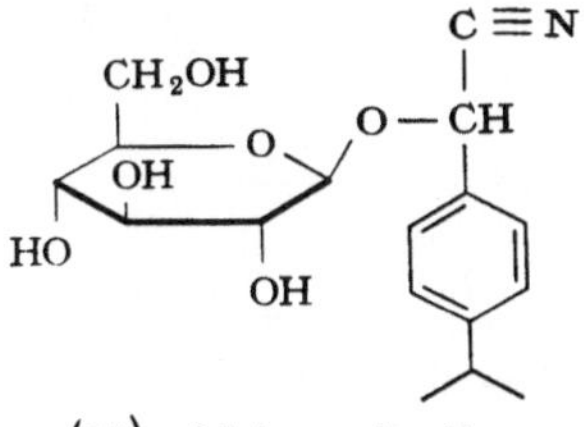

$C \equiv N$

CH_2OH

$O - CH$

OH   O

HO

OH

PALLARES (1946)   (*87*)

(**14**)   Polydesmus-glucoside

$C \equiv N$

$Ara - O - Ara - O - CH$

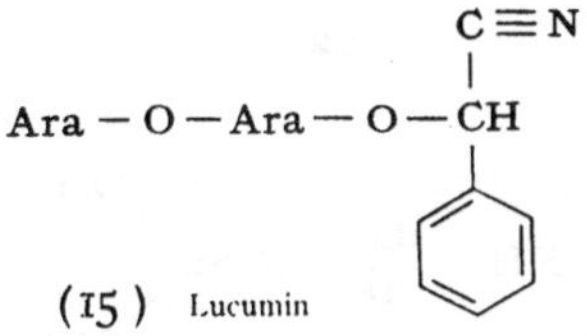

BACHSTEZ, PRIETO and   (*8*)
GAJA (1948)

(**15**)   Lucumin

$C \equiv N$

CH_2OH

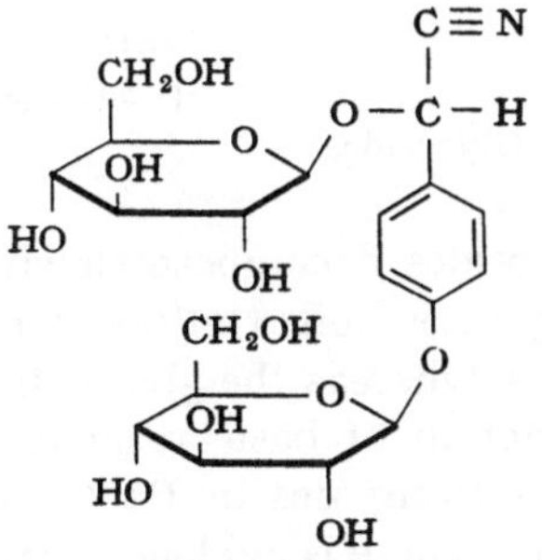

$O - C - H$

OH   O

HO

OH

CH_2OH O

OH   O

HO

OH

YOUNG and HAMILTON   (*122, 121*)
(1966)

(**16**)   Proteacin

$C_{14}H_{23}O_8N$?
(**17**)
Barterioside

This compound is suggested to be   PARIS, BOUQUET and PARIS   (*88*)
a β-D-glucoside containing a   (1969)
cyclopentene residue, thus
resembling gynocardin (**4**)

CH_2OH

$O - C - C \equiv N$

OH   O $\|$

CH

HO

OH CH

$\|$ EYJÓLFSSON (1969)   (*45*)

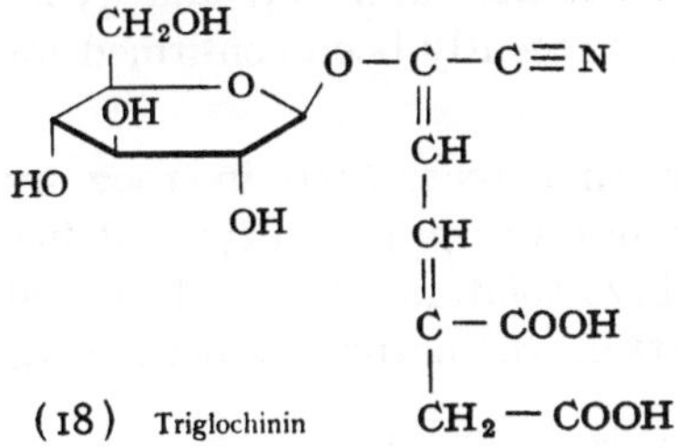

$C - COOH$

(**18**)   Triglochinin $CH_2 - COOH$

Chemical Properties

The most important chemical features that in principle apply to all cyanogenic glycosides (19) are summarized in *Chart 1*.

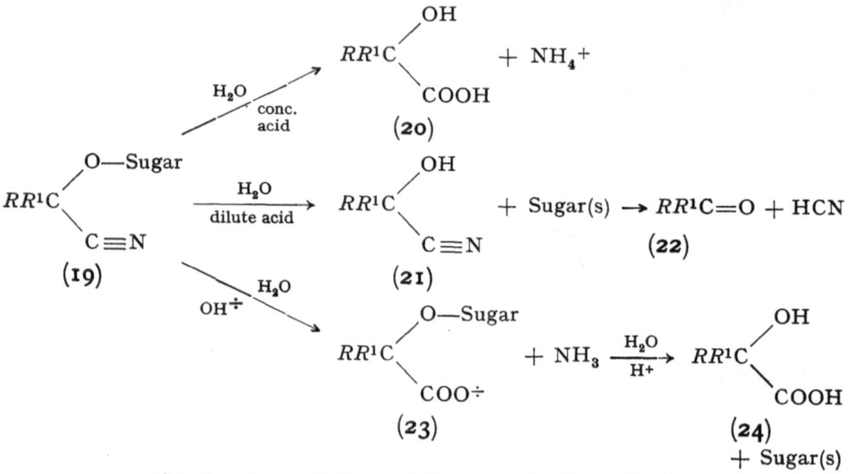

$$RR^1C\diagup{}^{OH}_{\diagdown COOH} \ + \ NH_4^+$$
(20)

$$\xrightarrow[\substack{conc.\\acid}]{H_2O}$$

$$RR^1C\diagup{}^{O-Sugar}_{\diagdown C\equiv N}$$
(19)

$$\xrightarrow[dilute\ acid]{H_2O}$$

$$RR^1C\diagup{}^{OH}_{\diagdown C\equiv N} \ + \ Sugar(s) \ \rightarrow \ RR^1C{=}O \ + \ HCN$$
(21)   (22)

$$\xrightarrow[OH\div]{H_2O}$$

$$RR^1C\diagup{}^{O-Sugar}_{\diagdown COO\div} \ + \ NH_3 \ \xrightarrow[H^+]{H_2O} \ RR^1C\diagup{}^{OH}_{\diagdown COOH}$$
(23)   (24)
                                    + Sugar(s)

Chart 1. Degradations of Cyanogenic Glycosides*

Hydrolysis is very easily effected; the glycosides decompose slowly in cold aqueous solutions and more or less rapidly when boiled with water. Hydrolytic cleavage is greatly catalyzed by acids, whereas the glycosidic bonds are generally quite stable towards the action of basic reagents. A notable exception is triglochinin (18) which decomposes in Ba(OH)$_2$ solutions at rather low temperature (45). Likewise, there is evidence that dhurrin (3) is unstable in the presence of basic ion-exchange resins or in basic solutions (82).

Glycosides possessing an asymmetric carbon atom at the point of attachment of sugar to aglycone (*i. e.* in the substituted α-hydroxynitrile grouping) may epimerize with great ease under the influence of bases, even in high dilution. This process must be operating in all glycosides of the amygdalin type, but is, of course, far less likely in glycosides derived from chiral, tertiary cyanohydrins such as lotaustralin (13) and gynocardin (4). The validity of this proposal has recently been confirmed for lotaustralin (15).

The alkali-induced epimerization was first recognized in 1907 by CALDWELL and COURTAULD (28) in their studies of prunasin (7), but has subsequently been confirmed by other investigators. It is of special interest here to point out that older reports on the natural co-occurrence

---

* References are to be found in *Table 1*, p. 76.

of the epimeric compounds prunasin (7) and sambunigrin (5) appear questionable and need reinvestigation. As an illustration, HÉRISSEY (68) reported the isolation of an equimolar mixture of (5) and (7) (named "prulaurasin" and still figuring in some modern handbooks) from leaves of *Prunus laurocerasus* L. Thirty years later the mixture was shown to be an artefact of isolation (90) arising as a result of the old practice of adding excess of basic reagents (such as CaCO₃) to crude plant extracts in order to neutralize plant acids.

On the other hand, very little or no epimerization seems to occur during hydrolysis of cyanogenic glycosides (19) with concentrated acids (HCl) if carried out under appropriate conditions. Hence, acid hydrolysis has been widely utilized for determination of absolute configurations of the aglycones in compounds of the amygdalin type (76).

Hydrolysis by dilute acids of the gynocardin- (4), (17) and acacipetalin type (8), (18) compounds, though formally following the scheme in *Chart 1*, does not afford oxo-compounds (22) as stable end products (33, 45, 98, 109). In the latter case, the expected oxo-compound will be a ketene, leading to formation of a carboxylic acid (see triglochinin, p. 87).

Enzymic Degradation

It is well established that the tissues of a plant containing cyanogenic glycosides almost invariably contain enzymes capable of decomposing them. If the cells of a living cyanophoric plant are injured mechanically or by the action of reagents (*e. g.* chloroform), the decomposition of the glycosides usually proceeds very rapidly, hydrocyanic acid being released almost instantaneously.

Reinvestigation of the enzymic degradation of amygdalin (1) by emulsin (60) showed that the decomposition proceeds stepwise as implied both from kinetic measurements and from the fact that the disaccharide (gentiobiose) was not present in hydrolysates. Amygdalin was first hydrolyzed to prunasin (7) and glucose; prunasin to D-mandelonitrile and glucose, and D-mandelonitrile was finally cleaved to benzaldehyde and hydrocyanic acid. The participation of three enzymes, *viz.* amygdalin hydrolase, prunasin hydrolase, and hydroxynitrile lyase was therefore proposed. In fact, the three enzymes were later shown to be present in bitter almond emulsin by electrophoretic methods (61). Further support for the consecutive degradation of amygdalin can be found in the isolation (9) from bitter almond emulsin of the crystalline enzyme that catalyzes the reaction:

$$\text{D-mandelonitrile} \rightleftharpoons \text{benzaldehyde} + \text{HCN}.$$

Moreover, two β-D-glucopyranosidases have been isolated in pure state from sweet almond emulsin (66), a preparation that also cleaves amygdalin.

Emulsin degrades all glycosides of the amygdalin- and acacipetalin type (45, 98), but has very little activity towards linamarin (2) and lotaustralin (13) (25). The ability of emulsin to decompose vicianin (6) is due to its content of an α-L-arabinosidase (80). By contrast, an enzyme system in the seeds of *Vicia angustifolia* Roth. cleaves vicianin to the disaccaride vicianose, benzaldehyde, and hydrogen cyanide (14).

During investigations of the enzymic degradation of dhurrin (3) it was ascertained that the parent plant *Sorghum vulgare* Pers. contains a hydroxynitrile lyase catalyzing the reaction:

$$p\text{-hydroxy-L-mandelonitrile} \rightleftharpoons p\text{-hydroxybenzaldehyde} + \text{HCN},$$

whilst its effect on L-mandelonitrile was negligible (21, 81, 104). Additionally, it was established that *S. vulgare* contains at least two β-D-glucopyranosidases, one of which has no effect on dhurrin, whereas the other is highly effective, but hydrolyzes other glycosides of the amygdalin type as well (taxiphyllin (10) was cleaved at an even higher rate than dhurrin) (81). Evidently, the enzymic decomposition of dhurrin takes place in two steps.

A β-D-glucopyranosidase fraction with pronounced activity towards linamarin (2) has been obtained from *Linum usitatissimum* L. (25). This glucosidase, called linamarase, also displays appreciable activity towards prunasin (7) and taxiphyllin (10). On the other hand, dhurrin (3) was hydrolyzed at a much lower rate, and amygdalin (1) not at all. To the knowledge of the author, higher plants have not been examined for enzymes catalyzing the reaction:

$$\alpha\text{-hydroxyisobutyronitrile} \rightleftharpoons \text{acetone} + \text{HCN}$$

but a fungal lyase has recently been reported catalyzing this reaction as well as that of the homologous 2-hydroxy-2-methylbutyronitrile (108).

III. Detection

Detection in Fresh Plant Specimens

The search for cyanogenic glycosides in plants is usually carried out in an indirect way, *i. e.* by detecting the hydrocyanic acid evolved on hydrolysis. In order to minimize loss of glycosides by enzymes present in the sample, care must be exercised to perform the test(s) as soon as possible after collection of the material. A vast number of publications dealing with the detection and quantitative assay of hydrogen cyanide is found in the chemical literature. In the present context, however, only qualitative methods currently used in the author's laboratory are considered. They are all based on the presence of enzymes in the plant material which are capable of decomposing the glycosides (autolysing technique).

1. *The picrate test* (Guignard-Mirande test) (*85*) is very rapid and simple and can be conveniently used as a field test: The material is placed in a micro test tube (3 ml) and treated with a drop of chloroform. A strip of filter paper, freshly moistened with solutions of picric acid and sodium carbonate, is fixed in the mouth of the tube which is then stoppered airtight. The tube is allowed to stand (preferably at approximately 35°) for 24 hr before estimating the result. The test paper turns orange to red or red-brown in the presence of hydrocyanic acid; the sensitivity is approximately 1 μg. Unfortunately, the test is not very specific as some volatile substances, such as hydrogen sulphide, sulphur dioxide, aldehydes, or ketones, may cause reddening of the paper (*46*).

2. *The nitrobenzaldehyde-dinitrobenzene test* (*58*) is performed analogously to 1, the test paper in this case being moistened with a buffer (pH 8.5) and inserted into the micro test tube. When the autolysing period has been completed the test paper is withdrawn and wetted with a solution of *p*-nitrobenzaldehyde and *o*-dinitrobenzene (0.05 M) in 2-methoxyethanol. A positive reaction is indicated by a purple colouration. The test is sensitive (less than 0.1 μg) and specific.

3. *The benzidine-cupric acetate test* (*47*) is accomplished in a slightly different manner: The material is first allowed to autolyze, whereupon a test paper dampened with a solution of benzidine acetate and cupric acetate is placed in the mouth of the micro test tube. A blue colouration signals the presence of hydrogen cyanide; the limit of detection is about 0.25 μg.

If only traces of hydrogen cyanide are present in the material, it is advisable to work with greater quantities and use steam distillation or aeration techniques in order to trap the liberated hydrocyanic acid by means of a suitable reagent. Recently, reports on the detection of hydrogen cyanide in nanogram quantities by chemical (*79*) or gas chromatographic (*31*) means have appeared. These methods may (with appropriate modifications) become effective tools in the future quest for cyanogenic plants.

Detection on Chromatograms

Cyanogenic glycosides are best detected on chromatograms by their ability to liberate hydrocyanic acid, but unspecific reagents may obviously also be used. The former approach demands the application of a proper enzyme reagent. No single enzyme preparation, however, can be claimed to possess sufficient decomposing activity towards all known cyanogenic glycosides (cf. p. 82). In experimental work carried out in the author's laboratory, chromatograms have generally been treated with an aqueous extract of the fresh parent plant and the results compared with those obtained by employing purified enzyme solutions (such as emulsin).

1. *Detection on paper chromatograms.* (i) The method of BUTLER and BUTLER (*26*): The dried chromatogram is sprayed lightly with an enzyme solution and held in a press adjacent to a sheet of paper that has been treated with picrate reagent. The two pieces of paper are separated by means of a perforated sheet and the hydrogen cyanide will be revealed as orange spots on the picrate paper. (ii) The method of BENNETT and TAPPER (*11*): The chromatogram is sprayed with an enzyme solution,

immediately covered with a thin sheet of plastic followed by a sheet of foam plastic and sandwiched between plane supports for 30 min. The chromatogram is then sprayed with a buffer (pH 8.5), followed by a spray with a solution of *p*-nitrobenzaldehyde and *o*-dinitrobenzene in 2-methoxyethanol. Hydrocyanic acid is revealed as purple spots.

2. *Detection on thin layer chromatograms*. Method (ii) described above is well suited for cellulose layers, whereas it cannot be applied satisfactorily to silica gel layers.

IV. Isolation

Before chromatographic methods had been developed, isolation of cyanogenic glycosides was generally beset with considerable difficulties. In fact, it has been proposed (*36*) that such difficulties may account for the relatively small number of plant species in which identity of the glycoside(s) has been established.

It must be emphasized that the plant material used for isolation should be as fresh as possible, and due precautions must be taken to arrest all enzymic activity, for instance by boiling the material with ethanol (*76*) or water (*45*).

Column Chromatography

1. *Chromatography on charcoal:* TOWERS et al. (*115*) isolated taxiphyllin (**10**) from leaves of *Taxus canadensis* by fractionation on a column of charcoal. Elution was performed successively with water, aqueous ethanol, absolute ethanol, and ethanol : benzene. The glucoside was obtained from the ethanol : benzene fractions.

2. *Gel filtration:* PIFFERI (*89*) obtained excellent separation of amygdalin (**1**) and prunasin (**7**) in deionized plant extracts using chromatography on Sephadex HL-20 and elution with ethanol : water (2/8, v/v). Sugars were eluted first, then amygdalin, and finally prunasin.

3. *Chromatography on polyvinylpyrrolidone:* REAY (*94*) has reported a method for the isolation of dhurrin (**3**) from leaves of *Sorghum vulgare*. Purified plant extracts (in isopropyl alcohol) were applied to a PVP column that was initially eluted with isopropyl alcohol. The glucoside was then recovered by irrigation with methanol : isopropyl alcohol mixtures containing a small amount of acetic acid.

4. *Chromatography on cellulose:* In the isolation of dhurrin (**3**) from *Sorghum vulgare*, MAO et al. (*82*) chromatographed purified extracts on a cellulose column with butanol : water (9/1, v/v) as the mobile phase. The isolation of triglochinin (**18**), an extremely polar glucoside, from the flowers of *Triglochin maritimum* L. was eventually accomplished by the following procedure (EYJÓLFSSON (*45*)): The aqueous plant extract was chromatographed in water on a column of polyamide, in order to remove phenolic substances. The eluate contained the glucoside together with large amounts of impurities (mainly carbohydrates), and was passed through a weakly basic ion-exchange resin withholding the glucoside along with other acidic compounds. The glucoside was recovered from the column by stepwise elution with increasing concentrations of acetic acid (up to 5 M), followed by 5 M formic acid. Final purification was attained on a cellulose column using ethyl acetate : acetic acid : water (8/1/1, v/v) as a solvent.

5. *Chromatography on silica gel:* The following examples may be cited, including solvent systems: Gynocardin (**4**) (COBURN and LONG (*33*)) from seeds of *Gynocardia odorata* (acetone : methanol; 9.5/0.5, v/v); linamarin (**2**) (CLAPP *et al.* (*32*)) from tubers of *Manihot esculenta* (chloroform : methanol; 5/1,·v/v); prunasin (**7**) (KOFOD and EYJÓLFSSON (*74, 76*)) from leaves of *Pteridium aquilinum* and *Cystopteris fragilis* (ethyl acetate : methanol; 9.5/0.5, v/v); and vicianin (**6**) (KOFOD and EYJÓLFSSON (*76*)) from leaves of *Davallia* species (ethyl acetate : methanol; 9/1, v/v).

Preparative Paper Chromatography

This method has been employed for the isolation of linamarin (**2**) (TSCHIERSCH (*116*)) from *Linum usitatissimum*, and of prunasin (**7**) from *Pteridium aquilinum* (BENNETT (*10*)).

Preparative Thin Layer Chromatography

TOWERS *et al.* (*115*) isolated dhurrin (**3**) from leaves of *Sorghum vulgare* by chromatography on silica gel, with methyl ethyl ketone : ethyl acetate : formic acid : water (5/3/2/1, v/v) as the mobile phase.

Gas Liquid Chromatography

BISSETT *et al.* (*15*) obtained a mixture of linamarin (**2**) and a methyllinamarin (**13**) from tubers of *Manihot esculenta* by column chromatography on silica gel. The dried mixture of the glucosides was fully trimethylsilylated and the ethers separated by repeated gas chromatography on columns packed with 10 or 20 per cent SE-30 on silanized Chromosorb W. The derivatives were collected in capillary tubes.

V. Structure Elucidation

Gynocardin

Although gynocardin has been known since 1904 (*92*), the structure of its aglycone remained unknown until 1966 when COBURN and LONG (*33*) published their results. Preliminary investigations by the present author (*41, 44*), mainly based on spectral data, are fully consistent with the structure proposed by COBURN and LONG for the glucoside (**25**, *Chart 2*). The following account of the structural assignment will be restricted to the latter study.

Elemental analysis of gynocardin yielded results indicating the composition $C_{12}H_{17}NO_8$ rather than, as earlier proposed, $C_{13}H_{19}NO_9$ (*93*). This was supported also by mass spectrometry. No parent ion was observable, but a peak at m/e 124, $C_6H_6NO_2^+$, was derived from the aglycone by elimination of a glucopyranosyloxy radical. Acetylation of gynocardin had earlier been suggested to give a heptaacetate (*93*), but was corrected to a hexaacetate (**28**) by elemental analysis and integration of its NMR spectrum. Similarly, fully methylated gynocardin was found to be a hexamethyl derivative (**29**), furnishing a molecular ion (at m/e 387) on electron impact. Hence, the presence of two hydroxyl groups in the aglycone moiety was strongly indicated. Verification was achieved by

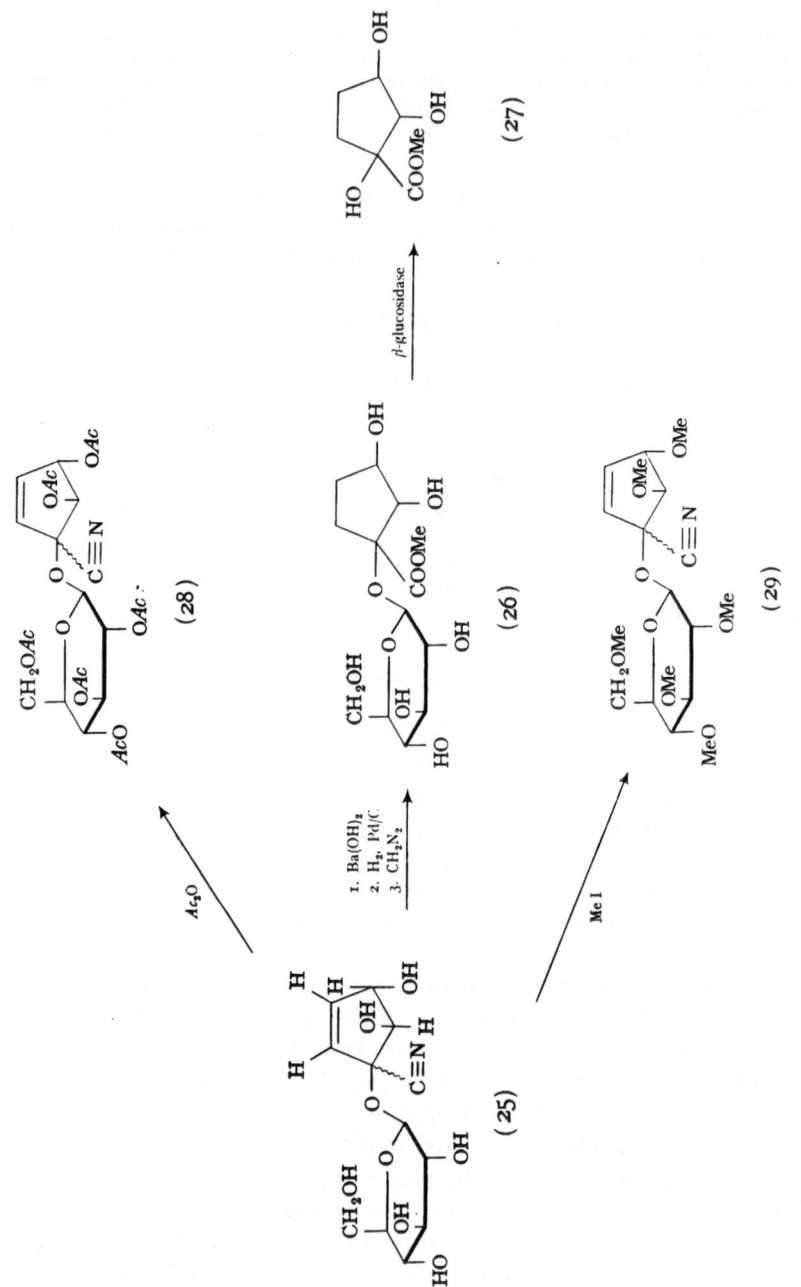

Chart 2. Conversions of Gynocardin

hydrolyzing dihydrogynocardinic acid methyl ester (26) with a β-gluco-sidase preparation from *Penicillium melanii* to a product shown (IR, NMR) to be 1,2,3-trihydroxycyclopentanecarboxylic acid methyl ester (27).

The NMR spectrum of gynocardin (in D_2O) exhibited two pairs of doublets, centered at δ 6.05 and δ 6.30 and representing two vinylic protons, with a coupling constant (6.2 c/s), suggestive of an unsymmetri-cally substituted cyclopentene ring. Signals at δ 4.31 (doublet) and δ 4.72 (multiplet) were assigned to two nonvinylic protons in the aglycone, coupled to each other with a coupling constant of 5.3 c/s indicating a *trans* arrangement. Besides, the low field proton was coupled to both of the vinylic protons.

The suggestion that barterioside (17) is a glycoside of the gynocardin type is supported chiefly by the presence in its NMR spectrum of signals corresponding to two vinylic cyclopentene protons resembling those encountered in the NMR spectrum of gynocardin (88).

Triglochinin

The structure (30) suggested for this new glucoside, apart from the stereochemistry of the aglycone, has been deduced mainly from spectro-scopic data, along with degradative work; the latter is summarized in *Chart 3* (45).

The presence of carboxylic acid groups in triglochinin was supported by a broad and intense IR absorption between 3.600 and 2.600 cm^{-1} as well as an intense band at 1.715 cm^{-1} (ν C=O). The nitrile ν C≡N band was located at 2.220 cm^{-1}. The low wave number position of this band, along with its high intensity when compared with similar bands in IR spectra of glycosides of the amygdalin-, linamarin-, or gynocardin type, suggested the presence of an α,β-unsaturated nitrile group. A medium strong band at 1.625 cm^{-1} is in accordance with the presence of con-jugated C=C bonds, and the conjugation shown in (30) is revealed by a broad UV absorption with λ_{max} at 275 nm (in water).

The NMR spectrum of the fully trimethylsilylated glucoside (in carbon tetrachloride) displayed signals characteristic of a β-D-glucopyranosyl residue, but, significantly, also two doublets centered at δ 5.93 and δ 6.69 with a coupling constant of 12.5 c/s attributable to the vinylic protons (H_A and H_B). In addition to these, only one signal was present, namely a two-proton singlet at δ 3.63 arising from a deshielded methylene group (H_C).

Hydrolysis of the compound, catalyzed by dilute acid or enzyme (β-glucosidase), afforded glucose, hydrocyanic acid, and a non-volatile acid, whereas oxo-compounds or volatile carboxylic acids were not

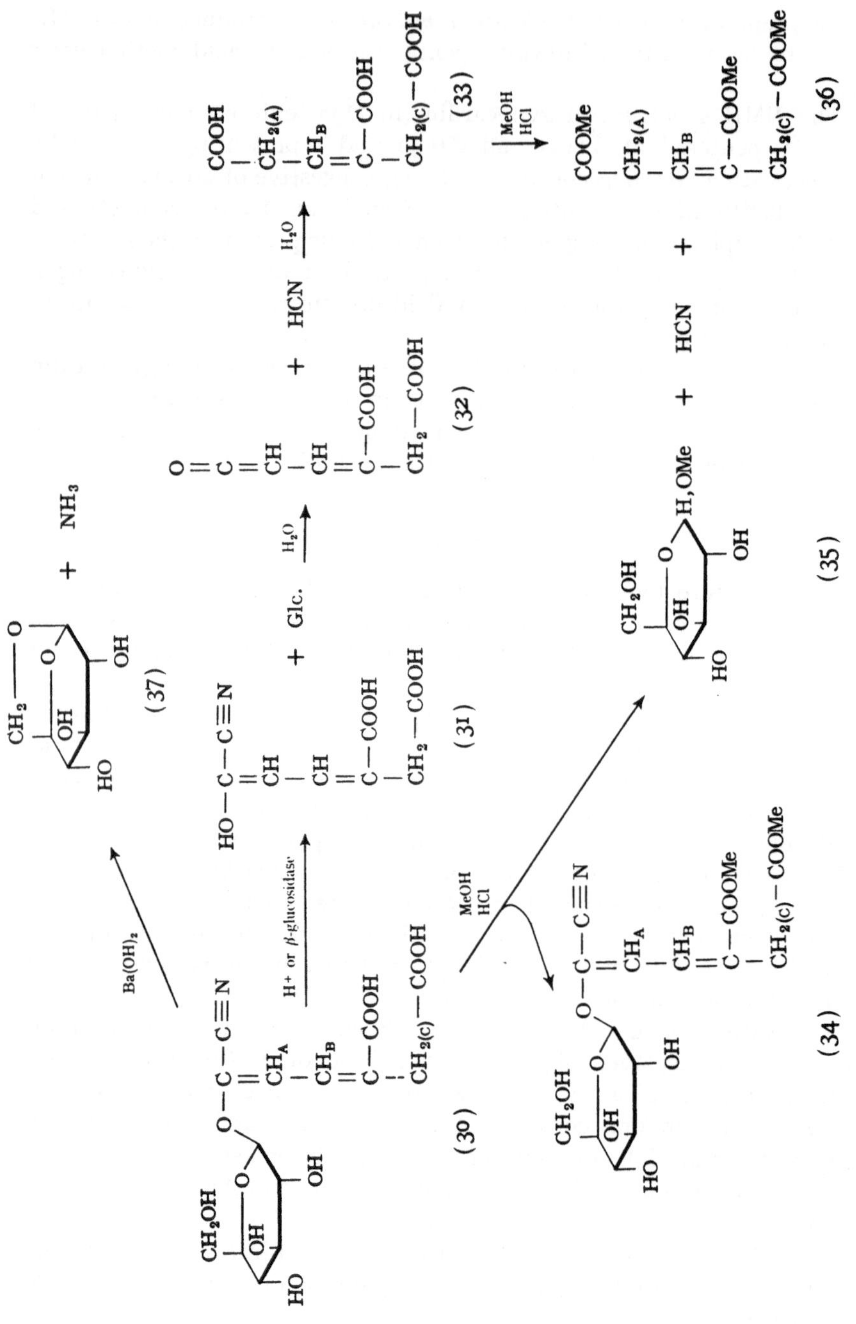

Chart 3. Degradations of Triglochinin

detectable. Elemental analysis of the acid, triglochinic acid (33), supported its composition as $C_7H_8O_6$. It was shown to be tribasic (integration of its NMR spectrum, titration, and conversion into the methyl ester (36) followed by NMR analysis) and α,β-unsaturated (λ_{max} = 209 nm in water; $\tilde{\nu}_{max}$ at 1.650 cm^{-1}). The trimethyl ester (36) showed a molecular ion peak of low intensity at m/e 230.

NMR analysis of the acid (33) (in $(CD_3)_2CO$) revealed a low field triplet at δ 7.21 (J_{AB} = 7 c/s) (H_B in (33)); a doublet at δ 3.38 (J_{AB} = 7 c/s) (H_A), and a singlet (H_C), slightly broadened due to allylic coupling, at δ 3.44. The NMR spectrum of the trimethyl ester (36) is in excellent agreement with that obtained for the free acid.

Titration of the glucoside indicated the presence of two carboxylic groups, and esterification with hydrochloric acid in methanol at 40° gave a dimethyl ester (34) (λ_{max} = 280 nm in methanol) as verified by its NMR spectrum. As expected, the esterification experiment also yielded a mixture of methylglucosides (35), hydrocyanic acid, and triglochinic acid trimethyl ester (36).

The 70 eV mass spectrum of triglochinin dimethyl ester (34) exhibited a very feeble signal at m/e 387, indicative of the molecular ion. A signal at m/e 225 corresponds to an ion formed by expulsion of glucose as the radical $C_6H_{10}O_5$, typical for O-glucopyranosides. A strong peak at m/e 198 can be interpreted as a fragment formed by ejection of HCN from the aglycone fragment (m/e 225).

In contrast to most other cyanogenic glycosides, triglochinin is sensitive to the action of aqueous barium hydroxide, even at relatively low temperature. On such treatment β-glucosan (37) is formed, but no glycosidic products resulting from hydrolysis of the nitrile group to a carboxylic group could be detected. This result is in agreement with the known lability of certain enolic glycosides towards basic reagents. Further studies on the stereochemistry of triglochinin are in progress in the author's laboratory.

Dhurrin and Taxiphyllin

The complete stereochemistry of these compounds, (3) and (10), was first established in 1964 by Towers et al. (115). Based on NMR studies the two glucosides were shown to be diastereomers with a β-D-glucopyranosyl residue linked to the cyanohydrin hydroxyl group in p-hydroxymandelonitrile.

Comparison of melting points of dhurrin- and taxiphyllin pentaacetates, as well as their specific rotations, with the corresponding constants for the tetraacetates of sambunigrin (5) and prunasin (7) strongly favoured L- and D-configurations for the aglycones of dhurrin and taxiphyllin, respectively. The argument, however, is weakened by the fact that dhurrin and taxiphyllin apparently possess almost identical specific rotations.

Analysis of the NMR spectra of dhurrin and taxiphyllin penta-acetates showed conspicuous differences in pertinent signals (*i. e.* the methine proton signals of the aglycones, and signals from the glucosyl residues), which were ascribed to effects from the aglycones. Comparison of the NMR spectra of taxiphyllin and prunasin acetates showed close similarity in relevant signals, and this information was used to support the postulated D-configuration of the aglycone of taxiphyllin. Unfortunately, similar comparisons between dhurrin and sambunigrin were not reported.

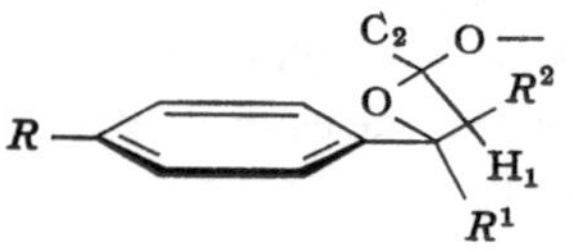

(38) Dhurrin pentaacetate ($R = OAc$; $R^1 = CN$; $R^2 = H$, $\delta = 5.69$; H_1, $\delta = 5.02$)

(39) Taxiphyllin pentaacetate ($R = OAc$; $R^1 = H$, $\delta = 5.54$; $R^2 = CN$; H_1, $\delta = 4.62$)

Fig. 1. (Based on Towers *et al.* (*115*))

Additional support for the assignment of L- and D-configuration to the aglycones of dhurrin and taxiphyllin, respectively, was derived from studies of molecular models of the respective pentaacetates. The most stable conformations of the compounds are depicted in *Fig. 1*. Inspection of the model suggests that a long range paramagnetic effect from the carbon-nitrogen triple bond in dhurrin (38) would shift the anomeric proton signal downfield relative to the anomeric proton in taxiphyllin (39). Again, the methine proton in the aglycone moiety of (38) would be expected to appear at a lower field than that of (39) due to a greater deshielding ring current effect from the aromatic nucleus in the former. The actual NMR spectra (see *Fig. 1*) were in agreement with both predictions.

Linamarin

The configuration of the anomeric carbon atom in the glucose moiety of linamarin (2) has been the subject of a great deal of discussion. Quite early it was argued (*39*) that linamarin might be an α-glucoside, primarily because of its resistance to hydrolysis by β-glucosidase preparations such as emulsin. This conclusion was soon challenged (*7*) on the basis of experiments which compared the properties of linamarase and emulsin. The weak activity of almond emulsin towards linamarin, however, has been repeatedly confirmed by various groups (*25, 35, 49*).

References, pp. 103—108

The chemical synthesis of linamarin by FISCHER and ANGER (52) greatly favours linamarin being a β-glucoside, which is further supported by recent IR studies (25) and careful investigation of the properties of linamarase (25). Finally, CLAPP et al. in 1966 proved unequivocally that linamarin is a β-glucoside by synthesizing both anomers and comparing their IR- and NMR spectra, and their optical rotations (32).

Lotaustralin

In contrast to linamarin (2), its homologue lotaustralin (13) contains a chiral centre in the aglycone. Hence two diastereomeric β-D-glucopyranosides are possible. Evidently, these diastereomers decompose to glucose, methyl ethyl ketone, and hydrogen cyanide on enzymic or acid catalyzed hydrolysis. A glucoside isolated from *Lotus australis* in 1938 (49) conforming to the above degradation was named lotaustralin as were all similar samples subsequently isolated or detected chromatographically in natural sources (1, 24, 108, see also 15 for review), in disregard of their stereochemistry which remained unestablished until very recently. In view of this, BISSETT et al. in 1969 (15) suggested that the pair of diastereomers be collectively named the methyllinamarins. The methyllinamarin occurring in *Trifolium repens* was named lotaustralin. Its stereochemistry was determined as follows:

Lotaustralin was treated with aqueous barium hydroxide converting the nitrile group into a carboxylic group. The product was then hydrolyzed with dilute hydrochloric acid to afford levorotatory 2-hydroxy-2-methylbutyric acid of known configuration (R). It was convincingly argued that the acid catalyzed hydrolysis of the intermediate carboxy glucoside must proceed predominantly by glucosyl-oxygen cleavage, whence the configuration of the asymmetric carbon atom in the aglycone of lotaustralin would be established as (R). Both epimeric methyllinamarins were prepared by chemical synthesis. The less levorotatory product was found to be identical with the natural glucoside in all respects.

It should be noted (15) that the constants recorded in the literature for samples of the methyllinamarins isolated prior to the work of BISSETT et al. do not allow conclusive identification of them as lotaustralin or its epimer (epilotaustralin). They all appear to be more or less contaminated with linamarin (2) which apparently has nearly the same botanical distribution (cf. *Table 2*, p. 93).

Proteacin

Degradation of this compound (16), recently obtained from *Macadamia integrifolia*, in the usual way yielded glucose, hydrocyanic acid, and p-hydroxybenzaldehyde in the molar ratio 2 : 1 : 1. This evidence, along with the facts that the glucoside is stable (cf. Nandina-glucoside below) and exhibits unchanged UV absorption in the presence of basic reagents, indicates that the glucosyl residues are linked both to the cyanohydrin and the phenolic hydroxyl groups. Comparison of the NMR spectrum of proteacin octaacetate with those of dhurrin and taxiphyllin pentaacetates (cf. p. 90) was used to adduce an L-configuration for the chiral centre in the aglycone of proteacin (*121*, *122*).

An apparently similar compound has been reported to occur in *Thalictrum aquilegifolium* (*106*). No constants are available as yet which would allow a comparison with proteacin.

Nandina-glucoside

The Nandina-glucoside (9), unique in containing a free cyanohydrin grouping, is unstable and decomposes rapidly to a glucoside of p-hydroxybenzaldehyde and hydrogen cyanide. It has not been isolated in crystalline form; the structure ascribed to it follows from results obtained by combined application of paper chromatography and quantitative determination of the cleavage products (5).

VI. Synthesis

The classical approach to synthesis of cyanogenic glycosides, by which several compounds of the amygdalin- and linamarin types have been prepared, involves condensation of an appropriate α-hydroxycarboxylic acid alkyl ester with an acetobromosaccharide. The product is subsequently converted into the corresponding amide which in turn is dehydrated to the desired nitrile. The overall procedure normally consists of at least four steps (*29, 52, 53*).

Preliminary studies in this laboratory (*75*) gave results which suggested that it would be possible to condense cyanohydrins directly with an acetobromosugar. Independently, CLAPP et al. (*32*) published an elegant synthesis of linamarin (2) essentially based on the same principle. Tetra-O-acetyl-α-D-glucopyranosyl bromide was condensed with acetone cyanohydrin in the presence of mercuric cyanide. The reaction mixture contained both anomers (as acetates) which were separated and deacetylated to linamarin and the anomeric isolinamarin. Lotaustralin (13) and its epimer, epilotaustralin, have also been prepared by this method (*15*).

VII. Distribution

Cyanogenesis in living organisms is generally ascribed to the presence of cyanogenic glycosides, an assumption which is apparently based mainly on the paucity of data pointing to other sources of hydrogen cyanide. Other possible sources, however, should not be overlooked. A pertinent example is found in the seeds of *Schleicera trijuga* Willd., known for almost fifty years as cyanogenic (*101*). Recently, KUNDU and BANDYOPADHYAY (*78*) reported on the isolation of two glyceride-like substances from the lipoid fraction of the seeds, suspected as the source of hydrocyanic acid. The exact chemical structure of these compounds is not yet known, but an IR analysis as well as high solubility in petroleum ether certainly indicates a non-glycosidic structure.

Table 2. *Distribution of Known Cyanogenic Glycosides in Higher Plants*

Compound	Species	Family	References
Acacipetalin (**8**)	*Acacia lasiopetala* Oliv. and *A. stolonifera* Burch.	Mimosaceae	(*98, 109*)
Acalyphin (**12**)	*Acalypha indica*	Euphorbiaceae	(*99*)
Amygdalin (**1**)	Species of *Cydonia, Eriobotrya, Prunus, Pyrus,* and other genera	Rosaceae	(*62*)
Barterioside (**17**)	*Barteria fistulosa* Mast.	Passifloraceae	(*88*)
Dhurrin (**3**)	*Sorghum vulgare* Pers.	Gramineae	(*82*)
Gynocardin (**4**)	*Gynocardia odorata* R. Br. and *Pangium edule* Reinw.	Flacourtiaceae	(*33, 71*)
Linamarin (**2**) and Methyllinamarin (**13**)	*Dimorphotheca berberiae* Harv., *D. ecklonis* D. C., and *Osteospermum jucundum* Norlindh	Compositae	(*24*)
	Cnidoscolus texanus (Muell. Arg.) Small ((**2**) only ?), *Hevea braziliensis* Muell. Arg. ((**2**) only ?), *Manihot esculenta* Crantz and other species in this genus	Euphorbiaceae	(*15, 24, 105*)
	Linum usitatissimum L. and other species in this genus	Linaceae	(*24*)
	Papaver nudicaule L.	Papaveraceae	(*1*)
	Nearly all investigated species of *Lotus; Phaseolus lunatus* L., and *Trifolium repens* L. (lotaustralin!)	Papilionaceae	(*15, 24*)
Lucumin (**15**)	*Lucuma mammosa* Gaertn.	Sapotaceae	(*8*)
Nandina-glucoside (**9**)	*Nandina domestica* Thunb.	Berberidaceae	(*5*)
	Goodia lotifolia Salisb. (?)	Papilionaceae	(*50*)
	Thalictrum aquilegifolium (?)	Ranunculaceae	(*106*)

(Table 2, continued)

Compound	Species	Family	References
Proteacin (16)	*Macadamia integrifolia*	Proteaceae	*(122)*
	Thalictrum aquilegifolium?	Ranunculaceae	*(106)*
Prunasin (7)	*Eremophila maculata* (Ker.) F. Muell.	Myoporaceae	*(62)*
	Eucalyptus cladocalyx F. Muell.	Myrtaceae	*(62)*
	Cystopteris fragilis Bernh. and *Pteridium aquilinum* (L.) Kuhn	Polypodiaceae	*(74, 76)*
	See amygdalin	Rosaceae	*(62)*
	Jamesia americana Torr et Gray	Saxifragaceae	*(62)*
	Linaria minor Desf. and *L. striata* D. C.	Schrophulariaceae	*(62)*
Sambunigrin (5)	*Sambucus nigra* L.	Caprifoliaceae	*(20)*
	Australian species of *Acacia*	Mimosaceae	*(49)*
	Ximenia americana L.	Olacaceae	*(49)*
	The co-occurrence of sambunigrin and prunasin in Rosaceae appears questionable	Rosaceae?	*(90)*
Taxiphyllin (10)	*Phyllanthus gasstroemi* Muell. Arg. (?)	Euphorbiaceae	*(51)*
	Macadamia ternifolia F. Muell. and *Stenocarpus sinuatus* Endl. (?)	Proteaceae	*(91)*
	Taxus species	Taxaceae	*(115)*
Triglochinin (18)	*Triglochin maritimum* L.	Juncaginaceae	*(45)*
Vicianin (6)	*Vicia angustifolia* Roth.	Papilionaceae	*(13)*
	Davallia species	Polypodiaceae	*(76)*
Zierin (11)	*Ziera laevigata* Sm.	Rutaceae	*(48)*

The classification of a given species as cyanogenic is subject to a host of uncertainties. Nothing general can be said about the cyanogenic properties in different species. The contents of hydrogen cyanide are extremely variable, ranging from traces to about 0.3–0.5 per cent (calculated on fresh weight basis) in leaves of *Nandina domestica* Thunb. (5) or 0.5–0.6 per cent in seeds of *Gynocardia odorata* R. Br. (*33*), the strongest cyanogenic plants known today. Besides, factors like age of plants, accumulation of cyanogens into different organs, and environmental variations (soil, climate, geographical location, *etc.*) have repeatedly been shown to play an important role.

Strictly speaking, classification of a given species as noncyanogenic is meaningless unless the problem is investigated on the enzymic level, since chemical tests suffer from lack of sensitivity as well as specificity. The present impression, based on current techniques, is that cyanogenic glycosides are not ubiquitous, but undoubtedly widely distributed in Nature.

Cyanogenesis in Higher Plants

The occurrence of cyanophoric species within the realm of higher plants has been dealt with in reviews by HEGNAUER (*62, 63, 64*) as well as in his well known treatise *Chemotaxonomie der Pflanzen*. A list covering the literature up to mid–1968 has been prepared by the present author (*42*). According to these surveys, at least 800 species of higher plants, representing more than 70 families, are known to be cyanogenic. For the present purpose, only the distribution of known cyanogenic glycosides will be considered and is enumerated in *Table 2*.

Cyanogenesis in Bacteria

Cyanogenesis has been detected in strains of the species *Pseudomonas aureofaciens, P. aeroginosa, B. chloroamphis*, and *Chromobacterium violaceum*, all belonging to the family Pseudomonadaceae, class Pseudomonadales (*83*). The exact mechanism of cyanogenesis is unknown although glycine appears to act as precursor in the formation of hydrogen cyanide, the methylene carbon of this amino acid being converted into the nitrile carbon atom (*22, 83, 84*).

Cyanogenesis in Fungi

This topic has been reviewed by HEGNAUER (*65*) according to whom approximately 30 species, mainly of the genera *Clitocybe* and *Marasmius*, are known as cyanophoric. STEVENS and STROBEL (*108*) have shown the presence of linamarin (**2**) and methyllinamarin (**13**) in an unidentified psychrophilic basidiomycete, explaining at least one possible pathway to cyanide production in this particular fungus.

Cyanogenesis in Animals

The reported instances of cyanogenic animals have recently been surveyed by the present author (*43*). The phenomenon seems to be widely prevalent in the class of Myriopoda of which eight species along with one species each of the orders Diplopoda (true millipedes) and Chilopoda (centipedes) have been shown to be cyanogenic. The mechanism of cyanogenesis is largely obscure, although PALLARES (87) has isolated a cyanogenic glucoside (14) from the millipede *Polydesmus vicinus* L. In millipedes, the cyanogen(s) are produced in special glands and may conceivably constitute a chemical defense and/or attack system (17, 102).

Cyanogenesis has also been observed in four species of moths, belonging to the sub-family Zygaeninae, class Insecta. These species, if disturbed, emit a smelly, colourless, volatile, and apparently non-cyanogenic fluid. On the other hand, crushed tissues of the moths, in all stages of evolution, contain cyanide, which is present also in the haemolymph (70).

VIII. Biosynthesis

The biosynthesis of cyanogenic glycosides has been a subject of intense research within the last decade, with BUTLER and CONN, together with their collaborators, as the leading groups. In view of the fact that an authoritative review has recently been published by CONN and BUTLER (34), only a brief summary will be presented here.

The aglycones of the cyanogenic glycosides (41, *Chart 4*) are generally derived from the closely related amino acids (40), as substantiated by numerous biosynthetic experiments leading to no less than seven glycosides of the amygdalin type ((1) (2), (3) (57, 77, 119), (6) (117), (7) (12), (9) (5), (10) (16), and (16) (106)) as well as the two linamarin type compounds (2) and (13) (27) in a number of plant species. It follows, that only L-amino acids can function as effective precursors and that the carboxylic carbon atom is lost at some stage during the *in vivo* synthesis. Moreover, administration of specifically labelled amino acids strongly indicates that all other linkages are retained during the conversions. Hence, all intermediates must be nitrogenous.

Much effort has been devoted to the problem of establishing the nature of the true intermediates on the biosynthetic pathway. Observations that the amino acid precursor does not fragment (except for the loss of a carboxylic group) during synthesis, and that α-oximino acids easily undergo conversion to the corresponding nitriles (6) have lead to experiments involving administration of ketoximes and aldoximes to cyanophoric plants. TAPPER et al. (114), in studies of the formation of linamarin ((45); $R = R^1 = CH_3$) in flax seedlings, found that isobutyraldoxime-U-^{14}C ((43); $R = R^1 = CH_3$) was converted virtually as effectively to the glucoside as was L-valine-U-^{14}C ((42); $R = R^1 = CH_3$) while incorporation of α-oximinoisovaleric acid-U-^{14}C ((44); $R = R^1 = CH_3$) was slightly less than half of valine. Similar results were obtained when α-oximinophenylpropionic acid-2-^{14}C ((44); $R = C_6H_5$, $R^1 = H$) and phenylacetaldoxime-U-^{14}C ((43); $R = C_6H_5$, $R^1 = H$) were administered to

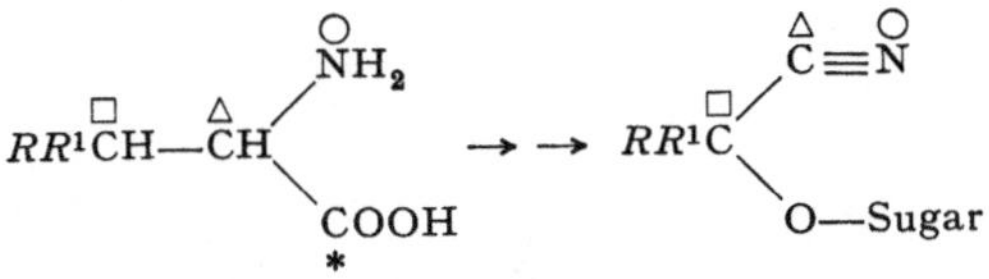

(40) (41)

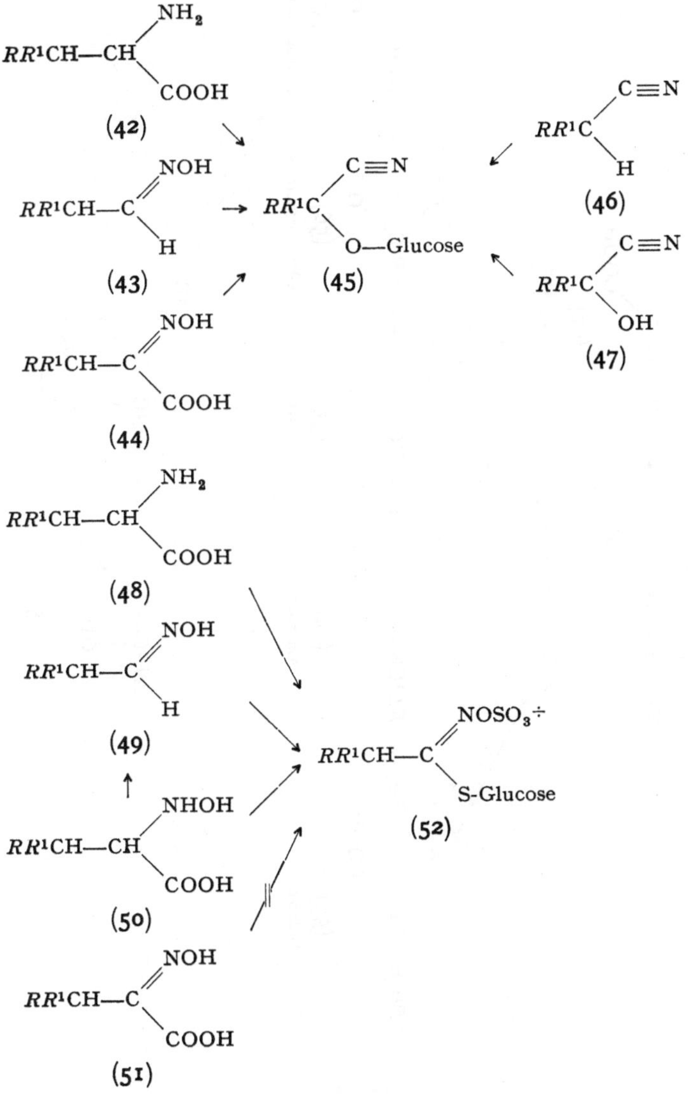

Chart 4. Summary of Experiments on the Biosynthesis of Cyanogenic Glycosides and Glucosinolates

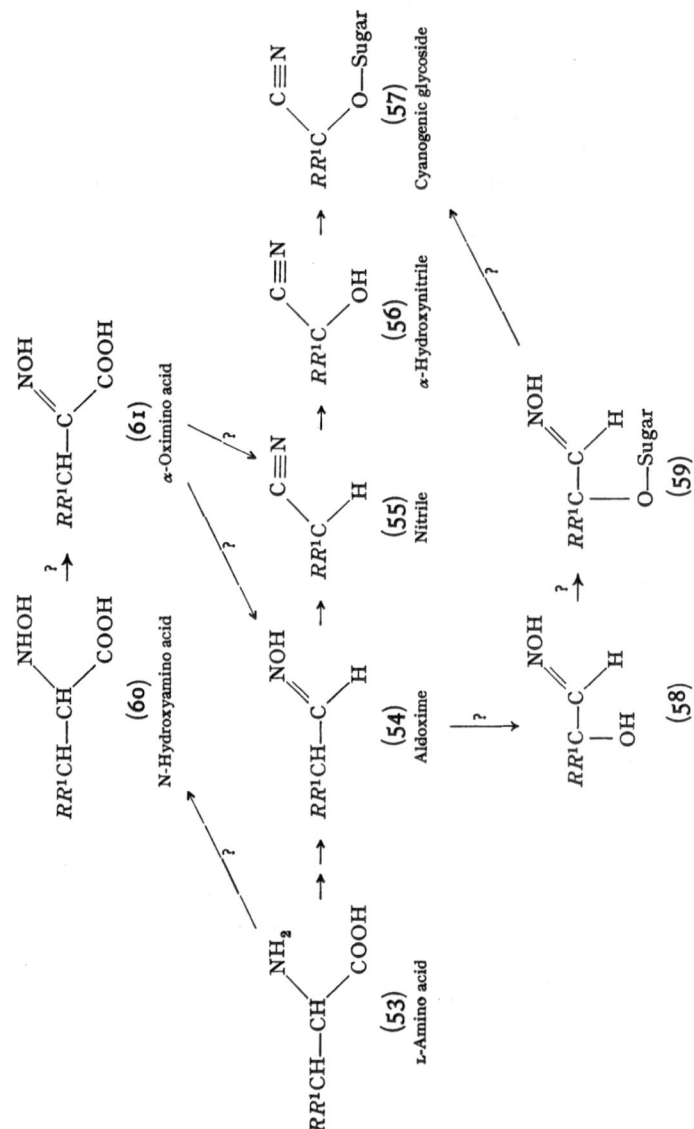

Chart 5. Biosynthesis of Cyanogenic Glycosides

shoots or leaves of cherry laurel (34, 59, 113). Incorporation of the keto-xime and the aldoxime into prunasin ((45); $R = C_6H_5$, $R^1 = H$) was five and about three times as effective, respectively, as that of L-phenyl-alanine-U-^{14}C ((42); $R = C_6H_5$, $R^1 = H$).

The evidence in favour of oximino acids as biosynthetic intermediates was not entirely conclusive. Thus, linamarin might arise from α-oximino-isovaleric acid by way of α-ketoisovaleric acid and valine, a possibility that was indeed supported by the observation (TAPPER et al. (114)) that valine isolated from flax was slightly labelled. Clearly, further experimentation was desirable to elucidate this point and evidence has now been obtained (34) which indicates approximately equal dilution of ^{14}C and ^{15}N in linamarin subsequent to administration of doubly labelled α-oximinoisovaleric acid to flax plants. Therefore, the oximino acid seems indeed to be a direct precursor in the biosynthesis.

If one accepts that aldoximes (54) are intermediates in the biosynthesis of cyanogenic glycosides, two routes appear possible (Chart 5). In the first, dehydration of the aldoxime to the corresponding nitrile (55) is followed by hydroxylation to the cyanohydrin (56) which is then glycosylated to the glycoside (57). Alternatively, the aldoxime (54) may undergo hydroxylation to give (58) which upon glycosylation (59) is converted further into glycoside (57). In accordance with the first proposal, HAHL-BROCK et al. (59) found that isobutyronitrile-1-^{14}C ((46); $R = R^1 = CH_3$; Chart 4) and α-hydroxyisobutyronitrile-1-^{14}C ((47); $R = R^1 = CH_3$) were both incorporated efficiently into linamarin in flax seedlings. Studies of the biosynthesis of prunasin in cherry laurel leaves gave analogous results.

The biosynthesis of glucosinolates (mustard oil glucosides) (52) (see ETTLINGER and KJÆR (40) for a recent review) seems strikingly parallel to that of the cyanogenic glycosides, amino acids (42), (48) and aldoximes (43), (49) in both series serving as highly effective precursors (113, 118). N-hydroxyamino acids (50) may be involved as direct intermediates for glucosinolates as well as in the production of the intermediate aldoximes (49), judging from investigations on the biosynthesis of glucotropaeolin ((52); $R = C_6H_5$, $R^1 = H$) by KINDL and UNDERHILL (73). It is somewhat disturbing at the moment that α-oximino acids (44), (51) seem to be true intermediates in the biosynthesis of cyanogenic glycosides, whilst KINDL and UNDERHILL (73) obtained no evidence supporting an analogous incorporation into glucotropaeolin. As a matter of fact, the seemingly high incorporation of α-oximino acids into cyanogenic glycosides may be due solely to their facile conversion into nitriles. Chart 5 presents a summary of the present concepts regarding the biosynthesis of cyanogenic glycosides.

IX. Metabolic Aspects

The Fate of Cyanogenic Glycosides in Living Organisms

As has been stated earlier, the cyanogenic glycosides are virtually always accompanied by enzymes capable of decomposing them. Hence, the important question arises as to whether these enzymes are continuously in action inside the intact organism or not, or, put in another way, whether the glycosides are to be considered metabolically active rather than inactive.

It is well established, that the content of cyanogenic glycosides in plant tissues generally reaches a maximum at a certain stage of growth, *viz.* while nitrogen metabolism is most extensive. Subsequently, the content diminishes or vanishes completely, indicating that enzymic degradation is really going on. In fact, evidence has been presented which indicates that the organs of certain living higher plants release hydrogen cyanide into the surrounding atmosphere (*3, 120*) similar to what has been observed in the cases of fungal or bacterial substrates (*65, 83*). Here again, however, the mechanism of cyanogenesis is poorly understood.

The most direct and unambiguous method for ascertaining participation of the glycosides in plant metabolism would be to feed a living cyanophoric plant radioactively labelled glycoside and then, after a suitable period of metabolism, search for labelled compounds. This approach has been tried by Tschiersch (*116*) but without success since, under the experimental conditions used, the plants did not take up the compounds administered. The metabolic fate of cyanogenic glycosides therefore must be inferred from results of experiments involving feeding with radioactive cyanide, or, less frequently, amino acids known to act as precursors of the glycosides in question.

Cyanide Metabolism

Administration of labelled hydrogen cyanide under various conditions to a number of higher plants, cyanogenic as well as noncyanogenic has consistently led to the production of labelled β-L-cyanoalanine ((66), *Chart 6*) (*18, 54, 55, 95*). Closely similar results have been obtained when appropriate labelled amino acids (62) were fed to plants known to produce cyanogenic glycosides (63) (*4, 5, 117*) indicating conversion of hydrocyanic acid, formed by decomposition of the glycosides, into cyanoalanine. L-serine (64) has been suggested as a plausible intermediate in this condensation (*37, 54, 86*) but L-cysteine (65) may be an even more likely candidate (*18, 67*).

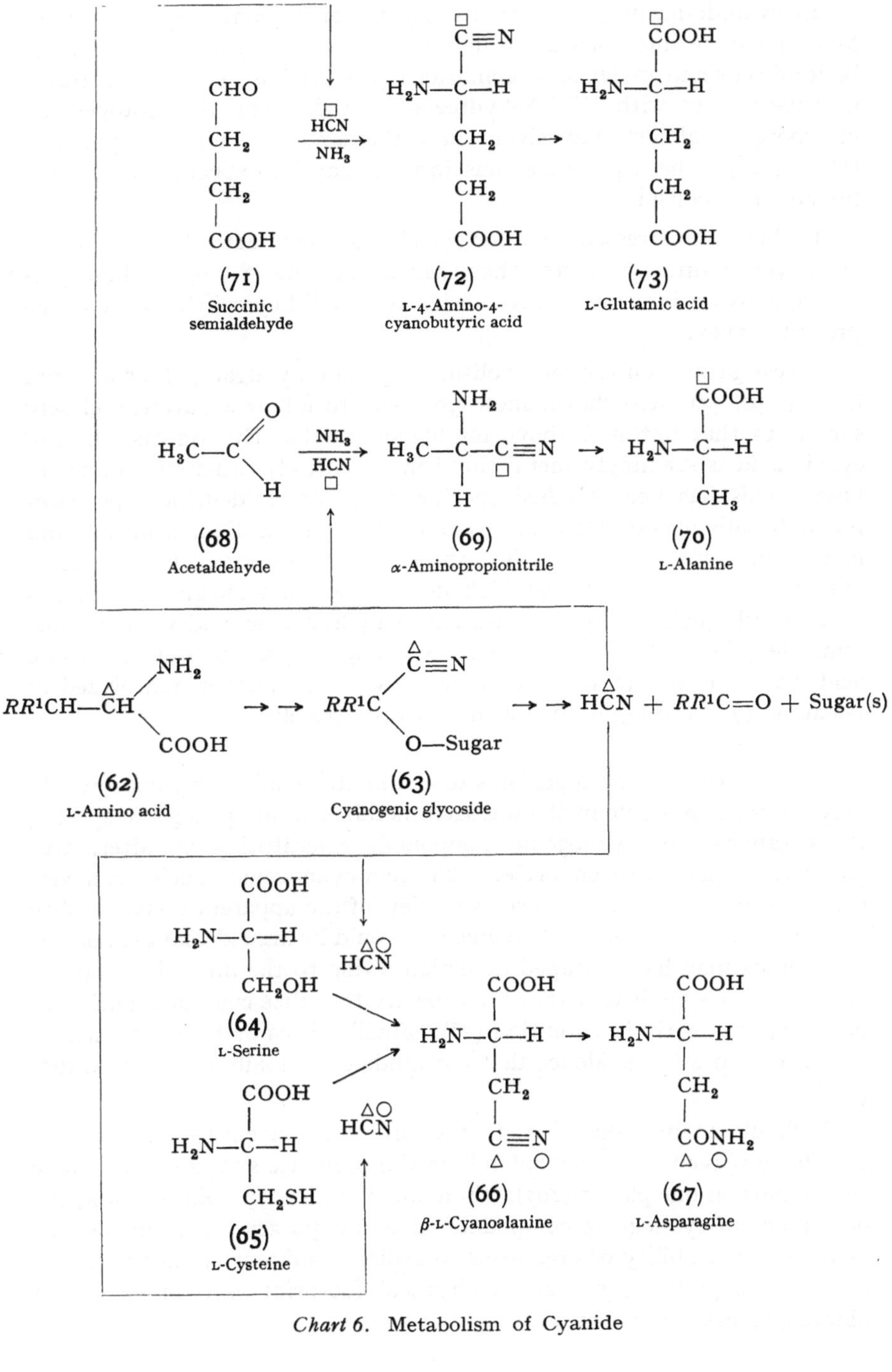

$$\underset{\substack{\text{(71)}\\ \text{Succinic}\\ \text{semialdehyde}}}{\overset{\text{CHO}}{\underset{\text{COOH}}{|\,\text{CH}_2\,|\,\text{CH}_2\,|}}} \xrightarrow[\text{NH}_3]{\text{HCN}} \underset{\substack{\text{(72)}\\ \text{L-4-Amino-4-}\\ \text{cyanobutyric acid}}}{\overset{\text{C}\equiv\text{N}}{\underset{\text{COOH}}{|\,\text{H}_2\text{N—C—H}\,|\,\text{CH}_2\,|\,\text{CH}_2\,|}}} \rightarrow \underset{\substack{\text{(73)}\\ \text{L-Glutamic acid}}}{\overset{\text{COOH}}{\underset{\text{COOH}}{|\,\text{H}_2\text{N—C—H}\,|\,\text{CH}_2\,|\,\text{CH}_2\,|}}}$$

(71) Succinic semialdehyde → (72) L-4-Amino-4-cyanobutyric acid → (73) L-Glutamic acid

$$\underset{\substack{\text{(68)}\\ \text{Acetaldehyde}}}{\text{H}_3\text{C—C}\overset{O}{\underset{H}{{}}}} \xrightarrow[\text{HCN}]{\text{NH}_3} \underset{\substack{\text{(69)}\\ \alpha\text{-Aminopropionitrile}}}{\text{H}_3\text{C—C—C}\equiv\text{N}\ \overset{\text{NH}_2}{\underset{H}{{}}}} \rightarrow \underset{\substack{\text{(70)}\\ \text{L-Alanine}}}{\overset{\text{COOH}}{\underset{\text{CH}_3}{|\,\text{H}_2\text{N—C—H}\,|}}}$$

(68) Acetaldehyde → (69) α-Aminopropionitrile → (70) L-Alanine

$$\underset{\substack{\text{(62)}\\ \text{L-Amino acid}}}{RR^1\text{CH—CH}\overset{\text{NH}_2}{\underset{\text{COOH}}{{}}}} \rightarrow\rightarrow \underset{\substack{\text{(63)}\\ \text{Cyanogenic glycoside}}}{RR^1\text{C}\overset{\text{C}\equiv\text{N}}{\underset{\text{O—Sugar}}{{}}}} \rightarrow\rightarrow \text{HCN} + RR^1\text{C}{=}\text{O} + \text{Sugar(s)}$$

(62) L-Amino acid → (63) Cyanogenic glycoside → HCN + RR¹C=O + Sugar(s)

$$\underset{\substack{\text{(64)}\\ \text{L-Serine}}}{\overset{\text{COOH}}{\underset{\text{CH}_2\text{OH}}{|\,\text{H}_2\text{N—C—H}\,|}}} \xrightarrow{\text{HCN}}$$

$$\underset{\substack{\text{(65)}\\ \text{L-Cysteine}}}{\overset{\text{COOH}}{\underset{\text{CH}_2\text{SH}}{|\,\text{H}_2\text{N—C—H}\,|}}} \xrightarrow{\text{HCN}}$$

$$\underset{\substack{\text{(66)}\\ \beta\text{-L-Cyanoalanine}}}{\overset{\text{COOH}}{\underset{\text{C}\equiv\text{N}}{|\,\text{H}_2\text{N—C—H}\,|\,\text{CH}_2\,|}}} \rightarrow \underset{\substack{\text{(67)}\\ \text{L-Asparagine}}}{\overset{\text{COOH}}{\underset{\text{CONH}_2}{|\,\text{H}_2\text{N—C—H}\,|\,\text{CH}_2\,|}}}$$

(64) L-Serine, (65) L-Cysteine → (66) β-L-Cyanoalanine → (67) L-Asparagine

Chart 6. Metabolism of Cyanide

In cyanide-feeding experiments, most higher plants rapidly convert β-cyanoalanine into asparagine (67) which was, in fact, the first metabolite discovered under such conditions (19). It has been demonstrated in experiments with ^{14}C-^{15}N-hydrogen cyanide that the isotopes are incorporated almost exclusively into the amide group of asparagine (18, 95, 96). The asparagine thus formed may be expected to undergo further metabolism.

It should be stressed that certain plants accumulate β-cyanoalanine, or convert it into products other than asparagine (86, 97). These processes, however interesting, are considered to fall beyond the scope of the present survey.

Investigations on the metabolism of cyanide by algae (55) or bacteria (22, 23, 30, 37) have shown these processes to follow a pathway closely similar to that outlined above for higher plants. By contrast, hydrocyanic acid is seemingly metabolized in an entirely different manner in fungi. This has been studied specifically in an unidentified, psychrophilic basidiomycete (110, 111) known to contain linamarin (2) and methyllinamarin (13) (108). Two routes were followed: (i) Condensation of hydrogen cyanide with acetaldehyde (68) and ammonia giving α-aminopropionitrile (69), or (ii) condensation of hydrocyanic acid with succinic semialdehyde (71) and ammonia yielding L-4-amino-4-cyanobutyric acid (72). The two metabolites were shown to be further hydrolyzed to L-alanine (70) and L-glutamic acid (73) respectively.

The ability of living organisms to assimilate cyanide may conceivably play an important role in the nitrogen metabolism of cyanogenic species, the catabolism of cyanogenic compounds constituting an alternative pathway to the nitrogen cycle. That non-cyanogenic species can also utilize cyanide seems more curious in view of the apparent unavailability of cyanide in these species. Obviously, it could be argued that cyanogenic compounds may have escaped detection owing to the limited sensitivity of the methods applied or that the capacity to utilize cyanide might have been acquired early in evolution (56). Finally, it cannot be excluded on the basis of present evidence that a cyanide cycle really exists in Nature (112).

It should be mentioned here that according to current theories cyanide may have played an important role in the prebiotic synthesis of organic compounds on our planet (107). From an evolutionary point of view, the occurrence of cyanogenic compounds in contemporary organisms as well as the distinct ability of organisms to utilize cyanide may therefore look less surprising, being perhaps a chemical footprint, a reminiscence of abiotic processes prevalent eons ago.

References

1. ABROL, Y. P.: Occurrence of Linamarin and Lotaustralin in Iceland Poppy (*Papaver nudicaule* Linn.). Indian J. Chemistry **4**, 251 (1966).
2. — Studies on the Biosynthesis of Amygdalin, the Cyanogenic Glycoside of Bitter Almonds (*Prunus amygdalus* Stokes). Indian J. Biochem. **4**, 54 (1967).
3. — Occurrence of Free Hydrocyanic Acid in Plants. Indian J. Exp. Biol. **5**, 191 (1967).
4. ABROL, Y. P. and E. E. CONN: Studies on Cyanide Metabolism in *Lotus arabicus* L. and *Lotus tenuis* L. Phytochem. **5**, 237 (1966).
5. ABROL, Y. P., E. E. CONN and J. R. STOKER: Studies on the Identification, Biosynthesis and Metabolism of a Cyanogenic Glucoside in *Nandina domestica* Thunb. Phytochem. **5**, 1021 (1966).
6. AHMAD, A. and I. D. SPENSER: The Conversion of α-Keto Acids and of α-Keto Acid Oximes to Nitriles in Aqueous Solution. Canad. J. Chem. **29**, 1340 (1961).
7. ARMSTRONG, H. E. and E. HORTON: Studies on Enzyme Action. XIII. Enzymes of the Emulsin Type. Proc. Roy. Soc. (London) **82 B**, 349 (1910).
8. BACHSTEZ, M., E. S. PRIETO y A. M. C. GAJA: Notas sombre drogas, plantas y alimentos mexicanos. X. Estudio de la lucumina, glucósido cianogenético del mamey (*Lucuma mammosa* G.). Ciencia (Mexico) **9**, 200 (1948).
9. BECKER, W. und E. PFEIL: Über das Flavinenzym D-Oxynitrilase. Biochem. Z. **346**, 301 (1966).
10. BENNETT, W. D.: Isolation of the Cyanogenic Glycoside Prunasin from Bracken Fern. Phytochem. **7**, 151 (1968).
11. BENNETT, W. D. and B. A. TAPPER: A Sensitive Method for Detecting Cyanoglycosides on Paper and Cellulose Thin Layers. J. Chromatogr. **34**, 428 (1968).
12. BEN-YEHOSHUA, S. and E. E. CONN: Biosynthesis of Prunasin, the Cyanogenic Glucoside of Peach. Plant Physiol. **39**, 331 (1964).
13. BERTRAND, G.: La vicianine, nouveau glucoside cyanhydrique contenu dans les graines de Vesce. C. R. hebd. séances Acad. Sci. **143**, 832 (1906).
14. BERTRAND, G. et G. WEISWEILLER: Sur la constitution du vicianose et de la vicianine. C. R. hebd. séances Acad. Sci. **151**, 884 (1910).
15. BISSETT, F. H., E. C. CLAPP, R. A. COBURN, M. G. ETTLINGER and L. LONG, Jr.: Cyanogenesis in Manioc: Concerning Lotaustralin. Phytochem. **8**, 2235 (1969).
16. BLEICHERT, E. F., A. C. NEISH and G. H. N. TOWERS: Biosynthesis of Taxiphyllin in *Taxus*. In: G. BILLEK (Edit.), Biosynthesis of Aromatic Compounds, Proceedings of the 2nd Meeting of the Federation of European Biochemical Societies, Vol. 3, p. 119. Oxford: Pergamon Press. 1966.
17. BLUM, M. S. and J. P. WOODRING: Secretion of Benzaldehyde and Hydrogen Cyanide by the Millipede *Pachydesmus crassicutis* (Wood). Science **138**, 512 (1962).
18. BLUMENTHAL, S. G., H. R. HENDRICKSON, Y. P. ABROL and E. E. CONN: Cyanide Metabolism in Higher Plants. III. The Biosynthesis of β-Cyanoalanine. J. Biol. Chem. **243**, 5302 (1969).
19. BLUMENTHAL-GOLDSCHMIDT, S., G. W. BUTLER and E. E. CONN: Incorporation of Hydrocyanic Acid Labelled with Carbon-14 into Asparagine in Seedlings. Nature **197**, 718 (1963).
20. BOURQUELOT, E. et E. DANJOU: Préparation du glucoside cyanhydrique du surreau à l'état cristallisé. J. Pharmac. Chim. **22**, 219 (1905).
21. BOVÉ, C. and E. E. CONN: Metabolism of Aromatic Compounds in Higher Plants. II. Purification and Properties of the Oxynitrilase of *Sorghum vulgare*. J. Biol. Chem. **236**, 207 (1961).

22. Brysk, M. M., W. A. Corpe and L. V. Hankes: β-Cyanoalanine Formation by *Chromobacterium violaceum*. J. Bacteriol. 97, 322 (1969).

23. Brysk, M. M., C. Lauinger and C. Ressler: Biosynthesis of Cyanide from [2-$^{14}C^{15}N$] Glycine in *Chromobacterium violaceum*. Biochim. Biophys. Acta 184, 583 (1969).

24. Butler, G. W.: The Distribution of the Cyanoglucosides Linamarin and Lotaustralin in Higher Plants. Phytochem. 4, 127 (1965).

25. Butler, G. W., R. W. Bailey and L. D. Kennedy: Studies on the Glucosidase "Linamarase". Phytochem. 4, 369 (1965).

26. Butler, G. W. and B. G. Butler: Biosynthesis of Linamarin and Lotaustralin in White Clover. Nature 187, 780 (1960).

27. Butler, G. W. and E. E. Conn: Biosynthesis of the Cyanogenic Glucosides Linamarin and Lotaustralin. I. Labelling Studies in Vivo with *Linum usitatissimum*. J. Biol. Chem. 239, 1674 (1964).

28. Caldwell, R. J. and S. L. Courtauld: Mandelonitrile Glucosides. Prulaurasin. J. Chem. Soc. (London) 91, 671 (1907).

29. Campbell, R. and W. N. Haworth: Synthesis of Amygdalin. J. Chem. Soc. (London) 125, 1337 (1924).

30. Castric, P. A. and G. A. Strobel: Cyanide Metabolism by *Bacillus megaterium*. J. Biol. Chem. 244, 4089 (1969).

31. Claeys, R. R. and H. Freund: Gas Chromatographic Separation of Hydrogen Cyanide on Porapak Q. Analysis of Trace Aqueous Solutions. Environ. Sci. Technol. 2, 458 (1968).

32. Clapp, R. C., F. H. Bissett, R. A. Coburn and L. Long, Jr.: Cyanogenesis in Manioc: Linamarin and Isolinamarin. Phytochem. 5, 1323 (1966).

33. Coburn, R. A. and L. Long, Jr.: Gynocardin. J. Organ. Chem. (U. S. A.) 31, 4312 (1966).

34. Conn, E. E. and G. W. Butler: The Biosynthesis of Cyanogenic Glycosides and Other Simple Nitrogen Compounds. In: J. B. Harborne and T. Swain (Edits.), Perspectives in Phytochemistry, p. 47. London: Academic Press. 1969.

35. Coop, I. E.: Cyanogenesis in White Clover (*Trifolium repens* L.). III. A Study of Linamarase, the Enzyme which Hydrolyses Lotaustralin. New Zealand J. Sci. Technol. 22B, 71 (1940).

36. Dillemann, G.: Composés Cyanogénétiques. In: W. Ruhland (Edit.), Handbuch der Pflanzenphysiologie, Vol. 8, p. 1050. Berlin-Göttingen-Heidelberg: Springer-Verlag. 1958.

37. Dunnill, P. M. and L. Fowden: Enzymatic Formation of β-Cyanoalanine from Cyanide by *Escherichia coli* Extracts. Nature 208, 1206 (1965).

38. Dunstan, W. R. and T. A. Henry: Cyanogenesis in Plant Parts. II. The Great Millet, *Sorghum vulgare*. Philos. Trans. Roy. Soc. (London) 199A, 399 (1902).

39. Dunstan, W. R., T. A. Henry and S. J. M. Auld: Cyanogenesis in Plants. Part IV. The Occurrence of Phaseolunatin in Common Flax *(Linum usitatissimum)*. Proc. Roy. Soc. (London) 78B, 152 (1906).

40. Ettlinger, M. G. and A. Kjær: Sulfur Compounds in Plants. In: T. J. Mabry, R. E. Alston and V. C. Runeckles (Edits.), Recent Advances in Phytochemistry, Proceedings of the Annual Symposium of the Phytochemical Society of North America, Vol. 1, p. 58. New York: Appleton-Century-Crofts. 1968.

41. Eyjólfsson, R.: Cyanogenic Glycosides in Nature, Chemistry and Distribution, A Review, p. 103. Thesis. The Royal Danish School of Pharmacy. Copenhagen. 1968.

42. EYJÓLFSSON, R.: Reference 41, p. 37.

43. — Reference 41) p. 56.

44. — Investigations on Cyanogenic Plants. Dansk Tidsskr. Farm. **42**, 301 (1968).

45. — Isolation and Structure Determination of Triglochinin, a New Cyanogenic Glucoside from *Triglochin maritimum*. Phytochem. (in press, 1969).

46. FARNSWORTH, N. R.: Biological and Phytochemical Screening of Plants. J. Pharm. Sci. **55**, 225 (1966).

47. FEIGL, F., V. GENTIL and E. JUNGREIS: Spot Test for Aliphatic and Aromatic Cyanides. Microchim. Acta 44 (1959).

48. FINNEMORE, H. and J. M. COOPER: Cyanogenetic Glucosides in Australian Plants. Part 4. *Ziera laevigata*. J. Proc. Roy. Soc. N. S. Wales **70**, 175 (1936).

49. — — The Cyanogenetic Constituents of Australian and Other Plants. Part VII. J. Soc. Chem. Ind. **57**, 162 (1938).

50. FINNEMORE, H. and D. K. LARGE: Cyanogenetic Glucosides in Australian Plants. Part 6. An Unstable Cyanogenetic Constituent in *Goodia lotifolia*. J. Proc. Roy. Soc. N. S. Wales **70**, 440 (1936).

51. FINNEMORE, H., S. K. REICHARD and D. K. LARGE: Cyanogenetic Glucosides in Australian Plants. Part 5. *Phyllanthus gasstroemi*. J. Proc. Roy. Soc. N. S. Wales **70**, 257 (1936).

52. FISCHER, E. und G. ANGER: Synthese des Linamarins und Glykolnitril-cellosids. Ber. dtsch. chem. Ges. **52**, 854 (1919).

53. FISCHER, E. und M. BERGMANN: Synthese des Mandelnitril-glucosids, Sambunigrins und ähnlicher Stoffe. Ber. dtsch. chem. Ges. **50**, 1047 (1917).

54. FLOSS, H. G., L. HADWIGER and E. E. CONN: Enzymatic Formation of β-Cyanoalanine from Cyanide. Nature **208**, 1207 (1965).

55. FOWDEN, L. and E. A. BELL: Cyanide Metabolism by Seedlings. Nature **206**, 110 (1965).

56. FRAENKEL, G. S.: The Raison d'Être of Secondary Plant Substances. Science **129**, 1466 (1959).

57. GANDER, J. E.: Incorporation of C^{14} into *p*-Hydroxymandelonitrile-β-glucose and Other Phenolic Substances in *Sorghum* Seedlings. J. Biol. Chem. **237**, 3229 (1962).

58. GUILBAULT, G. G. and D. N. CRAMER: Ultra Sensitive, Specific Method for Cyanide Using *p*-Nitrobenzaldehyde and *o*-Dinitrobenzene. Analyt. Chemistry **38**, 834 (1966).

59. HAHLBROCK, K., B. A. TAPPER, G. W. BUTLER and E. E. CONN: Conversion of Nitriles and α-Hydroxynitriles to Cyanogenic Glucosides in Flax Seedlings and Cherry Laurel Leaves. Arch. Biochem. Biophys. **125**, 1013 (1968).

60. HAISMAN, D. R. and D. J. KNIGHT: The Enzymic Hydrolysis of Amygdalin. Biochem. J. **103**, 528 (1967).

61. HAISMAN, D. R., D. J. KNIGHT and M. J. ELLIS: The Electrophoretic Separation of the β-Glucosidase of Almond "Emulsin". Phytochem. **6**, 1501 (1967).

62. HEGNAUER, R.: Die Verbreitung der Blausäure bei den Cormophyten. 3. Mitteilung. Die blausäurehaltigen Gattungen. Pharm. Weekblad **94**, 241 (1959).

63. — Chemotaxonomische Betrachtungen. 10. Die systematische Bedeutung des Blausäuremerkmales. Pharm. Zentralhalle **99**, 322 (1960).

64. — Die Verbreitung der Blausäure bei den Cormophyten. 4. Mitteilung. Neue Untersuchungen über die Verbreitung der Cyanogenese. Pharm. Weekblad **96**, 577 (1961).

65. — Chemotaxonomie der Pflanzen, Vol. 1, p. 97 and 135. Basel: Birkhäuser Verlag. 1962.

66. HELFERICH, B. und T. KLEINSCHMIDT: Zur Kenntnis des Süßmandel-Emulsins. Z. physiol. Chem. **349**, 25 (1968).

67. HENDRICKSON, H. R. and E. E. CONN: Cyanide Metabolism in Higher Plants. IV. Purification and Properties of the β-Cyanoalanine Synthase of Blue Lupine. J. Biol. Chem. **244**, 2632 (1969).

68. HÉRISSEY, H.: Sur la "prulaurasine" glucoside cyanhydrique cristallisé retiré des feuilles de laurier-cerise. J. Pharmac. Chim. **23**, 1 (1906).

69. — Présence de l'amygdonitrileglucoside dans le *Cerasus padus* Delarbe. J. Pharmac. Chim. **26**, 194 (1907).

70. JONES, D. A., J. PARSONS and M. ROTHSCHILD: Release of Hydrocyanic Acid from Crushed Tissues of All Stages in the Life-Cycle of Species of the Zygaeninae (Lepidoptera). Nature **193**, 52 (1962).

71. JONG DE, A. W. K.: L'acide cyanhydrique des feuilles du *Pangium edule*. Rec. trav. chim. Pays-Bas **28**, 24 (1909).

72. JORISSEN, A. et E. HAIRS: La linamarine. Nouveau glucoside, fournissant de l'acide cyanhydrique par dédoublement et retiré du *Linum usitatissimum*. Bull. Acad. Roy. Sci. Belges **21**, 529 (1891).

73. KINDL, H. and E. W. UNDERHILL: Biosynthesis of Mustard Oil Glucosides: N-Hydroxyphenylalanine, a Precursor of Glucotropaeolin and a Substrate for the Enzymatic and Nonenzymatic Formation of Phenylacetaldehyde Oxime. Phytochem. **7**, 745 (1968).

74. KOFOD, H. and R. EYJÓLFSSON: The Isolation of the Cyanogenic Glucoside Prunasin from *Pteridium aquilinum* (L.) Kuhn. Tetrahedron Letters 1289 (1966).

75. — — A New Approach to the Synthesis of Cyanogenic Glycosides. Tetrahedron Letters 5349 (1966).

76. — — Cyanogenesis in Species of the Fern Genera *Cystopteris* and *Davallia*. Phytochem. **8**, 1509 (1969).

77. KOUKOL, J., P. MILJANICH and E. E. CONN: The Metabolism of Aromatic Compounds in Higher Plants. VI. Studies on the Biosynthesis of Dhurrin, the Cyanogenic Glucoside of *Sorghum vulgare*. J. Biol. Chem. **237**, 3223 (1962).

78. KUNDU, M. K. and C. BANDYOPADHYAY: Studies on Some Chemical Aspects of Kusum Oil. J. Amer. Oil Chem. Soc. **46**, 23 (1969).

79. LUIS, P., C. N. CARDUCCI and A. SÁ: Ultramicro Detection of Cyanide and Cyano Compounds. Microchim. Acta 7 (1969).

80. MANNERS, D. J. and D. C. TAYLOR: Studies on Carbohydrate Metabolizing Enzymes. Part XVIII. The α-L-Arabinosidase Activity of Almond Emulsin. Carbohyd. Res. **7**, 497 (1968).

81. MAO, C.-H. and L. ANDERSON: Cyanogenesis in *Sorghum vulgare*. III. Partial Purification and Characterization of Two β-Glucosidases from *Sorghum* Tissues. Phytochem. **6**, 473 (1967).

82. MAO, C.-H., J. P. BLOCHER, L. ANDERSON and D. C. SMITH: Cyanogenesis in *Sorghum vulgare*. I. An Improved Method for the Isolation of Dhurrin; Physical Properties of Dhurrin. Phytochem. **4**, 297 (1965).

83. MICHAELS, R. and W. A. CORPE: Cyanide Formation by *Chromobacterium violaceum*. J. Bacteriol. **89**, 106 (1965).

84. MICHAELS, R., L. V. HANKES and W. A. CORPE: Cyanide Formation from Glycine by Nonproliferating Cells of *Chromobacterium violaceum*. Arch. Biochem. Biophys. **111**, 121 (1965).

85. MIRANDE, M.: Procéde rapide pour la recherche des plantes à acide cyanhydrique. C. R. hebd. séances Acad. Sci. **149**, 140 (1909).

86. NIGAM, S. N. and W. B. McCONNELL: Incorporation of Serine-U-^{14}C into β-Cyanoalanine and γ-Glutamyl-β-Cyanoalanine in *Vicia sativa.* Canad. J. Biochem. **46**, 1327 (1968).

87. PALLARES, E. S.: Note on the Poison Produced by the *Polydesmus (Fontaria) vicinus*, Lin. Arch. Biochem. **9**, 105 (1946).

88. PARIS, M., A. BOUQUET and R.-R. PARIS: Sur le bartérioside, nouvel hétéroside cyanogénétique des écorces de racine du *Barteria fistulosa* Mast. C. R. hebd. séances Acad. Sci. **268 D**, 2804 (1969).

89. PIFFERI, P. G.: Separazione di glucosidi cianogenetici per gel-filtrazione. Boll. Sci. Fac. Chim. Ind. (Bologna) **24**, 215 (1966).

90. PLOUVIER, V.: Recherches sur l'isomérisation d'hétérosides cyanogénétiques. C. R. hebd. séances Acad. Sci. **200**, 1985 (1935).

91. — Sur la recherche des polyacools et des hétérosides cyanogénétiques chez quelques Protéacées. C. R. hebd. séances Acad. Sci. **259**, 665 (1964).

92. POWER, F. B. and F. H. GORNALL: Gynocardin, a New Cyanogenetic Glucoside. Preliminary Note. Proc. Chem. Soc. (London) **20**, 137 (1904).

93. POWER, F. B. and F. H. LEES: Gynocardin, a New Cyanogenetic Glucoside. J. Chem. Soc. (London) **87**, 349 (1905).

94. REAY, P. F.: An Improved Procedure for the Isolation of Dhurrin. Phytochem. **8**, 2259 (1969).

95. RESSLER, C., Y.-H. GIZA and S. N. NIGAM: β-Cyanoalanine, Product of Cyanide Fixation and Intermediate in Asparagine Biosynthesis in Certain Species of *Lathyrus* and *Vicia.* J. Amer. Chem. Soc. **91**, 2766 (1969).

96. RESSLER, C., G. R. NAGARAJAN and C. LAUINGER: Biosynthesis of Asparagine from β-L-[^{14}C^{15}N] Cyanoalanine in *Lathyrus sylvestris* W. Seedlings: Origin of the Amide Nitrogen. Biochim. Biophys. Acta **184**, 578 (1969).

97. RESSLER, C., S. N. NIGAM and Y.-H. GIZA: Toxic Principle in Vetch. Isolation and Identification of γ-L-Glutamyl-L-β-Cyanoalanine from Common Vetch Seeds. Distribution in Some Legumes. J. Amer. Chem. Soc. **91**, 2758 (1969).

98. RIMINGTON, C.: The Occurrence of Cyanogenetic Glucosides in South African Species of *Acacia.* II. Determination of the Chemical Constitution of Acacipetalin. Its Isolation from *Acacia stolonifera* Burch. Onderstepoort J. Vet. Sci. Animal Ind. **5**, 445 (1935).

99. RIMINGTON, C. and G. C. S. ROETS: Chemical Investigation of the Plant *Acalypha indica.* Isolation of Triacetonamine, a Cyanogenetic Glucoside and Quebrachite. Onderstepoort J. Vet. Sci. Animal Ind. **9**, 193 (1937).

100. ROBIQUET et BOUTRON-CHARLARD: Nouvelles expériences sur les amandes amères, et sur l'huile volatile qu'elles fournissent. Ann. chim. phys. **44**, 352 (1830).

101. ROSENTHALER, L.: Über die Samen von *Schleichera trijuga.* Schweiz. Apoth.-Ztg. **58**, 17 (1920).

102. SCHILDKNECHT, H., U. MASCHWITZ und P. KRAUSS: HCN im Wehrsekret des Erdläufers *(Pachymerium ferrugineum).* Naturwiss. **55**, 230 (1968).

103. SCHRADER, J. C. C.: Neue Wahrnehmungen über die Blausäure. Gilbert Annalen **13**, 503 (1803).

104. SEELY, M. K., R. S. CRIDDLE and E. E. CONN: The Metabolism of Aromatic Compounds in Higher Plants. VIII. On the Requirement of Hydroxynitrile Lyase for Flavin. J. Biol. Chem. **241**, 4457 (1966).

105. SEIGLER, D. S. and J. J. BLOOMFIELD: Constituents of the Genus *Cnidoscolus.* Phytochem. **8**, 935 (1969).

106. SHARPLES, D. and J. R. STOKER: The Identification and Biosynthesis of Two Cyanogenic Glycosides in *Thalictrum aquilegifolium.* Phytochem. **8**, 597 (1969).

107. Smith, A. E., C. Galand and K. Bahadur: The Inevitable Appearance of Protocells on the Primitive Earth. Spaceflight 11, 325 (1969).
108. Stevens, D. L. and G. A. Strobel: Origin of Cyanide in Cultures of a Psychrophilic Basidiomycete. J. Bacteriol. 95, 1094 (1968).
109. Steyn, D. G. and C. Rimington: The Occurrence of Cyanogenetic Glucosides in South African Species of Acacia. I. Onderstepoort J. Vet. Sci. Animal Ind. 4, 51 (1935).
110. Strobel, G. A.: The Fixation of Hydrocyanic Acid by a Psychrophilic Basidiomycete. J. Biol. Chem. 241, 2618 (1966).
111. — 4-Amino-4-Cyanobutyric Acid as an Intermediate in Glutamate Biosynthesis. J. Biol. Chem. 242, 3265 (1967).
112. — Cyanide Utilization in Soil. Soil Sci. 103, 299 (1967).
113. Tapper, B. A. and G. W. Butler: Conversion of Oximes to Mustard Oil Glucosides (Glucosinolates). Arch. Biochem. Biophys. 120, 719 (1967).
114. Tapper, B. A., E. E. Conn and G. W. Butler: Conversion of α-Keto-Isovaleric Acid Oxime and Isobutyraldoxime to Linamarin in Flax Seedlings. Arch. Biochem. Biophys. 119, 593 (1967).
115. Towers, G. H. N., A. G. McInnes and A. C. Neish: The Absolute Configurations of the Phenolic Cyanogenetic Glucosides Taxiphyllin and Dhurrin. Tetrahedron 20, 71 (1964).
116. Tschiersch, B.: Über den Stoffwechsel der Blausäure. II. Zum Mechanismus der Blausäure-Assimilation. Flora (Jena) 154A, 445 (1964).
117. — Über den Stoffwechsel des Vicianins. Flora (Jena) 157A, 43 (1966).
118. Underhill, E. W.: Biosynthesis of Mustard Oil Glucosides: Conversion of Phenylacetaldehyde Oxime and 3-phenylpropionaldehyde Oxime to Glucotropaeolin and Gluconasturtin. Eur. J. Biochem. 2, 61 (1967).
119. Uribe, E. G. and E. E. Conn: The Metabolism of Aromatic Compounds in Higher Plants. VII. The Origin of the Nitrile Nitrogen Atom of Dhurrin (β-D-Glucopyranosyloxy-L-p-Hydroxymandelonitrile) J. Biol. Chem. 24, 92 (1966).
120. Weiss, M.: Die Blausäure in Apfelembryonen. Flora (Jena) 149A, 386 (1960).
121. Young, R. L.: Personal Communication.
122. Young, R. L. and R. A. Hamilton: A Bitter Principle in Macadamia Nuts. Proceedings Sixth Annual Meeting, Hawaii Macadamia Producers Association, p. 27, 1966.

(Received, December 31, 1969)

Naturstoffe mit Pyridinstruktur und ihre Biosynthese

Von D. GROSS, Halle (Saale)

Einführung

Bei den zahlreichen Pyridinverbindungen aus dem Tier- und Pflanzenreich handelt es sich entweder um einfache Abkömmlinge dieser Base, die sich durch verschiedenartige Substitution unterscheiden, oder um komplizierter gebaute Naturstoffe, die den Pyridinkern als ein charakteristisches Strukturelement enthalten. Vielfach sind die natürlich vorkommenden Pyridinverbindungen durch spezifische biochemische Funktionen oder durch besondere physiologische Wirkungen ausgezeichnet. Einige dieser Pyridinstrukturen treten im Grundstoffwechsel aller Organismen auf, andere sind nur Inhaltsstoffe einiger höherer Pflanzen oder Ausscheidungsprodukte bestimmter Mikroorganismen. Wegen dieser Vielgestaltigkeit kommt den Naturstoffen der Pyridinreihe besonderes biogenetisches Interesse zu.

Zwischen natürlichen Pyridin- und Piperidinverbindungen besteht eine enge strukturelle Verwandtschaft. Trotzdem liegt beiden Stoffklassen biogenetisch kein gemeinsames Entstehungsprinzip zugrunde, da ihre Biosynthese im allgemeinen von unterschiedlichen Vorstufen ausgeht und eine gegenseitige Umwandlung beider Ringsysteme nur in Einzelfällen möglich zu sein scheint.

In der nachfolgenden Übersicht werden natürlich vorkommende Pyridinstrukturen vorgestellt. Methodische Einzelheiten der Isolierung sowie der Struktur- und Konfigurationsaufklärung sind der zitierten Originalliteratur zu entnehmen.

Im zweiten Teil werden zum Pyridinringsystem führende Biosynthesewege besprochen und einige Hypothesen zur Biogenese experimentell noch nicht untersuchter Pyridinverbindungen diskutiert.

In der vorliegenden Arbeit bleiben Piperidinverbindungen weitgehend unberücksichtigt.

I. Natürlich vorkommende Pyridinverbindungen

1. Carbonsäuren und einfache Pyridinderivate

Von den in der Natur auftretenden Pyridincarbonsäuren besitzen Nicotinsäure (1) und Nicotinsäureamid (2) besondere Bedeutung im pflanzlichen und tierischen Stoffwechsel. In gebundener Form findet man sie als Bestandteil der Pyridinnucleotide, von denen die Coenzyme Nicotinsäureamid-adenin-dinucleotid (NAD) (3) und Nicotinsäureamid-adenin-dinucleotid-phosphat (NADP) an der biologischen Wasserstoffübertragung beteiligt sind.

(1)	(2)	(3)
Nicotinsäure	Nicotinsäureamid	Nicotinsäureamid-adenindinucleotid (NAD)

Literaturverzeichnis: SS. 146—161

Ungebunden vorliegende Nicotinsäure und ihr Amid dienen als Ausgangs- oder Reservestoffe für eine *de novo*- bzw. eine Resynthese der Pyridinnucleotide. Aus dem Nicotinsäurestoffwechsel leitet sich eine Reihe weiterer Pyridinstrukturen ab. Als tierische Endprodukte treten oxydierte und N-methylierte Verbindungen wie N^1-Methylnicotinsäureamid (4), N^1-Methyl-2-pyridon-3-carboxamid (5), N^1-Methyl-4-pyridon-3-carboxamid (6) und N^1-Methyl-6-pyridon-3-carboxamid (7) auf.

(4)	(5)	(6)	(7)
N^1-Methylnicotin-säureamid	N^1-Methyl-2-pyridon-3-carboxamid	N^1-Methyl-4-pyridon-3-carboxamid	N^1-Methyl-6-pyridon-3-carboxamid

Überschüssige Nicotinsäure kann außerdem mit Glucuronsäure oder mit Aminosäuren wie Glycin oder Ornithin zu Nicotinoylglucuronsäure, Nicotinursäure bzw. 2,5-Dinicotinoylornithin verknüpft werden und in dieser konjugierten Form entgiftet zur Ausscheidung gelangen.

Trigonellin (8), das N-Methylbetain der Nicotinsäure, findet sich im tierischen Organismus und in verschiedenen höheren Pflanzen wie *Trigonella foenum-graecum* L. *(30)*, *Quisqualis indica* L. *(73)* und *Pisum sativum* L. *(102)*. Es stellt möglicherweise eine Art Entgiftungsprodukt dar. Die Methylierung von Nicotinsäure (1) zu Trigonellin (8) ist bei höheren Pflanzen verschiedentlich beobachtet worden *(27, 126, 279)*.

Neben Nicotinsäure ist als natürliche Pyridincarbonsäure die beim oxydativen Tryptophanabbau auftretende Picolinsäure (9) zu nennen (vgl. S. 128, *Schema 1*). Einige substituierte Pyridinmonocarbonsäuren werden im Abschnitt I.4. aufgeführt.

(8)	(9)
Trigonellin	Picolinsäure

Unter den in der Natur vorkommenden Pyridindicarbonsäuren besitzt Chinolinsäure (10) eine Schlüsselstellung für alle diejenigen Pyridinverbindungen, deren Biosynthese über den Pyridinnucleotid-Cyclus (vgl.

S. 133, *Schema 3*) verläuft. Als weitere Pyridindicarbonsäure ist die von
LINGENS 1960 als Akkumulationsprodukt einer Nicotinsäure-Mangel-
mutante von *Escherichia coli* isolierte Cinchomeronsäure (11) zu nennen
(*151*). 2,6-Dipicolinsäure (12) ist ein Ausscheidungsprodukt von *Penicillium*-Arten wie *P. citreo-viride (193—196)* und findet sich außerdem in
den Sporen einiger *Bacillus*-Arten (*204, 208*).

| (10) | (11) | (12) |
| Chinolinsäure | Cinchomeronsäure | 2,6-Dipicolinsäure |

Pyridin soll nach älteren Angaben in verschiedenen höheren Pflanzen,
z. B. *Aplopappus hartwegi* Blake, *Atropa belladona* L. und *Quisqualis
indica* L. auftreten (*30*). Als einfachen pflanzlichen Pyridinabkömmling
hat MANSKE bereits 1942 das 3-Methoxypyridin (13) aus *Equisetum
arvense* L. und *Thermopsis rhombifolia* Richards. isoliert (*30*). Das aus
tierischen Organismen schon länger bekannte N-Methylpyridinium wurde
vor kurzem in *Vandopsis longicaulis* Schltr. *(Orchidaceae)* nachgewiesen
(*34a*).

| (13) | (14) | (15) | (16) |
| 3-Methoxypyridin | α-Picolin | 4-Aminonicotinsäuremethylester | 3-Hydroxymethylpyridin |

Von WILKINSON ist α-Picolin (14) in Rumex obtusifolius L.
aufgefunden worden (*269*). Der Arbeitsgruppe um SCHREIBER gelang
kürzlich die Isolierung von 4-Aminonicotinsäuremethylester (15) aus
Fontanesia phillyreoides Labill. (*36*). 3-Hydroxymethylpyridin (16) stellt
ein Ausscheidungsprodukt der Mykobakterien dar und wurde von MIURA
bei *Mycobacterium platypoecilus (170, 171)* und von uns bei *M. tuberculosis, M. bovis, M. phlei, M. smegmatis, M. fortuitum, M. avium* und
einigen unklassifizierten Arten nachgewiesen (*96*). Pyrimin (17), von
SHIMAN und NEILANDS 1965 als Ausscheidungsprodukt eines *Pseudomonas*-Stammes GH aufgefunden (*234*), enthält in α-Stellung des Pyridinringes einen δ-Glutamoylrest, der entweder cyclisiert als *Δ*[1]-Pyrrolin-
5-carbonsäure oder in offenkettiger Form vorliegt.

Literaturverzeichnis: SS. 146—161

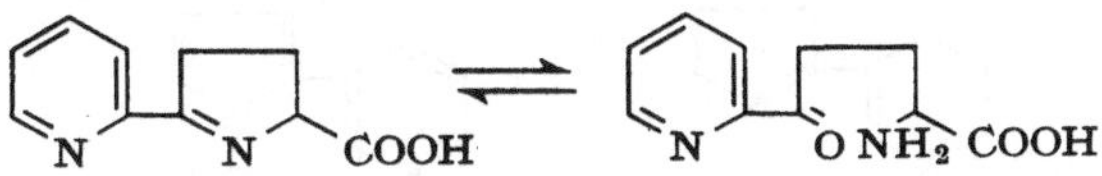

(17)
Pyrimin

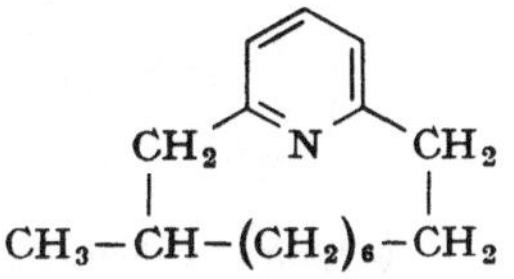

(18)
Muscopyridin

Als ein tierisches Pyridinderivat sei das 1957 von BIEMANN und Mitarb. (*26*) aus *Moschus moschiferus* isolierte Muscopyridin (**18**) genannt, bei dem die C-Atome 2 und 6 des Pyridins über eine methylverzweigte Dekamethylenbrücke miteinander verbunden sind.

Ebenfalls vom Pyridinringsystem leitet sich das Pyridoxalphosphat (**19**) ab, das als prosthetische Gruppe verschiedener Enzyme im pflanzlichen und tierischen Aminosäurestoffwechsel besondere Bedeutung besitzt. Es dürfte wie Pyridoxal (**20**), Pyridoxin (**21**) und Pyridoxamin (**22**) in allen lebenden Zellen anzutreffen sein (*74*).

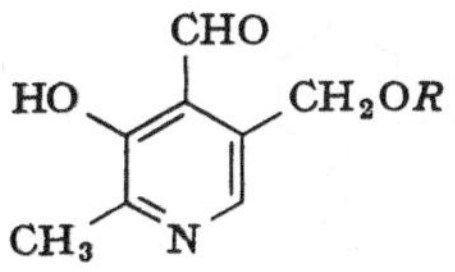

(19) $R = \circledP$ Pyridoxalphosphat
(20) $R = H$ Pyridoxal

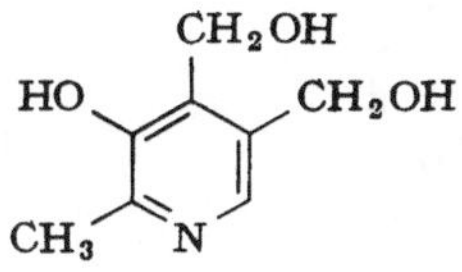

(21)
Pyridoxin

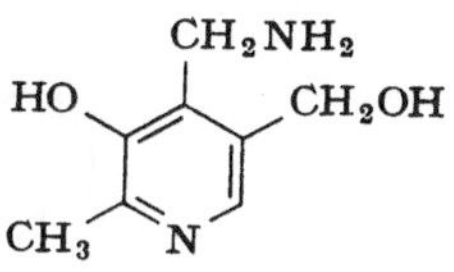

(22)
Pyridoxamin

2. Alkaloide

a) Nicotin und verwandte Alkaloide

Nicotin (**24**) und die ihm strukturähnlichen Nebenalkaloide sind β-substituierte Pyridinabkömmlinge, die bevorzugt in der Gattung *Nicotiana* sowie in verschiedenen nicht miteinander verwandten Pflanzenfamilien vorkommen (*30, 235, 270*). Als Substituenten finden sich Pyrrolidyl-, Piperidyl- und 2,3'-Pyridylpiperidylreste, die teilweise N-methyliert und unterschiedlich hydriert sind. Durch diese Variationen ergeben sich folgende Alkaloidstrukturen: Nornicotin (**23**), Myosmin (**25**), Nicotyrin (**26**), Anabasin (**27**), N-Methylanabasin (**28**), Anatabin (**29**), N-Methylanatabin (**30**), 2,3'-Dipyridyl (**31**), Anibin (**32**) und Nicotellin (**33**). Diese Alkaloide sind in anderen Zusammenfassungen bereits ausführlicher besprochen (*19, 30*) und sollen deshalb hier nur kurz abgehandelt werden.

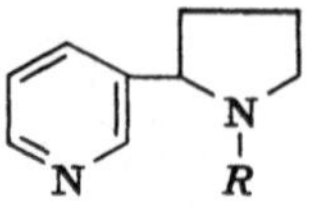

(23) R = H   Nornicotin

(24) R = CH₃   Nicotin

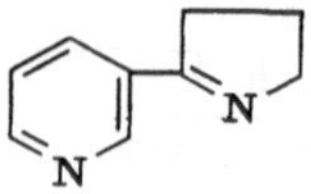

(25) Myosmin

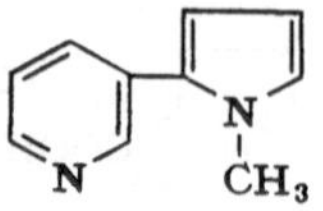

(26) Nicotyrin

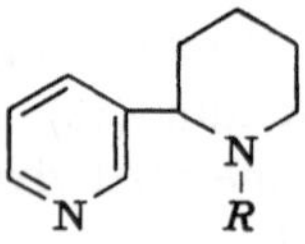

(27) R = H   Anabasin

(28) R = CH₃   N-Methylanabasin

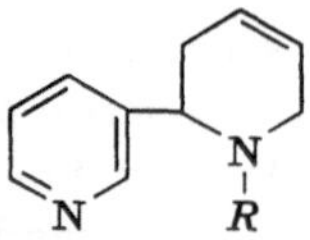

(29) R = H   Anatabin

(30) R = CH₃   N-Methylanatabin

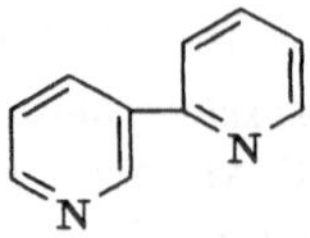

(31) 2,3'Dipyridyl

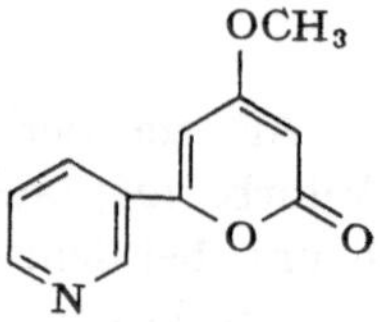

(32) Anibin

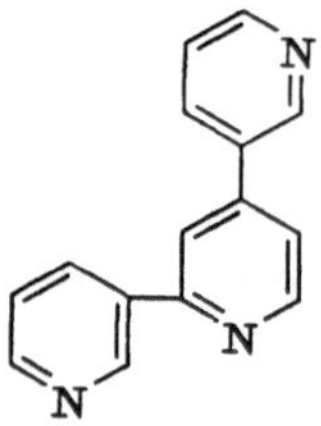

(33) Nicotellin

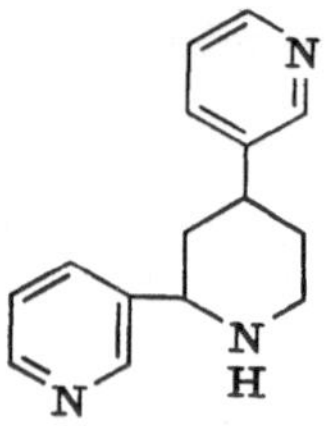

(34) Anatallin

Das linksdrehende natürliche Anabasin (**27**) ist oxydativ zu L(—)-N-Methylpipecolinsäure abgebaut worden, deren absolute Konfiguration bereits bekannt war. Somit konnte für das C-Atom 2′ des L(—)-Ana-

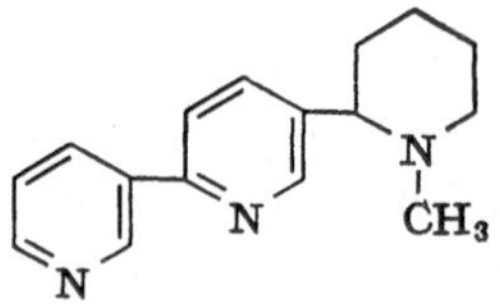

(35) Anabasamin

basins S-Konfiguration ermittelt werden (*155*). Die gleiche Konfiguration wurde für (—)-N-Methylanabasin (**28**), (—)-Anatabin (**29**) und (—)-N-Methylanatabin (**30**) abgeleitet (*155*). Zu diesen seit längeren bekannten Pyridinalkaloiden kommen das 1968 von KISAKI und Mitarb. (*133*) aus den Wurzeln von *Nicotiana tabacum* L. isolierte Anatallin (**34**) und das von der Taschkenter Arbeitsgruppe um SADYKOV (*178, 219*) in den Samen von *Anabasis aphylla* L. aufgefundene Anabasamin (35). Letzteres stellt wegen der besonderen Ringverknüpfung einen neuartigen Strukturtyp dar.·

b) Pyridonalkaloide

Als α-Pyridonalkaloide sind das Ricinin (36) aus *Ricinus communis* L. *(Euphorbiaceae)* (*30*) und das 1964 von MUKHERJEE und CHATTERJEE (*179, 180*) aus den Blättern von *Trewia nudiflora* L. *(Euphorbiaceae)* isolierte Nudiflorin (37) sowie das kürzlich von GANGULI (*81 a*) in der gleichen Pflanze gefundene Ricinidin (37 a) zu nennen. Diese Alkaloide weisen eine für Naturstoffe seltene Nitrilgruppe auf.

(36)	(37)	(37a)
Ricinin	Nudiflorin	Ricinidin

c) Terpenoide Alkaloide

Seit einiger Zeit kennt man aus 10 Kohlenstoffatomen aufgebaute Alkaloide, denen ein Pyridinring mit einem ankondensierten Methylcyclopentenring zugrunde liegt. Diese Alkaloide werden heute in einer gesonderten Gruppe als Pyrindanalkaloide zusammengefaßt. Ihre Biogenese scheint in einheitlicher Weise über Monoterpeneinheiten zu verlaufen.

Grundkörper dieser Verbindungsklasse ist das L(—)-Actinidin (38), ein 1959 von SAKAN und Mitarb. aus *Actinidia polygama* (Sieb. et Zucc.) Maxim. *(Actinidiaceae)* isoliertes Alkaloid (*80, 220, 221, 222*). Diese Base konnte kürzlich auch als Inhaltsstoff von *Valeriana officinalis* L. nachgewiesen werden (*70a, 259*). Aus der gleichen Pflanze isolierte FRANCK (*77*) (—)-Methoxyactinidin (39).

(38)	(39)
L(—)-Actinidin	(—)-Methoxyactinidin

Ein weiterer Vertreter des Actinidintyps ist das 1963 von HAMMOUDA
und LE MEN (*103*) in *Tecoma stans* Juss. *(Bignoniaceae)* und kürzlich
von Taschkenter Autoren (*4a*) auch in *Pedicularis rhinantoides* Schrenk
(Scrophulariaceae) aufgefundene Tecostidin (40). Strukturelle Ähnlichkeit
weist das 1967 im Laboratorium von SAKAN (*223*) aus *Boschniakia
rossica* (Cham. et Schldl.) B. Fedtsch. *(Orobanchaceae)* isolierte und nach
neueren Untersuchungen von DICKINSON und JONES (*64*) auch in *Tecoma
stans* enthaltene Boschniakin (41) auf. Formal unterscheiden sich
Actinidin, Tecostidin und Boschniakin durch ihren Oxydationsgrad.

(40)   (41)

Tecostidin   Boschniakin

Die Bestimmung der absoluten Konfiguration hat jedoch ergeben,
daß natürliches Actinidin und Tecostidin am C-Atom 7 gleiche Konfiguration besitzen (*46, 222*), während Boschniakin in dieser Position
R-Konfiguration aufweist (*223*).

Von DANILOVA und KONOVALOVA sind 1952 aus *Plantago indica* L.
(Plantaginaceae) die beiden C_{10}-Alkaloide Plantagonin und Indicain
isoliert worden (*59*). Letzteres trägt eine Aldehydgruppe und kann leicht
zu Plantagonin oxydiert werden (*58*). Diese damals in ihrer Konstitution
noch unbekannten Alkaloide wurden 1963 von YUNUSOV und Mitarb. auch
in *Pedicularis olgae* Rgl. *(Scrophulariaceae)* nachgewiesen (*262*). Darüber
hinaus ist Plantagonin auch in *Pedicularis dolichorrhiza* Schrenk. (*2*) und
in *P. rhinantoides* (*4a*) sowie neben Indicain in *Plantago albicans* L. (*7a*)
enthalten. Die von der Taschkenter Arbeitsgruppe 1965 für Plantagonin
und Indicain (*156*) und kürzlich für die aus *Pedicularis olgae* isolierten
Basen Pediculin (*3*) und Pedicularin (*2a*) angegebenen Strukturformeln
erscheinen aus chemischer und biochemischer Sicht unwahrscheinlich und
nur ungenügend gesichert. Durch neuere chemische und chemisch-physika-
lische Untersuchungen erfolgte 1968 durch die Taschkenter Autoren (*4*) und
in Schweden durch TORSSELL (*257*) eine Korrektur. Die nachstehend angegebenen Konstitutionsformeln für Plantagonin (42) und Indicain (43)
dürften heute weitgehend gesichert sein. Boschniakin (41) und Indicain
(43) besitzen formal dasselbe Kohlenstoffgerüst. Auf Grund der unterschiedlichen chemisch-physikalischen Daten dürfte es sich aber nicht um
identische Verbindungen handeln. Plantagonin zeigt weitgehende Übereinstimmung mit Boschniakinsäure, einem Oxydationsprodukt des

Boschniakins (41) (223), und könnte deshalb R-Konfiguration aufweisen. Möglicherweise kommt Pediculin und Pedicularin (2a, 3) ebenfalls eine Actinidin-ähnliche Grundstruktur zu.

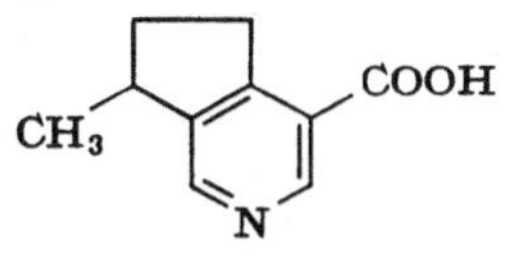

(42)
Plantagonin

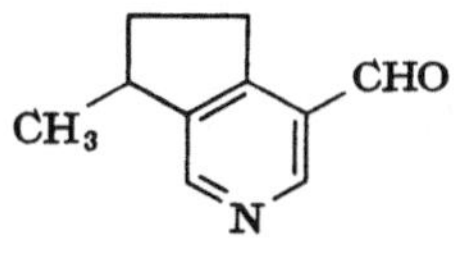

(43)
Indicain

Zu diesen monoterpenoiden Strukturtypen sind auch zwei quartäre Alkaloide (44) und (45) zu zählen, die 1966 von TORSSELL und WAHLBERG (258, 259) aus den Wurzeln von *Valeriana officinalis* L. isoliert und in ihrer Konstitution aufgeklärt worden sind. Diese beiden Alkaloide tragen am Heterostickstoff des Actinidinteils einen p-Hydroxyphenyläthylrest und dürften biogenetisch mit Actinidin (38) verwandt sein, zumal sie mit diesem und mit Methoxyactinidin (39) gemeinsam in *V. officinalis* vorkommen. Ob den anderen dünnschichtchromatographisch in *Valeriana officinalis* aufgefundenen Basen (70a) eine Actinidin-ähnliche Struktur zukommt, ist noch unbekannt.

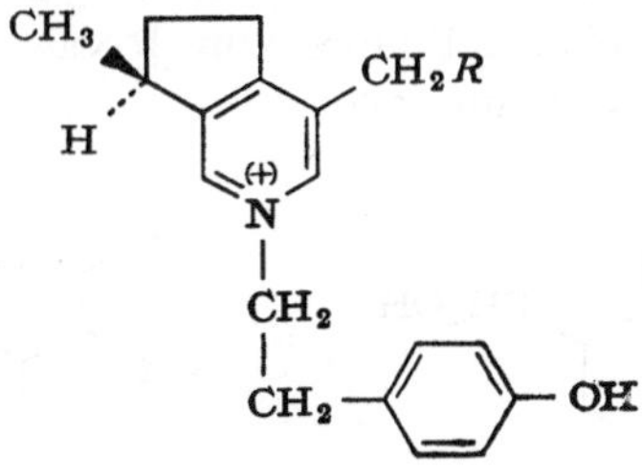

(44) R = H
(45) R = OH

Weiterhin sind heute verschiedene Piperidinalkaloide bekannt, die dasselbe Kohlenstoffskelett wie die des Actinidintyps aufweisen und daher mit aufgeführt werden sollen.

Es handelt sich um die diastereomeren α-, β- und δ-Skytanthine (46) aus *Skytanthus acutus* Meyen *(Apocynaceae)*, die 1961/1962 von DJERASSI und Mitarb. (65, 66), von APPEL und MÜLLER (13b) sowie von der italienischen Arbeitsgruppe um MARINI-BETTOLO (43) in ihrer Konstitution aufgeklärt wurden und deren absolute Konfiguration ebenfalls bekannt ist (41, 42, 71). Skytanthin wird von Dehydroskytanthin (47) und von zwei Hydroxyskytanthinen (48) und (49) (5, 13a, 40, 161a) sowie von β-Skytanthin-N-oxid (vgl. (46)) (238a) begleitet.

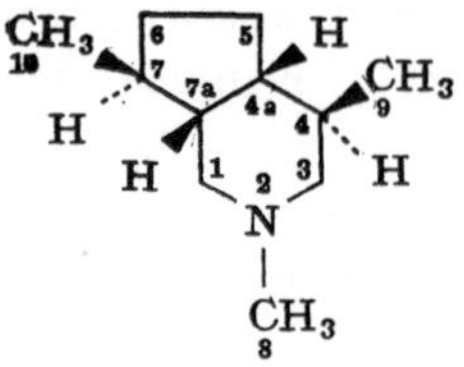

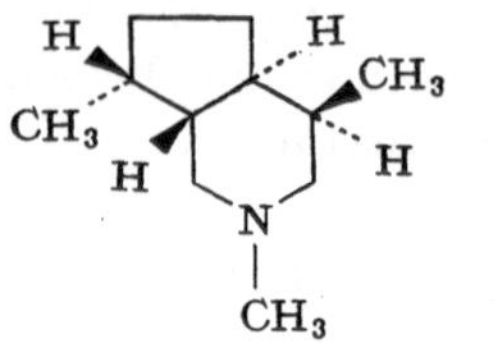

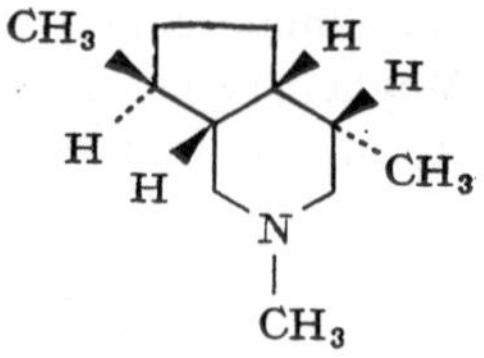

(46)

α-Skytanthin   β-Skytanthin   δ-Skytanthin

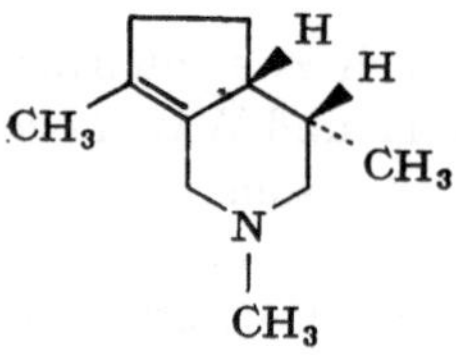

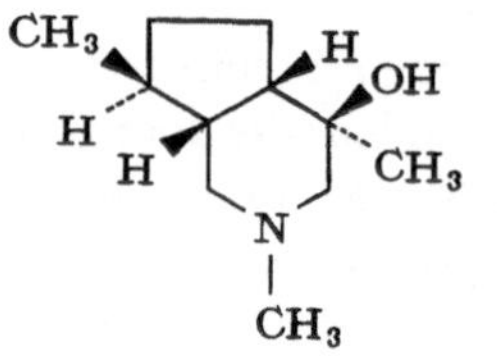

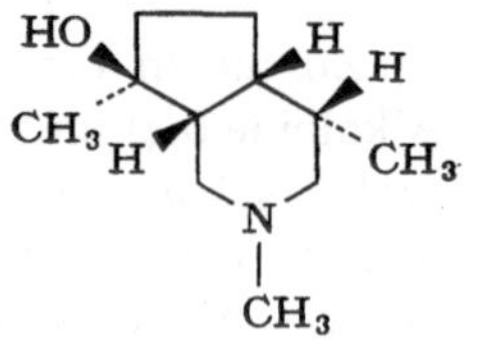

(47)   (48)   (49)

Δ⁷-Dehydroskytanthin   Hydroxyskytanthine

Strukturell analog gebaut sind Tecostanin (50), das 1963 von HAM-
MOUDA, PLAT und LE MEN aus *Tecoma stans* Juss. isoliert wurde (*104,
105*), und das in der gleichen Pflanze von JONES, FALES und WILDMAN
(*125*) aufgefundene Tecomanin (51).

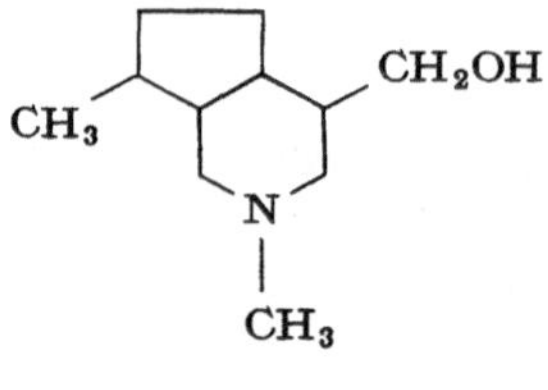

(50)   (51)

Tecostanin   Tecomanin

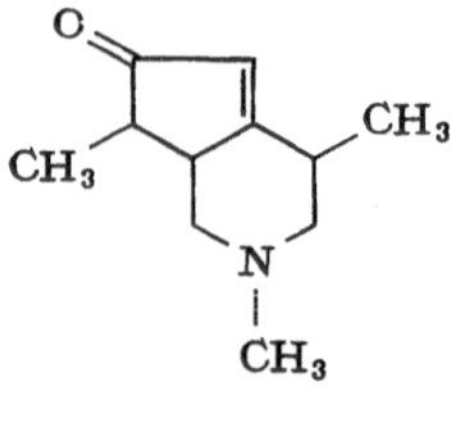

DICKENSON und JONES berichten 1969 über das Vorkommen und die
Identifizierung weiterer Alkaloide in *Tecoma stans*, wobei das Pflanzen-
material verschiedener Herkunft Unterschiede im Alkaloidspektrum auf-
wies (*64*). Neben dem aus *Boschniakia rossica* bekannten Boschniakin (41)
wurden als neue Verbindungen N-Norskytanthin (vgl. 46) und die beiden
Hydroxyskytanthine (52) und (53) beschrieben, die aber nicht mit den aus
Skytanthus acutus bereits isolierten 4- und 7-Hydroxyskytanthinen (48)
und (49) identisch sind.

Literaturverzeichnis: SS. 146—161

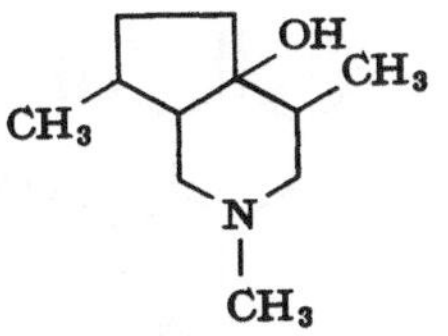

(52) 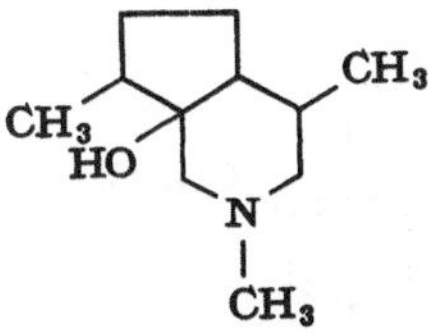   (53)

Hydroxyskytanthin

Neben diesen genannten Alkaloiden mit 10 Kohlenstoffatomen wurde in *Tecoma stans* überraschenderweise 4-Noractinidin (54) entdeckt *(64)*. Mit gewisser biogenetischer Berechtigung kann man dieses und die C_9- bzw. C_{10}-Basen (55) und (56) in die Gruppe der C_{10}-Monoterpenalkaloide einordnen, da diesen Verbindungen formal das Actinidingerüst zugrunde liegt. Das trifft für das von einer australischen Forschergruppe *(15, 16)* aus *Rauwolfia verticillata* (Lour.) Bail. isolierte Alkaloid mit der Arbeitsbezeichnung RW 47 (55) zu. Diese Base ist sehr wahrscheinlich mit dem 1968 in Indien von RAY und CHATTERJEE *(215)* in *Alstonia venenata* R. Br. *(Apocynaceae)* aufgefundenen Venoterpin (55) identisch *(106a)*. Das kürzlich aus *Jasminum*-Arten isolierte Alkaloid (56) besitzt ebenfalls Actinidinstruktur *(106a)*.

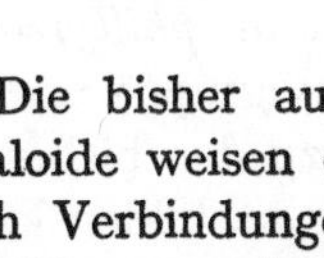

(54)   (55)   (56)

4-Noractinidin   Alkaloid RW 47
                    Venoterpin

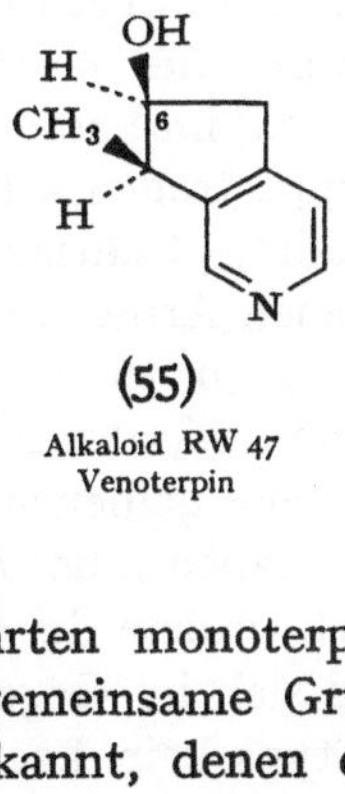

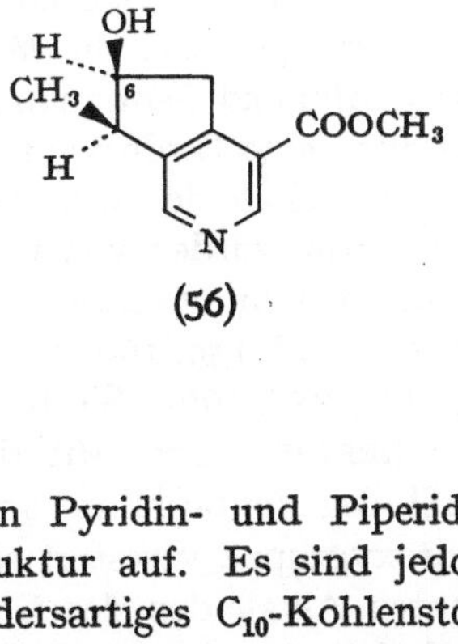

Die bisher aufgeführten monoterpenoiden Pyridin- und Piperidinalkaloide weisen eine gemeinsame Grundstruktur auf. Es sind jedoch auch Verbindungen bekannt, denen ein andersartiges C_{10}-Kohlenstoffskelett zugrunde liegt.

Dazu gehört das 1968 von HART, JOHNS und LAMBERTON *(106)* aus *Jasminum*-Arten und in *Ligustrum novoguineense* Lingelsh. *(Oleaceae)* entdeckte Jasminin (57). Dieses Alkaloid enthält einen Pyridinring mit einer Lactamgruppierung. Von den australischen Autoren werden für Jasminin zwei Strukturvorschläge diskutiert. Die endgültige Konstitutionsaufklärung steht noch aus, jedoch erscheint das linke Formelbild aus biogenetischen Gründen wahrscheinlicher.

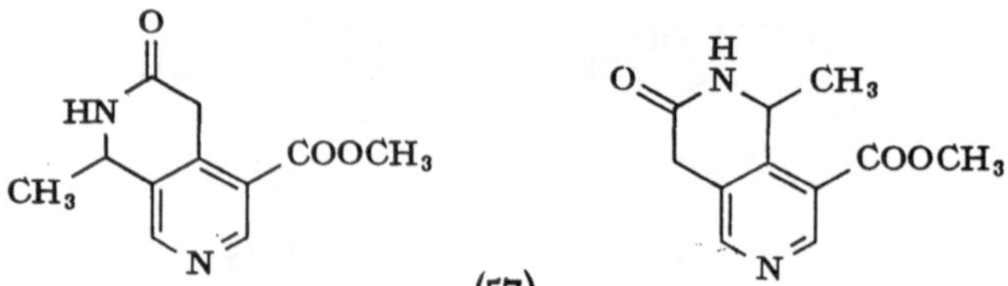

(57)
Jasminin

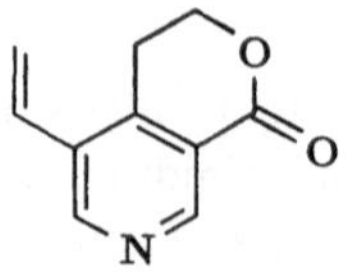

(58)
Gentianin

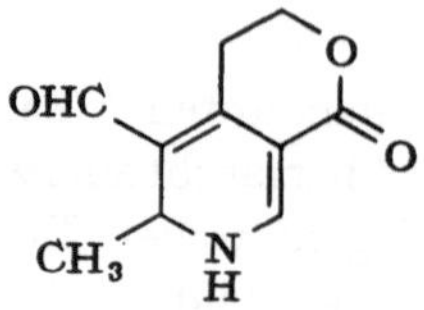

(59)
Gentioflavin

Die Alkaloide Gentianin (58) und Gentioflavin (59) sind ebenfalls aus 10 C-Atomen aufgebaut. Verbindungen dieses Strukturtyps enthalten im allgemeinen einen substituierten Pyridinring, der eine α,β-ungesättigte δ-Lactongruppierung trägt. Somit ergibt sich eine andere Grundstruktur als bei den bisher genannten Alkaloiden. Den strukturverwandten C_9-Basen Gentianidin (60) und Erythrocentaurin (61) fehlt die Formylgruppe bzw. die Methylgruppe des Gentioflavins (59), während das C_{11}-Alkaloid Gentianamin (62) (224) im Lactonring des Gentianins (58) zusätzlich eine Hydroxymethylgruppe tragen soll. Diese und einige verwandte Alkaloide wie Gentianadin, Gentiananin, Gentianain und Gentialutin wurden von verschiedenen Arbeitskreisen aus Gentianaceen, Loganiaceen und Oleaceen isoliert (7b, 30, 36, 39, 52, 79, 87, 124, 131, 134, 142, 147, 148, 159, 160, 161, 173, 203, 206, 211, 212, 212a, 217a, 218, 224, 225, 233, 252, 266). Sie treten teilweise gemeinsam vergesellschaftet auf.

SCHREIBER und Mitarb. (36) haben aus *Fontanesia phillyreoides* Labill. das Fontaphillin (63) isoliert, in dem die Carboxylgruppe und die Hydroxygruppe verestert sind und sich im Gegensatz zu den bisher genannten Alkaloiden des Gentianintyps kein Lactonring ausbilden kann.

Es ist umstritten, ob Gentianin und die ähnlich gebauten Alkaloide als native Verbindungen anzusehen sind oder ob sie künstlich entstehen. Verschiedene Befunde sprechen für eine Artefaktbildung (36, 76, 87, 134, 160, 206). Diese kann bei der ammoniakalischen Aufarbeitung des Pflanzenmaterials auftreten oder durch aus Glutamin stammendem „Ammoniak" zustande kommen. Trotzdem bleibt eine Restkonzentration an pflanzlichem Gentianin, die nicht durch Artefaktbildung zu erklären ist.

Literaturverzeichnis: SS. 146—161

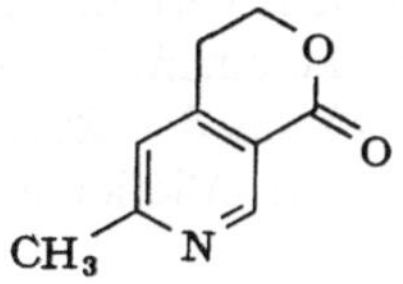

$$CH_2OH$$

(60)
Gentianidin

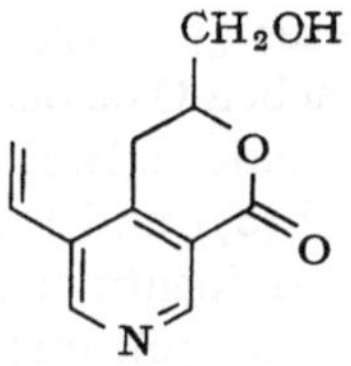

(62)
Gentianamin

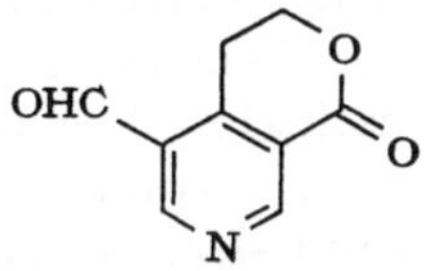

(61)
Erythrocentaurin

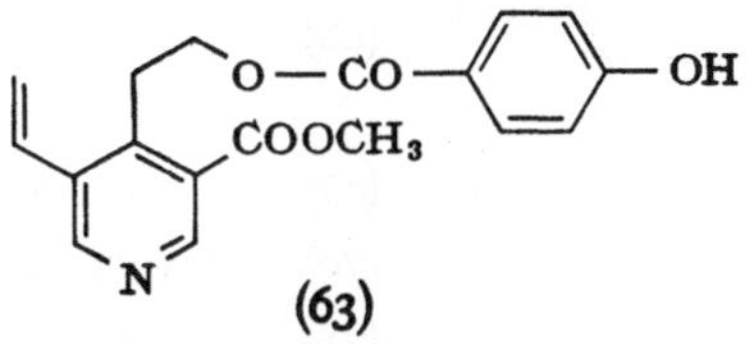

(63)
Fontaphillin

Abschließend seien als sesquiterpenoide Pyridinalkaloide die von BÜCHI und Mitarb. aus *Pogostemon patchouli* Pellet. *(Lamiaceae)* isolierten C_{15}-Basen Epiguaipyridin (64) und Patchoulipyridin (66) genannt *(37)*. Diese Gruppe wird durch das kürzlich aufgefundene Guaipyridin (65) ergänzt *(82)*.

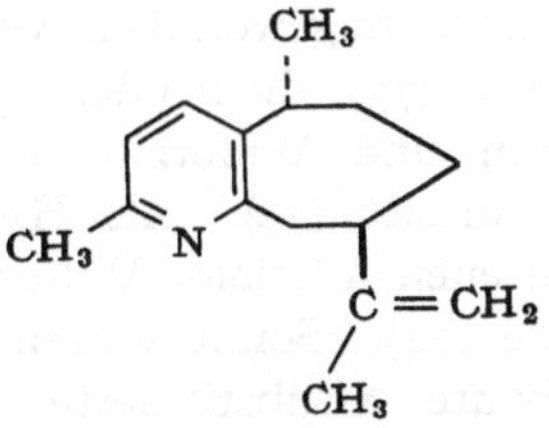

(64)
Epiguaipyridin

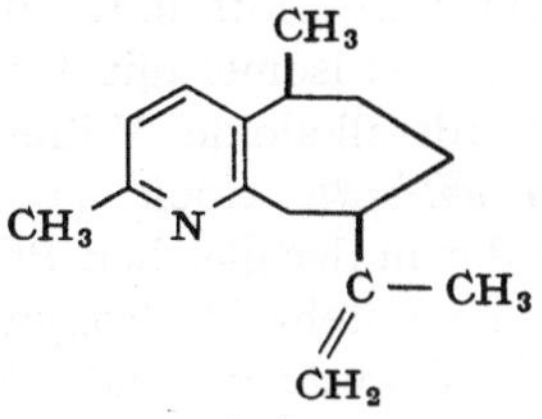

(65)
Guaipyridin

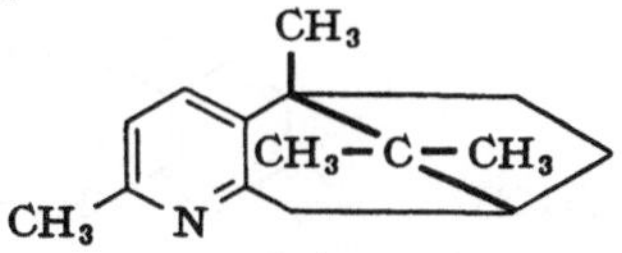

(66)
Patchoulipyridin

d) Verschiedenartige Pyridinalkaloide

Von der australischen Arbeitsgruppe um CROW und HODGKIN wurden aus der Rutacee *Halfordia scleroxyla* F. Muell. mehrere 4-Phenyl-2-pyridyloxazolalkaloide isoliert *(54, 55, 57)*. Dazu gehören Halfordinol (67), Halfordin (68) und Halfordinon (69) sowie als quartäre Base N-Methylhalfordiniumchlorid. Diese Alkaloide wurden auch in *Halfordia kendack*

(Montr.) Guillaumin nachgewiesen (*56*). Biogenetisch interessant erscheint dabei, daß in diesen Pflanzen neben Oxazolalkaloiden auch Verbindungen aus der Reihe der Chinolinalkaloide, z. B. Dictamin vorkommen. DREYER (*70*) isolierte später Halfordinol (**67**) und Halfordin (**68**) aus der Rutacee *Aeglopsis chevalieri* Swing. und identifizierte zwei neue Alkaloide als isomere O-Isopentenylhalfordinole (**70**), (**71**).

(**67**) $R = H$ Halfordinol
(**68**) $R = CH_2CHOH\ C(CH_3)_2$ Halfordin
          $\quad\quad\quad\quad\quad | $
          $\quad\quad\quad\quad\quad OH$
(**69**) $R = CH_2CO\ CH(CH_3)_2$ Halfordinon
(**70**) $R = CH_2CH=C(CH_3)_2$
(**71**) $R = CH_2CH_2C=CH_2$
          $\quad\quad\quad\quad | $
          $\quad\quad\quad\quad CH_3$

Das von PAILER und LIBISELLER (*199*) bearbeitete Evonin aus *Evonymus europaea* L. enthält als Säurekomponente eine Pyridin-3-carbonsäure, die in 2-Stellung eine C_5-Seitenkette mit endständiger Carboxylgruppe trägt. Diese Seitenkette der Evoninsäure (**72**) ist für eine biogenetische Verwandtschaft zur Mevalonsäure ungewöhnlich verzweigt. Evoninsäure ist isomer mit der Wilfordinsäure (**73**), die bei der Hydrolyse der Wilfordiaalkaloide Wilforgin, Wilforin und Wilforzin aus *Tripterygium wilfordii* Hook. *(Celastraceae)* auftritt (*24*). Als Hydrolyseprodukt der in der gleichen Pflanze enthaltenen Alkaloide Wilfordin und Wilfortrin entsteht Hydroxywilfordinsäure (**74**). Somit weisen Evonin und die Wilfordiaalkaloide ähnlich gebaute α-substituierte Pyridin-3-carbonsäuren auf.

(**72**)
Evoninsäure

(**73**) $R = H$ Wilfordinsäure
(**74**) $R = OH$ Hydroxywilfordinsäure

Im Clivimin aus *Clivia miniata* Rgl., das von der Arbeitsgruppe um DÖPKE in Struktur und Stereochemie untersucht worden ist, findet sich als Pyridinkörper die 2,6-Lutidin-3,5-dicarbonsäure (**75**), über deren Carboxylgruppen esterartig das Alkaloid Clivimin aufgebaut ist (*69, 163*).

Literaturverzeichnis: SS. 146—161

Aus der Gruppe der Steroidalkaloide soll das von Kikuchi und Mitarb. strukturell aufgeklärte Epipachysamin B (76) aus *Pachysandra terminalis* Sieb. et Zucc. aufgeführt werden, in dem Nicotinsäure amidartig an die in 3-Stellung befindliche Aminogruppe gebunden ist (*132*).

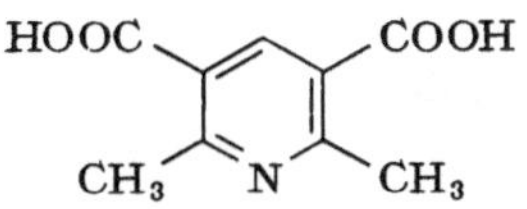

(75)

2,6-Lutidin-3,5-dicarbonsäure

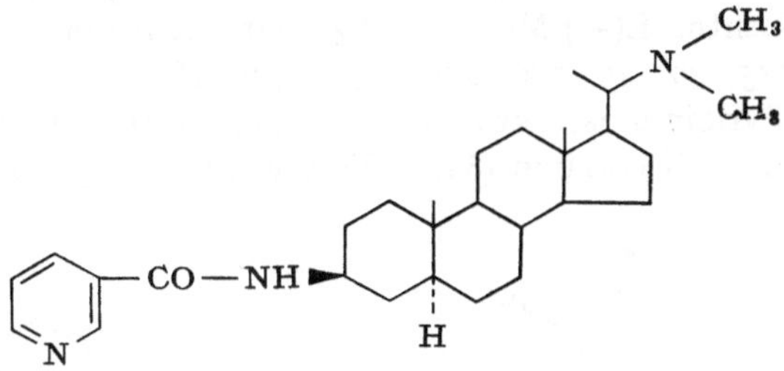

(76)

Epipachysamin B

Während die Lobeliaalkaloide im allgemeinen in die Gruppe der Piperidine gehören (*149*), sind die von Tschesche und Mitarb. aus *Lobelia syphilitica* L. isolierten Alkaloide Syphilobin A (77) und Syphilobin F (78) Pyridinring-haltig (*260*).

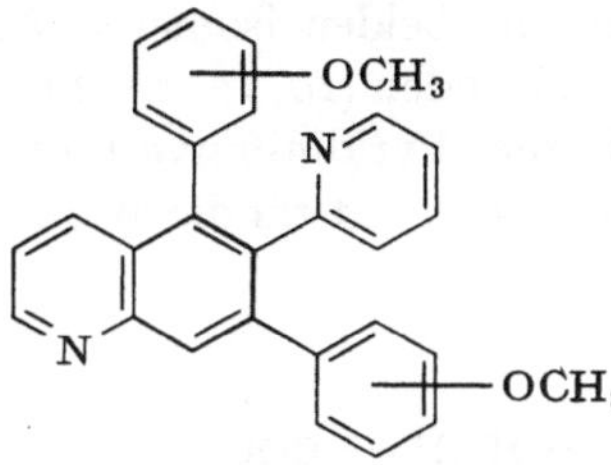

(77)

Syphilobin A

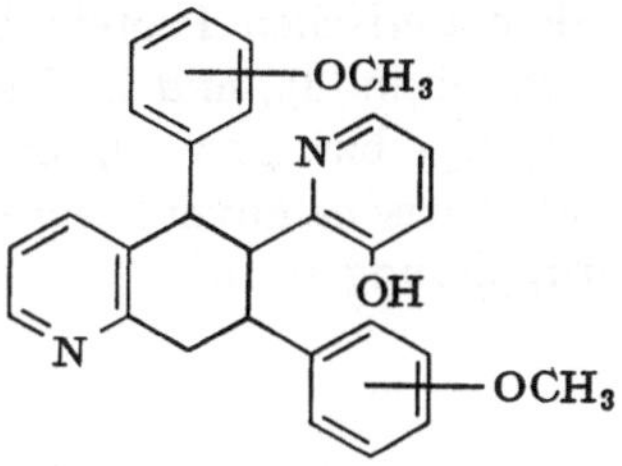

(78)

Syphilobin F

Zum Abschluß sollen die in *Areca catechu* L. vorkommenden Tetrahydropyridinalkaloide Guvacin (79), Guvacolin (80), Arecaidin (81) und Arecolin (82) genannt werden (*30*), die möglicherweise eine biogenetische Verwandtschaft zu den von der Nicotinsäure abzuleitenden Alkaloiden zeigen.

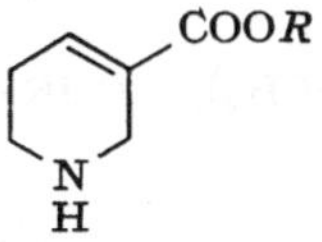

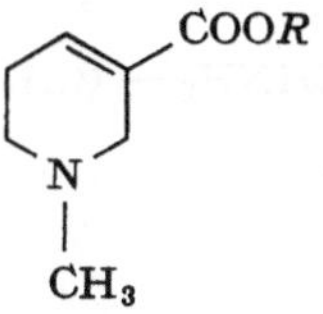

(79) R = H Guvacin

(80) R = CH₃ Guvacolin

(81) R = H Arecaidin

(82) R = CH₃ Arecolin

3. Aminosäuren mit Pyridinstruktur

Unter den ungewöhnlichen, nichtproteinogenen Aminosäuren mit heterocyclischer Grundstruktur finden sich auch einige Vertreter aus der Pyridinreihe. So enthalten *Mimosa pudica* L. und *Leucaena glauca* Benth. L(—)-Mimosin (83), eine α-Aminosäure mit γ-Pyridonstruktur (*25, 30, 35, 109, 117, 150, 236, 268*). Als pflanzliche Aminosäure mit Pyridiniumstruktur ist von NOGUCHI und Mitarb. aus Tabakblättern L(+)-Nicotianin (84) isoliert worden (*185, 186*).

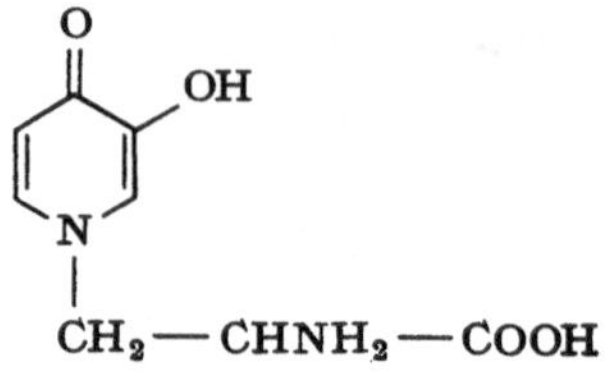

(83)

Mimosin

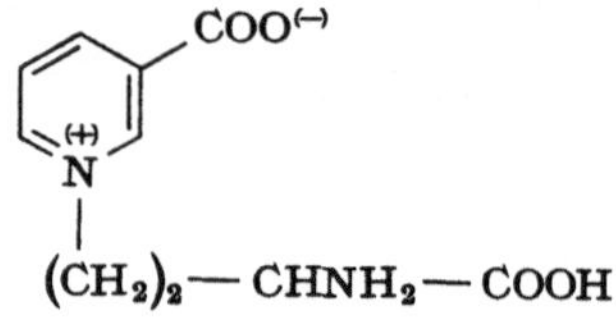

(84)

Nicotianin

In dem tierischen Protein Elastin sind die beiden isomeren Aminosäuren Desmosin (85) und Isodesmosin (86) zu finden (*10, 12, 22, 200, 254*). Diese 1,3,4,5- bzw. 1,2,3,5-tetrasubstituierten Pyridiniumverbindungen weisen aliphatische unverzweigte Seitenketten mit endständigen α-Aminosäuregruppierungen auf.

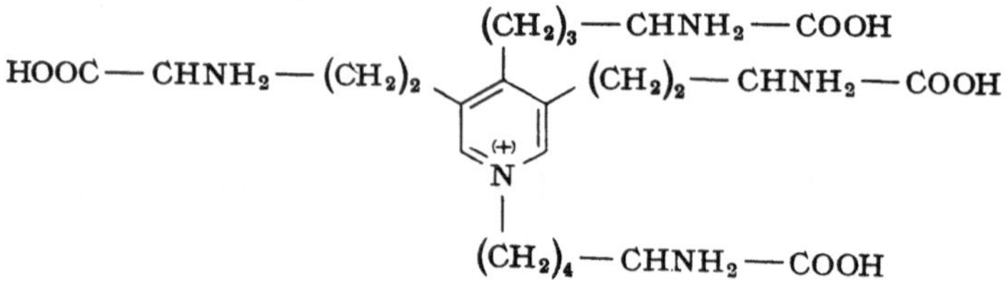

(85)

Desmosin

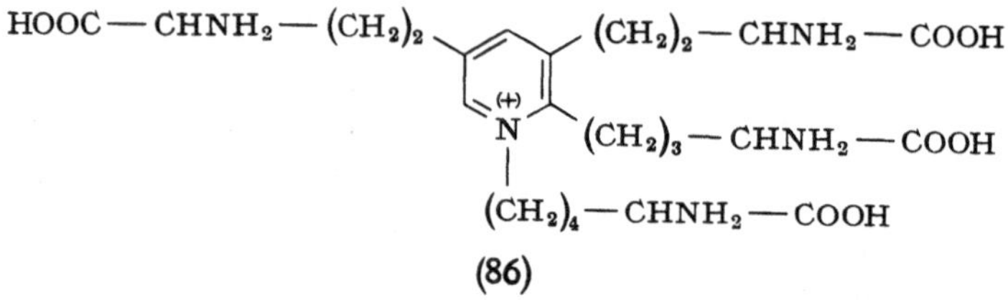

(86)

Isodesmosin

Literaturverzeichnis: SS. 146—161

4. Mikrobielle Stoffwechselprodukte

Einige Mikroorganismen sind imstande, ungewöhnliche Verbindungen mit Pyridinstruktur zu bilden und auszuscheiden, wobei diese Substanzen oftmals Antibiotica-Wirkung besitzen.

Die α-Pyridincarbonsäuren Fusarinsäure (87) und Dehydrofusarinsäure (88), Ausscheidungsprodukte einiger *Gibberella*- und *Fusarium*-Arten (*48, 81, 184, 237, 238, 272*), haben sich als pflanzliche Welketoxine erwiesen.

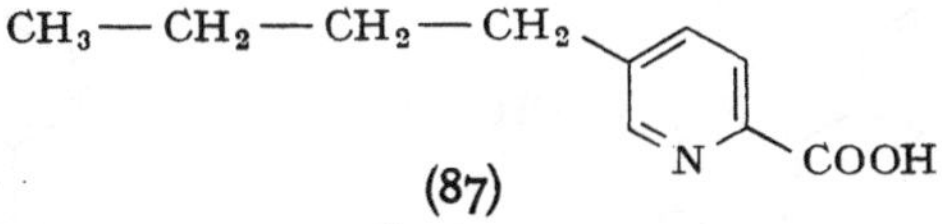

CH_3 — CH_2 — CH_2 — CH_2

(87)

Fusarinsäure

CH_3 — CH = CH — CH_2

COOH

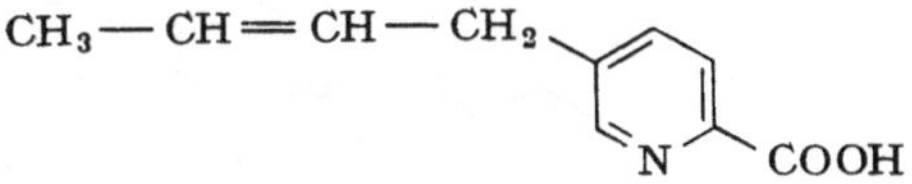

(88)

Dehydrofusarinsäure

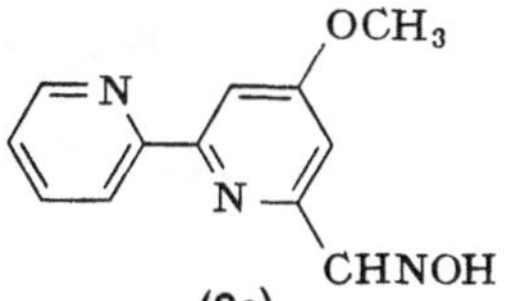

(89)

Caerulomycin

Von *Streptomyces caeruleus* Baldacci wird Caerulomycin (89) produziert, das 1967 von Divekar und Mitarb. isoliert und aufgeklärt wurde (*67*). Caerulomycin ist ein wegen der α,α'-Dipyridylstruktur und wegen der Aldoximgruppe interessanter Naturstoff.

Takahashi und Mitarb. haben in Piericidin A (90) und Piericidin B (91) aus *Streptomyces mobaraensis* durchsubstituierte Pyridinverbindungen mit einer mehrfach methylverzweigten Seitenkette aufgefunden (*242* bis *250*). Für die C-Atome 9 und 10 der Seitenkette wurde S-Konfiguration ermittelt, die Doppelbindungen sind *trans*-ständig (*250*).

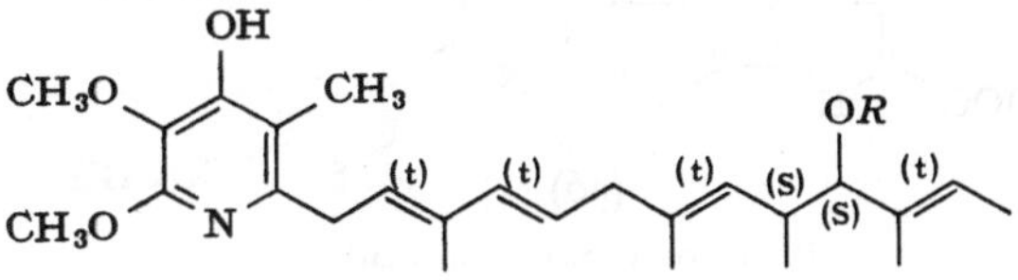

(90) R = H Piericidin A
(91) R = CH₃ Piericidin B

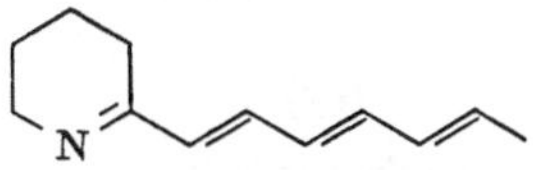

(92)

Nigrifactin

Das kürzlich aus *Streptomyces* Stamm FFD-101 isolierte Nigrifactin (92) erwies sich als 2-Heptatrienyl-Δ^1-piperidein (*253*), soll aber wegen seiner strukturellen Ähnlichkeit mit aufgeführt werden.

Streptonigrin (93) aus *Streptomyces flocculus*, von Rao, Biemann und Woodward in seiner Konstitution aufgeklärt, enthält einen 5-Amino-3-methylpicolinsäurerest, der in den Positionen 4 und 6 weitere Substituenten trägt (*213*). Im makrocyclisch gebauten Pyridomycin (94) aus *Streptomyces albidofuscus* finden sich ein 3-Hydroxypicolinsäurerest und eine 3-Pyridylmethylgruppe, wie japanische Autoren nachgewiesen haben (*136*, *190*).

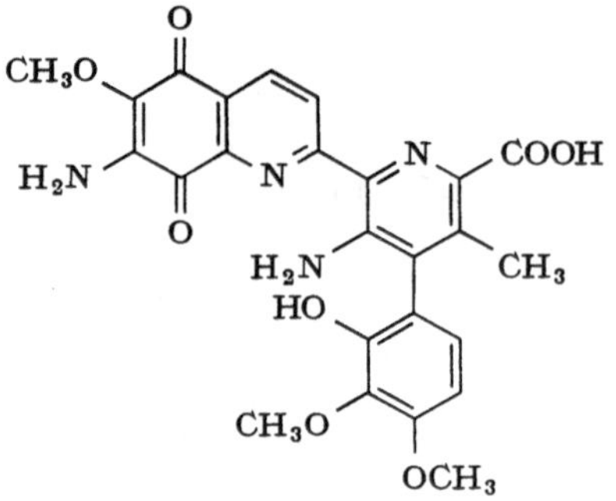

(93)
Streptonigrin

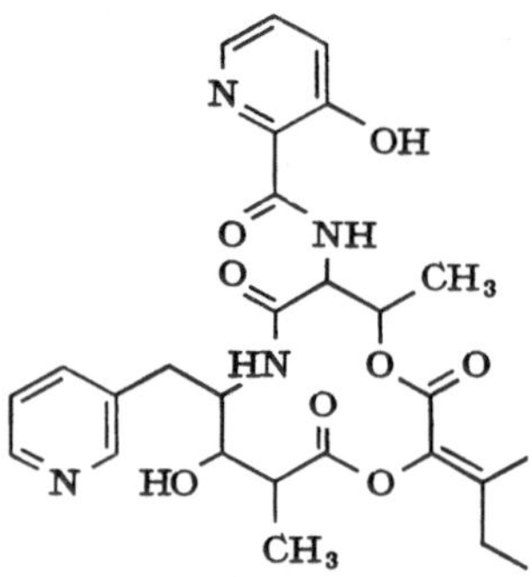

(94)
Pyridomycin

Einige Antibiotica der Streptogramin B-Gruppe enthalten an cyclische Peptide gebundene 3-Hydroxypicolinsäure (95). Zu diesen Verbindungen gehören Staphylomycin S, Vernamycin B_δ, Ostreogrycin B, B_1, B_2 und B_3 sowie Viridogrisein und Doricin (*263*). Das von *Bacillus pumilus* produzierte Micrococcin P besitzt als Säurekomponente (96) einen mit Thiazolylresten substituierten Pyridinring (*31*).

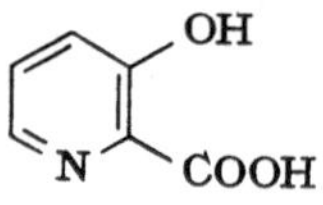

(95)
3-Hydroxypicolinsäure

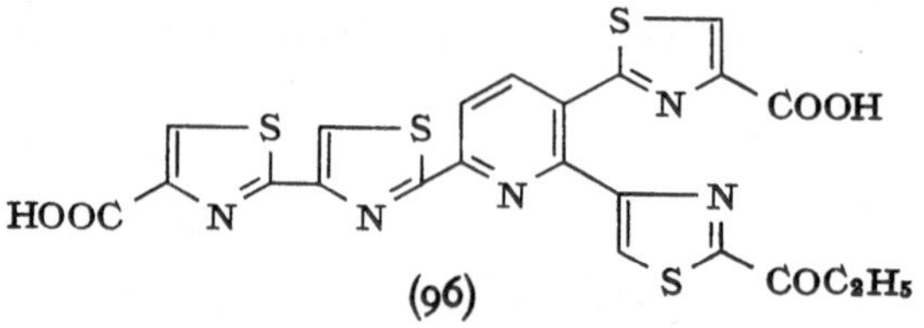

(96)
Micrococcin P (Säurekomponente)

II. Biosynthese des Pyridinringes

Mit der Auffindung und Strukturaufklärung eines Naturstoffes ergibt sich die Frage nach seiner Biosynthese. Im vorhergehenden Kapitel ist gezeigt worden, daß das heterocyclische Ringsystem des Pyridins in mannigfaltiger Form im Tier- und Pflanzenreich auftritt. Es war zu

Literaturverzeichnis: SS. 146—161

untersuchen, aus welchen Bausteinen des Grundstoffwechsels der Pyridinring entsteht und ob der Pyridinring in allen Organismen in einheitlicher Weise aus gemeinsamen Vorstufen aufgebaut wird oder ob Alternativwege angelegt sind. Als Resultat vielfältiger Experimente hat sich ergeben, daß der Pyridinring in der Natur auf unterschiedlichen Bildungswegen entsteht, wobei sich Tryptophan, Asparaginsäure und Lysin sowie Acetat, Propionat, Glycerin und Mevalonat als spezifische Precursoren erwiesen haben. Diese Verbindungen des Primärstoffwechsels bauen in verschiedenartiger Weise das Ringsystem des Pyridins auf, wie nachfolgend besprochen wird.

1. Entstehung der Nicotinsäure

Eine Reihe wichtiger Naturstoffe mit Pyridinstruktur leitet sich biogenetisch von der Nicotinsäure (1) ab, so daß mit der Biosynthese dieser Pyridin-3-carbonsäure begonnen werden soll.

Für die Bildung der Nicotinsäure (1) sind heute zwei Hauptwege bekannt. Diese führen zu Chinolinsäure (10), die das Bindeglied beider Synthesewege darstellt und nach einem offenbar allen Organismen gemeinsam zugrunde liegenden Bildungsprinzip, dem Pyridinnucleotid-Cyclus (vgl. S. 133), zu Nicotinsäure (1) decarboxyliert und zu verschiedenartig gebauten Pyridinverbindungen umgewandelt wird.

a) Oxydativer Tryptophanabbau

Im Säugetier-Organismus wurde schon vor etwa 20 Jahren ein enger Zusammenhang zwischen Tryptophanstoffwechsel und Nicotinsäurebildung beobachtet. Durch intensive Tracerexperimente an Säugetieren und vor allem durch biochemisch-genetische Studien an dem Pilz *Neurospora crassa* hat sich der in *Schema 1* zusammengestellte und heute strukturell wie enzymatisch weitgehend gesicherte Abbauweg des Tryptophans ergeben (*165, 267*).

Aus Tryptophan (97) entsteht durch oxydative Aufspaltung des Pyrrolringes N-Formylkynurenin (98), das zu Kynurenin (99) hydrolysiert und zu 3-Hydroxykynurenin (100) hydroxyliert wird. Seitenkettenabspaltung führt zu Alanin (101) und 3-Hydroxyanthranilsäure (103), deren oxydative Ringöffnung α-Amino-β-carboxymuconsäure-ε-semialdehyd (104) ergibt. Dieses kürzlich in seiner Struktur aufgeklärte acyclische Spaltprodukt (*138—141*) ist sehr instabil und kann folgenden Reaktionen unterliegen:

1. Spontane Cyclisierung und Wasserabspaltung führen zu Chinolinsäure (10).

2. Decarboxylierung, Cyclisierung und Dehydratisierung ergeben in einer enzymatisch gesteuerten Reaktion Picolinsäure (9).

3. Durch Decarboxylierung, Desaminierung und weitere Reaktionsschritte erfolgt ein Abbau über γ-Oxalylcrotonsäure, α-Ketoadipinsäure
und Glutaryl-CoA zu Acetyl-CoA und Kohlendioxid, den Endprodukten
des vollständigen Tryptophanabbaues (Glutarate Pathway).

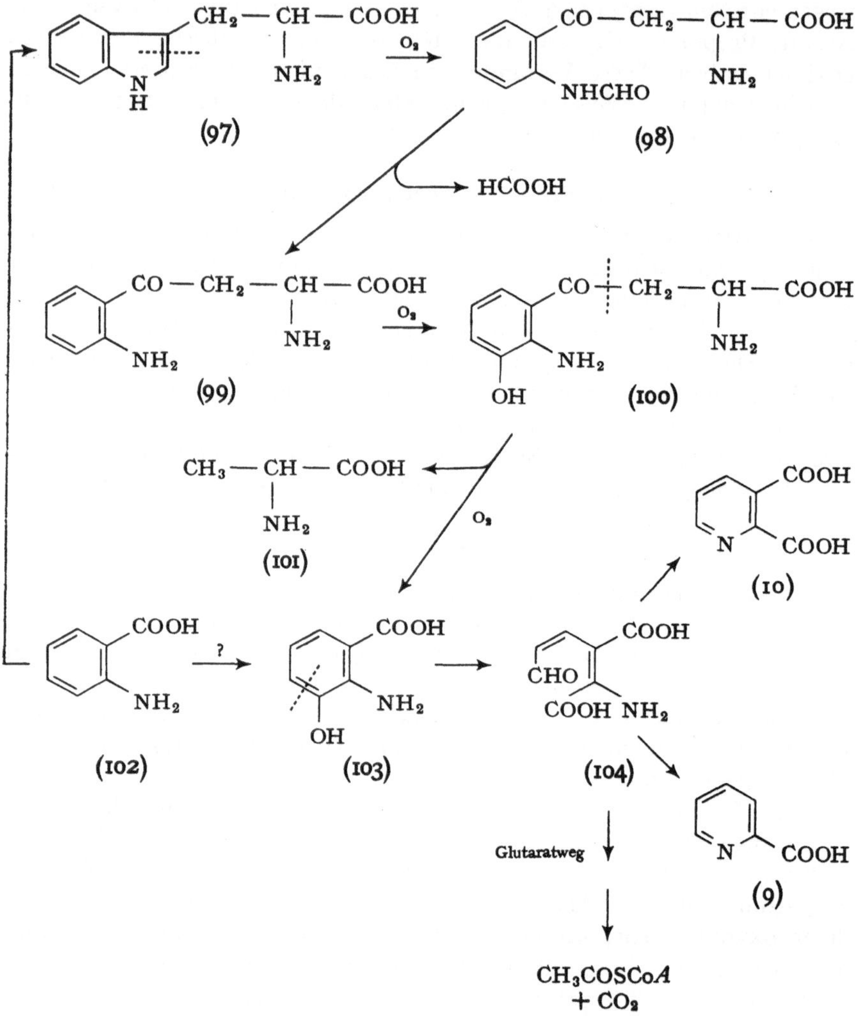

Schema 1. Oxydativer Tryptophanabbau

Der über Chinolinsäure (10) zu Nicotinsäure (1) führende Tryptophanabbau *(Schema 1)* ist von den Arbeitsgruppen um BONNER und um
YANOFSKY für *Neurospora crassa* (32, 275) und von anderen Autoren für

die Pilze *Claviceps purpurea* (*72, 261*), *Karlingia rosea* (*152*), *Tricho-phyton equinum* (*83*), *Fusarium oxysporum* (*63*) und aerob wachsende *Saccharomyces*-Arten (*6, 7, 152, 239, 240, 241*) sowie für den Säugetier-organismus (*275*) sichergestellt. Auch einige *Streptomyces*- und *Xantho-monas*-Arten (*60, 152, 153, 271*) sowie die Alge *Cyanidium caldarium* (*152*) verwerten Tryptophan (97) zur Nicotinsäuresynthese.

Darüber hinaus wird von PONTECORVO (*207*) für *Aspergillus*-Arten eine direkte Hydroxylierung der Anthranilsäure (102) zu 3-Hydroxy-anthranilsäure (103) diskutiert. Dieser kurzgeschlossene Biosyntheseweg würde unter Umgehung des Tryptophans (97) zu Nicotinsäure (1) führen (vgl. *Schema 1*), er ist aber noch nicht mit Sicherheit nachgewiesen.

b) C_3- und C_4-Kondensation

Die Annahme, daß Nicotinsäure (1) auch bei anderen Mikroorganismen und in der höheren Pflanze generell durch oxydativen Tryptophanabbau entsteht, ließ sich experimentell nicht bestätigen. YANOFSKY hat 1954 nachgewiesen, daß *Escherichia coli* und *Bacillus subtilis* Tryptophan nicht zur Nicotinsäuresynthese verwenden (*274*). In späteren Untersuchungen wurde der zur Nicotinsäure führende Tryptophanabbau für eine Reihe weiterer Mikroorganismen ausgeschlossen. Das gilt für *Mycobacterium tuberculosis* (*135, 172, 217*), *M. bovis* Stamm BCG (*174*), *M. phlei, Pseudo-monas aeruginosa, P. jodinum* und *P. trifolii* (*153*) sowie für *Anacystis nidulans, Chlorella pyrenoidosa, Saprolegnia ferax* und *Streptomyces venecuela* (*152*). Auch die im Pyridomycin (94) enthaltenen Pyridinringe gehen nicht aus Tryptophan hervor, wie Untersuchungen an *Strepto-myces pyridomyceticus* gezeigt haben (*191*).

In der höheren Pflanze findet ebenfalls keine Umwandlung von Tryptophan in Nicotinsäure (*112*) oder biogenetisch von ihr abgeleiteter Verbindungen wie Trigonellin (8) (*14, 146*), Nicotin (24) (*33, 88, 112, 143*) oder Ricinin (36) (*264*) statt. In allen Fällen zeigten radioaktiv mar-kiertes Tryptophan oder seine Folgeprodukte keinen Einbau in den Pyridinring dieser Alkaloide.

Interessanterweise wurde kürzlich aus Tabakblättern 6-Hydroxy-kynurensäure isoliert (*157*). Diese Chinolin-2-carbonsäure entsteht im tierischen Organismus aus dem Tryptophanstoffwechsel. Das Auf-treten dieser Verbindung in höheren Pflanzen spricht dafür, daß be-stimmte Stufen des tierischen Tryptophanmetabolismus offenbar auch in der Tabakpflanze realisiert sein müssen.

Parallel zu den genannten Untersuchungen wurde an Bakterien und in der höheren Pflanze intensiv nach einem Bildungsweg der Nicotin-säure gesucht.

ORTEGA und BROWN erhielten 1959/1960 erste Hinweise, daß Glycerin und Dicarbonsäuren des Tricarbonsäurecyclus bei *Escherichia coli* wirksame Nicotinsäurevorstufen darstellen (*197, 198*). Diese Befunde wurden später durch andere Autoren gesichert und weiter ausgebaut (*9, 47, 189*).

Durch Tracerexperimenten an *Mycobacterium bovis* Stamm BCG wurde gefunden, daß Asparaginsäure (106) unter Verlust ihrer C-1-Carboxylgruppe in Nicotinsäure inkorporiert wird und die C-Atome 2 und 3 sowie die Carboxylgruppe liefert, wobei der Heterostickstoff der Nicotinsäure dem Aminostickstoff der Asparaginsäure entstammt (*90, 99, 174, 175*). Die C-Atome 4, 5 und 6 der Nicotinsäure gehen aus dem Glycerinstoffwechsel hervor (*95*). Weiterhin konnte nachgewiesen werden, daß die Nicotinsäurebiosynthese bei *M. bovis* Stamm BCG über Chinolinsäure (10) verläuft (*93*) (vgl. *Schema 2*).

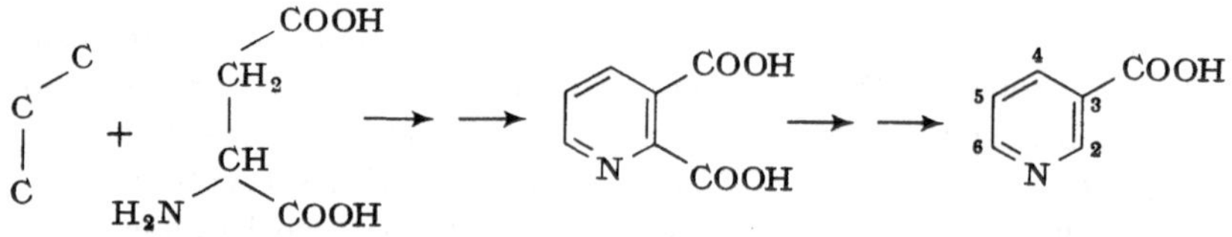

Schema 2. Biosynthese von Nicotinsäure durch C_3- + C_4-Kondensation

Ähnliche Befunde liegen von ALBERTSON und MOAT (*8, 172*) für die Nicotinsäurebiosynthese bei *Mycobacterium tuberculosis* vor. Demgegenüber stehen Befunde von RIO ESTRADA und PATINO (*217*), die keinen Einbau von ^{14}C-markiertem Glycerin in Nicotinsäure nachweisen konnten.

Spätere Untersuchungen an *Streptomyces pyridomyceticus* (*191*) und an anaeroben *Saccharomyces*-Kulturen (*7, 110*) bestätigen das in *Schema 2* dargestellte Bildungsprinzip des Pyridinringes auch für diese Mikroorganismen.

Somit synthetisieren diese Organismen, denen offensichtlich bestimmte Enzyme des oxydativen Tryptophanabbaues fehlen, das Pyridinringsystem durch Kondensation eines C_3-Körpers wie Glycerin oder Glycerinaldehyd (105) mit einer C_4-Dicarbonsäure wie Aspartat (106) *(Schema 2)*, wobei über noch unbekannte Zwischenstufen Chinolinsäure (10) entsteht. Dabei ergibt der C_3-Baustein die C-Atome 4, 5 und 6 der Chinolinsäure, während die C-Atome 2 und 3 sowie die Carboxylgruppen aus den vier C-Atomen der Dicarbonsäure hervorgehen. Chinolinsäure wird dann wie beim Tryptophanabbau in den Pyridinnucleotid-Cyclus (vgl. S. 133, *Schema 3*) eingeschleust und zu verschiedenartigen Pyridinstrukturen umgewandelt.

Literaturverzeichnis: SS. 141—161

Darüber hinaus liegen für Bakterien experimentelle Befunde vor, die auf ein andersartiges Nicotinsäurebildungsvermögen hinweisen.

Der von SCOTT und HUSSEY (231) bei *Serratia marcescens* gefundene Einbau von Asparaginsäure-[3-^{14}C] und -[4-^{14}C] ist mit dem in *Schema 2* dargestellten Bildungsweg in Übereinstimmung. Daher überrascht, daß nach Applikation von Asparaginsäure-[2-^{14}C] 87% der ^{14}C-Aktivität in der Carboxylgruppe der Nicotinsäure nachgewiesen wurden, denn das C-Atom 2 der Asparaginsäure sollte ja dem C-Atom 2 der Nicotinsäure entsprechen.

Clostridium butylicum verwertet *in vivo* und im zellfreien Rohextrakt Glycerin und Asparaginsäure zur Nicotinsäuresynthese, wie ISQUITH und MOAT gezeigt haben (122, 123). Dagegen werden im fraktionierten *in vitro*-System Formiat und Acetat an Stelle von Glycerin in Nicotinsäure inkorporiert (123). SCOTT und Mitarb. diskutieren, daß bei *C. butylicum* N-Formylasparaginsäure als Nicotinsäurevorstufe anzusehen ist, wobei die Formylgruppe das C-Atom 6 der Nicotinsäure ergibt (229, 229a, 232). Allerdings ist durch Doppelmarkierung noch nicht gesichert, daß N-Formylasparaginsäure tatsächlich intakt in Nicotinsäure inkorporiert wird.

Es ist noch offen, ob auch bei anderen Bakterien das C-Atom 6 der Nicotinsäure aus einem Formiatrest hervorgehen kann oder ob der bei *C. butylicum* gefundene Formiateinbau einen Sonderfall darstellt.

Interessant erscheint in diesem Zusammenhang auch die Tatsache, daß Hefen unter anaeroben Kulturbedingungen Nicotinsäure über den C$_3$- + C$_4$-Kondensationsweg bilden (7, 110), während aerob der Tryptophanabbau beschritten wird (6, 7, 152, 239, 240, 241). Offenbar sind in der Hefezelle die Enzymsysteme für beide Synthesewege angelegt, die je nach Kulturbedingungen reguliert werden.

Parallel zu diesen an Mikroorganismen durchgeführten Experimenten wurden an höheren Pflanzen insbesondere Nicotin (24) und Ricinin (36) auf ihre Biosynthese untersucht. Diese Arbeiten stammen im wesentlichen aus den Laboratorien um R. U. BYERRUM, R. F. DAWSON, E. LEETE, L. MARION, H. RAPOPORT und G. R. WALLER.

Es ist schon länger bekannt, daß sich Pyridinalkaloide wie Nicotin (24) und Ricinin (36) biogenetisch von der Nicotinsäure bzw. Chinolinsäure ableiten. Darüber hinaus wurde später nachgewiesen, daß die C-Atome 2 und 3 dieser Alkaloide aus den C-Atomen 2 und 3 einer C$_4$-Dicarbonsäure wie Aspartat (106) oder Succinat hervorgehen. Der Heterostickstoff des Pyridinringes dürfte auch in der höheren Pflanze dem Aminostickstoff des Aspartats entsprechen, wenn das auch wegen der hohen Transaminaseaktivität nur schwierig nachzuweisen ist.

Als wirksamer Precursor der C-Atome 4, 5 und 6 des Pyridinringes hat sich Glycerin (105) erwiesen. Kürzlich konnte gezeigt werden, daß nach Applikation von Glycerinaldehyd-[3-^{14}C] an *Nicotiana rustica* die

Radioaktivität zu über 50% im C-Atom 4 lokalisiert ist. Es wird postuliert, daß Glycerinaldehyd-phosphat mit Asparaginsäure kondensiert und anschließend zum Pyridinringsystem cyclisiert.

Einzelheiten dieser zahlreichen Biosynthesearbeiten sind neueren Übersichten (*91, 144, 145, 176, 214*) und der dort zitierten Originalliteratur zu entnehmen. Bei zusammenfassender Betrachtung dieser Untersuchungen ergibt sich, daß der Pyridinring des Nicotins (**24**), Anabasins (**27**) und Ricinins (**36**) nach dem in *Schema 2* dargestellten C_3- + + C_4-Kondensationsmechanismus entsteht. Dieses Bildungsprinzip ist auch für einige Strukturverwandte des Nicotins anzunehmen. Vor der Chinolinsäure liegende Zwischenstufen sind derzeitig nicht bekannt.

Von der Arbeitsgruppe um WALLER konnte für Nicotin (**24**) und Ricinin (**36**) wahrscheinlich gemacht werden, daß sich diese Alkaloide aus dem Pyridinnucleotid-Cyclus ableiten (*78, 265*). Dieser Cyclus scheint für den Stoffwechsel zahlreicher Pyridinverbindungen außerordentlich bedeutsam zu sein und wird im folgenden Abschnitt eingehend besprochen.

c) Pyridinnucleotid-Cyclus

Die NAD-Biosynthese wird seit Jahren intensiv bearbeitet. Die an verschiedenartigen biologischen Objekten durchgeführten Untersuchungen haben zur Aufklärung des Pyridinnucleotid-Cyclus geführt, der in seiner Gesamtheit oder in Einzelschritten für eine Vielzahl von Organismen weitgehend sichergestellt ist. Wegen der Fülle an vorliegenden Untersuchungen sei auf neuere Übersichtsarbeiten verwiesen (*50, 84, 91, 118*). Die nach einem der beiden Bildungsmechanismen (Weg A oder Weg B in *Schema 3*) entstandene Chinolinsäure (**10**) wird in Gegenwart von Mg-Ionen und 5-Phosphoribosyl-1-pyrophosphat (PRPP) über intermediär gebildetes Chinolinsäuremononucleotid (**107**) unter Decarboxylierung zu Nicotinsäuremononucleotid (**108**) umgesetzt *(Schema 3)*. In einer ATP-abhängigen Reaktion entstehen entweder Nicotinsäureadenin-dinucleotid (**109**), das durch Glutamin oder NH_3 zu NAD (**3**) amidiert wird, oder freie Nicotinsäure (**1**). Ausgehend von Tryptophan (**97**) oder von Aspartat (**106**) plus Glycerinanaloga (**105**) erfolgt auf diese Weise eine *de novo* NAD-Synthese, offenbar bei allen Organismen nach gleichem Prinzip.

Der NAD-Abbau führt entweder direkt oder über Nicotinsäureamidmononucleotid (**110**) zu Nicotinsäureamid (**2**), das zu Nicotinsäure (**1**) desamidiert werden kann. Nicotinsäure bildet mit 5-Phosphoribosyl-1-pyrophosphat und ATP erneut Nicotinsäuremononucleotid (**108**), so daß der Cyclus für eine NAD-Resynthese geschlossen wird (Nicotinate Pathway). Nicotinsäureamid kann auch über Nicotinsäureamidmononucleotid (**110**) direkt zu NAD (**3**) umgesetzt werden (Nicotinamide Pathway).

Literaturverzeichnis: SS. 146—161

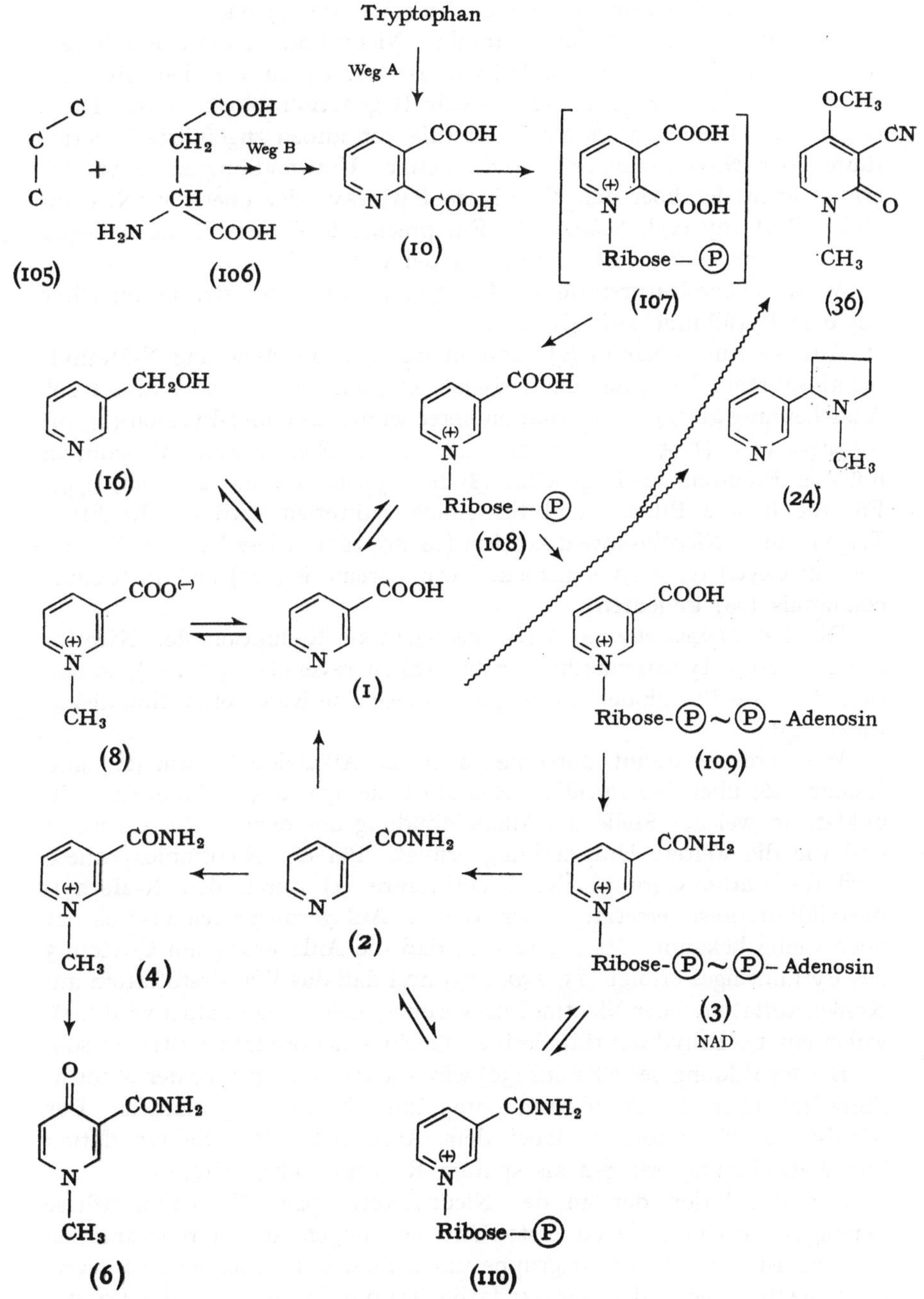

Tryptophan

Weg A

Weg B

(105)   (106)

(10)

(107)   (36)

(16)

(108)

(24)

(8)   (1)   (109)

(4)   (2)   (3)

NAD

(6)   (110)

Schema 3. Biosynthese von Pyridinverbindungen über den Pyridinnucleotid-Cyclus

Für einige Organismen besitzen Nicotinsäure (1) oder Nicotinsäureamid (2) den Charakter eines Vitamins (Nicotinsäureheterotrophe Organismen), weil die *de novo* NAD-Synthese auf einem vor dem Pyridinnucleotid-Cyclus liegenden Syntheseschritt genetisch blockiert ist. Diese Organismen benötigen als NAD-Vorstufe von außen zugeführte Nicotinsäure oder Nicotinsäureamid. Die weitere Umwandlung zu NAD (3) erfolgt entweder über den Nicotinate Pathway oder über den Nicotinamide Pathway (vgl. *Schema 3*). Entsprechende Untersuchungen liegen z. B. für verschiedene *Lactobacillus*-Arten vor (*192*).

Verschiedene Naturstoffe mit Pyridinstruktur leiten sich unmittelbar aus dem Pyridinnucleotid-Cyclus ab.

Aus Nicotinsäureamid (2) entsteht durch N-Methylierung N-Methylnicotinsäureamid (4), das im tierischen Organismus entweder direkt zur Ausscheidung gelangt oder zum entsprechenden 2- und 4-Pyridon (5), (6) oxydiert wird (*1, 51, 75, 86, 116, 181, 210*). Nicotinsäure (1) kann in höheren Pflanzen zu Trigonellin (8) methyliert werden (*27, 126, 279*). Für die höhere Pflanze und für einige Bakterien wird die Reaktion Trigonellin → Nicotinsäure diskutiert (*28, 29, 127*). Diese Demethylierung verläuft oxydativ, wie Versuche an Torula cremoris (*127*) und an Ricinus communis (*98*) andeuten.

Die bei *Mycobacterium*-Arten beobachtete Reduktion der Nicotinsäure (1) zu 3-Hydroxymethylpyridin (16) ist reversibel (*97, 280*), so daß sich für dieses Pyridinderivat möglicherweise eine Reservefunktion diskutieren läßt.

Wie bereits erwähnt, entstehen auch die Alkaloide Nicotin (24) und Ricinin (36) über den Pyridinnucleotid-Cyclus (*78, 265*). Dabei ist noch unklar, an welcher Stelle die Alkaloidbildung aus dem Cyclus abzweigt und wie die weitere Umwandlung erfolgt. Bei der Nicotinbiosynthese wird die Carboxylgruppe der Nicotinsäure (1) durch den N-Methylpyrrolidiniumrest ersetzt. Über diesen Anlagerungsmechanismus ist noch wenig bekannt. Man weiß nur, daß die Anlagerung am C-Atom 3 des Pyridinringes erfolgt (*85, 230, 273*) und daß das Wasserstoffatom am Kohlenstoffatom 6 der Nicotinsäure vorübergehend abgespalten wird (*61*), wobei ein 1,6-Dihydropyridinderivat als Zwischenprodukt auftreten soll.

Bei der Bildung des Ricinins (36) wird die Carboxylgruppe der Nicotinsäure (1) über die Amidgruppierung zum Nitril umgewandelt. Der Nitrilstickstoff entspricht dabei dem Amidstickstoff. Die Oxydation und Methylierung erfolgen als spätere Reaktionsschritte (*92*).

Zur Regulation der an der Nicotinsäure- bzw. NAD-Biosynthese beteiligten Enzyme liegen erste Untersuchungen an Mikroorganismen vor. So ist von der Arbeitsgruppe um LINGENS für *Saccharomyces cerevisiae* nachgewiesen, daß der oxydative Tryptophanabbau durch Tryptophan (97) induziert wird, aber durch Nicotinsäure (1), Chinolinsäure (10),

Literaturverzeichnis: SS. 146—161

Nicotinsäureamid (2) oder NAD (3) nicht beeinflußt wird (*110*). Die 3-Hydroxyanthranilsäure-Oxydase [(103) → (104)] wird durch diese Verbindungen nicht reprimiert, feed back gehemmt oder durch Tryptophan induziert (*111*).

Bei *Xanthomonas pruni* werden die Tryptophan-abbauenden Enzyme nicht durch Tryptophan reprimiert; zur Regulation der Tryptophan-Pyrrolase [(97) → (98)] wird von SAXTON und Mitarb. ein feed back-Mechanismus diskutiert (*226*). Nach Untersuchungen von KAHN und BLUM (*128, 129*) erfolgt in zellfreien Extrakten der Alge *Astasia longa* durch Nicotinsäure oder Chinolinsäure keine Repression der Chinolinsäurephosphoribosyl-Transferase [(10) → (108)] und der Nicotinsäuremononucleotid-Pyrophosphorylase [(1) → (108)]. Bei *Bacillus subtilis* und *B. megaterium* ist von SAXTON et al. eine Repression der Chinolinsäurephosphoribosyl-Transferase durch Nicotinsäure nachgewiesen, während Nicotinsäurephosphoribosyl-Transferase durch Nicotinsäure nicht reprimiert wird (*226*). *Escherichia coli* zeigt umgekehrte Verhältnisse, d. h. Repression der Nicotinsäurephosphoribosyl-Transferase, aber nicht der Chinolinsäurephosphoribosyl-Transferase (*226*). Nach Arbeiten von NAMBARU und ASAHINA hat sich Pyridazin-3-carbonsäure bei *E. coli* als ein nichtkompetitiver Hemmstoff erwiesen, der eine Akkumulation von Chinolinsäure bewirkt (*182, 183*).

Die Regulation der Alkaloidsynthese bei höheren Pflanzen ist noch wenig untersucht. Wuchsstoffe wie 3-Indolylessigsäure, 2,4-Dichlorphenoxyessigsäure und andere Auxine hemmen nach Angaben von YASUMATSU zwar die Nicotinbildung bei *Nicotiana tabacum* und *N. glutinosa*, der Wirkungsmechanismus ist jedoch noch unbekannt (*276*).

Die Nicotinsäurebiosynthese soll bei *N. tabacum* durch Helminthosporinsäure gehemmt werden, während die Bildung des Pyrrolidinringes nicht beeinflußt wird (*277*).

Das vorliegende Material gestattet noch keine endgültige Aussage über die Regulation der Pyridinnucleotid-Biosynthese, zumal diese Mechanismen bei verschiedenartigen Organismen unterschiedlich zu sein scheinen.

2. Biosynthese von Fusarinsäure und 2,6-Dipicolinsäure

Die Biosynthese der Fusarinsäure (87) ist von HILL und Mitarb. (*113*) an *Gibberella fujikuroi* und von der Arbeitsgruppe um VINING (*62, 63, 68*) an *Fusarium oxysporum* Schlecht. untersucht worden. Diese Arbeiten haben ergeben, daß Fusarinsäure nach einem Kondensationsmechanismus aufgebaut wird, der sich vom Nicotinsäure-Bildungsweg der Bakterien und höheren Pflanzen *(Schema 2)* unterscheidet.

Nach den vorliegenden Tracerexperimenten gehen die C-Atome 2, 3 und 4 sowie die Carboxylgruppe der Fusarinsäure (87) aus den vier

C-Atomen der Asparaginsäure (106) oder des Asparaginsäure-β-semial-
dehyds hervor, während die n-Butylseitenkette und die C-Atome 5 und 6
des Pyridinringes aus drei Molekülen Acetat durch Kopf-Schwanz-
Kondensation gebildet werden *(Schema 4)*. Der Heterostickstoff der
Fusarinsäure entstammt mit hoher Wahrscheinlichkeit dem Aminostickstoff der Asparaginsäure.

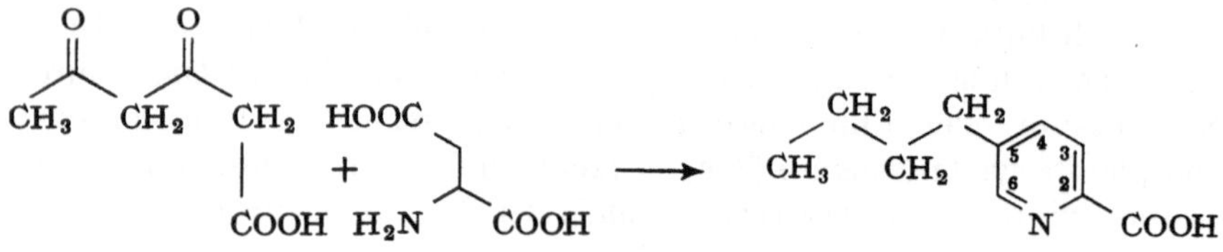

Schema 4. Biosynthese der Fusarinsäure *(87)*

Im Gegensatz zur Nicotinsäurebiosynthese durch C_3- + C_4-Kon-
densation (vgl. *Schema 2*) werden bei der Fusarinsäure drei C-Atome der
Asparaginsäure in den Pyridinring eingebaut, während das vierte als
Carboxylgruppe erhalten bleibt. Interessanterweise entsteht aber bei
Fusarium oxysporum Nicotinsäure (1) nicht wie Fusarinsäure (87) aus
Asparaginsäure (106), sondern aus Tryptophan (97) *(63)*. Somit werden
in diesem Organismus die Pyridinringe von Nicotinsäure (1) und von
Fusarinsäure (87) auf unterschiedlichen Bildungswegen aufgebaut.

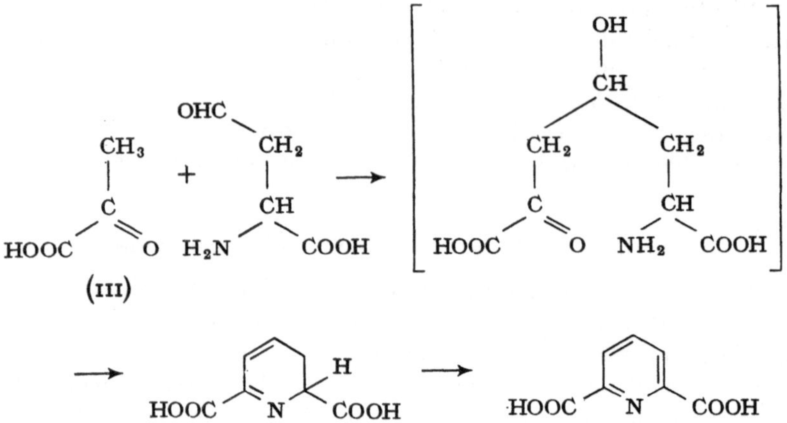

Schema 5. Biosynthese von 2,6-Dipicolinsäure *(12)*

Literaturverzeichnis: SS. 146—161

Eine analoge Asparaginsäureinkorporation wie bei der Fusarinsäure findet man auch bei der 2,6-Dipicolinsäure (12). Die Biosynthese dieser Pyridindicarbonsäure ist intensiv von verschiedenen Arbeitskreisen an *Bacillus*-Arten (*20, 23, 49, 162, 204, 209*) und an *Penicillium citreo-viride* (*114, 130, 251*) untersucht worden. Es hat sich gezeigt, daß offensichtlich Asparaginsäure-β-semialdehyd mit Pyruvat (111) zu einem Pimelin-säurederivat kondensiert, das zu 2,3-Dihydropicolinsäure (112) cyclisiert und durch Dehydrierung in 2,6-Dipicolinsäure (12) übergeht *(Schema 5)*.

Die Bildung von 2,6-Dipicolinsäure ist bei verschiedenen *Bacillus*-Arten eng mit der Versporung gekoppelt und läuft nicht im vegetativen System ab (*20, 23, 208*).

3. Lysin als Pyridinvorstufe

Die Aminosäure Lysin (113) stellt im allgemeinen den Precursor für das Piperidinringsystem dar (*100, 149*). Darüber hinaus leiten sich aber auch einige Naturstoffe mit Pyridinstruktur von dieser Aminosäure ab, wie nachfolgend aufgezeigt wird.

a) Biosynthese von Mimosin sowie Desmosin und Isodesmosin

Die Biosynthese des Mimosins (83) ist im wesentlichen von der Arbeits-gruppe um SPENSER (*187, 236, 255, 256*) sowie von HYLIN (*117*) unter-sucht worden. Die Experimente an *Mimosa pudica* und *Leucaena glauca* haben ergeben, daß der γ-Pyridonring des Mimosins nicht aus Asparagin-säure (106) und einem Glycerinanalogen (105) entsteht, sondern aus Lysin (113) gebildet wird. Als Vorstufe der aliphatischen Seitenkette ist Serin nachgewiesen (*255*) *(Schema 6)*.

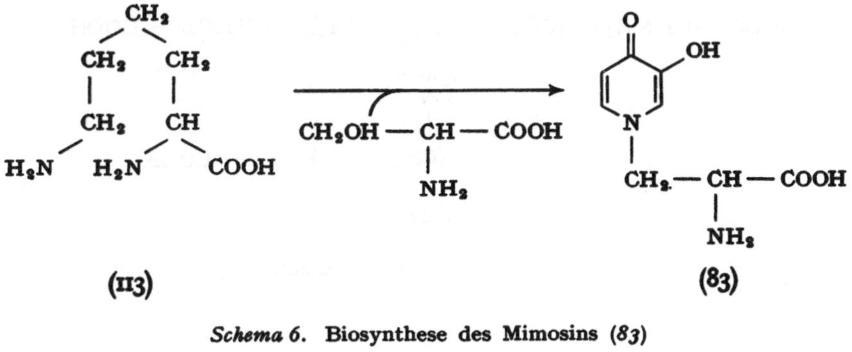

Schema 6. Biosynthese des Mimosins (*83*)

Der Heterostickstoff des Mimosins könnte der α- oder der ε-Amino-gruppe des Lysins entstammen. Entsprechende Untersuchungen liegen noch nicht vor. Der negative Einbau von δ-Hydroxylysin-[6-14C] spricht gegen eine zeitig erfolgende Hydroxylierung im Verlaufe der Biosynthese-kette (255). Die mehrfach nachgewiesene Inkorporation von ¹⁴C-mar-kierter Asparaginsäure in den Pyridonkern des Mimosins (187, 236, 255) erfolgt offenbar indirekt über die Lysinbiosynthese. Dafür spricht auch der Einbau von α-Aminoadipinsäure-[6-¹⁴C] (255). Zwischenstufen der Mimosinbiosynthese sind noch nicht bekannt.

Nach Biosyntheseexperimenten von ANWAR und ODA (11—13) und von MILLER und Mitarb. (168, 169) an embryonaler Kükenaorta sowie von der Arbeitsgruppe um PARTRIDGE (201, 202) an Ratten entstehen die Aminosäuren Desmosin (85) und Isodesmosin (86) ebenfalls aus Lysin. Die Hypothese, daß vier Moleküle dieser Aminosäure das Desmosin- bzw. Isodesmosinmolekül aufbauen, ließ sich durch den Einsatz ¹⁴C-markierten Lysins und durch chemischen Abbau des Desmosins zur Ermittlung der ¹⁴C-Verteilung bestätigen.

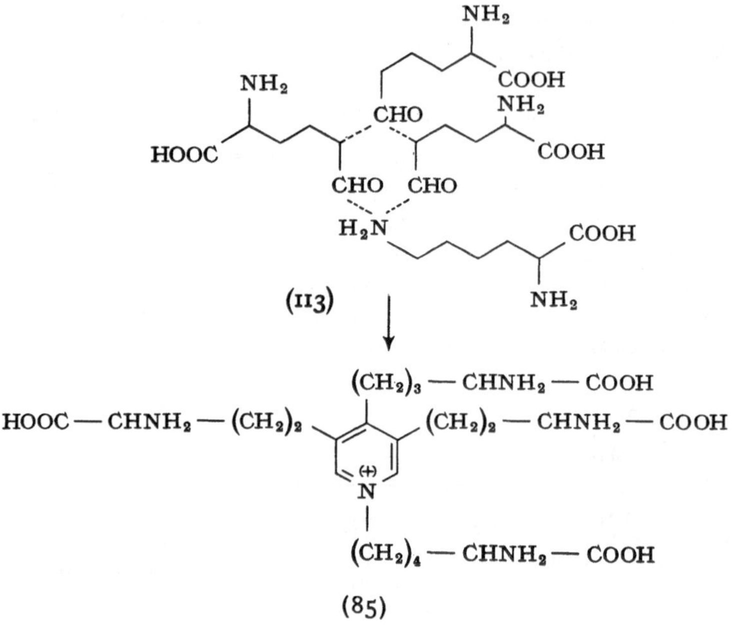

Schema 7. Biosynthese von Desmosin (*85*)

Als Zwischenstufe der Desmosinbiosynthese wurde α-Aminoadipin-säure-δ-semialdehyd nachgewiesen (169). Hemmstoffe der Desmosin-bzw. Isodesmosinbiosynthese sind Penicillamin und β-Aminopropio-nitril; Kupfermangel wirkt stark inhibierend (167, 188, 205).

Literaturverzeichnis: SS. 146—161

b) In vitro-Biosynthese von Anabasin

Der Pyridinring des Anabasins (27) entsteht in *Anabasis aphylla* und *Nicotiana* durch C_3- $+$ C_4-Kondensation (vgl. *Schema 2*), während der Piperidinring aus Lysin (113) aufgebaut wird (*100, 149*). Darüber hinaus existiert nach in vitro-Untersuchungen, die an Erbsen- und Lupinenextrakten durchgeführt wurden, ein Biosyntheseweg des Pyridinringes des Anabasins (27) aus Cadaverin (114), dem Decarboxylierungsprodukt des Lysins (113).

In Erbsen und Lupinen findet sich kein Anabasin. Sie enthalten aber nach HASSE und Mitarb. (*107, 108*) sowie nach MANN und SMITHIES (*158*) eine Diaminoxydase, die diesen Pflanzen verabreichtes Cadaverin (114) zum ω-Aminovaleraldehyd (115) oxydiert. Wie SCHÖPF und Mitarb. zeigen konnten, dimerisiert das nach spontaner Cyclisierung entstandene Δ^1-Piperidein (116) leicht zu Tetrahydroanabasin (117) (*227*). Schließlich bewirkt ein von HASSE und BERG (*107*) aufgefundener niedermolekularer Mn-abhängiger Faktor die weitere Dehydrierung zu Anabasin (27). Diese interessante *in vitro*-Anabasinsynthese ist von MOTHES und Mitarb. mit Cadaverin-[1,5-^{14}C] gesichert (*177*), konnte aber noch nicht *in vivo* in höheren Pflanzen nachgewiesen werden.

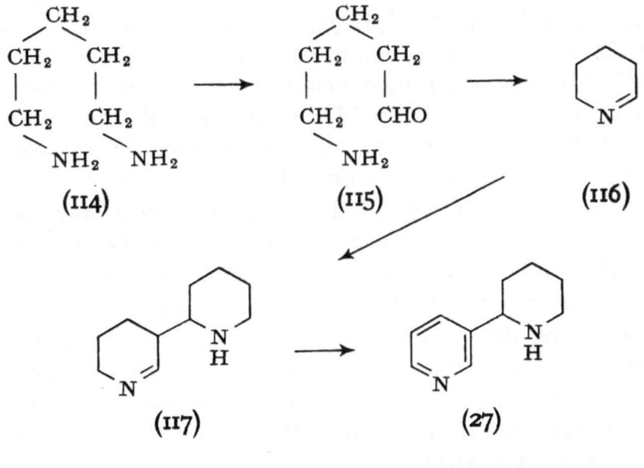

Schema 8. In vitro-Biosynthese von Anabasin (27)

Möglicherweise läuft sie in *Haloxylon salicornicum* (Moq.-Tand) Boiss. nach der in *Schema 8* dargestellten Reaktionsfolge ab, da in dieser Pflanze Anabasin (27) neben Aldotripiperidein vorkommt (*166, 228*), das als spontanes Umwandlungsprodukt des Tetrahydroanabasins (117) bekannt ist.

Das gemeinsame Auftreten von strukturverwandten Pyridin- und Piperidinalkaloiden in einer Pflanze, z. B. bei *Nicotiana* oder *Lobelia*, könnte aber als Hinweis angesehen werden, daß die Dehydrierung des Piperidinringes zum Pyridin in Einzelfällen möglich sein kann. Als derartige „Alkaloidpaare" seien Anabasin (27) und 2,3'-Dipyridyl (31), Anatallin (34) und Nicotellin (33) sowie Syphilobin A bzw. F (77), (78) und die entsprechenden Piperidinalkaloide aus *Lobelia* genannt.

4. Entstehung des Pyridinringes auf dem Monoterpenweg

Wie in den vorhergehenden Abschnitten gezeigt, gehen bei den bisher beschriebenen Biosynthesewegen die C-Atome und der Heterostickstoff des Pyridinringes ganz oder teilweise aus dem Aminosäurestoffwechsel [Tryptophan (97), Aspartat (106), Lysin (113)] und aus einfachen Verbindungen des Grundstoffwechsels wie Acetat, Glycerin (105) und Pyruvat (111) hervor. Seit kurzem kennt man ein weiteres Bildungsprinzip, bei dem das Pyridinring-haltige Grundgerüst verschiedener Alkaloide aus zwei Isoprenbausteinen entsteht.

Die ersten experimentellen Angaben zu diesem Biosyntheseweg stammen von der italienischen Arbeitsgruppe um MARINI-BETTOLO. Diese Autoren haben 1964 an *Skytanthus acutus* nachgewiesen, daß Mevalonat-[2-14C] (118) wesentlich besser in Skytanthin (46) eingebaut wird als Phenylalanin oder Acetat (44, 45, 154). Auf Grund dieses Befundes wurde für Skytanthin ein monoterpenoider Biosyntheseweg postuliert. AUDA und Mitarb. konnten 1967 den Einbau von Mevalonat-[2-14C] in Skytanthin bestätigen und die 14C-Verteilung im isolierten Alkaloid durch chemischen Abbau ermitteln (17). Bei dreijährigen *Skytanthus*-Pflanzen wurde die Radioaktivität zu 40% im C-Atom 3 und zu 45% in Position 6 gefunden. Das entspricht der Inkorporation von zwei Mevalonsäureeinheiten (Weg A in Schema 9). Überraschenderweise fanden die Autoren bei jüngeren *Skytanthus*-Pflanzen in den C-Atomen 3 und 9 des Skytanthins je 25% und im C-Atom 6 45% der Gesamtradioaktivität (Weg B in Schema 9). Diese 14C-Verteilung deutet auf eine Randomerisierung, bei der die C-Atome 2 und 6 des einen Mevalonatrestes im Verlauf der Biosynthesefolge gleichwertig werden (17), oder auf das Vorliegen eines Radioisomeren-Gemisches (13a).

Darüber hinaus haben AUDA und Mitarb. nachgewiesen, daß Mevalonat-[2-14C] (118), Acetat-[2-14C] und Geranylpyrophosphat-[1-14C] (119) mit guten Einbauraten in Actinidin (38) inkorporiert werden, während andere als Pyridinvorstufen bekannte Verbindungen wie Chinolinsäure (10), Aspartat (106) oder Lysin (113) als Actinidinprecursoren ausgeschlossen werden konnten (18).

Diese Versuche sprechen dafür, daß der Pyridinring des Actinidins und der Piperidinring des Skytanthins in offenbar ähnlicher Weise aus Mevalonsäure (118) über Geranylpyrophosphat (119) aufgebaut

Literaturverzeichnis: SS. 146—161

werden. In analoger Weise könnten auch die monoterpenoiden Pyridinalkaloide gebildet werden, wobei der p-Hydroxyphenyläthylrest der
Valeriana-Alkaloide (44) und (45) möglicherweise aus einem Tyraminrest
hervorgeht (*70a*). Es ist noch unbekannt, ob die Ringsysteme des Pyridins
und des Piperidins in dieser Alkaloidgruppe im Verlauf der Biosynthese
ineinander übergehen können oder ob sie auf parallel angelegten Bildungswegen entstehen.

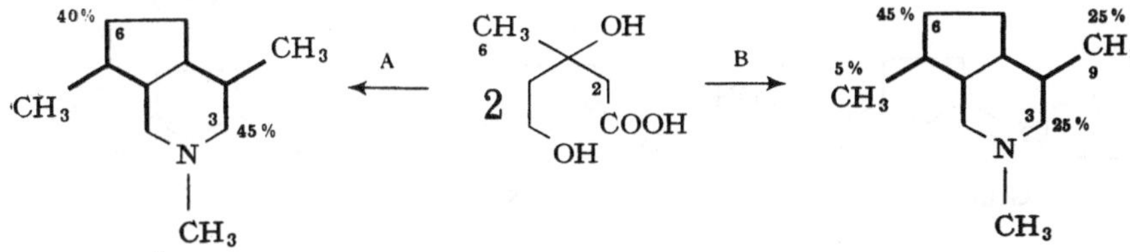

Schema 9. ^{14}C-Verteilung in Skytanthin nach Gabe von Mevalonsäure-[2-^{14}C] an
Skytanthus acutus verschiedenen Alters

Die Biosynthese der monoterpenoiden Pyridinalkaloide steht im
engen Zusammenhang mit der Bildung der Iridoide und der als Secoiridoide bezeichneten Bitterstoffe der *Gentianaceen*, da sich diese stick-
stofffreien Verbindungen ebenfalls von der Mevalonsäure (118) ableiten
(*91a*). Das ist u. a. nachgewiesen für Nepetalacton (*164, 216*), Verbenalin
(*114a, 115*), Aucubin (*115*), Loganinsäure (*122*) und Loganin (*123*) (*21,
34b, 52a, 53, 53b, 99a, 153a*), Plumierid (*278*) sowie für Swertiamarin,
Swerosid und Gentiopikrosid (*125*) (*52a, 53, 53a, 53b, 89a, 99a, 119, 120,
121*). Die Einführung eines Stickstoffatoms durch Ersatz des Hetero-
sauerstoffs ist in der Klasse der Iridoide und Secoiridoide wegen ihrer
labilen cyclischen Acetalgruppierung chemisch leicht durchzuführen. So
gibt die Umsetzung von Gentiopikrosid (125) und von Swertiamarin mit
Ammoniak unter milden Bedingungen das Alkaloid Gentianin (58) (*38,
137*). Eine derartige Reaktion, dürfte prinzipiell auch für die höhere
Pflanze möglich sein, so daß die Entstehung monoterpenoider Alkaloide
vom Typ des Actinidins (38), des Skytanthins (46) und des Gentianins
(58) über iridoide Zwischenstufen entsprechend Schema 10 anzunehmen
ist. Für ein gemeinsames Bildungsprinzip spricht auch das Vorkommen
iridoider und secoiridoider Verbindungen neben strukturähnlichen, stick-
stoffhaltigen Basen in einer Pflanze, z. B. Matatabilacton und Actinidin
(38) in *Actinidia polygama* (*220, 221*), Valepotriate (*253a*) und monoterpenoide Pyridinalkaloide (38), (39), (44) und (45) in *Valeriana officinalis* (*70a, 77, 259*), Boschnialacton und Boschniakin (41) in *Bosch-*

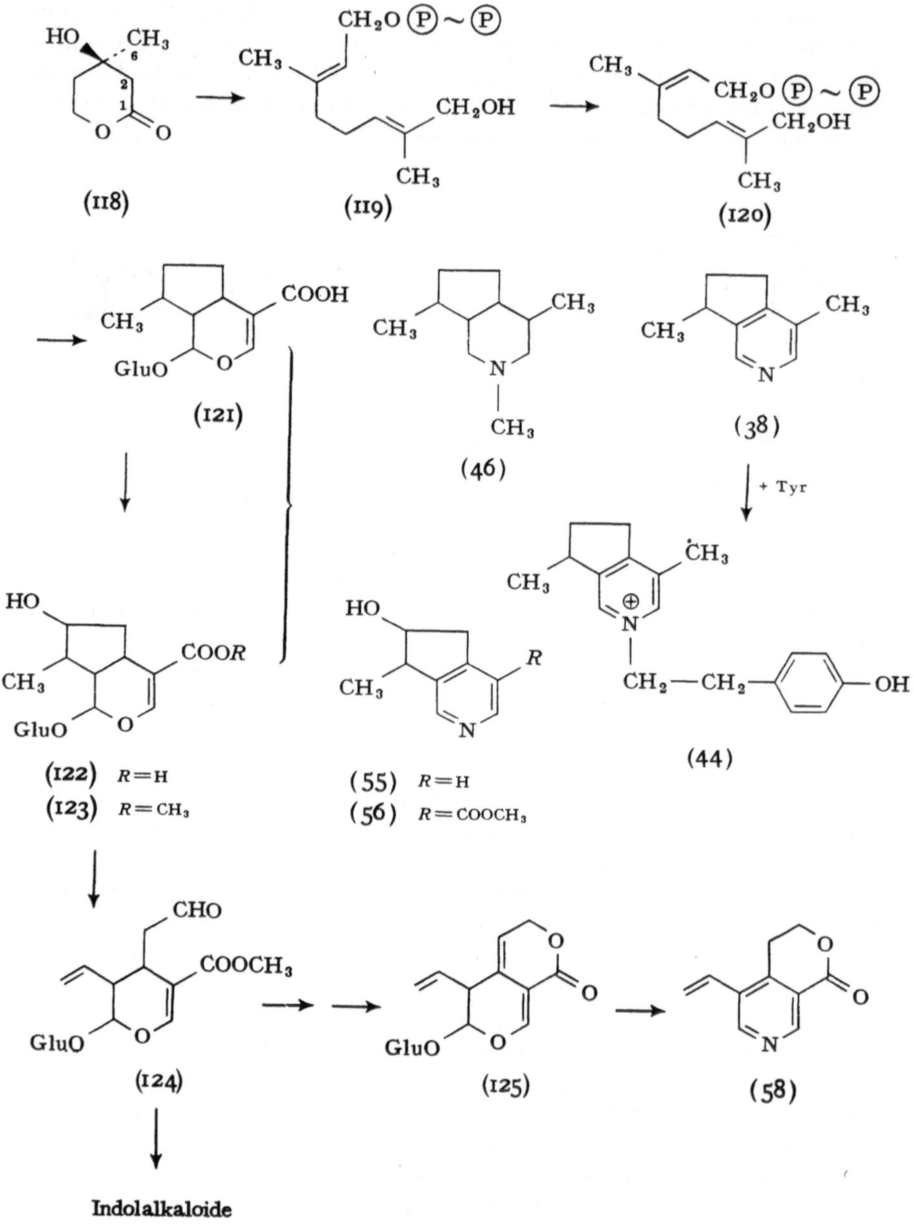

Schema 10. Möglicher Biosyntheseweg monoterpenoider Pyridin- und Piperidinalkaloide

Literaturverzeichnis: SS. 146—161

niakia rossica (223). Einzelheiten dieser Biosynthesewege sind noch nicht bekannt, auch nicht die Abzweigstellen der zu den einzelnen Verbindungstypen führenden Biosyntheseketten.

Darüber hinaus ist die Biogenese der Iridoide, Secoiridoide und der monoterpenoiden Pyridin- und Piperidinalkaloide in enger Verbindung zum Ablauf der Biosynthese zahlreicher Indolalkaloide zu sehen. Nach Untersuchungen, die insbesondere von D. ARIGONI in Zürich, A. R. BATTERSBY in Liverpool, H. INOUYE in Kyoto und A. I. SCOTT in Brighton durchgeführt wurden, entsteht der Nicht-Tryptamin-Teil der Indolalkaloide aus zwei Molekülen Mevalonsäure (118), die über Geranyl- und Nerylpyrophosphat (119), (120) und iridoide Intermediärprodukte wie Loganinsäure (122) und Loganin (123) zu Secologanin (124) umgesetzt werden (vgl. *89, 91a, 145a, 145b*). Über weitere Zwischenstufen und kompliziert erscheinende Umlagerungsreaktionen wird die große Palette der Indolalkaloide vom *Corynanthe-, Aspidosperma-, Iboga-* und *Strychnos-*Typ gebildet. Das Auftreten typischer C_{19}-Indolalkaloide neben der monoterpenoiden Pyridinbase Venoterpin (55) in *Rauwolfia-* und *Alstonia-*Arten (*15, 16, 215*) weist möglicherweise auch auf ein einheitliches Entstehungsprinzip aus gemeinsamen Ausgangs- oder Zwischenstufen.

Die Bedeutung dieses isoprenoiden Biosyntheseweges, der erst in seinen Anfängen untersucht ist, liegt offensichtlich in seiner Vielgestaltigkeit.

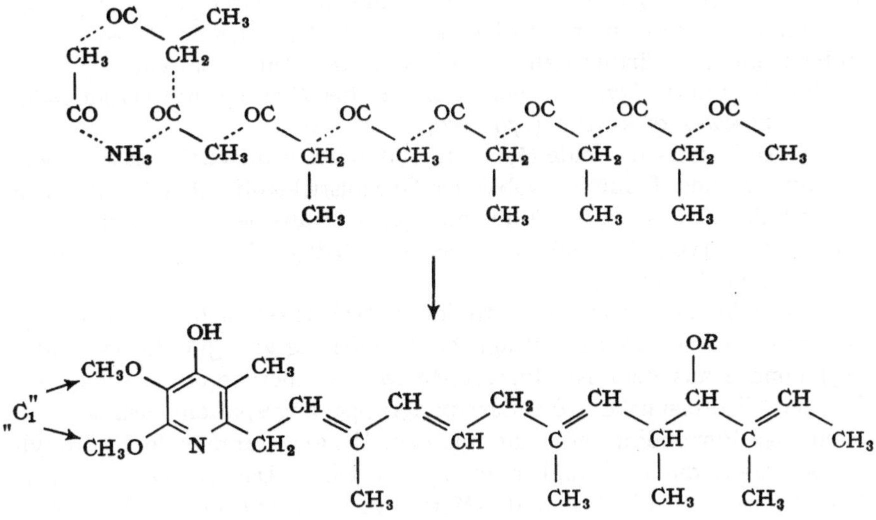

(90) $R = H$ Piericidin A

(91) $R = CH_3$ Piericidin B

Schema 11. Biosynthese der Piericidine A und B

5. Biosynthese der Piericidine A und B

Nach der Isolierung und Konstitutionsaufklärung von Piericidin A (90) und Piericidin B (91) haben Takahashi und Mitarb. (*132a, 244*) kürzlich auch Ergebnisse zur Biosynthese dieser Pyridinverbindungen veröffentlicht. Inkorporationsversuche mit ^{14}C-markierten Precursoren wie Acetat, Formiat, Propionat, Mevalonat, Methionin und Aspartat an *Streptomyces mobaraensis* ergaben, daß die beiden Methoxygruppen über den C_1-Pool (Methionin) geliefert werden, während der Pyridinring und die Seitenkette aus 4 Acetat- und 5 Propionateinheiten aufgebaut werden *(Schema 11)*. Dabei soll sich zunächst eine C_{23}-Kette ausbilden, deren eines Ende mit „Ammoniak" reagiert und zum Pyridinring cyclisiert. Dieses zu den Piericidinen führende Bildungsprinzip unterscheidet sich stark von den bisher aufgeführten Entstehungswegen des Pyridinringes.

III. Hypothesen zur Biogenese weiterer Pyridinverbindungen

Bei Betrachtung biogenetisch bisher noch nicht untersuchter Pyridinstrukturen lassen einige von ihnen Beziehungen zum Nicotinsäure-Stoffwechsel erkennen.

So könnten sich die *Areca*-Alkaloide (79), (80), (81) und (82) von der Nicotinsäure (1) ableiten. Dabei ist zu diskutieren, ob der Pyridinring nachträglich zum Tetrahydropyridin hydriert wird oder ob möglicherweise ein nach C_3- + C_4-Kondensation entstandenes Zwischenprodukt nicht vollständig aromatisiert werden kann. Entsprechende Biosyntheseuntersuchungen dürften daher von besonderem Interesse sein.

Für Nudiflorin (37) wird ebenfalls ein über Nicotinsäure verlaufender Biosyntheseweg diskutiert (*179, 180*).

Die Halfordia-Alkaloide (67)—(71) entstehen wahrscheinlich aus Nicotinsäure (1) und Tyramin, wobei der Oxazolstickstoff entweder aus dem Amidstickstoff des Nicotinsäureamids (2) oder aus der Aminogruppe des Tyramins hervorgehen könnte. Der O-Substituent ist möglicherweise isoprenoider Herkunft.

Der 4-Aminonicotinsäuremethylester (15) leitet sich eventuell von Nicotinsäure (1) durch nachträgliche Aminierung ab. 3-Methoxypyridin (13) könnte aus dem Nicotinsäurestoffwechsel hervorgehen, wobei wie bei der Nicotinsynthese die Carboxylgruppe abgespalten wird und an Stelle des Pyrrolidinringes eine aus dem Wasser stammende Hydroxylgruppe tritt, die nachträglich methyliert wird. Das gemeinsame Vorkommen von Nicotin (24) und 3-Methoxypyridin (13) deutet diese Möglichkeit an. α-Picolin (14) und 3-Hydroxypicolinsäure (95) dürften aus dem Chinolinsäurestoffwechsel hervorgehen.

Bei den Säurekomponenten (72) (73) (74) des Evonins und der *Wilfordia*-Alkaloide ist in Position 2 der Nicotinsäure eine C_5-Seitenkette angeknüpft, deren Herkunft aus Mevalonsäure denkbar ist, wobei allerdings bei der Evoninsäure (72) eine ungewöhnliche Isoprenverknüpfung vorliegen würde.

Pyrimin (17) entsteht vielleicht aus Chinolinsäure (10) unter Decarboxylierung und Anknüpfung eines Glutaminsäurerestes in α-Stellung.

Für Micrococcinsäure (96) wird von HALL *et al.* eine Bildung aus 4 Molekülen Cystein, einem Molekül α-Aminobuttersäure und einem Molekül α-Aminoadipinsäure diskutiert (*101*), wobei sich zunächst eine Peptidsequenz mit eingebauter α-Aminoadipinsäure ausbilden soll. Aus den Cysteinresten könnten sich durch Cyclisierung und Dehydrierung die Thiazolringe bilden, während der Pyridinring durch Ringschluß und Wasserstoffabspaltung aus α-Aminoadipinsäure entstehen soll.

Für einige weitere Pyridinverbindungen wie Pyridoxalphosphat (19), Muscopyridin (18) und die 2,6-Lutidin-3,5-dicarbonsäure (75) des Clivimins ist vielleicht ein modifizierter Polyacetat-Weg zu erwägen. Endgültige Aussagen lassen sich jedoch erst nach Abschluß entsprechender Biosyntheseexperimente machen.

Schlußbetrachtung

Man kennt heute etwa hundert Naturstoffe mit Pyridinstruktur. Diesen Verbindungen liegt ein einfach oder mehrfach substituierter Pyridinring zugrunde, wobei die α- und β-Positionen zur Substitution bevorzugt sind. Als Substituenten finden sich vor allem Methyl-, Hydroxymethyl-, Carboxy-, Carboxamid-, Hydroxy-, Methoxy- und Aminogruppen sowie verzweigte und unverzweigte Seitenketten. Über die Carboxylgruppen einiger Pyridincarbonsäuren bauen sich ester- oder amidartig verschiedene höher molekulare Verbindungen auf. Vielfach ist in β-Stellung des Pyridinringes ein weiteres heterocyclisches Ringsystem angefügt, z. B. Pyrrolidin-, Pyridin-, Piperidin-, Oxazol- oder Thiazolreste. Darüber hinaus findet man auch Pyridon- und Pyridiniumstrukturen.

Man darf mit großer Sicherheit das Auffinden weiterer natürlicher Pyridinstrukturen erwarten, vor allem als Stoffwechselprodukte bestimmter Mikroorganismen oder als Inhaltsstoffe höherer Pflanzen. Bei der Isolierung der Alkaloide sind bei den bisher allgemein üblichen Aufarbeitungsmethoden die quartären Verbindungen meist nicht miterfaßt worden, so daß in dieser Richtung chemisch und biochemisch interessante Naturstoffe auch aus der Pyridinreihe zu erwarten sein dürften.

Nach unseren heutigen Kenntnissen erfolgt die Biosynthese des Pyridinringsystems auf verschiedenartigen Wegen, die vom biologischen Objekt, von der Struktur der zu synthetisierenden Verbindungen und unter Umständen von äußeren Lebensbedingungen abhängig sind. In

der Pyridinreihe ist besonders auffällig, daß eine Verbindung, wie z. B.
die Nicotinsäure (1), von verschiedenartigen Organismen (Mikroorganismen, höhere Pflanze und Säugetier) nach unterschiedlichen Mechanismen
synthetisiert werden kann. Anderseits sind in einem Organismus, wie
z. B. der Hefezelle, verschiedenartige Biosynthesewege für die Nicotinsäure angelegt. Weiterhin können in einem Organismus parallel zwei
zum Pyridinringsystem führende Bildungswege beschritten werden,
wie das bei *Fusarium oxysporum* für Nicotinsäure (1) und Fusarinsäure
(87) bewiesen ist.

Nach den derzeitig vorliegenden Ergebnissen kennt man zwar in
etlichen Fällen das Grundprinzip der Pyridinbiosynthese, man weiß aber
noch sehr wenig über die intermediär auftretenden Produkte, über
Reaktionsmechanismen und den zeitlichen Ablauf der Biosynthesekette
sowie über die Enzymologie, Regulation und Kinetik der einzelnen
Syntheseschritte. Der stereochemische Ablauf der Alkaloidbiogenese ist
ebenfalls noch weitgehend unbekannt. Die Klärung dieser noch offenen
Fragen muß weiteren Untersuchungen vorbehalten bleiben, wobei *in-
vitro*-Experimente immer mehr in den Vordergrund treten werden.

Literaturverzeichnis

1. ABELSON, D., A. BOYLE and H. SELIGSON: Identification of N'-Methyl-4-pyri-
done-3-carboxamide in Human Plasma. J. Biol. Chem. **238**, 717 (1963).

2. ABDUSAMATOV, A., S. KHAKIMDZHANOV and S. Y. YUNUSOV: The Alkaloids
of *Pedicularis*. Khim. Prir. Soedin. **4** (3), 195 (1968) [Chem. Abstr. **69**, 57450a
(1968)].

2a. — — — Structure of Pedicularine. Khim. Prir. Soedin. **5** (5), 457 (1969)
[Chem. Abstr. **72**, 67156k (1970)].

3. ABDUSAMATOV, A., K. UBAEV and S. Y. YUNUSOV: The Alkaloids of *Pedicularis olgae*. Khim. Prir. Soedin. **4** (2), 136 (1968) [Chem. Abstr. **69**, 67572f
(1968)].

4. ABDUSAMATOV, A., M. R. YAGUDAEV and S. Y. YUNUSOV: Structure of Plantagonine and Indicaine. Khim. Prir. Soedin. **4**(4), 265 (1968) [Chem. Abstr.
70, 68580g (1969)].

4a. ABDUSAMATOV, A. and S. Y. YUNUSOV: Pedicularis Alkaloids. Khim. Prir.
Soedin. **5**, (4), 334 (1969) [Chem. Abstr. **72**, 51805k (1970)].

5. ADOLPHEN, G., H. H. APPEL, K. H. OVERTON and W. D. C. WARNOCK: Hydroxy-
skytanthines I and II. Two Minor Alkaloids of *Skytanthus acutus* Meyen.
Tetrahedron **23**, 3147 (1967).

6. AHMAD, F., and A. G. MOAT: Nicotinic Acid Biosynthesis in Prototrophs and
Certain Yeast Auxotrophs. Federat. Proc. (Amer. Soc. Exp. Biol.) **24**, 480
(1965).

7. — — Nicotinic Acid Biosynthesis in Prototrophs and Tryptophan Auxotrophs
of *Saccharomyces cerevisiae*. J. Biol. Chem. **241**, 775 (1966).

7a. AHMED, Z. F., A. M. RIZK and F. M. HAMMOUDA: Phytochemical Studies of
Egyptian *Plantago* Species (Alkaloids). J. pharm. pharmacol. **17** (Suppl.),
39 S (1965).

7b. AKRAMOV, S. T., M. R. YUGATAEV, T. U. RAKHMATULLAEV, A. SAMATOV and S. Y. YUNUSOV: Structures of *Gentiana* Alkaloids (Gentianadine, Gentianamine and Gentianaine). Khim. Prir. Soedin. **5** (1), 14 (1969) [Chem. Abstr. **71**, 3518b (1969)].

8. ALBERTSON, J. N. and A. G. MOAT: Biosynthesis of Nicotinic Acid by *Mycobacterium tuberculosis*. J. Bacteriol. **89**, 540 (1965).

9. ANDREOLI, A. J., M. IKEDA, Y. NISHIZUKA and O. HAYAISHI: Quinolinic Acid: a Precursor to Nicotinamide Adenine Dinucleotide in *Escherichia coli*. Biochem. Biophys. Res. Comm. **12**, 92 (1963).

10. ANWAR, R. A.: Comparison of Elastins from Various Sources. Canad. J. Biochem. **44**, 725 (1966).

11. ANWAR, R. A. and G. ODA: The Biosynthesis of Desmosine and Isodesmosine. J. Biol. Chem. **241**, 4638 (1966).

12. — — Structure of Desmosine and Isodesmosine. Nature **210**, 1254 (1966).

13. — — The Biosynthesis of Desmosine and Isodesmosine II. Incorporation of Specifically Labeled Lysine into Desmosine and Isodesmosine. Biochim. Biophys. Acta **133**, 151 (1967).

13a. APPEL, H. H.: Sobre la Existencia de Esquitantinas „Radioisomeras". Scientia (Valparaiso) **35**, 128 (1968) [Chem. Abstr. **71**, 124733b (1969)].

13b. APPEL, H. H. and B. MÜLLER: Alkaloids from *Skytanthus acutus*. Scientia (Valparaiso) **28** (115), 5 (1961) [Chem. Abstr. **57**, 2332s (1962)].

14. ARONOFF, S.: Experiments on the Biogenesis of the Pyridine Ring in Higher Plants. Plant Physiol. **31**, 355 (1956).

15. ARTHUR, H. R., S. R. JOHNS, J. A. LAMBERTON and S. N. LOO: A New Monoterpenoid Alkaloid (RW 47) from *Rauwolfia verticillata* (Lour.) Bail. of Hong Kong. Austral. J. Chem. **20**, 2505 (1967).

16. ARTHUR, H. R. and S. N. LOO: An Examination of *Rauwolfia verticillata* of Hong Kong II. Phytochem. **5**, 977 (1966).

17. AUDA, H., H. R. JUNEJA, E. J. EISENBRAUN, G. R. WALLER, W. R. KAYS and H. H. APPEL: Biosynthesis of Methylcyclopentane Monoterpenoids. I. *Skytanthus* Alkaloids. J. Amer. chem. Soc. **89**, 2476 (1967).

18. AUDA, H., G. R. WALLER and E. J. EISENBRAUN: Biosynthesis of Methylcyclopentane Monoterpenoids. III. Actinidine. J. Biol. Chem. **242**, 4157 (1967).

19. AYER, W. A. and T. E. HABGOOD: The Pyridine Alkaloids. In: R. H. F. MANSKE, The Alkaloids, Chemistry and Physiology, Vol. XI, p. 459, Academic Press, New York-London 1968.

20. BACH, M. L. and C. GILVARG: Biosynthesis of Dipicolinic Acid in Sporulating *Bacillus megaterium*. J. Biol. Chem. **241**, 4563 (1966).

21. BATTERSBY, A. R., E. S. HALL and R. SOUTHGATE: Alkaloid Biosynthesis. Part XIII. The Structure, Stereochemistry, and Biosynthesis of Loganin. J. Chem. Soc. (London) C **1969**, 721.

22. BEDFORD, G. R. and A. R. KATRITZKY: Proton Magnetic Resonance Spectra of Degradation Products from Elastin. Nature **200**, 652 (1963).

23. BENGER, H.: Zur Biosynthese der Dipicolinsäure. Ein Beitrag zur Biologie der Sporenbildung. Z. Hyg. Infektionskrankheiten **148**, 318 (1962) [Chem. Zbl. **136**, 9623 (1965)].

24. BEROZA, M.: Alkaloids from *Tripterygium wilfordii* Hook. The Chemical Structure of Wilfordic and Hydroxywilfordic Acids. J. Organ. Chem. (USA) **28**, 3562 (1963).

25. BEYERMAN, H. C., L. MAAT and M. P. HEGARTY: The Absolute Configuration of Mimosine. Rec. trav. chim. Pays-Bas **83**, 1078 (1964).

26. Biemann, K., G. Büchi and B. H. Walker: The Structure and Synthesis of Muscopyridine. J. Amer. Chem. Soc. **79**, 5558 (1957).

27. Blake, C. O.: Niacin Metabolism in the Corn Seedling: The Biosynthesis of Trigonelline. Amer. J. Bot. **41**, 231 (1954).

28. Blaim, K.: Der Abbau von Trigonellin zu Nicotinsäure in keimenden Samen. Naturwiss. **47**, 332 (1960).

29. — Untersuchungen über biochemische Prozesse und die physiologische Bedeutung von Trigonellin. Roczniki Nauk rolniczych, Ser. A **85**, 307 (1962) [Chem. Zbl. **135** (31), 1208 (1964)].

30. Boit, H.-G.: Ergebnisse der Alkaloid-Chemie bis 1960. S. 125ff., Akademie-Verlag Berlin, 1961.

31. Bolton, M. and J. Walker: Chemistry of Micrococcin P. Part XI. Application of a Simple Micromethod for the Identification of Lower Volatile Fatty Acids to the Structural Study of 2-Acylthiazoles. J. Chem. Soc. (London) C **1967**, 2095.

32. Bonner, D. M. and C. Yanofsky: The Biosyntheses of Tryptophan and Niacin and their Relationships. J. Nutrit. **44**, 603 (1951).

33. Bowden, K.: Biogenesis of Nicotine. Nature **172**, 768 (1953).

34. Bowman, R. M. and E. Leete: Observations on the Administration of Iridodial-7-^{14}C to *Vinca rosea*. Phytochem. **8**, 1003 (1969).

34a. Brandänge, S. and B. Lüning: Studies on *Orchidaceae* Alkaloids. XVII. Alkaloids from *Vandopsis longicaulis* Schltr. Acta Chem. Scand. **24**, 353 (1970).

34b. Brechbühler-Bader, S., C. J. Coscia, P. Loew, Ch. von Szczepanski and D. Arigoni: The Chemistry and Biosynthesis of Loganin. Chem. Comm. **1968**, 136.

35. Brewbaker, J. L. and J. W. Hylin: Variation of Mimosine Content among Leucaena Species and Related Mimosaceae. Crop. Sci. **5**, 348 (1965) [Chem. Abstr. **64**, 3958 (1966)].

36. Budzikiewicz, H., C. Horstmann, K. Pufahl und K. Schreiber: Isolierung von Fontaphillin, Gentianin und 4-Amino-nicotinsäuremethylester aus Blattextrakten von *Fontanesia phillyreoides* Labill. Chem. Ber. **100**, 2798 (1967).

37. Büchi, G., I. M. Goldman and D. W. Mayo: The Structures of two Alkaloids from Patchouli Oil. J. Amer. Chem. Soc. **88**, 3109 (1966).

38. Canonica, L., F. Pelizzoni and G. Jommi: Transformation of Gentiopicroside to Gentianine. Gazz. chim. ital. **92**, 298 (1962) [Chem. Abstr. **57**, 7330g (1962)].

39. Casanova, C. and A. G. Gonzales: Alkaloids of Canary Island Plants. IX. *Ixanthus viscosus*. An. Real. Soc. espan. Fisica Quim. (Madrid), Ser. B **60**, 607 (1964) [Chem. Abstr. **62**, 15067f (1965)].

40. Casinovi, C. G., F. Delle Monache, G. Grandolini, G. B. Marini-Bettolo and H. H. Appel: Two New Alkaloids from *Skytanthus acutus* Meyen. Chem. and Ind. **1963**, 984.

41. Casinovi, C. G., F. Delle Monache, G. B. Marini-Bettolo, E. Bianchi e J. Garbarino: Ricerche nella serie della Skytanthina. Nota II. Sintesi di tre stereoisomeri a configurazione nota della Skytanthina. Gazz. chim. ital. **92**, 479 (1962).

42. — — — — — Syntheses in the Skytanthin Series. Sci. Rept. Ist. Super Sanita **1**, 588 (1961) [Chem. Abstr. **63**, 4349d (1965)].

43. Casinovi, C. G., J. A. Garbarino and G. B. Marini-Bettolo: Structure of the Alkaloid of *Skytanthus acutus* Meyen. Chem. and Ind. **1961**, 253.

44. Casinovi, C. G., G. Giovannozzi-Sermanni e G. B. Marini-Bettolo: Studi preliminari sulla biosintesi delle skitantine. Gazz. chim. ital. **94**, 1356 (1964).

45. — — — The Biosynthesis of Skytanthines. Rend. Accad. Naz. XL **16—17**, 89 (1966) [Chem. Abstr. **68**, 10289 (1968)].

46. CAVILL, G. W. K. and A. ZEITLIN: Synthesis of D-(+)-Tecostidine and related Actinidine Derivatives. Austral. J. Chem. **20**, 349 (1967).

47. CHANDLER, L. R., N. OGASAWARA, R. K. GHOLSON and A. J. ANDREOLI: In vitro Biosynthesis of Quinolinic Acid. Federat. Proc. (Amer. Soc. Exp. Biol.) **25**, 217 (1966).

48. CHANDRAMOHAN, D. and A. MAHADEVAN: Detection of Fusaric Acid in the Mycelium and Conidia of *Fusarium oxysporum* f. *vasinfectum*. Experientia **24**, 427 (1968).

49. CHASIN, L. A. and J. SZULMAJSTER: Biosynthesis of Dipicolinic Acid in *Bacillus subtilis*. Biochem. Biophys. Res. Comm. **29**, 648 (1967).

50. CHAYKIN, S.: Nicotinamide Coenzymes. Annu. Rev. Biochem. **36** (I), 149 (1967).

51. CHAYKIN, S., M. DAGANI, L. JOHNSON and M. SAMLI: The Fate of Nicotinamide in the Mouce. J. Biol. Chem. **240**, 932 (1965).

52. CIESLAK, J., J. KUDUK and F. RULKO: Alkaloidy *Gentiana lutea*. Acta polon. pharm. **21**, 265 (1964) [Chem. Abstr. **62**, 13507d (1965)].

52a. COSCIA, C. J., L. BOTTA and R. GUARNACCIA: On the Mechanism of Iridoid and Secoiridoid Monoterpene Biosynthesis. Arch. Biochem. Biophys. **136**, 498 (1970).

53. COSCIA, C. J. and R. GUARNACCIA: Biosynthesis of Gentiopicroside, a Novel Monoterpene. J. Amer. Chem. Soc. **89**, 1280 (1967).

53a. — — Natural Occurrence and Biosynthesis of a Cyclopentanoid Monoterpene Carboxylic Acid. Chem. Comm. **1968**, 138.

53b. COSCIA, C. J., R. GUARNACCIA and L. BOTTA: Monoterpene Biosynthesis. I. Occurrence and Mevalonoid Origin of Gentiopicroside and Loganic Acid in *Swertia caroliniensis*. Biochemistry **8**, 5036 (1969).

54. CROW, W. D. and J. H. HODGKIN: Alkaloids of the Australian *Rutaceae*: *Halfordia scleroxyla*. I. The Structure of N-Methylhalfordinium chloride. Tetrahedron Letters **1963**, 85.

55. — — Alkaloids of the Australian *Rutaceae: Halfordia scleroxyla*. II. Isolation and Structure of the Alkaloids. Austral. J. Chem. **17**, 119 (1964).

56. — — Alkaloids of the Australian *Rutaceae: Halfordia scleroxyla* and *Halfordia kendack*. IV. Co-occurrence of Oxazole and Quinoline alkaloids. Austral. J. Chem. **21**, 3075 (1968).

57. CROW, W. D., J. H. HODGKIN and J. S. SHANNON: Alkaloids of the Australian *Rutaceae: Halfordia scleroxyla*. III. Mass Spectrometry of Halfordinol and related Oxazoles. Austral. J. Chem. **18**, 1433 (1965).

58. DANILOVA, A. V.: Alkaloids of *Plantago ramosa* II. Zhurn. Obshchei Khimii (USSR) **26**, 2069 (1956) [Chem. Abstr. **51**, 5098 (1957)].

59. DANILOVA, A. V. and R. A. KONOVALOVA: Alkaloids from *Plantago indica*. Zhurn. Obshchei Khimii (USSR) **22**, 2237 (1952) [Chem. Abstr. **48**, 691a (1954)].

60. DAVIS, D., L. M. HENDERSON and D. POWELL: The Niacin-Tryptophan Relationship in the Metabolism of *Xanthomonas pruni*. J. Biol. Chem. **189**, 543 (1951).

61. DAWSON, R. F., D. R. CHRISTMAN, A. D'ADAMO, M. L. SOLT and A. P. WOLF: The Biosynthesis of Nicotine from Isotopically Labeled Nicotinic Acids. J. Amer. Chem. Soc. **82**, 2628 (1960).

62. DESATY, D., A. G. McINNES, D. G. SMITH and L. C. VINING: Use of ^{13}C in Biosynthetic Studies. Incorporation of Isotopically Labeled Acetate and Aspartate into Fusaric Acid. Canad. J. Biochem. **46**, 1293 (1968).

63. DESATY, D. and L. C. VINING: Biosynthesis of Nicotinic Acid by *Fusarium oxysporum* Schlecht. Canad. J. Biochem. **45**, 1953 (1967).

64. DICKINSON, E. M. and G. JONES: Pyrindane Alkaloids from *Tecoma stans*. Tetrahedron **25**, 1523 (1969).
65. DJERASSI, C., J. P. KUTNEY and M. SHAMMA: Alkaloid Studies. XXXII. Studies on *Skytanthus acutus* Meyen. The Structure of the Monoterpenoid Alkaloid Skytanthine. Tetrahedron **18**, 183 (1962).
66. DJERASSI, C., J. P. KUTNEY, M. SHAMMA, J. N. SHOOLERY and L. F. JOHNSON: Alkaloid Studies. XXVII. The Structure of Skytanthine. Chem. and Ind. **1961**, 210.
67. DIVEKAR, P. V., G. READ and L. C. VINING: Caerulomycin, a new Antibiotic from *Streptomyces caeruleus* Baldacci. II. Structure. Canad. J. Chem. **45**, 1215 (1967).
68. DOBSON, T. A., D. DESATY, D. BREWER and L. C. VINING: Biosynthesis of Fusaric Acid in Cultures of *Fusarium oxysporum* Schecht. Canad. J. Biochem. **45**, 809 (1967).
69. DOEPKE, W., M. BIENERT, A. L. BURLINGAME, A. L. SCHNOES, P. W. JEFFS and D. S. FARRIER: Structures and Stereochemistry of Clivonine and Clivimine. Tetrahedron Letters **1967**, 451.
70. DREYER, D. L.: Chemotaxonomy of the *Rutaceae*. III. Isolation of Halfordinol Derivatives from *Aeglopsis chevalieri* Swing. J. Organ. Chem. (USA) **33**, 3658 (1968).
70a. EDNER, G., D. GROSS, und H. R. SCHÜTTE: Über monoterpenoide Valeriana-Alkaloide. Arch. Pharmaz. (Im Druck.)
71. EISENBRAUN, E. J., A. BRIGHT and H. H. APPEL: The Absolute Configuration and Partial Synthesis of α-, β-, γ- and δ-Skytanthine. Chem. and Ind. **1962**, 1242.
72. ERGE, D., D. GROEGER and K. MOTHES: Zum Tryptophan-Stoffwechsel und zur Nicotinsäuresynthese bei zentralasiatischem Mutterkorn. Arch. Pharm. **295**, 474 (1962).
73. FANG, SHEN-TIN and JEN-HUNG CHU: The Chemical Constituents of Leaves of *Quisqualis indica*. Acta chimica sinica (Hua Hsueh Hsueh Pao) **30**, 226 (1964) [Chem. Abstr. **61**, 7359f (1964)].
74. FASELLA, P.: Pyridoxal Phosphate. Annu. Rev. Biochem. **36** I, 185 (1967).
75. FELSTED, R. L. and S. CHAYKIN: N^1-Methylnicotinamide Oxidation in a Number of Mammals. J. Biol. Chem. **242**, 1274 (1967).
76. FLOSS, H. G., U. MOTHES und A. RETTIG: Biosyntheseversuche mit Gentianaceen II. Die Beziehung zwischen Gentianin und Gentiopikrosid. Z. Naturforsch. **19b**, 1106 (1964).
77. FRANCK, B.: Synthesen in Anlehnung an die Alkaloid-Biogenese. Abh. dtsch. Akad. Wiss. Berlin, **1970**, im Druck.
78. FROST, G. M., K. S. YANG and G. R. WALLER: Nicotinamide Adenine Dinucleotide as a Precursor of Nicotine in *Nicotiana rustica* L. J. Biol. Chem. **242**, 887 (1967).
79. FU, FENG-YUNG and NAN-CHUN SUN: The Chemical Constituents of *Gentiana macrophylla*. Acta pharm. sinica (Yao Hsüeh Hsüeh Pao) **6**, 198 (1958) [Chem. Abstr. **53**, 8310 (1959)].
80. FUJINO, A.: III. Chemical Structure of Actinidine. J. Chem. Soc. Japan **81**, 1327 (1960).
81. GÄUMANN, E., S. NAEF-ROTH und H. KOBEL: Über Fusarinsäure, ein zweites Welketoxin des *Fusarium lycopersici* Sacc. Phytopatholog. Z. **20**, 1 (1952).
81a. GANGULI, S. N.: Isolation of Ricinidine from Plant Source. Phytochemistry **9**, 1667 (1970).
82. GEN, VAN DER, A. and L. M. VAN DER LINDE: Isolation and Synthesis of Guaipyridine, a new Basic Constituent of Patchouli Oil. IUPAC 5th International Symposium on the Chemistry of Natural Products, Abstracts, London, 1968, p. 293.

83. Georg, L. K.: Conversion of Tryptophane to Nicotinic Acid by *Trichophyton equinum*. Proc. Soc. Exp. Biol. Med. **72**, 653 (1949).

84. Gholson, R. K.: The Pyridine Nucleotide Cycle. Nature **212**, 933 (1966).

85. Gholson, R. K., J. L. R. Chandler, K. S. Yang and G. R. Waller: Nicotine Biosynthesis. Federat. Proc. (Amer. Soc. Exp. Biol.) **23**, 528 (1964).

86. Gluecksohn-Waelsch, S., P. Greengard, G. P. Quinn and L. S. Teicher: Genetic Variations of an Oxidase in Mammals. J. Biol. Chem. **242**, 1271 (1967).

87. Govindachari, T. R., S. S. Sathe and N. Viswanathan: Gentianine, an Artefact in *Enicostema littorale*. Indian J. Chem. **4**, 201 (1966).

88. Grimshaw, J. and L. Marion: The Pyridine Ring and the Problem of its Biosynthesis. Nature **181**, 112 (1958).

89. Groeger, D.: Indolalkaloide. In K. Mothes und H. R. Schütte, Biosynthese der Alkaloide, S. 459 und 699, VEB Deutscher Verlag der Wissenschaften, Berlin 1969.

89a. Gröger, D. und P. Simchen: Über den Einbau von Loganin in Gentiopikrosid. Z. Naturforsch. **24b**, 356 (1969).

90. Gross, D.: Untersuchungen zur Biosynthese der Nicotinsäure bei *Mycobacterium tuberculosis*. Abh. dtsch. Akad. Wiss. Berlin, Klasse Chem., Geol. Biol. **1966** (3), 151.

91. — Pyridinalkaloide. In: K. Mothes und H. R. Schütte, Biosynthese der Alkaloide, S. 215, VEB Deutscher Verlag der Wissenschaften, Berlin 1969.

91a. — Die Biosynthese iridoider Naturstoffe. Fortschr. Bot. **32**, 93 (1970).

92. Gross, D., P. Banditt, J. W. Kurbatov und H. R. Schütte: Zur Biosynthese des Ricinins. Z. Pflanzenphysiol. **58**, 410 (1968).

93. Gross, D., P. Banditt, A. Zureck und H. R. Schütte: Chinolinsäure als Zwischenstufe in der Nicotinsäure-Biosynthese. Z. Naturforsch. **23b**, 390 (1968).

94. Gross, D., G. Edner und H. R. Schütte, in Vorbereitung.

95. Gross, D., A. Feige, R. Stecher, A. Zureck und H. R. Schütte: Untersuchungen zur Biosynthese der Nicotinsäure bei Mycobacterium tuberculosis. Z. Naturforsch. **20b**, 1116 (1965).

96. Gross, D., A. Feige, A. Zureck und H. R. Schütte: 3-Hydroxymethylpyridin, ein neues Stoffwechselprodukt der Mycobakterien. Z. Naturforsch. **22b**, 835 (1967).

97. — — — — Die Biosynthese von 3-Hydroxymethylpyridin bei *Mycobacterium bovis* Stamm BCG. European J. Biochem. **4**, 28 (1968).

98. Gross, D., D. Müller und H. R. Schütte: Zur Methylierung bei der Ricininbiosynthese. Z. Naturforsch. **24b**, 705 (1969).

99. Gross, D., H. R. Schütte, G. Hübner und K. Mothes: The Biosynthesis of Nicotinic Acid by *Mycobacterium tuberculosis*. Tetrahedron Letters **1963**, 541.

99a. Guarnaccia, R., L. Botta and C. J. Coscia: Mechanism of Secoiridoid Monoterpene Biosynthesis. J. Amer. Chem. Soc. **91**, 204 (1969).

100. Gupta, R. N.: Biosynthesis of the Piperidine Alkaloids. Lloydia **31**, 318 (1968).

101. Hall, G. E., N. Sheppard and J. Walker: Chemistry of Micrococcin P. Part X. Proton Magnetic Resonance Spectrum of Dimethyl Micrococcinate, and the Probable Mode of Biosynthesis of Micrococcinic Acid. J. Chem. Soc. (London) **1966**, 1371.

102. Hammouda, F. M. and Z. F. Ahmed: A Phytochemical Study of the Seeds of *Pisum sativum*. I. Congr. Sci. Farm., Conf. Comun., (Pisa) **21**, 551 (1962) [Chem. Abstr. **59**, 903 (1963)].

103. Hammouda, Y. and J. Le Men: Sur la técostidine: nouvel alcaloïde du *Tecoma stans* Juss. (4e mémoire). Monoterpénoide (4e mémoire). Bull. soc. chim. France **1963**, 2901.

104. Hammouda, Y., M. Plat and J. Le Men: Isolement de la técostanine. Alcaloïdes du *Tecoma stans* (Juss.) *Bignoniacées* (2e mémoire) Ann. pharm. franc. **21**, 699 (1963).

105. — — — Structure de la técostanine: alcaloïde du *Tecoma Stans* Juss. (3e mémoire). Monoterpénoides (2e mémoire). Bull. soc. chim. france **1963**, 2802.

106. Hart, N. K., S. R. Johns and J. A. Lamberton: A New Monoterpenoid Alkaloid from *Jasminum* Species and from *Ligustrum novoguineense* (Family Oleaceae). Austral. J. Chem. **21**, 1321 (1968).

106a. — — — Alkaloids of *Jasminum* Species (Family *Oleaceae*). II. Isolation of a New Monoterpenoid Alkaloid and Other Constituents. Austral. J. Chem. **22**, 1283 (1969).

107. Hasse, K. und P. Berg: Oxydation von Cadaverin zu Anabasin. Naturwiss. **44**, 584 (1957).

108. Hasse, K. und H. Maisack: Δ^1-Pyrrolin und Δ^1-Piperidein aus Putrescin und Cadaverin durch enzymatische Oxydation. Naturwiss. **42**, 627 (1955).

109. Hegarty, M. P.: The Isolation and Identification of 5-Hydroxypiperidine-2-carboxylic Acid from *Leucaena glauca* Benth. Austral. J. Chem. **10**, 484 (1957).

110. Heilmann, H.-D. und F. Lingens: Zur Regulation der Nicotinsäure-Biosynthese in *Saccharomyces cerevisiae*. Hoppe Seyler's Z. physiol. Chem. **349**, 231 (1968).

111. — — Reinigung und Eigenschaften der 3-Hydroxyanthranilat-Oxygenase aus *Saccharomyces cerevisiae*. Hoppe Seyler's Z. physiol. Chem. **349**, 223 (1968).

112. Henderson, L. M., J. F. Someroski, D. R. Rao, P. L. Wu, T. Griffith and R. U. Byerrum: Lack of a Tryptophan-Niacin Relationship in Corn and Tobacco. J. Biol. Chem. **234**, 93 (1959).

113. Hill, R. D., A. M. Unrau and D. T. Canvin: The Biosynthesis of Fusaric Acid from ^{14}C-labelled Acetate in *Gibberella fujikuroi*. Canad. J. Chem. **44**, 2077 (1966).

114. Hodson, P. H. and J. W. Foster: Dipicolinic Acid Synthesis in *Penicillium citreo-viride*. J. Bacteriol. **91**, 562 (1966).

114a. Horodysky, A. K., G. R. Waller and E. J. Eisenbraun: Biosynthesis of Methylcyclopentane Monoterpenoids. IV. Verbenalin. J. Biol. Chem. **244**, 3110 (1969).

115. Hüni, J. E. S., H. Hiltebrand, H. Schmid, D. Gröger, S. Johne und K. Mothes: Zur Biosynthese des Verbenalins und Aucubins. Experientia **22**, 656 (1966).

116. Huff, S. D. and S. Chaykin: Genetic and Androgenic Control of N[1]-Nethyl-nicotinamide Oxidase Activity in Mice. J. Biol. Chem. **242**, 1265 (1967).

117. Hylin, J. W.: Biosynthesis of Mimosine. Phytochem. **3**, 161 (1964).

118. Ichiyama, A., S. Nakamura and Y. Nishizuka: Studies on the Biosynthesis of Nicotinamide Adenine Dinucleotide (NAD) in Mammals and its Regulatory Mechanism. I and II. Arzneimittel-Forsch. **17**, 1346, 1525 (1967).

119. Inouye, H., S. Ueda, Y. Aoki and Y. Takeda: Zur Biosynthese der Iridoidglucoside. Tetrahedron Letters **1969**, 2351.

120. Inouye, H., S. Ueda und Y. Nakamura: Zur Biosynthese der bitteren Glucoside der Gentianazeen, des Gentiopicrosids, des Swertiamarins und des Swerosids. Tetrahedron Letters **1967**, 3221.

121. Inouye, H., S. Ueda and Y. Takeda: The Biological Conversion of Sweroside into Gentiopicroside and Vindoline and a Biogenetic Aspect of some Indole Alkaloids. Tetrahedron Letters **1968**, 3453.

122. Isquith, A. J. and A. G. Moat: Biosynthesis of Nicotinic Acid and NAD by *Clostridium butylicum*. Bacteriol. Proc. **1965**, P 6, 74.

123. — — Biosynthesis of NAD and Nicotinic Acid by *Clostridium butylicum*. Biochem. Biophys. Res. Comm. **22**, 565 (1966).

124. Iyer, R. S. R., B. Pathak and A. K. Bose: Gentianine from an Indian Medicinal Plant. Naturwiss. **43**, 251 (1956).

125. Jones, G., H. M. Fales and W. C. Wildman: The Structure of Tecomanine. Tetrahedron Letters **1963**, 397.

126. Joshi, J. G. and P. Handler: Biosynthesis of Trigonelline. J. Biol. Chem. **235**, 2981 (1960).

127. — — Metabolism of Trigonelline. J. Biol. Chem. **237**, 3185 (1962).

128. Kahn, V. and J. J. Blum: Studies of Nicotinic Acid Metabolism in Astasia longa I. Incorporation of Nicotinic Acid into Pyridine Derivatives in Exponentially Growing and in Synchronized Cultures. J. Biol. Chem. **243**, 1441 (1968).

129. — — Studies of Nicotinic Acid Metabolism in Astasia longa II. Pathway and Regulation of NAD-Biosynthesis in Cell-free Preparations. J. Biol. Chem. **243**, 1448 (1968).

130. Kanie, M., S. Fujimoto and J. W. Foster: Chemical Degradation of Dipicolinic Acid-C^{14} and its Application in Biosynthesis by *Penicillium citreoviride*. J. Bacteriol. **91**, 570 (1966).

131. Karpovich, V. N.: Phytochemical Investigation of Transbaikal Species of Gentian. Tr. Leningr. Khim.-Farmatsevt. Inst. **12**, 201 (1960) [Chem. Abstr. **57**, 12631 (1962)].

132. Kikuchi, T., S. Uyeo and T. Nishinaga: Studies on the Alkaloids of *Pachysandra terminalis* Sieb. et Zucc. (4): Structure of Epipachysamine-B, -C and Terminaline. Tetrahedron Letters **1965**, 1993.

132 a. Kimura, Y., N. Takahashi and S. Tamura: Biosynthesis of Piericidins A and B by *Streptomyces mobaraensis*. Agric. Biol. Chem. (Tokyo) **33**, 1507 (1969).

133. Kisaki, T., S. Mizusaki and E. Tamaki: Phytochemical Studies on Tobacco Alkaloids. XI. A New Alkaloid in *Nicotiana tabacum* Roots. Phytochemistry **7**, 323 (1968).

134. Koch, M.: Gentianine and Swertiamarin from *Anthocleista procera*. Trav. Lab. Matiere Med. Pharm. Galenique **50**, 94 (1965) [Chem. Abstr. **67**, 64594 (1967)].

135. Konno, K., K. Oizumi, Y. Shimizu and S. Oka: Niacin Metabolism in Mycobacteria. Amer. Rev. Respirat Dis. **91**, 383 (1965).

136. Koyama, G., Y. Iitaka, K. Maeda and H. Umezawa: The Structure of Pyridomycin. Tetrahedron Letters **1967**, 3587.

137. Kubota, T. and Y. Tomita: Derivation of Gentianin from Swertiamarin. Tetrahedron Letters **1961**, 453.

138. Kuss, E.: Identifizierung des Zwischenproduktes der Nicotinsäure-Biosynthese aus 3-Hydroxy-anthranilsäure. Hoppe Seyler's Z. physiol. Chem. **345**, 195 (1966).

139. — Identifizierung des acyclischen Spaltproduktes der Nicotinsäure-Biosynthese, II. Synthese der Diastereomeren des ε-Hydroxy-β-carboxy-norleucin. Hoppe Seylers's Z. physiol. Chem. **348**, 1589 (1967).

140. — Identifizierung des acyclischen Spaltproduktes der Nicotinsäure-Biosynthese, III. Synthese der Diastereomeren des β-Carboxy-lysins. Hoppe Seylers's Z. physiol. Chem. **348**, 1596 (1967).

154   D. Gross:

141. Kuss, E., Identifizierung des acyclischen Spaltproduktes der Nicotinsäure-Biosynthese, IV. Umwandlung des Zwischenproduktes in ε-Hydroxy-β-carboxy-
norleucin und β-Carboxylysin. Hoppe Seyler's Z. physiol. Chem. **348**, 1602
(1967).
142. Lavie, D. and R. Taylor-Smith: The Isolation of Gentianine from *Anthocleista procera (Loganiaceae)*. Chem. and Ind. **1963**, 781.
143. Leete, E.: The Biogenesis of the Pyridine Ring of Nicotine. Chem. and Ind.
1957, 1270.
144. — Biosynthesis of Alkaloids. Science **147**, 1000 (1965).
145. Leete, E.: Alkaloid Biogenesis. In: P. Bernfeld, Biogenesis of Natural
Compounds, second Ed., Pergamon Press, Oxford-London-Edinburgh-New York-
Toronto-Sydney-Paris-Braunschweig, p. 953, 1967.
145 a. — Biosynthesis of Quinine and Related Alkaloids. Accounts chem. Res. **2**,
59 (1969).
145 b. — Alkaloid Biosynthesis. Advances in Enzymol. **32**, 373 (1969).
146. Leete, E., L. Marion and I. D. Spenser: The Biogenesis of Alkaloids. XIV.
A Study of the Biosynthesis of Damascenine and Trigonelline. Canad. J. Chem.
33, 405 (1955).
147. Liang, Hsiao-Tien, Te-Chuan Yu and Feng-Yung Fu: Investigation of the
Chemical Constituents of *Gentiana macrophylla*. II. The Structure and Synthesis of Gentianidine. Acta pharm. Sinica (Yao Hsue Hsue Pao) **11**, 412
(1964) [Chem. Abstr. **62**, 5309g (1965)].
148. — — — Scientia sinica **14**, 869 (1965) Chem. Zbl. **1968** (31), 1180.
149. Liebisch, H. W.: Piperidinalkaloide. In: K. Mothes und H. R. Schütte;
Biosynthese der Alkaloide, S. 275, VEB Deutscher Verlag der Wissenschaften,
Berlin 1969.
150. Lin, Jung-Yao, and Kuo-Huang Lin: Free Amino Acids in the Seeds of
Leucaena glauca I. Isolation and Identification of Mimosine. J. Formosan
Med. Assoc. **60**, 651 (1961) [Chem. Abstr. **57**, 7577 (1962)].
151. Lingens, F.: Cinchomeronsäure, Zwischenprodukt der Nicotinsäure-Biosynthese bei Bakterien. Angew. Chem. **72**, 920 (1960).
152. Lingens, F. und P. Vollprecht: Zur Biosynthese der Nicotinsäure in Strepto-
myceten, Algen, Phycomyceten und Hefe. Hoppe Seyler's Z. physiol. Chem.
339, 64 (1964).
153. Lingens, F., P. Vollprecht und V. Gildemeister: Zur Biosynthese der
Nicotinsäure in *Xanthomonas-* und *Pseudomonas*-Arten, *Mycobacterium phlei*
und Rotalgen. Biochem. Z. **344**, 462 (1966).
153 a. Loew, P., and D. Arigoni: The Biological Conversion of Loganin into Indole
Alkaloids. Chem. Comm. **1968**, 137.
154. Luchetti, M. A.: Biosynthesis of Skytanthine *in vitro*. Ann. Ist. Super.
Sanita **1**, 563 (1965) [Chem. Abstr. **65**, 9349b (1966)].
155. Lukes, R., A. A. Arojan, J. Kovar und K. Blaha: Zur Konfiguration stick-
stoffhaltiger Verbindungen XV. Bestimmung der absoluten Konfiguration von
Anabasin und Anatabin. Collect. Czech. Chem. Comm. **27**, 751 (1962).
156. Lutfullin, K. L., P. Kh. Yuldashev and S. Y. Yunusov: Investigation of the
Alkaloids of Pedicularis olgae. Structure of Plantagonine and Indicaine. Khim.
Prir. Soedin. **1965** (5), 365 [Chem. Abstr. **64**, 3620c (1966)].
157. Macnicol, P. K.: Isolation of 6-Hydroxykynurenic Acid from the Tobacco
Leaf. Biochem. J. **107**, 473 (1968); vgl. auch Phytochem. **7**, 1779 (1968).
158. Mann, P. J. G. and W. R. Smithies: Plant Enzyme Reactions Leading to
the Formation of Heterocyclic Compounds. I. The Formation of Unsaturated
Pyrrolidine and Piperidine Compounds. Biochem. J. **61**, 89 (1955).

159. MAREKOV, N., N. MOLLOV and S. POPOV: Minor Alkaloids of *Gentiana* Species. C. R. Acad. bulg. Sci., math. natur. 18, 999 (1965).

160. MAREKOV, N. and S. POPOV: Minor Alkaloids of *Erythracea centaurium*. C. R. Acad. bulg. Sci., math. natur. 20, 441 (1967).

161. MAREKOV, N. L. and S. S. POPOV: The Structure of Gentioflavine, a New Alkaloid of Some *Gentiana* Species. Tetrahedron 24, 1323 (1968).

161 a. MARINI-BETTOLO, G. B.: Skytanthines: A New Group of Natural Alkaloids. Ann. Ist. Super Sanita 4, 489 (1968) [Chem. Abstr. 71, 91703m (1969)].

162. MARTIN, H. H. and J. W. FOSTER: Biosynthesis of Dipicolinic Acid in Spores of *Bacillus megaterium*. J. Bacteriol. 76, 167 (1958).

163. MEHLIS, B.: Zur Struktur von Clivimin und Clivonin. Naturwiss. 52, 33 (1965).

164. MEINWALD, J., G. M. HAPP, J. LABOWS and T. EISNER: Cyclopentanoid Terpen Biosynthesis in a Phasmid Insect and in Catmint. Science 151, 79 (1966).

165. MEISTER, A.: Biochemistry of the Amino Acids, Vol. 2, p. 849, sec. Edit., Academic Press, New York-London 1965.

166. MICHEL, K.-H., F. SANDBERG, F. HAGLID and T. NORIN: Alkaloids of *Haloxylon salicornicum* (Moq.-Tand.) Boiss. Acta pharm. Suecica 4, 97 (1967).

167. MILLER, E. J., G. R. MARTIN, C. E. MECCA and K. A. PIEZ: The Biosynthesis of Elastin Cross-links. J. Biol. Chem. 240, 3623 (1965).

168. MILLER, E. J., G. R. MARTIN and K. A. PIEZ: The Utilization of Lysine in the Biosynthesis of Elastin Crosslinks. Biochem. Biophys. Res. Comm. 17, 248 (1964).

169. MILLER, E. J., S. R. PINNEL, G. R. MARTIN and E. SCHIFFMANN: Investigation of the Nature of the Intermediates Involved in Desmosine Biosynthesis. Biochem. Biophys. Res. Comm. 26, 132 (1967).

170. MIURA, Y.: On an Novel Metabolite of Acid-Fast Bacilli; Produktion of 3-Pyridinemethanol by Mycobacteria. Kosankinbyo Kenkyu-Zassi 18, 133 (1966).

171. — 3-Pyridinemethanol, a Novel Metabolite of Acid-Fast Bacilli. Sci. Rep. Res. Inst. Tohoku Univ., C, 14 (3—4), 114 (1967).

172. MOAT, A. G. and J. N. ALBERTSON: Niacin Biosynthesis by *Mycobacterium tuberculosis*. Federat. Proc. (Amer. Soc. Exp. Biol.) 23, 528 (1964).

173. MOLLOV, N., N. MAREKOV, S. POPOV and B. KOUZMANOV: Alkaloids of Some *Gentiana* Species. C. R. Acad. bulg. Sci., math. natur. 18, 947 (1965).

174. MOTHES, E.: Zinkmangel und Nicotinsäure-Biosynthese bei *Mycobacterium tuberculosis* (Stamm BCG). Z. allg. Mikrobiol., Morphol., Physiol., Ökol. Mikroorganismen 4, 42 (1964).

175. MOTHES, E., D. GROSS, H. R. SCHÜTTE und K. MOTHES: Die Biosynthese der Nicotinsäure bei *Mycobacterium tuberculosis* (Stamm BCG). Naturwiss. 48, 623 (1961).

176. MOTHES, K. und H. R. SCHÜTTE: Die Biosynthese von Alkaloiden I. Angew. Chem. 75, 265 (1963).

177. MOTHES, K., H. R. SCHÜTTE, H. SIMON und F. WEYGAND: Die Bildung von Anabasin aus Cadaverin-[1,5-^{14}C] mit Hilfe von Extrakten aus Erbsenkeimlingen. Z. Naturforsch. 14b, 49 (1959).

178. MUKHAMEDZHANOV, S. Z., K. A. ASLANOV, A. S. SADYKOV, V. B. LEONTEV and V. K. KIRYUKHIN: Structure of Anabasamine. Khim. Prir. Soedin. 4, (3), 158 (1968) [Chem. Zbl. 1969 (14), 1313] [Chem. Abstr. 69, 87277e (1968)].

179. MUKHERJEE, R. and A. CHATTERJEE: Structure of Nudiflorine, a Naturally Occuring Isomer of Ricinidine. Chem. and Ind. 1964, 1524.

180. — — Structure and Synthesis of Nudiflorine, a New Pyridone Alkaloid. Tetrahedron 22, 1461 (1966).

181. Murashige, K., D. Mc Daniel and S. Chaykin: N-Methylnicotinamide Oxidation in Hog Liver. Biochim. Biophys. Acta **118**, 556 (1966).

182. Nambaru, S. and M. Asahina: Action of Pyridazine-3-carboxylic Acid, an Analogue of Nicotinic Acid, to Bacteria (I). Antagonism of Pyridazine-3-carboxylic Acid to Nicotinic Acid by *Lactobacillus arabinosus* 17-5 and *Escherichia coli* B. Vitamins (Kyoto) **34**, 527 (1966).

183. — — Action of Pyridazine-3-carboxylic Acid, an Analogue of Nicotinic Acid, to Bacteria (II). Accumulation of Quinolinic Acid in the Culture Media of *Escherichia coli* B grown in the Presence of Pyridazine-3-carboxylic Acid. Vitamins (Kyoto) **34**, 533 (1966).

184. Nishimura, S.: Fusaric Acid Production of Fusarium. Nippon Shokubutsu Byori Gakkaiho **22**, 274 (1957) [Chem. Abstr. **54**, 4771 (1960)].

185. Noguchi, M., H. Sakuma and E. Tamaki: N-(3-Amino-3-carboxypropyl)-β-carboxy-pyridinium Betain: a New Amino Acid from Tobacco Leaves. Arch. Biochem. Biophys. **125**, 1017 (1968).

186. — — — The Isolation and Identification of Nicotianine: a New Amino Acid from Tobacco Leaves. Phytochemistry **7**, 1861 (1968).

187. Notation, A. D. and I. D. Spenser: Biosynthesis of Mimosine. Incorporation of Aspartic Acid into the Pyridone Nucleus. Canad. J. Biochem. **42**, 1803 (1964).

188. O'Dell, B. L., D. F. Elsden, J. Thomas, S. M. Partridge, R. H. Smith and R. Palmer: Inhibition of the Biosynthesis of the Cross-links in Elastin by Lathyrogen. Nature **209**, 401 (1966).

189. Ogasawara, N., J. L. R. Chandler, R. K. Gholson, R. J. Rosser and A. J. Andreoli: Biosynthesis of Quinolinic Acid in a Cell-free System. Biochim. Biophys. Acta **141**, 199 (1967).

190. Ogawara, H., K. Maeda, G. Koyama, H. Naganawa and H. Umezewa: The Chemistry of Pyridomycin. Chem. Pharm. Bull. (Japan) **16**, 679 (1968).

191. Ogawara, H., K. Maeda and H. Umezawa: The Biosynthesis of Pyridomycin. I. Biochemistry **7**, 3296 (1968).

192. Ohtsu, E., A. Ichiyama, Y. Nishizuka and O. Hayaishi: Pathways of Nicotinamide Adenine Dinucleotide Biosynthesis in Nicotinic Acid or Nicotinamide Requiring Microorganisms. Biochem. Biophys. Res. Comm. **29**, 635 (1967).

193. Ooyama, J.: Biosynthesis of Dipicolinic Acid by Molds. II. Dipicolinic Acid Producers. Rept. Ferment Res. Inst. **20**, 95 (1961).

194. — Biosynthesis of Dipicolinic Acid by Molds. V. Dipicolinic Acid Producers. Rept. Ferment Res. Inst. **21**, 103 (1962).

195. Ooyama, J., N. Nakamura and O. Tanabe: Biosynthesis of Dipicolinic Acid by a *Penicillium* sp. Bull. Agr. Chem. Soc. **24**, 743 (1960).

196. — — — Biosynthesis of Dipicolinic Acid by Molds. I. Isolation and Identification of Dipicolinic Acid from Culture Filtration of a *Penicillium* sp. Rept. Ferment Res. Inst. **19**, 75 (1961).

197. Ortega, M. V. and G. M. Brown: Precursors of Nicotinic Acid in *Escherichia coli*. J. Amer. Chem. Soc. **81**, 4437 (1959).

198. — — Precursors of Nicotinic Acid in *Escherichia coli*. J. Biol. Chem. **235**, 2939 (1960).

199. Pailer, M. und R. Libiseller: Über Evonymus-Alkaloide, I. Mitt.: Zur Reindarstellung und Konstitution des Evonins (aus *Evonymus europaea* L.). Monatsh. Chem. **93**, 403 (1962).

200. Partridge, S. M., D. F. Elsden and J. Thomas: Constitution of the Cross-linkages in Elastin. Nature **197**, 1297 (1963).

201. PARTRIDGE, S. M., D. F. ELSDEN, J. THOMAS, A. DORFMAN, A. TELSER and P.-L. HO: Biosynthesis of the Desmosine and Isodesmosine Cross-Bridges in Elastin. Biochem. J. **93**, 30 C (1964).

202. — — — — — — Incorporation of Labelled Lysine into the Desmosine Cross-Bridges in Elastin. Nature **209**, 399 (1966).

203. PERNET, R., M. DUPIOL and G. COMBES: Sur la présence de Gentianine dans *Anthocleista rhizophoroides* Bak. Bull. soc. chim. France **1964**, 281.

204. PERRY, J. J. and J. W. FOSTER: Studies on the Biosynthesis of Dipicolinic Acid in Spores of *Bacillus cereus* var. *mycoides*. J. Bacteriol. **69**, 337 (1955).

205. PINNEL, S. R., G. R. MARTIN and E. J. MILLER: Desmosine Biosynthesis: Nature of Inhibition by D-Penicillamine. Science **161**, 475 (1968).

206. PLAT, M., M. KOCH, A. BOUQUET, J. LE MEN and M.-M. JANOT: Présence d'un hétéroside générateur de gentianine dans l'*Anthocleista procera* Leprieur ex-Bureau *(Loganiacées)*. Monoterpènoïdes I. Bull. soc. chim. France **1963**, 1302.

207. PONTOCORVO, G.: New Fields in the Biochemical Genetics of Microorganisms. Biochem. Soc. Symposia **4**, 40 (1950).

208. POWELL, J. F.: Isolation of Dipicolinic Acid (Pyridine-2 : 6-dicarboxylic Acid) from Spores of *Bacillus megatherium*. Biochem. J. **54**, 210 (1953).

209. POWELL, J. F. and R. E. STRANGE: Synthesis of Dipicolinic Acid from α, ε-Diketopimelic Acid. Nature **184**, 878 (1959).

210. QUINN, G. P. and P. GREENGARD: The Pathway for the Biosynthesis of N^1-Methyl-4-pyridone-3-carboxamide. Arch. Biochem. Biophys. **115**, 146 (1966).

211. RAI, J. and K. A. THAKAR: Chemical Investigation of *Enicostemma littorale*, Blume. Current Sci. (India) **35**, 148 (1966).

212. RAKHMATULLAEV, T. U., S. T. AKRAMOV and S. Y. YUNUSOV: Alkaloids from *Swertia marginata*, *S. graciliflora*, and *Dipsacus azureus*. Khim. Prir. Soedin. **5** (1), 64 (1969) [Chem. Abstr. **70**, 112376 (1969)].

212a. — — — *Gentiana* Alkaloids. Khim. Prir. Soedin. **5** (1), 32 (1969) [Chem. Abstr. **71**, 13247b (1969)].

213. RAO, K. V., K. BIEMANN and R. B. WOODWARD: The Structure of Streptonigrin. J. Amer. Chem. Soc. **85**, 2532 (1963).

214. RAPOPORT, H., The Biosynthesis of the Pyridine and Piperidine Alkaloids. The Tobacco Alkaloids. Abh. dtsch. Akad. Wiss. Berlin, Klasse Chem., Geol. Biol. **1966** (3), 111.

215. RAY, A. B. and A. CHATTERJEE: Venoterpine, a New Monoterpenoid Alkaloid from the Fruits of *Alstonia venenata* R. Br. Tetrahedron Letters **1968**, 2763.

216. REGNIER, F. E., G. R. WALLER, E. J. EISENBRAUN and H. AUDA: The Biosynthesis of Methylcyclopentane Monoterpenoids. II. Nepetalactone. Phytochem. **7**, 221 (1968).

217. RIO-ESTRADA, C. DEL and H. PATINO: Nicotinic Acid Biosynthesis by *M. tuberculosis*. J. Bacteriol. **84**, 871 (1962).

217a. RULKO, F.: Alkaloids of Gentianaceae. V. Alkaloids of *Menyanthes trifoliata* L. Rocz. Chem. (Ann. Soc. chim. polonorum) **43**, 1831 (1969) [Chem. Abstr. **72**, 67158n (1970)].

218. RULKO, F., L. DOLEJS, A. D. CROSS, J. W. MURPHY and T. P. TOUBE: The Structure of Alkaloid from *Gentiana tibetica* King. Rozniki Chem. (Ann. Soc. chim. polonorum) **41**, 567 (1967) [Chem. Abstr. **67**, 117012e (1967)].

219. SADYKOV, A. S., S. Z. MUKHAMEDZHANOV and K. A. ASLANOV: Structure of Anabasamine, a New Base from Anabasis aphylla Seeds. Doklady Akad. Nauk (UzbSSR) **24**, 34 (1967) [Chem. Abstr. **68**, 78473e (1968)].

220. Sakan, T., A. Fujino and F. Murai: Chemische Untersuchung der Bestandteile von *Actinidia polygama* (Miq.). I. Isolierung von Matatabilacton und Actinidin. J. Chem. Soc. Japan **81**, 1320 (1960).

221. Sakan, T., A. Fujino, F. Murai, Y. Butsugan and A. Suzui: On the Structure of Actinidine and Matatabilactone, the Effective Components of *Actinidia polygama*. Bull. Chem. Soc. Japan **32**, 315 (1959).

222. Sakan, T., A. Fujino, F. Murai, A. Suzui, Y. Butsugan and Y. Terashima: The Absolute Structure of Actinidine. Bull. Chem. Soc. Japan **33**, 712 (1960).

223. Sakan, T., F. Murai, Y. Hayashi, Y. Honda, T. Shono, M. Nakajima and M. Kato: Structure and Stereochemistry of Boschniakine, Boschnialactone, and Boschnialinic Acid, an Oxidation Product of Boschnialactone. Tetrahedron **23**, 4635 (1967).

224. Samatov, A., S. T. Akramov ans S. Y. Yunusov: Alkaloids of Gentiana. The Structure of Gentianadine and Gentianamine. Khim. Prir. Soedin. **3**, 182 (1967) [Chem. Abstr. **67**, 117007g (1967)].

225. Sargazakov, D. S.: The Alkaloids of *Gentiana tianschanica*. Izvest. Akad. Nauk. Kirgiz. SSR. Ser. Estestren. i. Tekh. Nauk. **2** (5), 103 (1960) [Chem. Abstr. **55**, 2020 (1961)].

226. Saxton, R. E., V. Rocha, R. J. Rosser, A. J. Andreoli, M. Shimoyama, A. Kosaka, J. L. R. Chandler and R. K. Gholson: A Comparative Study of the Regulation of NAD Biosynthesis. Biochim. Biophys. Acta **156**, 77 (1968).

227. Schöpf, Cl., F. Braun und A. Komzak: Der Übergang von Δ^1-Piperidein in Tetrahydro-anabasin unter zellmöglichen Bedingungen. Chem. Ber. **89**, 1821 (1956).

228. Schütte, H. R. und G. Siegel, unveröffentlichte Versuche.

229. Scott, T. A., E. Bellion and M. Mattey: The Synthesis of Nicotinic Acid from N-Formyl-L-aspartate by Extracts of *Clostridium butylicum*. Biochem. J. **107**, 23 P (1968).

229a. — — — The Conversion of N-Formyl-L-aspartate into Nicotinic Acid by Extracts of *Clostridium butylicum*. European J. Biochem. **10**, 318 (1969).

230. Scott, T. A. and J. P. Glynn: The Incorporation of [2,3,7-^{14}C] Nicotinic Acid into Nicotine by *Nicotiana tabacum*. Phytochem. **6**, 505 (1967).

231. Scott, T. A. and H. Hussey: The Biosynthesis of Nicotinic Acid in *Serratia marcescens*. Biochem. J. **96**, 9 C (1965).

232. Scott, T. A. and M. Mattey: The Incorporation of Formate into Nicotinic Acid by *Clostridium butylicum*. Biochem. J. **107**, 606 (1968).

233. Shibata, S., M. Fujita and H. Igeta: Detection and Isolation of an Alkaloid Gentianine from Japanese Gentianaceous Plants. J. Pharm. Soc. Japan **77**, 116 (1957) [Chem. Abstr. **51**, 6089 (1957)].

234. Shiman, R. and J. B. Neilands: Isolation, Characterization, and Synthesis of Pyrimine, an Iron (II)-Bindung Agent from *Pseudomonas* GH. Biochemistry **4**, 2233 (1965).

235. Smith, H. H. and D. V. Abashian: Chromatographic Investigations on the Alkaloids Content of *Nicotiana* Species and Interspecific Combinations. Amer. J. Bot. **50**, 435 (1963).

236. Spenser, I. D. and A. D. Notation: A Synthesis of Mimosine. Canad. J. Chem. **40**, 1374 (1962).

237. Stoll, C.: Über Stoffwechsel und biologisch wirksame Stoffe von *Gibberella fujikuroi* (Saw.) Woll., dem Erreger der Bakanaekrankheit. Phytopatholog. Z. **22**, 233 (1954) [Chem. Zbl. **127**, 2788 (1956)].

238. Stoll, C., J. Renz und E. Gäumann: Über die Bildung von Furarinsäure und Dehydrofusarinsäure durch das *Fusarium lycopersici* Sacc. in saprophytischer Kultur. Phytopathol. Z. **29**, 388 (1957) [Chem. Zbl. **129**, 763 (1958)].

238a. Streeter, P. M., G. Adolphen and H. H. Appel: β-Skytanthine N-Oxide in *Skytanthus acutus* Meyen. Chem. and Ind. **1969**, 1631.

239. Suomalainen, H.: Changes in the Cell Constituents of Bakers' Yeast in Changing Growth Conditions. Pure Appl. Chem. **7**, 639 (1963) [Chem. Abstr. **60**, 3457c (1964)].

240. Suomalainen, H., A. Bjorklund, K. Vihervaara and E. Oura: Nicotinic Acid and NAD Contents of Bakers' Yeast in Changing Culture Conditions. J. Inst. Brewing **71**, 221 (1965) [Chem. Abstr. **63**, 6282 (1965)].

241. Suomalainen, H., T. Nurminen, K. Vihervaara and E. Oura: Effect of Aeration on the Synthesis of Nicotinic Acid and NAD by Bakers' Yeast. J. Inst. Brewing **71**, 227 (1965) [Chem. Abstr. **63**, 6283a (1965)].

242. Suzuki, A., N. Takahashi and S. Tamura: Chemical Structure of Piericidin A. Part IV. Structural Confirmation for Pyridine Ring in Piericidin A through Synthesis. Agric. Biol. Chem. (Tokyo) **30**, 13 (1966).

243. — — — Chemical Structure of Piericidin A. Part V. Mass Spectrometrical Confirmation for the Side Chain Structure of Piericidin A. Agric. Biol. Chem. (Tokyo) **30**, 18 (1966).

244. Takahashi, N., Y. Kimura and S. Tamura: Biosynthesis of Piericidins A and B. Tetrahedron Letters **1968**, 4659.

245. Takahashi, N., A. Suzuki, Y. Kimura, S. Miyamoto and S. Tamura: Structure of Piericidin B and Stereochemistry of Piericidins. Tetrahedron Letters **1967**, 1961.

246. Takahashi, N., A. Suzuki, Y. Kimura, S. Miyamoto, S. Tamura, T. Mitsui and J. Fukami: Isolation, Structure and Physiological Activities of Piericidin B, Natural Insecticide Produced by a *Streptomyces*. Agric. Biol. Chem. (Tokyo) **32**, 1115 (1968).

247. Takahashi, N., A. Suzuki, S. Miyamoto, R. Mori and S. Tamura: Chemical Structure of Piericidin A. Part I. Functional Groups. Agric. Biol. Chem. (Tokyo) **27**, 583 (1963).

248. Takahashi, N., A. Suzuki and S. Tamura: Structure of Piericidin A. J. Amer. Chem. Soc. **87**, 2066 (1965).

249. — –— — Chemical Structure of Piericidin A. Part III. Structure of Piericidin A and Octahydropiericidin A. Agric. Biol. Chem. (Tokyo) **30**, 1 (1966).

250. Takahashi, N., S. Yoshida, A. Suzuki and S. Tamura: Chemical Structure of Piericidin A. Part VI. Stereochemistry. Agric. Biol. Chem. (Tokyo) **32**, 1108 (1968).

251. Tanenbaum, S. W. and K. Kaneko: Biosynthesis of Dipicolinic Acid and of Lysine in *Penicillium citreo-viride*. Biochemistry **3**, 1314 (1964).

252. Taylor-Smith, R. E.: Investigations on Plant of West Africa. II. Isolation of Anthocleistin from *Anthocleista procera (Loganiaceae)*. Tetrahedron **21**, 3721 (1965).

253. Terashima, T., Y. Kuroda and Y. Kaneko: Studies on a New Alkaloid of *Streptomyces*. Structure of Nigrifactin. Tetrahedron Letters **1969**, 2535.

253a. Thies, P. W.: Die Konstitution der Valepotriate. Mitteilung über die Wirkstoffe des Baldrians. Tetrahedron **24**, 313 (1968).

254. Thomas, J., D. F. Elsden and S. M. Partridge: Partial Structures of Two Major Degradation Products from the Cross-linkages in Elastin. Nature **200**, 651 (1963).

255. Tiwari, H. P., W. R. Penrose and I. D. Spenser: Biosynthesis of Mimosine: Incorporation of Serine and α-Aminoadipic Acid. Phytochemistry 6, 1245 (1967).

256. Tiwari, H. P. and I. D. Spenser: Precursors of Mimosine in Mimosa pudica. Canad. J. Biochem. 43, 1687 (1965).

257. Torssell, K.: The Structures of Alkaloids form Pedicularis olgae Regel (Scrophulariaceae) and Plantago indica (P. ramosa, Plantaginaceae). Acta Chem. Scand. 22, 2715 (1968).

258. Torssell, K. and K. Wahlberg: The Structure of the Principal Alkaloids from Valeriana officinalis L. Tetrahedron Letters 1966, 445.

259. — — Isolation, Structure and Synthesis of Alkaloids from Valeriana officinalis L. Acta Chem. Scand. 21, 53 (1967).

260. Tschesche, R., D. Klöden und H. W. Fehlhaber: Über die Alkaloide aus Lobelia syphilitica L. II. Syphilobin A und Syphilobin F. Tetrahedron 20, 2885 (1964).

261. Tyler, V. E. and A. E. Schwarting: The Culture of Claciceps purpurea III. Tryptophan Metabolism. J. Amer. Pharm. Assoc. Sci. Edit. 43, 207 (1954).

262. Ubaev, K., P. Kh. Yuldashev and S. Y. Yunusov: Pedicularis olgae Alkaloids. Uzbeksk. Khim. Zhurn. 7 (3), 33 (1963) [Chem. Abstr. 59, 15602a (1963)].

263. Vazquez, D.: The Streptogramin Family of Antibiotics. In: D. Gottlieb and P. D. Shaw: Antibiotics, Vol. I, p. 387, Springer-Verlag Berlin-Heidelberg-New York 1967.

264. Waller, G. R. and L. M. Henderson: Biosynthesis of the Pyridine Ring of Ricinine. J. Biol. Chem. 236, 1186 (1961).

265. Waller, G. R., K. S. Yang, R. K. Gholson and L. A. Hadwiger: The Pyridine Nucleotide Cycle and its Role in the Biosynthesis of Ricinine by Ricinus communis. J. Biol. Chem. 241, 4411 (1966).

266. Wan, A. S. C. and Y. L. Chow: Alkaloids of Fagraea fragrans Roxb. J. Pharm. Pharmacol. 16, 484 (1964).

267. Weber, F. und O. Wiss: Am Tryptophan-Stoffwechsel beteiligte Enzyme. In: Hoppe-Seyler/Tierfelder: Handbuch der physiologisch- und pathologisch-chemischen Analyse, Band 6, S. 806, Springer-Verlag Berlin-Heidelberg-New York 1966.

268. Wibaut, J. P.: Some Recent Development in Pyridine Chemistry. Progress in Organic Chemistry 2, 179 (1953).

269. Wilkinson, S.: α-Picoline from Rumex obtusifolius L. Nature 181, 636 (1958).

270. Willaman, J. J. and H. L. Li: General Relationships among Plants and their Alkaloids. Econom. Bot. 17, 180 (1963).

271. Wilson, R. G. and L. M. Henderson: Tryptophan-Niacin Relationship in Xanthomonas pruni. J. Bacteriol. 85, 221 (1963).

272. Yabuta, T., K. Kambe and T. Hayaishi: Biochemistry of Bakanae-Fungus. I. Fusarinic Acid, a New Product of the Bakanae-Fungus. J. Agr. Chem. Soc. Japan 10, 1059 (1934) [Chem. Abstr. 29, 1132 (1937)].

273. Yang, K. S., R. K. Gholson and G. R. Waller: Studies on Nicotine Biosynthesis. J. Amer. Chem. Soc. 87, 4184 (1965).

274. Yanofsky, C.: The Absence of a Tryptophan-Niacin Relationship in E. coli and Bacillus subtilis. J. Bacteriol. 68, 577 (1954).

275. — Tryptophan and Niacin Synthesis in Various Organisms. In: W. D. McElroy and H. B. Glass: The Amino Acid Metabolism, p. 930, The Johns Hopkins Press, Baltimore 1955.

276. Yasumatsu, N.: Studies on the Chemical Regulation of Alkaloid Biosynthesis in Tobacco Plants. Part II. Inhibition of Alkaloid Biosynthesis by Exogenous Auxins. Agric. Biol. Chem. (Tokyo) **31**, 1441 (1967).

277. Yasumatsu, N., A. Sakurai and S. Tamura: Studies on the Chemical Regulation of Alkaloid Biosynthesis in Tobacco Plants. Part. I. Inhibition of Nicotine Biosynthesis by Helminthosporol and Helminthosporic Acid. Agric. Biol. Chem. (Tokyo) **31**, 1061 (1967).

278. Yeowell, D. A. und H. Schmid: Zur Biosynthese des Plumierids. Experientia **20**, 250 (1964).

279. Zeijlemaker, F. C. J.: The Metabolism of Nicotinic Acid in the Green Pea and its Connection with Trigonelline. Acta Bot. Neerl. **2**, 123 (1953).

280. Zureck, A.: 3-Hydroxymethyl-Pyridin im Stoffwechsel von Mycobakterien. Zbl. Bakteriol., Parasitenkunde Infektionskrankheiten Abt. I., Orig. **206**, 522 (1968).

(Eingelaufen am 1. Oktober 1969)

Peptide Alkaloids

By E. W. WARNHOFF, London, Canada

With 6 Figures

Contents

I. Introduction

The term *peptide alkaloids*, first proposed by Goutarel and co-workers (*32*) in 1964, has been applied to a rapidly expanding group of closely related polyamide plant bases composed largely of simple amino acids. Although these compounds have also been called *basic peptides* (*12*), it seems generally accepted that the emphasis of the first term is more appropriate in view of their relatively small size and the fact that each incorporates a fragment which is not (or at least is no longer) a common amino acid (*61*). For the purpose of the present review a peptide alkaloid is defined as a *plant* base containing two or more amide groups. By this definition are included certain polyamide bases (*e. g.* lunarine, homaline) whose amino acid origin is less obvious*. This review follows the suggestion of Goutarel and co-workers (*32*) that, in spite of their polyamide nature, the ergot alkaloids logically belong with the indole alkaloids because of the importance of lysergic acid in their structures. Moreover, the basic lysergic acid portion is distant from the polyamide part and has usually been cleaved from it during chemical investigation. Therefore, these compounds are not treated here. The peptide compounds isolated by Wieland and co-workers from *Amanita* species are also excluded since most of them are not basic and they have been reviewed recently (*58*).

Peptide alkaloids have so far been found in fifteen plants from seven plant families (Table 3). The amounts of alkaloid present vary from *ca.* 0.02 to 0.9% of the dried plant weight and are distributed throughout the plant although the seeds and roots appear to be the richest parts from the limited information available. The presence of alkaloidal material in *Ceanothus americanus* (*8, 10, 11*) and *Lunaria biennis* (*14*) had long been known, and occasional investigations into the bases of *Ceanothus americanus* and/or *Ceanothus velutinus* were made earlier (*1, 6, 7, 23, 40, 41*). These studies showed that the ceanothus alkaloid fraction was a complex mixture of closely related compounds, as has since been found to be true of the alkaloid fraction of most of the other plants examined. Little real progress in the chemistry of these compounds was possible until the development of column and thin layer chromatography, nuclear magnetic resonance and organic mass spectroscopy. Once chromatographic techniques permitted solution of the separation and purification problems, the structures of the first few bases were determined by a combination of chemical degradation and application of spectroscopic data. More recently, automatic amino acid and mass spectroscopic analysis of these alkaloids has been developed to the point where the structures of new bases can often be deduced from these measurements alone.

---

* On the other hand the definition excludes compounds such as pithecolobine (*59, 60*) which bears a resemblance to homaline.

II. General Properties of Peptide Alkaloids

The alkaloids belong to three basically different types: the *p*-aryloxy-macrocyclic peptides (1), the lunarine group (2), and the homaline type (3). The first class is by far the largest and of the thirty five compounds listed in Table 2, twenty seven have the basic formula (1) or a ring opened modification of it. There are seven members of the lunarine type (2), and homaline is so far the only representative of structure (3).

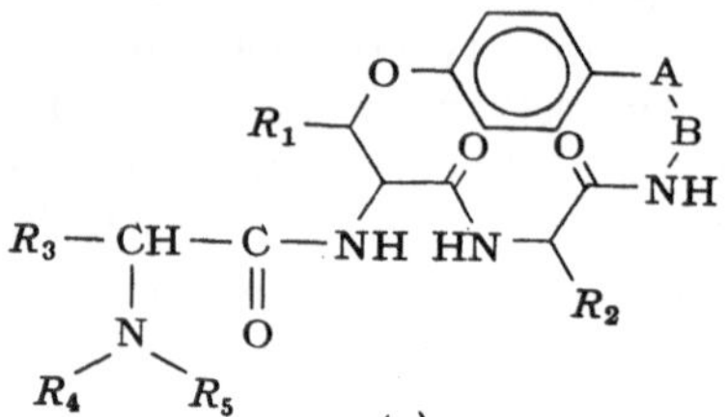

A—B = CH=CH

$$\begin{array}{c} CH-CH_2 \\ | \\ OH \end{array}$$

$$\begin{array}{c} C-CH_2 \\ \parallel \\ O \end{array}$$

R_1, R_2, R_3 = amino acid residues
R_4, R_5 = H, Me, or another amino acid

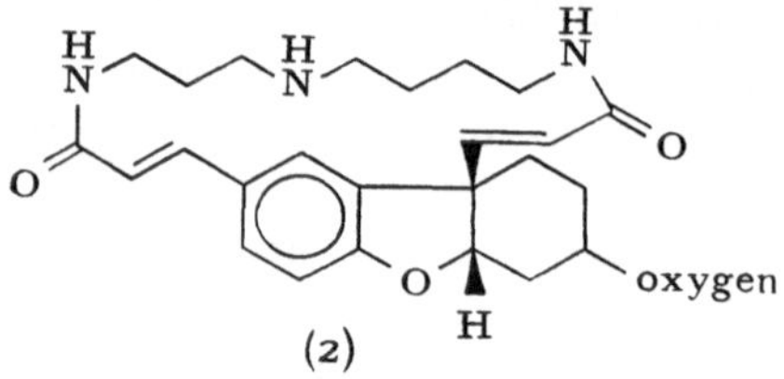

(1)

(2)

or

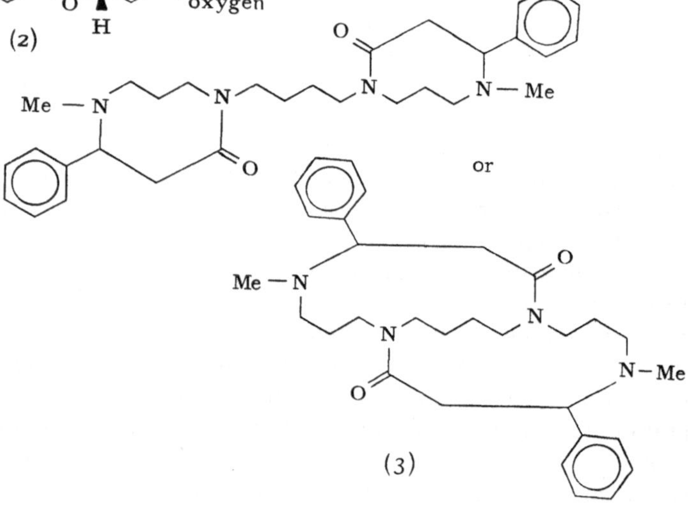

(3)

The *p*-aryloxymacrocyclic alkaloids (1) are secondary or tertiary amines composed of three or four simple amino acids plus a modified tyramine unit, and they generally have the high melting points (most > 200°) expected of polyamides. The amino acids found so far are leucine,

isoleucine, valine, phenylalanine, proline, tryptophan and their N-methy-
lated and β-hydroxylated derivatives. All of the aminoacids whose
stereochemistry has been determined belong to the L-series with the single
exception of D-*threo*-phenylserine in lasiodine-A. The β-hydroxyamino
acids have all been found to belong to the *threo* series so far. The few
alkaloids of this class whose basicity has been determined exhibit the
expected base weakening effect of the adjacent carboxamido group
(Table 1). Most members of this class are styrylamides [(1), CH=CH for
A–B] whose conjugated double bond is rapidly reduced by hydrogen in
the presence of palladium. The dihydrocompounds [(1), CH_2—CH_2 for
A–B] usually melt 20–80° higher than the parent alkaloid and are *more*
polar than the parent alkaloid on thin layer chromatography. These
alkaloids generally have large negative optical rotations ($[\alpha]_D \sim -200°$
to $-400°$), much of which results from the inherently dissymmetric
styrene chromophore since they undergo a large positive shift ($\Delta[\alpha]_D \sim$
$+100°$ to $+250°$) on hydrogenation of the double bond. The large
negative rotations of the bases of this group are in contrast to the large
positive rotations (or else no rotation) of the lunarine alkaloids.

The lunarine and homaline alkaloids have in common that both types
are amides of long chain polyamines and β-phenylpropionic acid derivatives.

III. General Methods and Techniques
of Structural Determination

The crude alkaloid mixtures have been separated by column and/or
thick layer chromatography on silicic acid, alumina, Florisil and Kieselgel.
Repeated chromatography with different solvent systems is often required to separate the mixtures of closely related compounds and to
isolate pure alkaloids. The best means of assessing purity has been tlc
behaviour in several solvent systems in conjunction with the melting
point and NMR spectrum.

UV spectra and the changes which occur in them on hydrogenation
have been very valuable in structural work with these compounds since
they permit ready recognition of the styrylamide, *p*-alkoxyphenyl-
carbonyl, tryptophan and cinnamamide chromophores, and they also
reveal whether the styrylamide is part of a macrocyclic ring or not (see
Section V). IR spectra have been chiefly useful for identifying carbonyl
functions and for determining the number of amide carbonyl groups by
integration of these peaks. Although there are occasional solubility
problems, particularly with the dihydro aryloxymacrocyclic bases, NMR
spectroscopy has been very valuable as a criterion of purity as well as a
source of structural information. The number of methyl groups on the

basic nitrogen can be determined by their large downfield shift in acid solution (CF_3COOH or CD_3COOD) (*22*).

Hydrolysis has been the most generally useful chemical technique, particularly the comparison of the hydrolysis products of both the alkaloid and its hydrogenated derivative. Acid hydrolysis of alkaloids containing the styrylamide unit destroys this group with the formation of reddish purple insoluble material whereas the hydrogenated alkaloids lacking the styrylamide group remain colorless on acid hydrolysis and give β-phenylethylamine derivatives. Hydrolysis is not without complications since amino acids (*e. g.* leucine, glycine) or other artifacts (*e. g.* β-phenyl-naphthalene) not present in the alkaloid may be formed during the hydrolysis. Moreover, even though the amino acid analysis may be automated, there still remains the problem of putting Humpty Dumpty together again. Ozonization of the styrylamide group has been helpful in determining the relative orientation of the carbon chain and the oxygen substituents on the aromatic ring.

Finally, high resolution mass spectrometry has become the most powerful tool for structural work with these alkaloids because of the characteristic way in which peptides break down on ionization (*9, 16, 17, 45*) (see Section VII). Even in the earlier work on the first members of the peptide alkaloid family before these breakdown patterns were established, low resolution mass spectra served usefully for determination of molecular weight and the basic terminal end (4) of the alkaloid whose cleavage always gives by far the strongest peak in the mass spectrum at m/e corresponding to (5). In the future high resolution mass spectroscopy will probably be the exclusive technique for structure determination of peptide alkaloids which will become routine unless and until new types of peptide alkaloids are found.

$$
\begin{array}{ccc}
 & \overset{\displaystyle O}{\overset{\displaystyle \|}{R_1\!-\!CH\!-\!C\!-\!NH\!-\!}} & \\
 & \underset{\displaystyle R_2 \quad\; R_3}{\overset{\displaystyle |}{N}} & \\
 & (4) &
\end{array}
\quad \rightarrow \quad
\begin{array}{c}
R_1\!-\!CH \\
\| \\
\underset{R_2 \quad\; R_3}{N^{\oplus}} \\
(5)
\end{array}
$$

In the discussion of individual alkaloids which follows, the first seven are covered in sufficient detail to provide complete evidence for the structure, both in order to give the flavor of the chemistry and to illustrate the transition that has taken place from chemical to mass spectroscopic methods of structure determination of the aryloxymacrocyclic bases. Moreover, much of the organic chemistry of these bases was developed in the work on the first four compounds which took place in different laboratories at about the same time. For the remaining alkaloids, all of which are listed in Table 2, only those with special points of interest are mentioned.

IV. Structure and Properties of Peptide Alkaloids

1. Pandamine (6 a)

Pandamine, $C_{31}H_{44}N_4O_5$, m. p. 256°, $[\alpha]_D - 103°$ ($CHCl_3$), was the first of the p-aryloxymacrocyclic alkaloids to have its structure determined (27, 32), chiefly by chemical methods. The base could be crystallized from the crude alkaloid fraction of *Panda oleosa*, and final purification was achieved by chromatography of the O-acetate on silicic acid. The mono-basic compound was a tertiary amine since only an O-acetate was formed with acetic anhydride pyridine. Saponification regenerated the alkaloid. The large downfield NMR shift of two methyl groups in acid solution $[\delta^{CDCl_3} 2.15 \rightarrow \delta^{CF_3COOH} 3.10$ ($J \sim 3$ Hz)] meant that the basic nitrogen atom was an N,N-dimethylamino group. The action of chromic acid-acetone reagent on pandamine produced a conjugated ketone whose UV spectrum was that of a p-alkoxyaromatic ketone. Alkaline hydro-lysis of pandamine in refluxing methanolic sodium hydroxide gave a basic, an acidic, and an amphoteric fraction. The basic fraction consisted of N,N-dimethyl-L-isoleucinamide (7) identical with a synthetic specimen. The amphoteric fraction was a mixture of racemized phenylalanine (8), racemized leucine (9), 2,p-dihydroxyphenylethylamine (10) and 2-meth-oxy-p-hydroxyphenylethylamine (11). The first four of the five hydrolysis products accounted for all of the carbon atoms of the original alkaloid.

Acid hydrolysis of pandamine in aqueous hydrochloric-acetic acid solution revealed more information about the linking of the pieces. Chromatography of the amphoteric products on an ion exchange column yielded L-phenylalanine and a fragment, $C_{20}H_{32}N_2O_4$, proved to be (12) by a double irradiation study of the NMR spectrum of its methyl ester. The sum of these two cleavage products has two fewer carbon atoms than the original alkaloid, presumbly the two carbon atoms of the phenyl-ethyl group. Incorporation of these two carbon atoms and reconstitution of the alkaloid structure provides the unique structure (6 a).

Further work confirmed this structural proposal (27). Desoxy-pandamine (6 b) was prepared by palladium catalyzed hydrogenolysis of chloropandamine (6 c) obtained from the action of hydrochloric-acetic acid mixtures on pandamine. Acid hydrolysis of desoxypandamine (6 b) under the conditions employed for pandamine now yielded the amino acid (13) containing the two carbon atoms lost in the hydrolysis of pandamine.

The optical rotations of the phenylalanine and N,N-dimethylleucine obtained from acid hydrolysis permitted assignment of the L-stereochemistry to these acids. The hydroxyl configuration was determined by Horeau's phenylbutyric anhydride method. However, the stereochemistry at the two asymmetric carbon atoms of the β-hydroxyleucine was not decided since the free amino acid did not survive either the acidic or the basic hydrolytic conditions used.

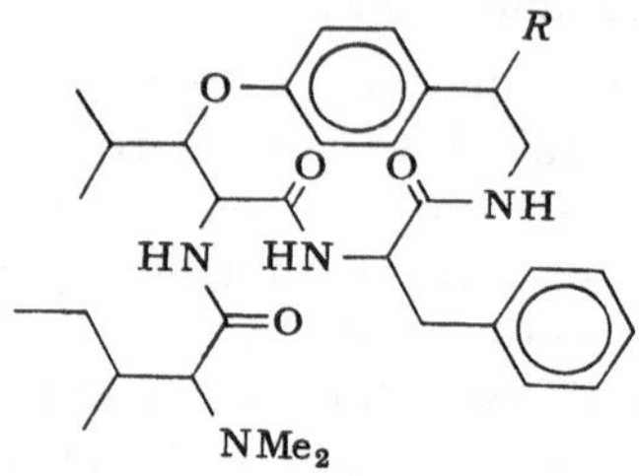

a $R = OH$ Pandamine
b $R = H$ Desoxypandamine
c $R = Cl$
d $R = =O$ Pandaminone

(6)

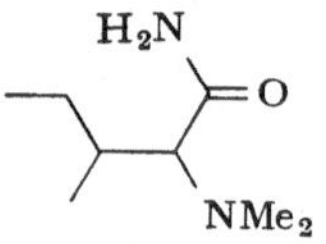

(7)

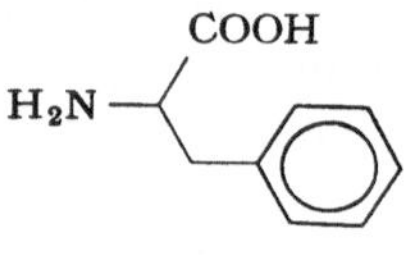

(8)

Phenylalanine

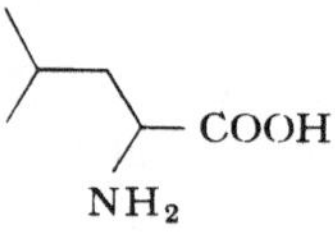

(9)

Leucine

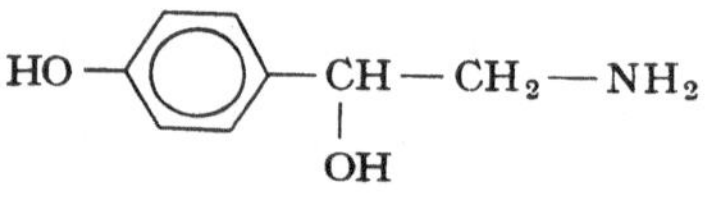

(10)

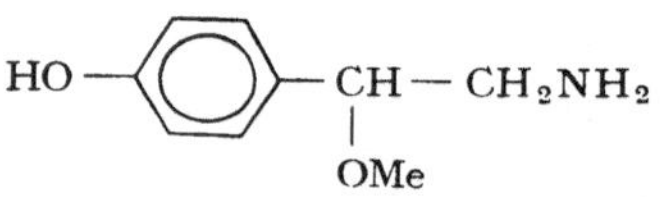

(11)

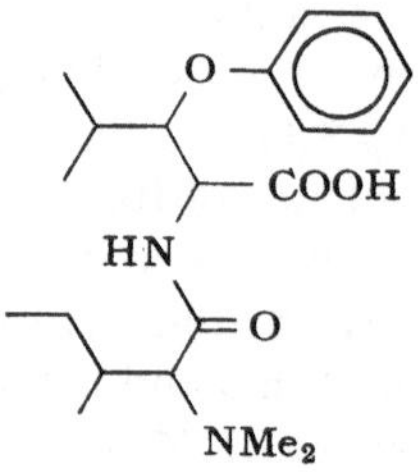

(12)

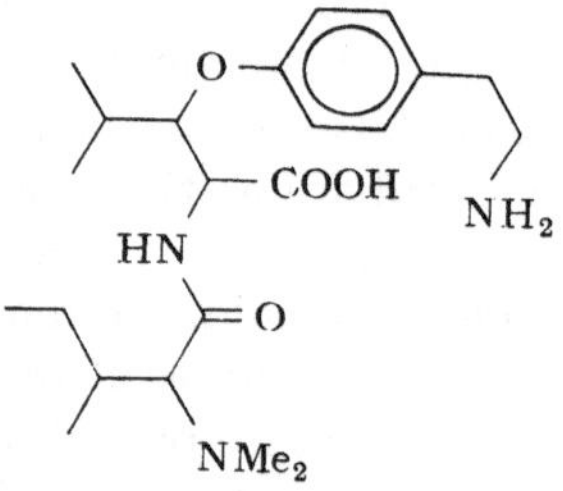

(13)

Although the structural study was now complete, there were three anomalous points of chemical interest to be rationalized: the formation of the methoxy amine (11), the formation of the monosubstituted phenyl ring of (12), and the formation of leucine (9). The methoxyl group is probably introduced by conjugate addition of methoxide to the quinone methine (**14**) derived by elimination of the benzylic hydroxyl from the phenolate anion (**15**). The corresponding model hydroxyaminophenol (**10**) undergoes the same reaction.

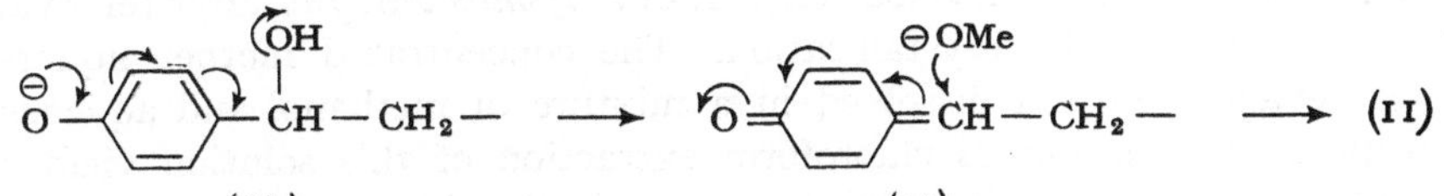

(15)   (14)   →   (11)

The rupture of the bond between the aromatic ring and the benzylic carbon atom during acid hydrolysis is probably a reverse aldehyde-phenol ether condensation (**16**) as outlined below:

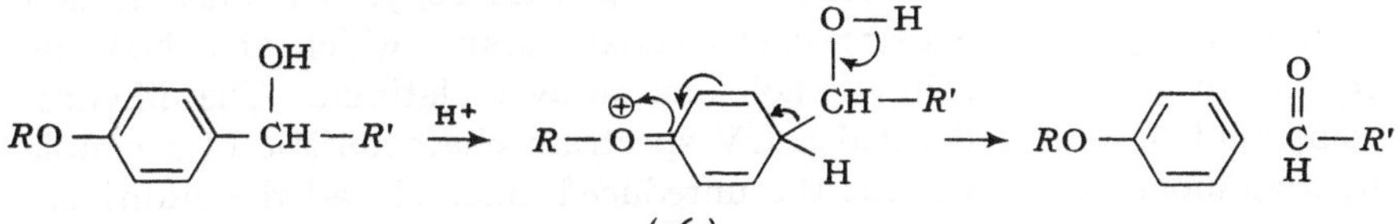

(16)

The leucine, which is not a constituent of the alkaloid, must be formed from β-hydroxyleucine, probably by hydrolysis to α-keto isocaproic acid and subsequent transamination [(**17**) → (**9**) or its equivalent]. In agreement with this idea, the alkaline hydrolysis of (**12**) does not give leucine unless phenylalanine is added to the reaction medium. The formation of leucine in this way has been observed with other alkaloids of this series (*e. g.* scutianine, frangulanine, americine, ceanothine-B).

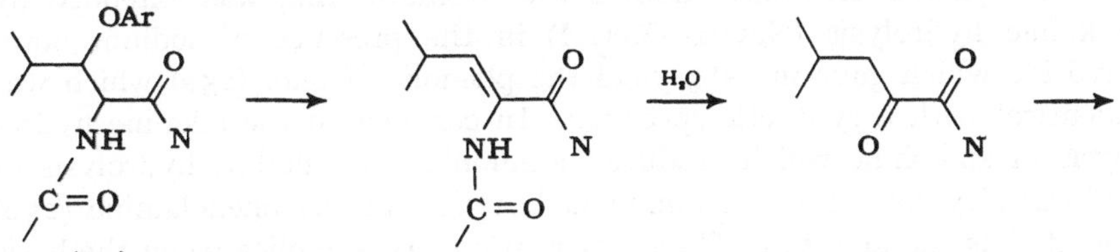

(17)

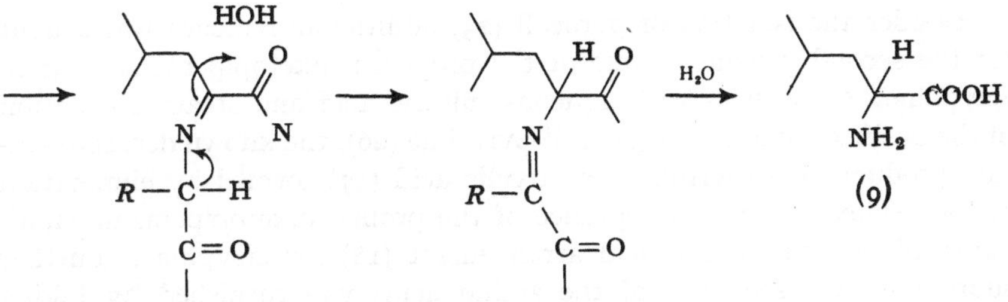

(9)

2. Zizyphine (18)

Zizyphine, $C_{33}H_{49}N_5O_6$, m. p. 121° dec., $[\alpha]_D - 464°$ (CHCl$_3$), was the second peptide alkaloid to have its structure determined (*61*). It was isolated (*26*) from the alcoholic extract of *Zizyphus oenoplia* after removal of betulinic acid by crystallization. The concentrated mother liquors were evaporated and dissolved in a mixture of methanol and aqueous acetic acid. Continuous chloroform extraction of this solution yielded crude zizyphine and zizyphinine acetates in the chloroform extract. The crude alkaloid was purified by precipitation as the oxalate, regeneration and chromatography on alumina (*26*).

The structure of zizyphine was worked out very systematically by classical chemical degradation in conjunction with IR and UV spectra, and then confirmed by NMR and mass spectroscopy. The alkaloid is a weakly monobasic (pKa 5.77–6.23) tertiary amine which absorbed one equivalent of hydrogen in ethanol solution over platinum. The dihydro-compound, $[\alpha]_D - 188°$, had a UV spectrum characteristic of a simple hydroquinone ether, whereas the unreduced alkaloid had the additional conjugation of a styrenoid chromophore. Hydrochloric acid hydrolysis gave L-proline (**19**) and L-isoleucine (**20**). Shorter hydrolysis time yielded proline and a larger fragment, L-(N,N-dimethylisoleucyl)-L-isoleucine (**21**), isolated as its crystalline ethyl ester. The identity of this fragment was confirmed by hydrolysis and synthesis. Destructive distillation of zizyphine at 260–330° produced 3-methyl-4-methoxyphenol (**22**) and pyrocoll (**23**). The phenol was oxidized to *p*-toluquinone (**24**) by ferric chloride and was found to be different from an authentic sample of 2-methyl-4-methoxyphenol.

The partial structure around the aromatic ring was extended by alkaline hydrolysis (NaOH—MeOH) in the presence of sodium boro-hydride which gave in 65% yield the phenolic alcohol (**25a**) which was identical with a synthetic specimen. In contrast to the alkaline hydro-lysis of zizyphine which produced ammonia, the alkaline hydrolysis of dihydrozizyphine gave no ammonia but instead the phenolamine (**25b**) isolated as the diacetate. These observations were explicable on the basis of the styrylamide chromophore suspected from the UV data.

Besides the isolation of pyrocoll (**23**) additional evidence was sought for the second proline residue of the proposed diketopiperazine unit in zizyphine. Hydrolysis with aqueous sulfuric acid and steam distillation of the basified hydrolysate gave Δ^1-pyrroline (**26**), the known decarboxyla-tion product of Δ^1-pyrroline carboxylic acid (**27**) formed by elimination of the aryloxy group. Acceptance of the proline diketopiperazine elimi-nates all but one amino acid arrangement (**18**) for zizyphine. Further support for the sequence of the amino acids was furnished by taking advantage of the known periodate cleavage of 1,2-diamines provided neither is tertiary. Lithium aluminium hydride reduction of dihydro-zizyphine yielded a polyamine, $C_{33}H_{59}N_5O_2$ (28), with no amide groups

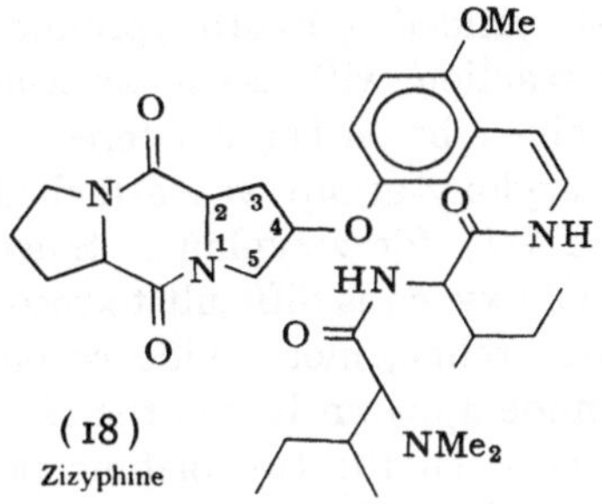

(18)
Zizyphine

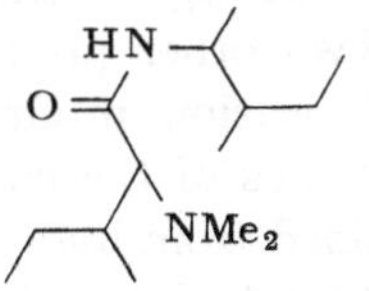

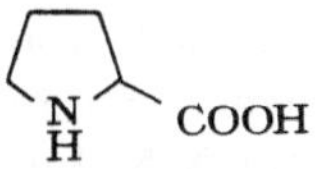

(19)
Proline

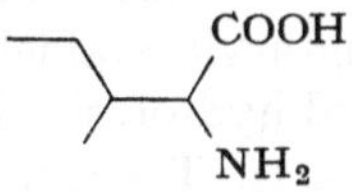

(20)
Isoleucine

(21)

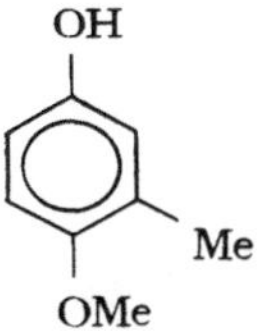

(22)

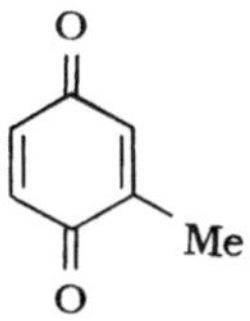

(24)

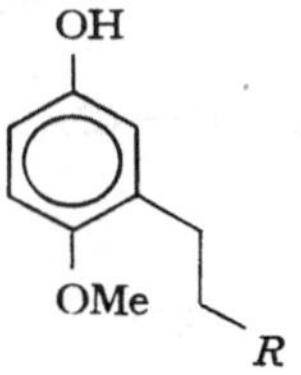

(25) a R = OH
b R = NH₂

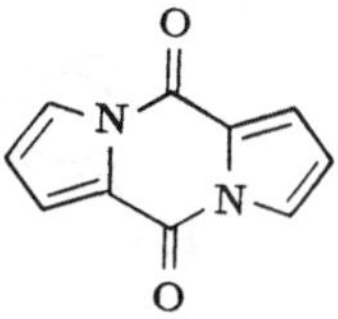

(23)
Pyrocoll

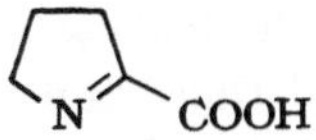

(27)

(26)

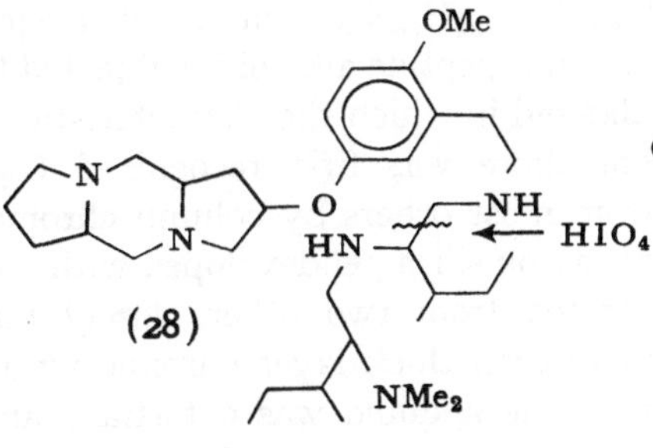

(28)

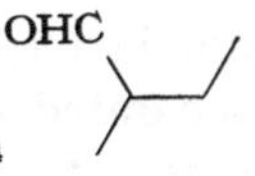

(29)

remaining. Since the polyamine reacted with one equivalent of periodic acid to give formaldehyde and 2-methylbutanal (29), isolated as the 2,4-dinitrophenylhydrazones, the amino acids must be linked as shown in (18) for zizyphine. The one remaining point of uncertainty was the

position of attachment of the aryloxy group to the proline diketopiperazine. Although the earlier mentioned formation of the phenolic alcohol
(25a) on basic hydrolysis suggests mild β-elimination from the 3-position,
ozonization of dihydrozizyphine to destroy the aromatic ring and subsequent acid hydrolysis yielded a hydroxyproline which was found to
give the same colour reaction with isatin as 4-hydroxyproline, but a
different colour than given by either diastereomeric 3-hydroxyproline.
On these grounds the aryloxy group was attached at the 4-position to
complete the structure (18) for zizyphine. Since the base catalyzed
elimination of the phenol oxygen is difficult to reconcile with attachment
at the 4-position of the proline, more evidence on this point would be
desirable. The styrylamide apparently has the *cis* configuration because
the coupling constant between the two olefinic protons is only 8.5 Hz.

Two other products isolated from hydrolysis were the nitrogen-free
condensation product (30) derived from alkaline condensation of two
molecules of the corresponding arylacetaldehyde, and the azlactone (31)
obtained from hydrochloric acid hydrolysis and pyridine-acetic anhydride
treatment of the ethyl ester of (21). The elimination of the N,N-dimethyl-
amino group is remarkable.

(30)   (31)

3. Ceanothine-B (32)

Ceanothine-B, $C_{29}H_{36}N_4O_4$, m. p. 238.5–240.5°, $[\alpha]_D - 293°$ (CHCl$_3$),
was one of the first peptide alkaloids subjected to structural determination
and is the alkaloid in which the characteristic warped macrocyclic styrylamide chromophore was first recognized (54, 55). The alkaloid was
separated from most others by column chromatography and thick layer
chromatography on silica gel developed with chloroform-methanol. It was
further separated from two other closely related bases, ceanothine-A
and -C, by silica gel thick layer chromatograms developed with chloro-
form-acetone. The alkaloid was a tertiary amine (not acetylated) with
one N_{basic}-methyl group whose NMR absorption shifted from δ^{CDCl_3}
1.97 to δ^{CD_3COOD} 3.02. Ceanothine-B absorbed one equivalent of hydrogen
in the presence of palladium with the loss of an enamide chromophore (33)
($\lambda_{max} \sim 250$ nm, $\varepsilon \sim 4000$) to produce dihydroceanothine-B, m. p.
272–278°, $[\alpha]_D - 87°$ (MeOH). Quantitative comparison of the amide

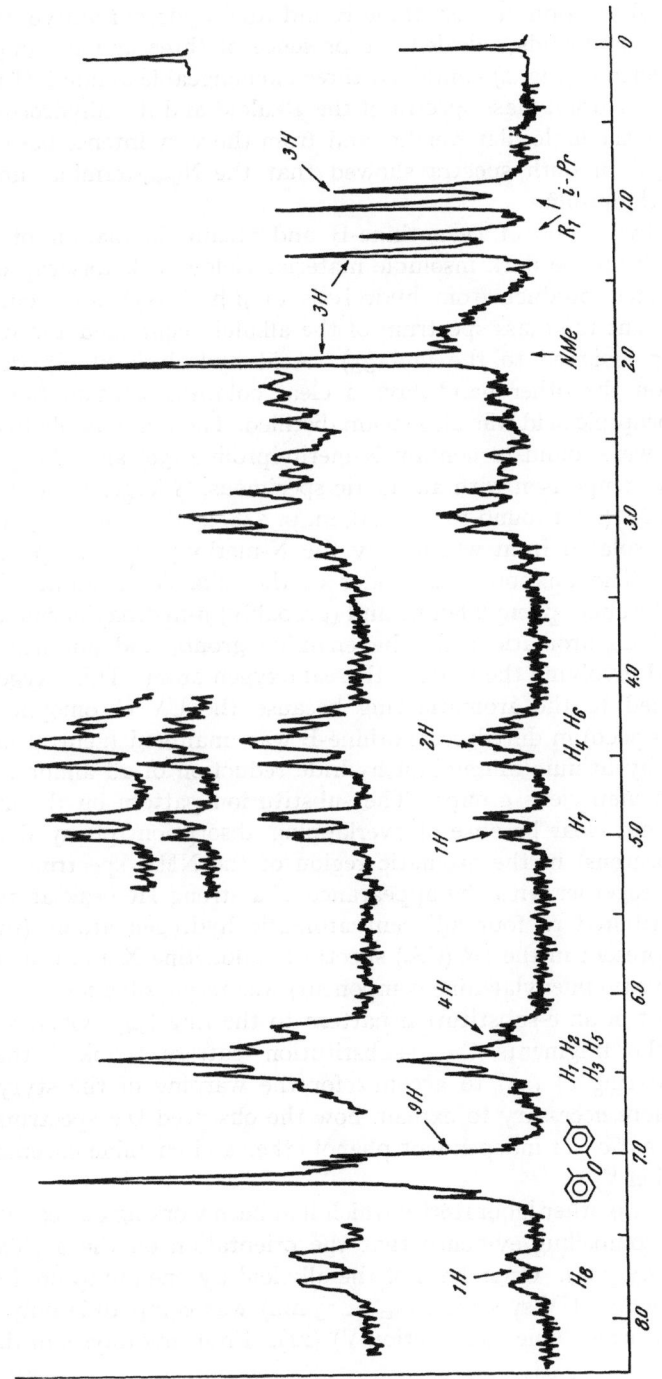

Fig. 1. NMR spectrum of ceanothine-B

carbonyl absorption of ceanothine-B and its dihydroderivative with that of model compounds revealed the presence of three amide groups. The NMR spectrum (Fig. 1) contained three exchangeable amide NH protons. The low resolution mass spectra of the alkaloid and its dihydrocompound confirmed the molecular weights and from the very intense base peak at m/e 84 (34) in both spectra showed that the N_{basic}-terminal unit must be N-methylproline.

Acid hydrolysis of ceanothine-B and steam distillation of volatile products from the dark insoluble material yielded α-ketoisocaproic acid, the expected product from hydrolysis of β-hydroxyleucyl compounds. In agreement the mass spectrum of the alkaloid contained a strong peak at m/e 97 assigned to the ion (35). Acid hydrolysis of dihydroceanothine-B on the other hand gave a clear colorless solution from which α-ketoisocaproic acid was also steam distilled. The non-volatile hydrolysis products were found to contain N-methylproline (36) and phenylalanine (8) by tlc comparison with authentic specimens. A larger basic fragment, $C_{23}H_{29}N_3O_3$ [(later found to be (37)], m. p. 248–254°, $[\alpha]_D + 63°$ (EtOH), was also isolated from which only the N-methylprolyl group had been removed. The unknown remainder of the alkaloid in addition to the N-methylproline, phenylalanine and (probably) β-hydroxyleucine residues contained an aromatic ring, the enamide group, and another ring of some kind involving the fourth ethereal oxygen atom. This oxygen atom was bonded to the aromatic ring because the UV chromophore (λ_{max} 232 nm, ε 9000) in dihydroceanothine-B was unaltered (neutral or acidic solution) by lithium aluminium hydride reduction of all amide carbonyl groups to methylene groups. The substitution pattern on the aromatic ring was not clear because of overlapping absorption (phenylalanyl and olefinic protons) in the aromatic region of the NMR spectrum (Fig. 1). This fact together with the appearance of a strong IR peak at 757 cm⁻¹ (CS₂) attributed to four adjacent aromatic hydrogen atoms (the peak was also present in the IR (CS₂) spectra of adouétine-X and frangulanine which have no phenylalanine component) was responsible for an incorrect assignment of an o-substitution pattern to the ring (54). On recombination of the fragments the o-substitution pattern required the nine-membered ring of (38) to account for the warping of the styrylamide chromophore necessary to explain how the observed UV spectrum could be a summation of independent phenol ether and enamide chromophores (see Section V).

Later, two other laboratories which had been working on ceanothine-B provided compelling evidence that the orientation on the aryloxy ring was actually *para*. Ozonolysis of the alkaloid by one group produced an aldehyde whose UV spectrum (λ_{max} 275 nm) was compatible only with a p-alkoxybenzaldehyde (see Section V) (21). From hydrolysis of dihydroceanothine-B the other group was able to isolate tyramine (39) (42). This result requires incorporation of another amino acid into the macrocyclic ring of ceanothine-B to make it geometrically possible, and therefore the correct structure of ceanothine-B is (32). More recently accumulated knowledge of the ion fragmentation of these alkaloids (9) also allowed assignments (21, 42) of peaks in the mass spectrum of the alkaloid which are only consistent with the revised formula (32). The failure of the infrared correlation remains unexplained.

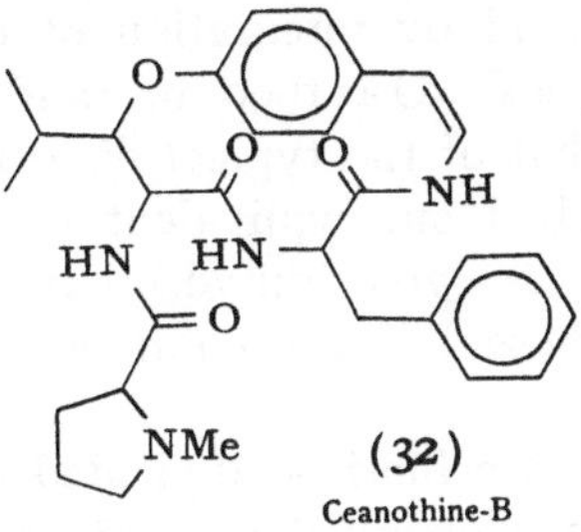

(32)
Ceanothine-B

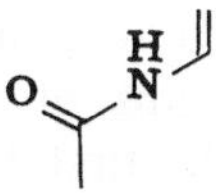

(33) 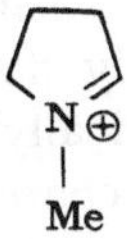　　　(34) 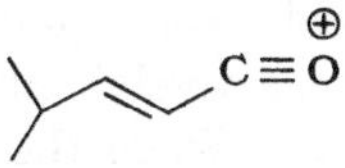　　　(35) 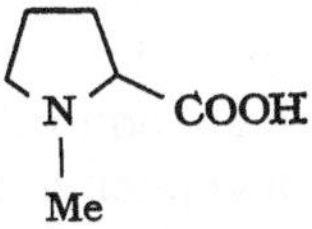　　　(36)
N-Methylproline

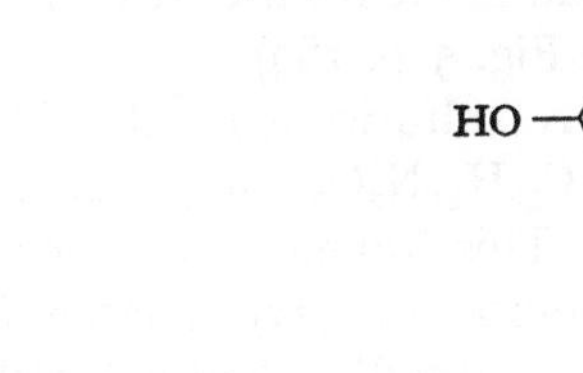

(37) 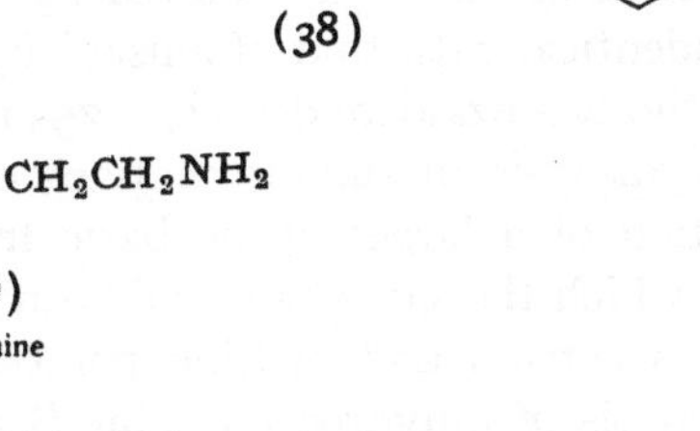　　　　　　(38)

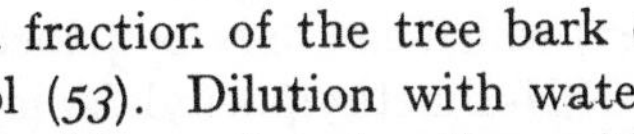

$$HO - \underset{}{\bigcirc} - CH_2CH_2NH_2$$

(39)
Tyramine

4. Scutianine (40)

The crude alkaloid fraction of the tree bark of *Scutia buxifolia* was extracted by methanol (53). Dilution with water, adjustment to pH 9 with ammonium hydroxide, and extraction with benzene yielded the mixture of bases which was further freed from extraneous matter by re-extraction into aqueous citric acid. Separation of scutianine, m. p. 186–187°, $[\alpha]_D - 399°$ (CHCl$_3$), was achieved by a two stage thick layer chromatography procedure. Although the chemical evidence adduced was alone sufficient to establish the structure of this alkaloid, extensive use was made of high resolution mass spectroscopy during the structural work (53). The molecular formula was determined by this means in conjunction with elemental analysis to be $C_{39}H_{47}N_5O_5$. The alkaloid was a monobasic tertiary amine (pKa 5.94) in agreement with the presence of four amide groups found by integration of the area under the IR amide carbonyl peaks (three C=O at 1690 cm^{-1} and one C=O at 1640 cm^{-1}). The UV spectrum was that of the typical distorted styrylamide. Moreover, the alkaloid absorbed one equivalent of hydrogen in methanol solution over platinum to give dihydroscutianine, m. p. 239–240°, $[\alpha]_D - 158°$ (CHCl$_3$), whose UV spectrum was now that of a phenol ether.

Hydrolysis with 6N hydrochloric acid yielded six amino acids detected by paper chromatography: phenylalanine (8), proline (19), β-hydroxyleucine (41), N,N-dimethylphenylalanine (42), leucine (9) and glycine. The first three amino acids were quantitatively determined (0.9, 0.88 and 0.60 equivalents, respectively) photometrically by means of the N-dinitrophenyl derivatives. Since the carbon content of the first four acids combined with the eight carbon atoms of the missing styrenoid piece accounts for all of the carbon atoms, the leucine and glycine (present in only slight amount) must be artifacts derived from β-hydroxyleucine by transamination (see pandamine discussion) and retroaldol processes, respectively. The missing piece was found by 6N hydrochloric acid hydrolysis of dihydroscutianine which yielded tyramine (39) identified by tlc. In view of the earlier misassignment of an o-orientation of the oxygen atom and the two carbon chain in ceanothine-B, further evidence was provided for the p-orientation in scutianine by ozonization of the double bond which gave an aldehyde whose UV absorption (λ_{max} 275 mμ) was identical with that of anisaldehyde but different from that of m- and o-methoxybenzaldehydes (λ_{max} 253 mμ) (see Fig. 5 p. 189).

Hydrolysis of dihydroscutianine with 6N sulfuric acid allowed the isolation of a larger cyclic basic fragment $C_{23}H_{29}N_3O_3$, m. p. 246–247°, from which the side chain had been cleaved. This fragment has the same mass spectrum and melting point as the compound (37) isolated from hydrolysis of dihydroceanothine-B and is undoubtedly identical with it. Hydrolysis of this fragment yielded tyramine (39), phenylalanine (8) and β-hydroxyleucine (41) which can only be joined as in (37). Reattachment of the missing, N,N-dimethylphenylalanine and proline to the unsaturated macrocycle in the only way possible gives the complete structure (40) for scutianine in complete agreement with the high resolution mass spectroscopic data. Paper chromatographic comparison of the β-hydroxyleucine from hydrolysis with the authentic *threo* and *erythro* compounds showed the alkaloid to have the *threo* configuration of this acid. Although the absolute configuration of the amino acids was not determined, the negative rotation of dihydroscutianine argues for the L-configuration.

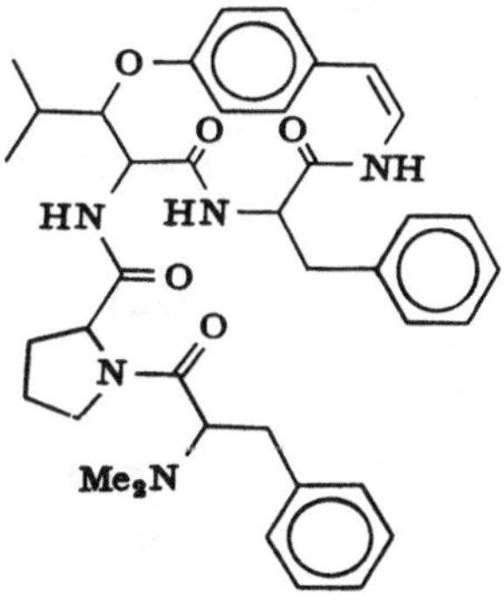

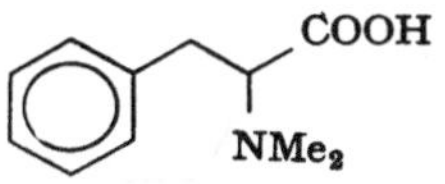

(42)

N,N-Dimethylphenylalanine

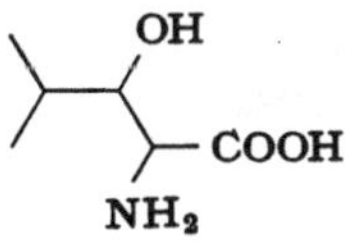

(40)

Scutianine

(41)

β-Hydroxyleucine

5. Integerrine (43)

This alkaloid is one whose entire structure was deduced from its mass spectrum alone (*48*) (see Section VII). The total amount of alkaloid isolated by extensive chromatography was 5.5 mg. High resolution mass spectroscopy gave the molecular formula $C_{35}H_{39}N_5O_4$ and the formulas of many of the fragmentation ions. The very intense base peak was at m/e 100 and the ion $C_6H_{14}N$ corresponding to it therefore is (44) in consideration of peaks at m/e 85 ($C_5H_{11}N$) (45) and m/e 550 ($M-C_3H_7$). The basic terminal amino acid is therefore N,N-dimethylvaline. The *p*-hydroxystyrylamine peak (46) appeared at m/e 135. The remaining two amino acids were tryptophan (47)-peaks at m/e 130 (C_9H_8N) (48), 159 ($C_{10}H_{11}N_2$) (49) and 170 ($C_{11}H_8NO$) (50) and β-phenylserine (51)-peak at m/e 131 (C_9H_7O) (52).

The joining of these four units was evident from other fragmentation peaks. Ions at m/e 224 ($C_{15}H_{14}NO$) (53), 347 ($C_{20}H_{17}N_3O_3$) (54) and 410 ($C_{26}H_{24}N_3O_2$) (55) showed that the phenolic oxygen was etherified with the β-hydroxyphenylalanine (51) and the styrylamine was acylated by the tryptophan carboxyl. Other ions at m/e 317 ($C_{20}H_{17}N_2O_2$) (56) and 289 ($C_{19}H_{17}N_2O$) (57) showed that the β-hydroxyphenylalanine and the tryptophan were directly linked to complete the macrocyclic ring which is present in the ion at m/e 451 ($C_{28}H_{25}N_3O_3$) (58). Final proof of the

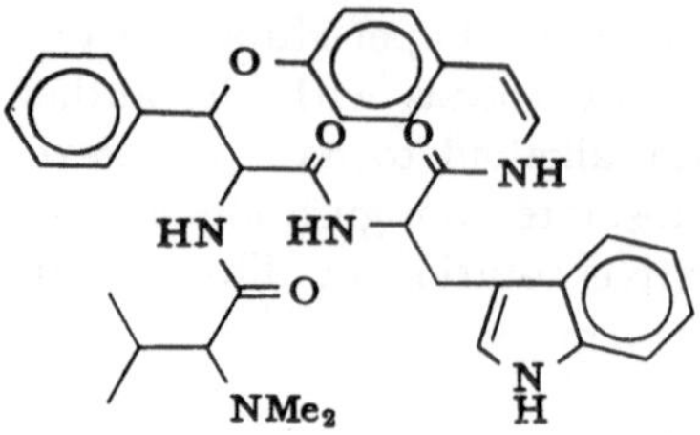

(43)
Integerrine

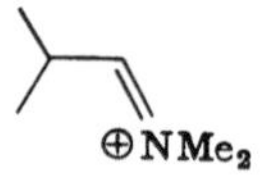

(44)

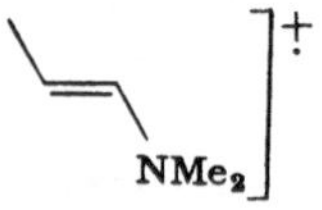

(45)

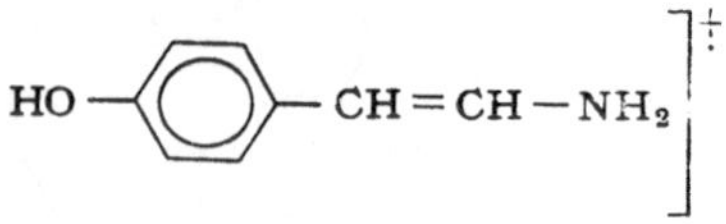

(46)

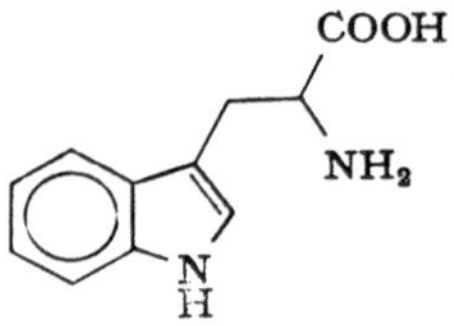

(47)
Tryptophan

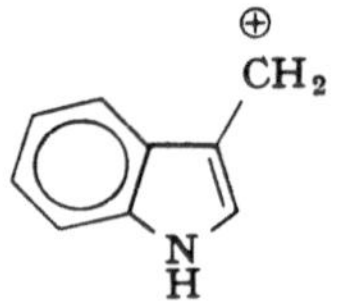

(48)

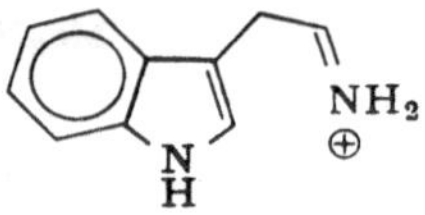

(49)

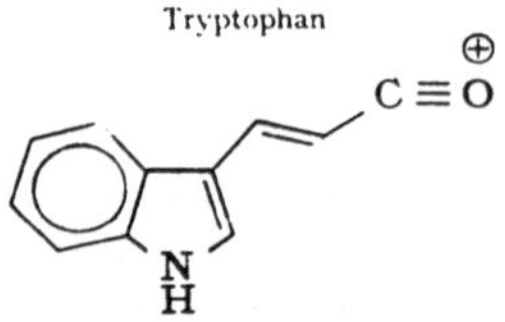

(50)

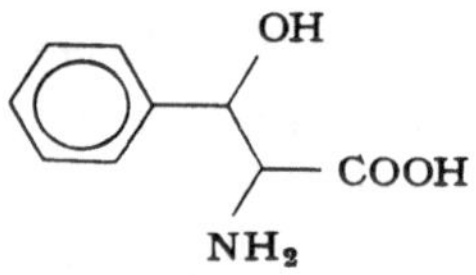

(51)
β-Phenylserine or β-Hydroxyphenylalanine

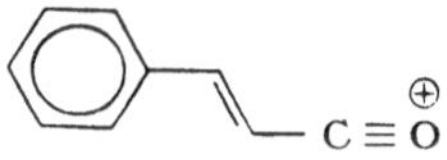

(52)

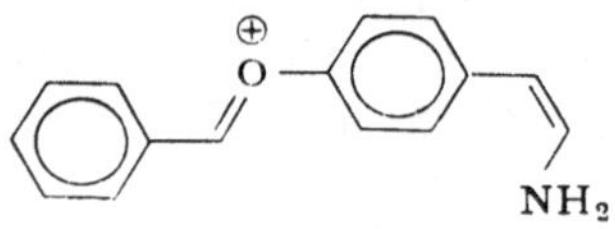

(53)

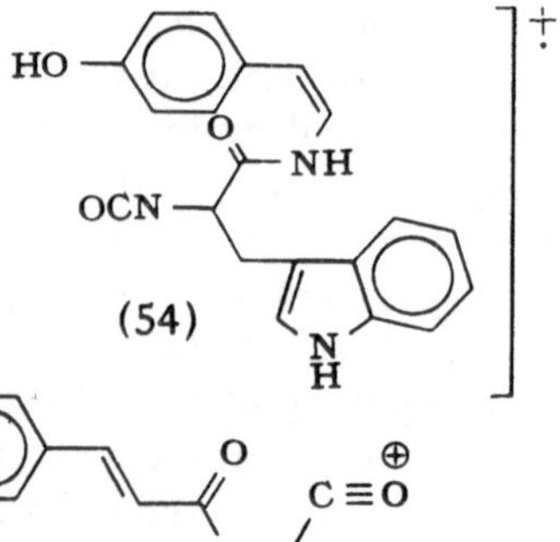

(54)

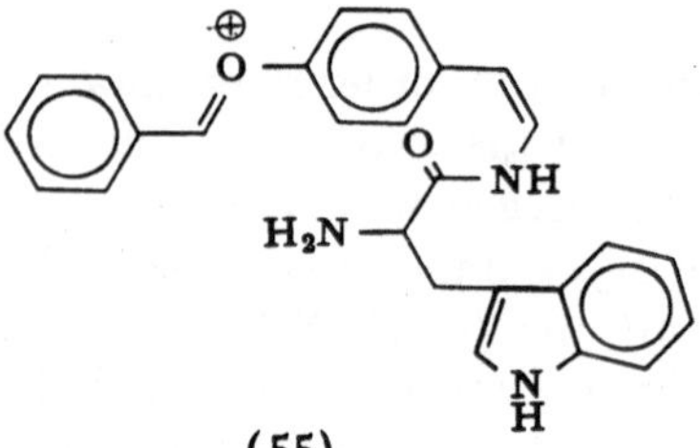

(55)

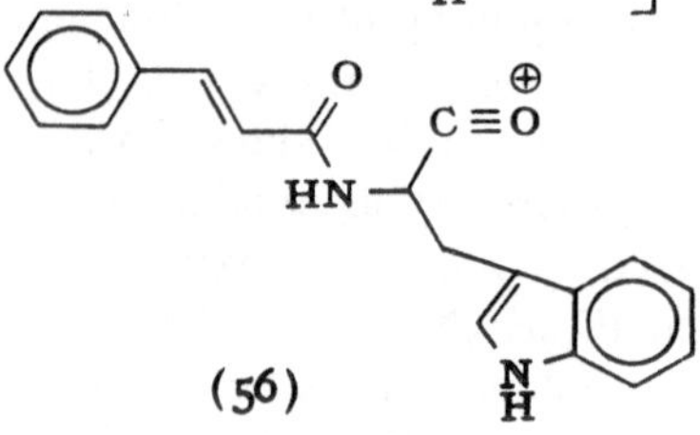

(56)

attachment of the N,N-dimethylvaline to the nitrogen of the β-hydroxy-phenylalanine comes from ions at m/e 229 ($C_{13}H_{13}N_2O_2$) (59) and 201 ($C_{12}H_{13}N_2O$) (60).

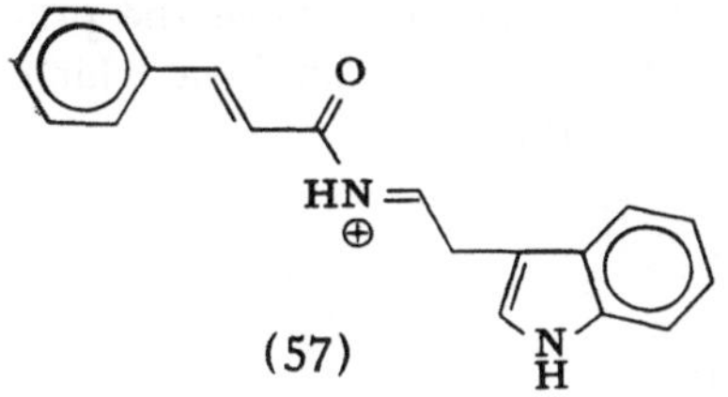

(57)

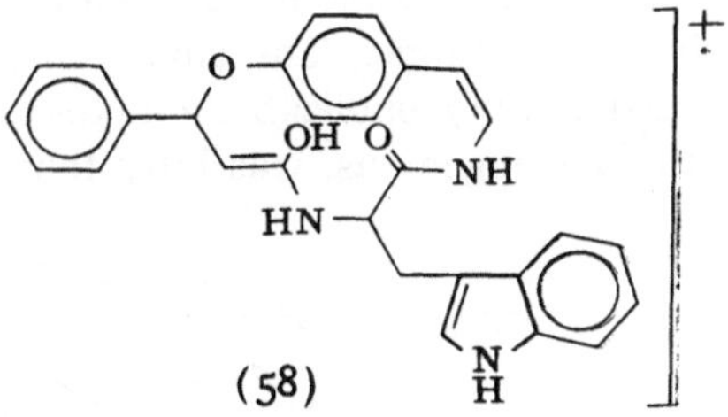

(58)

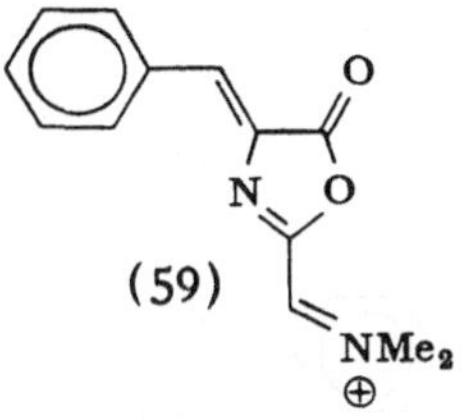

(59)

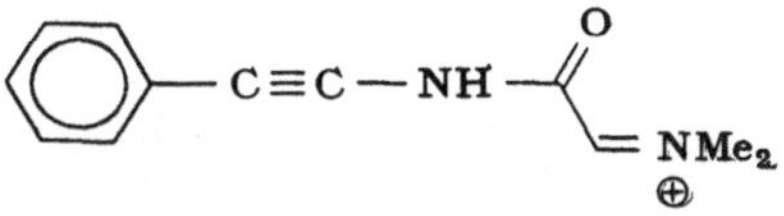

(60)

6. Lunarine (61)

Lunarine, $C_{25}H_{31}N_3O_4$, m. p. 227–231°, $[\alpha]_D + 300°$ (CHCl₃), the most important alkaloid of its class, was isolated by several groups (5, 14, 15, 18, 19, 44) before serious work on its structure began. It is a monobenzenoid, monobasic amine with two secondary amide groups. (IR 1650 cm⁻¹). A third oxygen atom was present in a ketone (IR 1700 cm⁻¹) which could be reduced to an alcohol, lunarinol, by boro-hydride. The fourth oxygen atom was an ether function. The molecule absorbed three equivalents of hydrogen with retention of the ketone function (35). The changes in the UV spectrum after hydrogenation were those to be expected of a cinnamamide whose double bond had been reduced. The amine function was secondary since an N-acetate and an N-tosylate were prepared (3).

A significant observation was that the UV spectrum of lunarine changed reversibly in acid and base and that these changes did not involve the basic nitrogen function (3). The UV spectrum of lunarinol changed in acidic but not in basic solution.

Alkali fusion of lunarine produced spermidine (62) (37) (also obtained from aqueous acidic hydrolysis), and the biphenyl acid (63), as well as the lactone (64) (2). These fragments accounted for 23 of the 25 carbon atoms, and a pair of structures (66) was proposed* which was in agreement.

---

* Structure II in reference (46) is different (presumably a misprint) from structure III given in reference (38).

with the substituent orientation in the biphenyl derivatives (63) and (64). The changes in the UV spectrum were attributed to two different chromophores in the molecule. In acid solution it is the alkoxycinnamamide which is responsible for the change; the model compound (65) underwent the same change. The change in base was thought to involve the proposed readily enolizable δ-keto-α,β-unsaturated amide, an idea which, although ingenious, was later found to be incorrect.

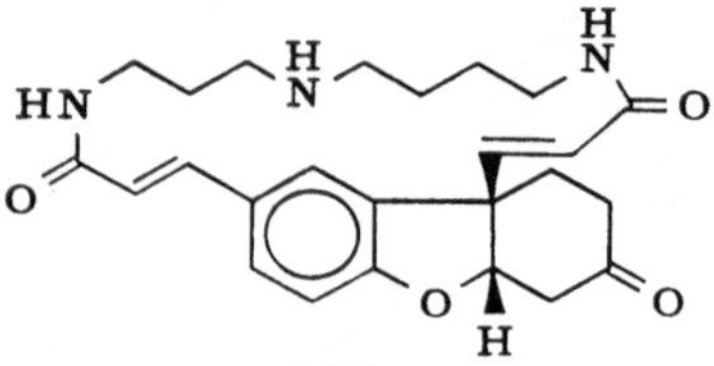

(61)
Lunarine

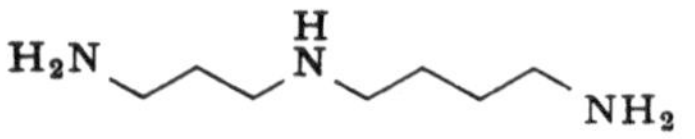

(62)
Spermidine

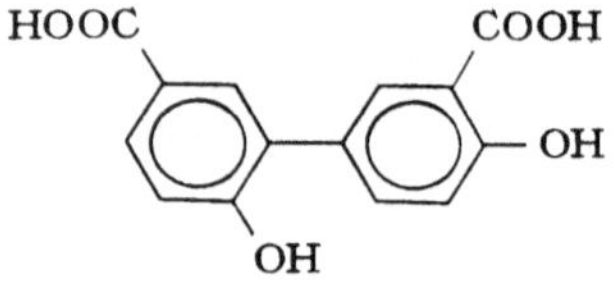

(63)

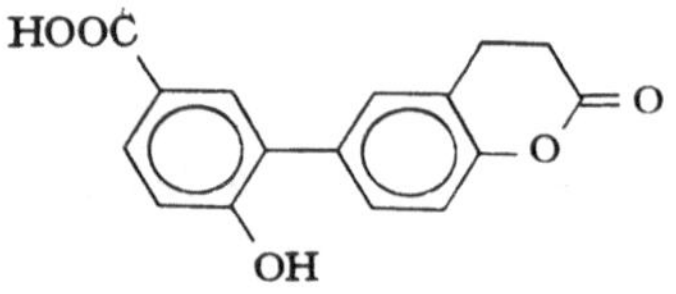

(64)

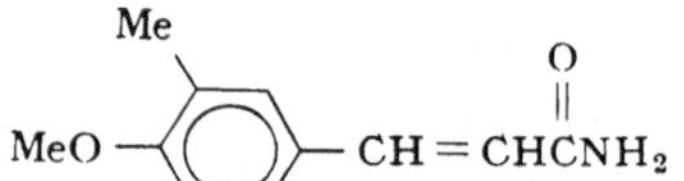

(65)

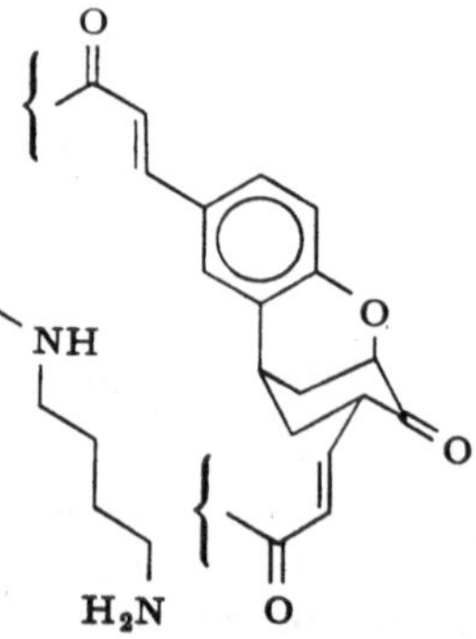

(66)

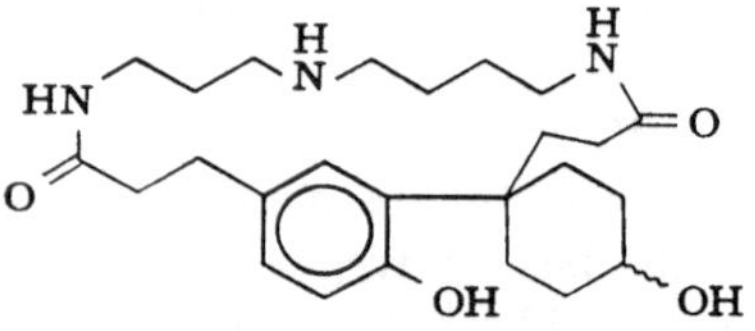

(67)

The molecule was finally bludgeoned into submission by two simultaneous and independent X-ray crystal structure analyses (46) which not only decided the orientation of the spermidine unit but also revealed the true structure and stereochemistry of the rest of the molecule to be (61). This revised structure allows ready explanation of the UV change in base as being due to the reversible elimination of phenolate ion from the β-position of the ketone which change cannot occur with lunarinol. The third molecule of hydrogen absorbed by lunarine presumably saturated this conjugated double bond. Thus hexahydrolunarine should be optically inactive although there is no information on this point*. The true structure of lunarine (61) requires the biphenyl products (63) and (64) which are formed during the alkali fusion of the alkaloid to arise from a rearrangement perhaps by some path such as depicted in Chart 1.

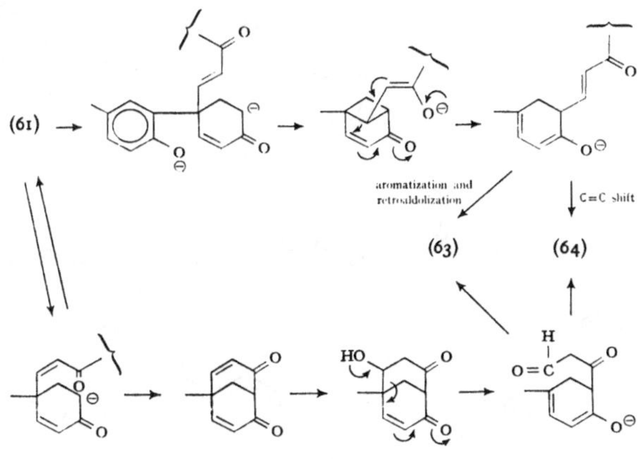

Chart 1. Conceivable Rearrangements of Lunarine on Alkaline Fusion

7. Homaline (68 β) or (69 β)

This unique peptide alkaloid, $C_{30}H_{42}N_4O_2$, m. p. 134°, $[\alpha]_D - 34°$ (CHCl$_3$), is dibasic, having two amide groups and two tertiary amine functions. Acid hydrolysis yields *trans*-cinnamic acid. The alkaloid forms a *single* monomethiodide and a bis methiodide. A double Hofmann elimination from the bis-methiodide took place at room temperature to give the optically inactive bis-cinnamamide (70). A second successive double Hofmann reaction with (70) followed by catalytic hydrogenation of the four double bonds in (71) led to the bis β-phenylpropionic acid amide of N,N'-di-*n*-propylputrescine (72) which was further hydrogenolyzed to di-propylputrescine (73). Of the four structures fitting these facts [(68) α and β and (69) α and β], the pair with the nitrogen atom β

---

* It is difficult to formulate the hexahydrolunarinol obtained by hydrogenation of lunarinol (the analysis given is consistent with a tetrahydro compound), or to formulate an optically active hexahydrolunarinol from hydrogenation of lunarine since the optically inactive (67) should be the product.

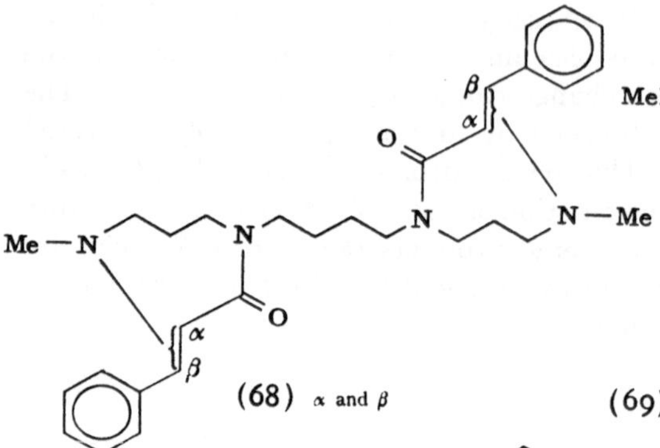

(68) α and β

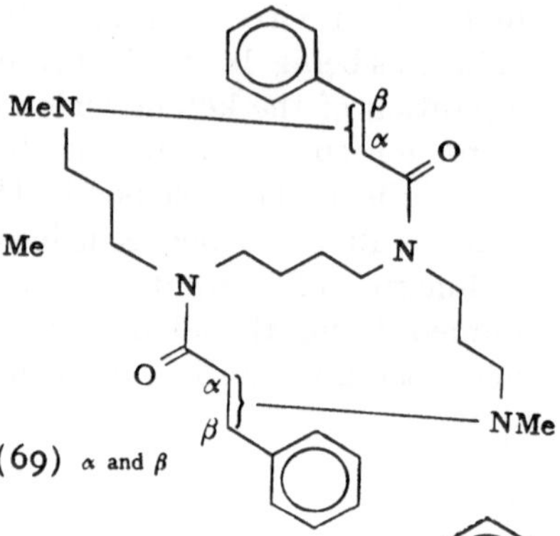

(69) α and β

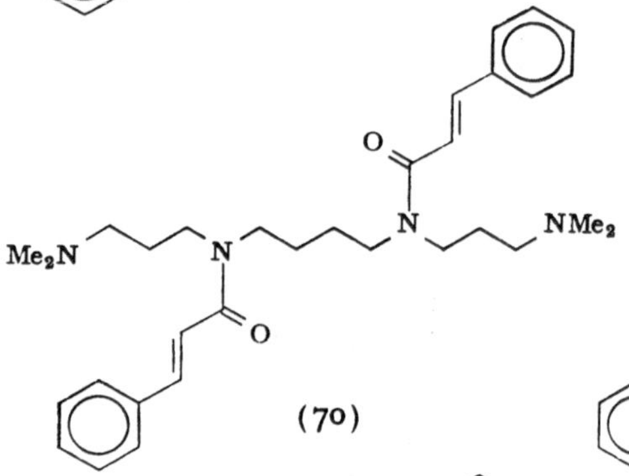

(70)

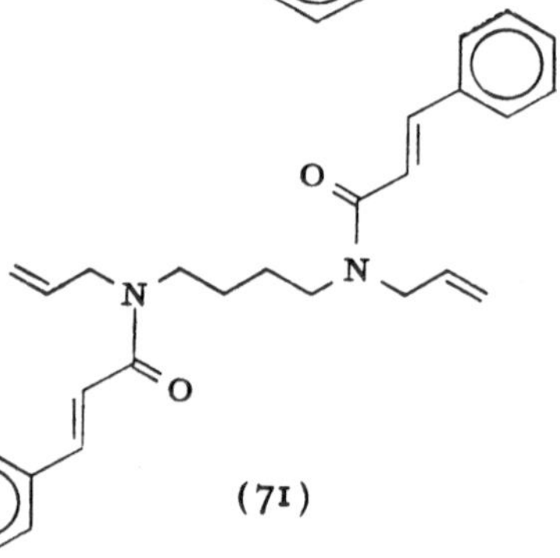

(71)

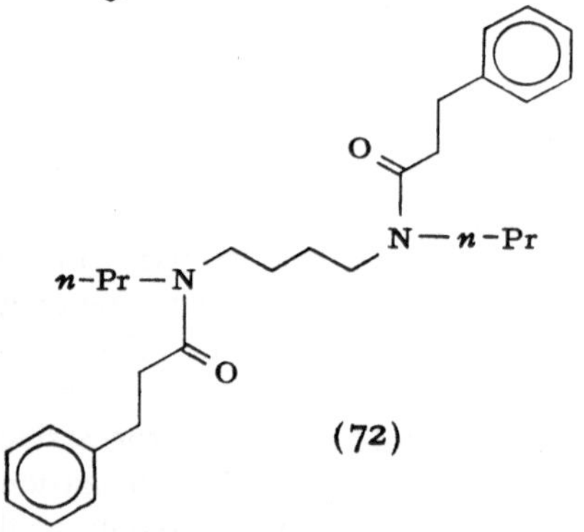

(72)

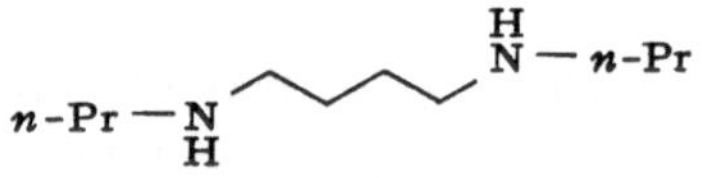

(73)

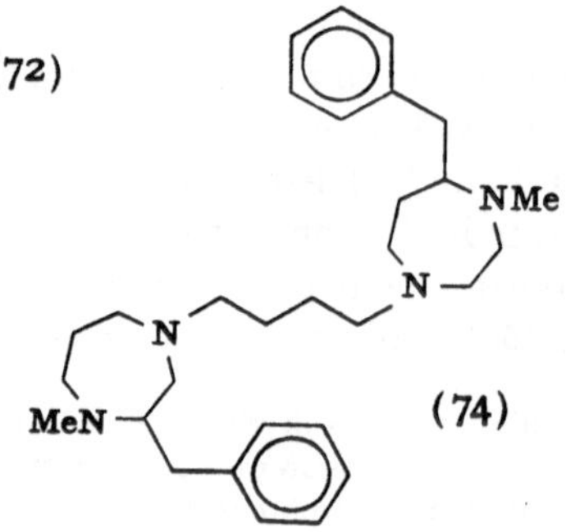

(74)

to the amide carbonyl for facile elimination during hydrolysis and Hofmann reactions is more likely than the alternative pair with the α-nitrogen atoms of phenylalanine. Moreover, one of these latter formulas (68) α has been excluded by synthesis of the racemic form of (74) which was different from the lithium aluminum hydride reduction product of homaline. A distinction between the two possibilities (68) β and (69) β has not yet been published (*33, 34*).

8. Adouétine-X (75)

The alkaloids ceanothamine-A and -B (*55*) were originally thought to be different from the isomeric alkaloid adouétine-X (*28, 29*), because of apparent differences in the UV spectra and because ceanothamine-B exhibited a large difference in optical rotation. Recently, a higher rotation was reported for adouétine-X (*43*) and ceanothamine-B has been found to be identical with this alkaloid by direct comparison (*43, 56*), although the discrepancy in optical rotations ($- 338°$ *vs.* $- 370°$) is still unresolved. The UV spectra of both samples are identical. Ozonization of ceanothamine-B ($\equiv$ adouétine-X) gave a p-alkoxybenzaldehyde as chemical proof of the p-orientation on the aromatic ring (see Fig. 5 p. 189) (*56*).

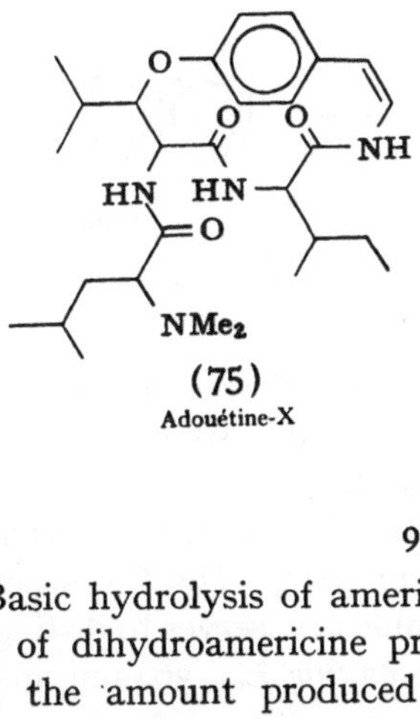

(75)
Adouétine-X

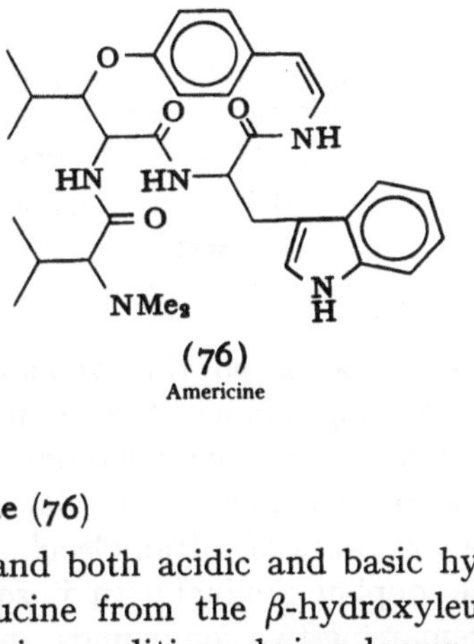

(76)
Americine

9. Americine (76)

Basic hydrolysis of americine (76) and both acidic and basic hydrolysis of dihydroamericine produced leucine from the β-hydroxyleucine unit, the amount produced under basic conditions being larger presumably because the necessary transamination reaction [(17) → (9)] is more efficient in basic medium. The reciprocal nature of the tyramine and ammonia yields from hydrolysis of the alkaloid and its dihydro derivative was pointed out (*cf.* the analogous result with zizyphine), indicating that most of the large yield of ammonia from the alkaloid arose from the enamide group (*20*).

10. Aralionine (77)

This alkaloid (77) has an extra carbonyl group in the form of a C-benzoylglycine unit, whose UV spectrum, being that of a β-keto amide, undergoes a bathochromic shift (λ_{max} 247 → λ_{max} 283 nm) when the solution is made alkaline. The benzoyl group is readily cleaved by methanolic potassium hydroxide leaving behind the des-benzoyl alkaloid. On the other hand 6N hydrochloric acid hydrolysis gave ω-aminoacetophenone as one product from loss of the carboxyl group instead. A second, neutral product of the acid hydrolysis was identified as β-phenylnaphthalene (78) formed by self-condensation of two units of β-phenylacetaldehyde (or its imine or imide) derived from the β-phenylserine unit in (77).

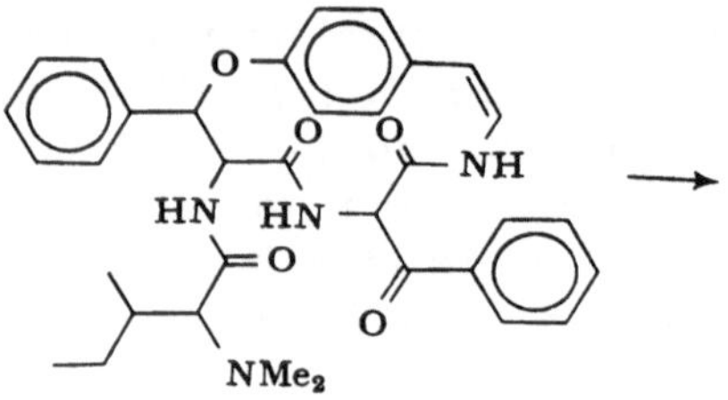

(77)

Aralionine

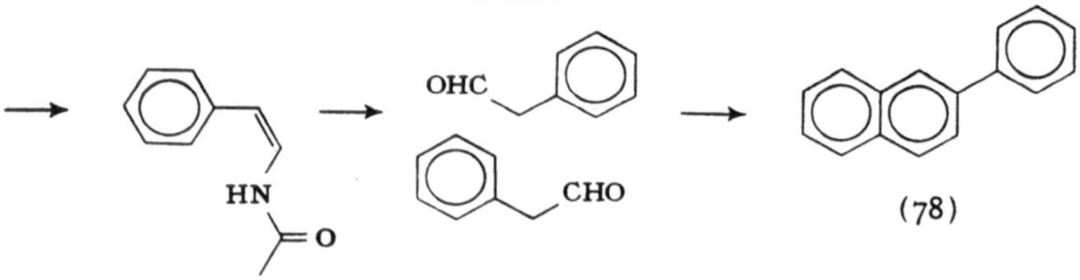

(78)

In the work on this alkaloid another method was developed for distinguishing among the position isomers of the small amount of hydroxyphenylethylamine produced by hydrolysis of the reduced enamide. The o-isomer is much less polar than the other two on paper chromatography because of intramolecular hydrogen bonding. The UV spectrum of the coupling product with diazotized sulfanilic acid serves to distinguish the remaining two compounds, the m-isomer having λ_{max} 362 nm and the p-isomer having λ_{max} 323 nm (47).

11. Ceanothine-A (79)

Ceanothine-A, $C_{30}H_{40}N_4O_4$, is a secondary amine. Since its N-basic terminal unit is N-methylphenylalanine, and since it has the macrocyclic styrylamide UV chromophore and a β-hydroxyleucyl group [peak in mass spectrum at m/e 97 corresponding to (35)], the alkaloid must have structure (79) with either a leucyl or isoleucyl group completing the macrocyclic ring (55).

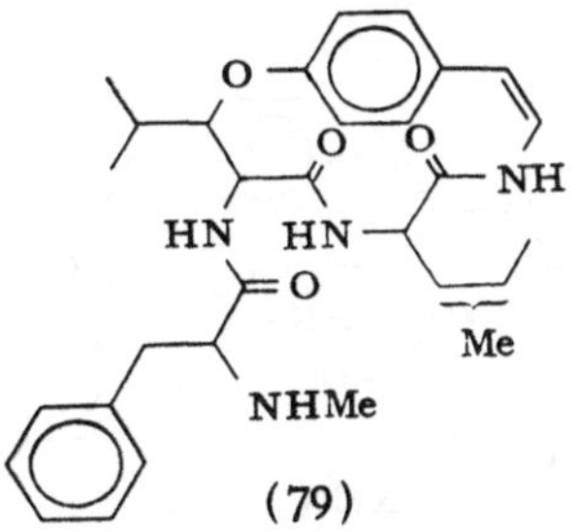

(79)
Ceanothine-A

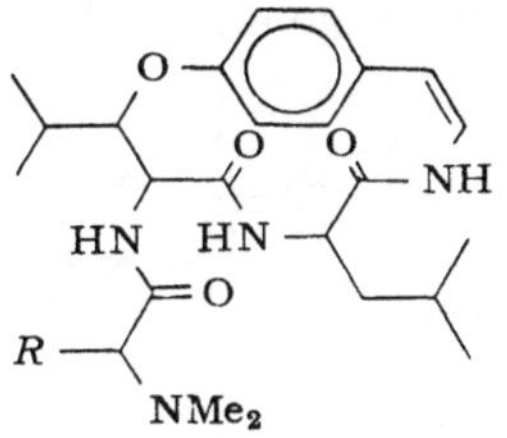

(80)

Franganine
Frangufoline
Frangulanine

a $R = (CH_3)_2CH-CH_2-$
b $R = Phenyl-CH_2-$
c $R = CH_3-CH_2-CH-$
       $\qquad\qquad\qquad |$
       $\qquad\qquad\qquad CH_3$

12. Franganine (80a), Frangufoline (80b) and Frangulanine (80c)

All three of these closely related alkaloids (49, 50) yield some glycine on hydrolysis with 6N hydrochloric acid. This artifact presumably arises from retroaldol cleavage of the β-hydroxyleucine unit common to all three bases.

13. Hymenocardine (81)

Hymenocardine (81) is so far unique both in having a β-hydroxy-valine unit to which the aromatic ring is attached and in having an aryl conjugated ketone instead of the usual enamide group. Basic hydrolysis (NaOH—MeOH) of the alkaloid resulted in opening of the macrocyclic ring by β-elimination of the phenolate to produce (82) (30, 31). This hitherto unprecedented mild ring opening is undoubtedly aided by the ketone function conjugated with the phenol ether which makes the phenolate ion in this molecule a better leaving group than in the other alkaloids.

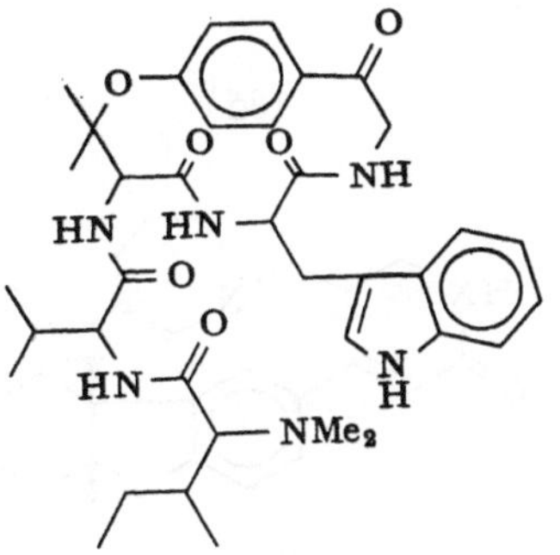

(81)
Hymenocardine

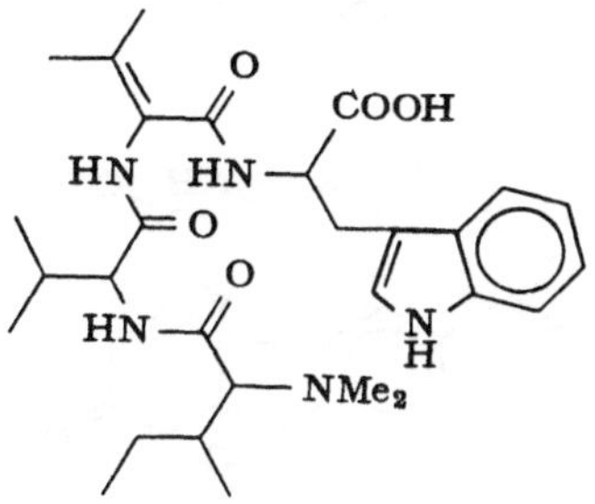

(82)

14. Integerrenine (83) and Integerressine (84)

Acid hydrolysis of both of these alkaloids (52) and their dihydro derivatives yield β-phenylnaphthalene from the β-phenylserine unit as discussed for aralionine (47).

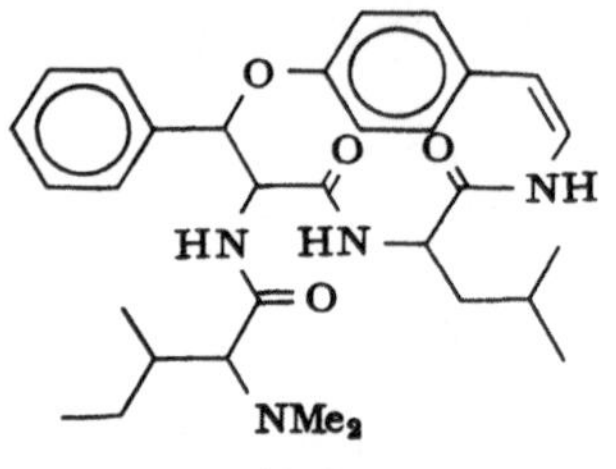

(83)
Integerrenine

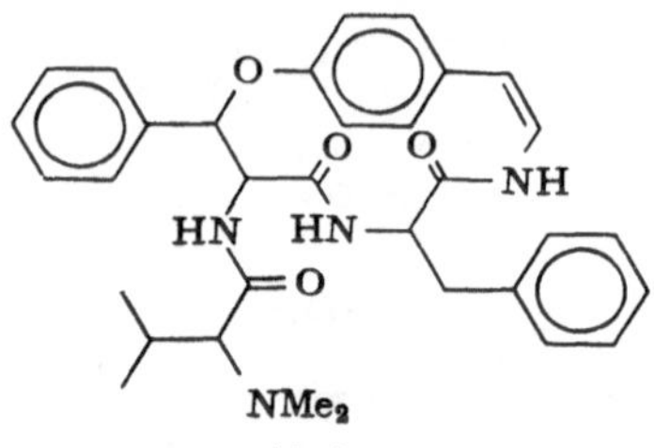

(84)
Integerressine

15. LBX and LBZ

Alkaloid LBX, whose molecular formula has one more CH_2 unit than lunarine, can be readily prepared by reaction of formaldehyde and lunarine. However, the structure (85) proposed for this alkaloid is unlikely to be correct because serious repulsive non-bonded interactions arise as the α-carbon atom of the ketone approches within bonding distance of the imonium ion (86) so long as the double bonds retain the *trans* configuration. Moreover, the ease with which LBX is formed from lunarine and formaldehyde suggests that addition of one of the amide nitrogen atoms to the imonium ion (86) has occurred to produce (87) or (88) as possible formulas for LBX*. Alkaloid LBZ, an alcohol from borohydride reduction of LBX, would then be one of the epimeric alcohols corresponding to (87) or (88) (39).

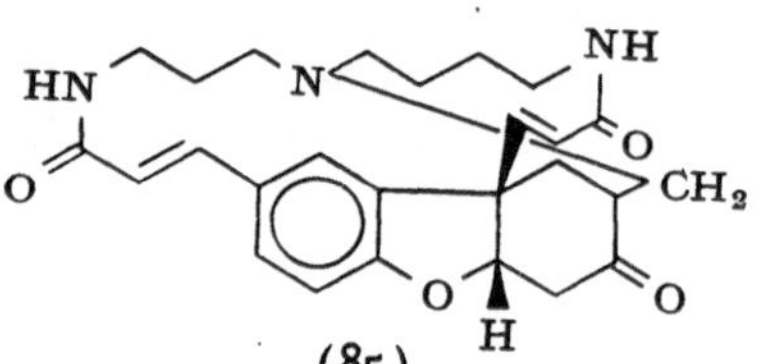

(85)

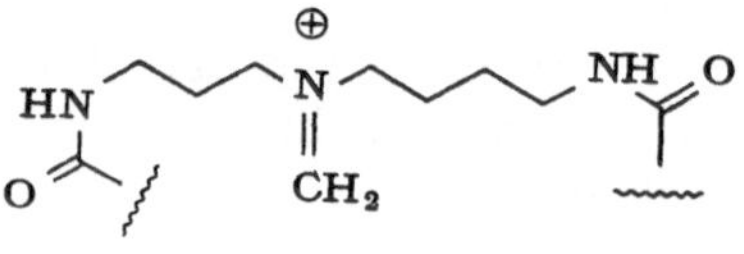

(86)

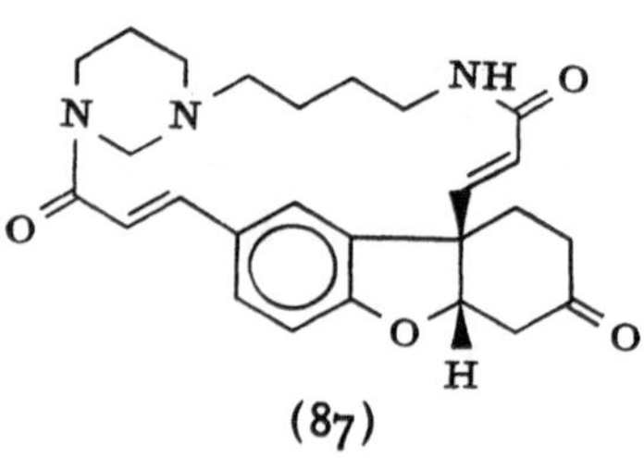

(87)

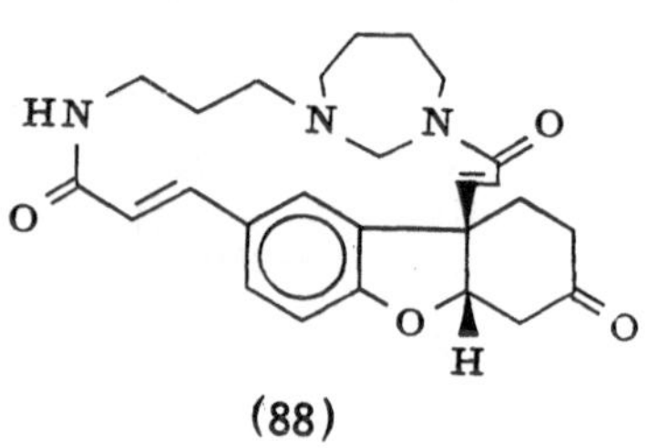

(88)

* Personal communication from Dr. P. POTIER.

16. Lasiodine-A (89)

Lasiodine-A (89) is so far the only naturally occurring peptide alkaloid derivative in which the usual macrocyclic ring is not present but in which are preserved the free phenolic hydroxyl and α,β-unsaturated amide unit that would be involved in opening or closing the ring (25). As mentioned earlier, opening of the macrocyclic ring by β-elimination of the phenol has been observed with hymenocardine (81) during alkaline hydrolysis. Even though no macrocyclic ring is present in lasiodine-A, the styryl-amide would still appear to be *cis* from the coupling constant of 9.5 Hz for the olefinic hydrogens. Lasiodine-A is dibasic, the second basic function residing in an N-methylvaline residue esterifying the hydroxyl of a β-phenylserine unit. Palladium catalyzed hydrogenation of the alkaloid reduces both double bonds on (89) and produces N-methylvaline by hydrogenolysis of the benzylic ester. Lasiodine-A with methanolic sodium hydroxide undergoes facile β-elimination of the N-methylvaline with introduction of a third double bond in (90). One of the amino acid components of lasiodine-A belongs to the D-series, D-*threo*-β-phenyl-serine.

(89)
Lasiodine-A

(90)

V. Ultraviolet Spectra of Peptide Alkaloids

In the initial structural work on the aryloxymacrocyclic alkaloids the UV spectra were a puzzle since in the 200–400 nm range the macro-cyclic enamides exhibit only a gradually decreasing absorption which ends ~ 300 nm. Once this composite absorption pattern was dissected, the UV spectra became very useful. The most interesting feature is the difference in the aryloxyenamide chromophore, depending on whether it is part of a ring or not. If it is not, there appears a long wavelength maximum due to the entire chromophore as observed in zizyphine and lasiodine-A (Fig. 2). Inclusion of the group within a fourteen-membered macrocyclic ring as in ceanothine-B (Fig. 2) reduces conjugation. In fact molecular models [see (91)] show that in the macrocycle the p-orbitals of the aryloxy and enamide chromophores can not overlap to any extent, and each group must therefore absorb independently. In Fig. 3 the summation curve of model enamide and *p*-alkoxyalkylbenzene chromophores is compared with the UV absorption of adouétine-X, and the two are found to be almost identical. Conversely, if the spectrum of a dihydro macrocyclic enamide alkaloid is subtracted from the spectrum of the unhydrogenated alkaloid, there remains the simple enamide chromophore (Fig. 4). However, if the same subtraction is done for a nonmacrocyclic aryloxyenamide base, the remainder is not the simple enamide absorption. As another example of the effect of the macrocyclic ring on conjugation, even the more flexible ketone (6d) derived from pandamine shows evidence of only partial conjugation (λ_{max} 266 nm, ε 7400) with the aromatic ring when compared with that of anisaldehyde given in Fig. 5.

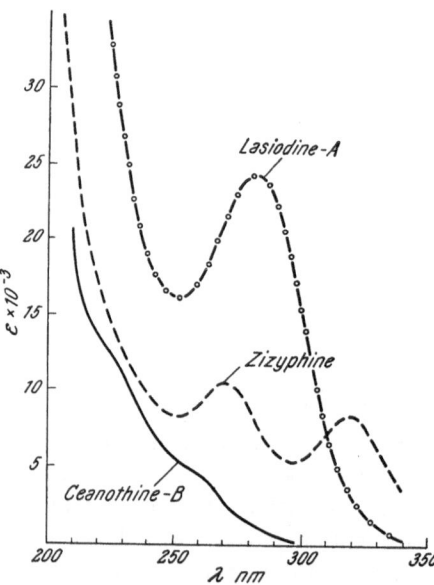

Of the other UV absorbing groups found in peptide alkaloids, the indole system is revealed by the maxima in the 270—290 nm region, and the C-benzoyl group of aralionine (77) by the maximum at 247 nm (ε 10,600), but the phenyl group of phenylalanine and phenylserine and the α,β-unsaturated amide group of lasiodine-A (89) serve merely to increase the rising end absorption near 210 nm. Careful quantitative measurements, however, reveal their presence.

The reversible changes in the UV absorption of lunarine (61) in neutral, basic and acidic medium are represented in Fig. 6. The long wavelength maximum at 355 nm in basic solution is that of the *p*-hydroxycinnamamide anion (92). In acid the bathochromic shift fortuitously has the same long wavelength maximum at 354 nm, but it must have a different origin in this case, particularly since the spectrum of the alcohol, lunarinol, undergoes the same change as lunarine in acid but does not

Fig. 2. UV spectra of ceanothine-B, zizyphine and lasiodine-A

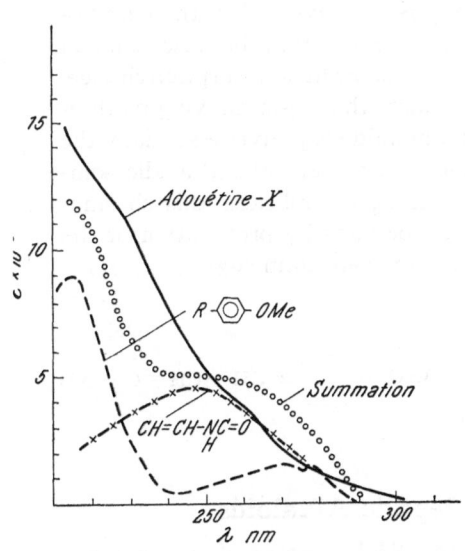

Fig. 3. UV spectrum of adouétine-X compared with summation curve from enamide and *p*-methoxybenzene chromophores

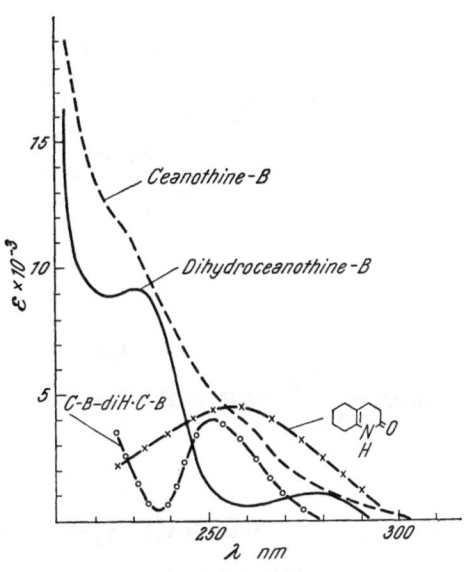

Fig. 4. Enamide chromophore obtained by subtraction of UV spectrum of dihydroceanothine-B from that of ceanothine-B

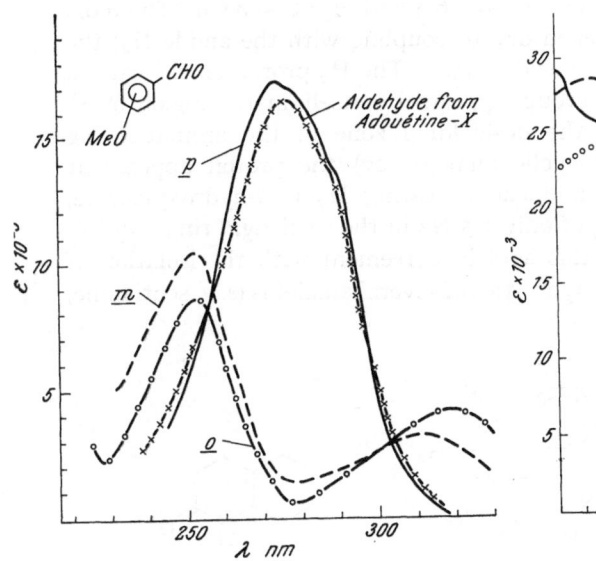

Fig. 5. UV spectra of methoxybenzaldehydes and the aldehyde from adouétine-X

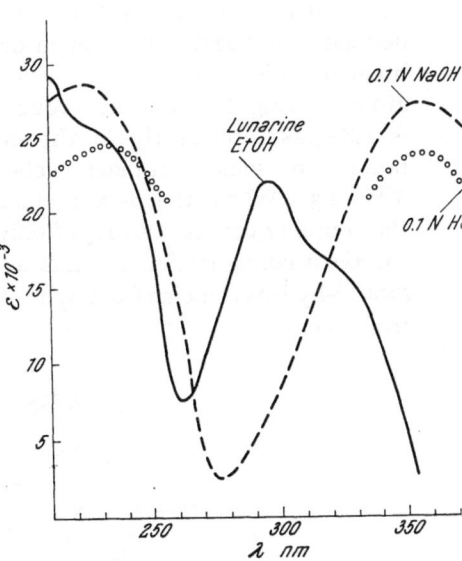

Fig. 6. UV spectrum of lunarine in neutral, acid an solution

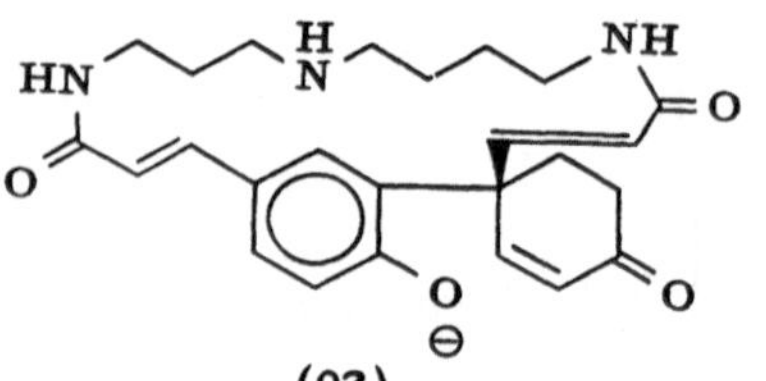

(92)

change in base. The secondary amino group is not involved in the chromophore modification because salts of lunarine show little absorption change. The fact that 4-methoxy-3-methyl-cinnamamide (65) gives essentially the same shift in neutral and acidic solution strongly implicates this chromophore, perhaps by protonation of the amide oxygen atom (93).

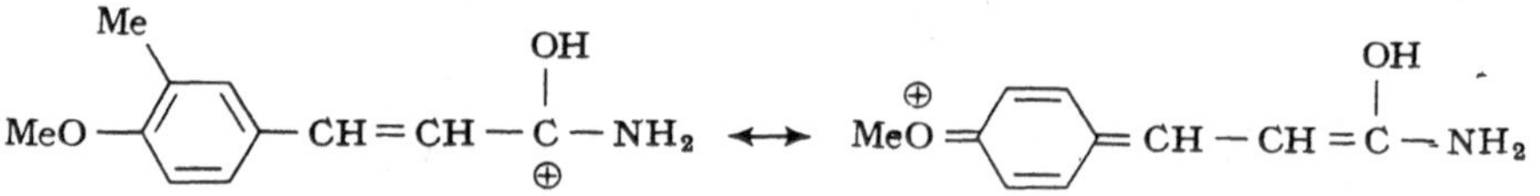

(93)

VI. NMR Spectra of Peptide Alkaloids

There are a number of features in the NMR spectra of the aryloxy-macrocyclic alkaloids (94) that are of interest. The pH dependence of the basic N-methyl groups has already been mentioned in Section III. The chemical shift of the two protons (H_1 and H_2) of the enamide group is also very pH dependent. In deuteriochloroform both protons are near δ 6.4 but in acetic acid H_1 appears $\sim \delta$ 6.8 and H_2 at $\sim \delta$ 6.0. The δ 6.0 doublet is noticeably broadened due to coupling with the amide H_3; the broadening is absent in acetic acid-d_4 (56). The H_7 proton on the carbon atom bearing the aryloxy group appears by itself in the region δ^{CDCl_3}, $\sim$ 4.8–5.25, apparently in the deshielding zone of the aromatic ring because in some non-macrocyclic analogs (27) the proton appears at δ^{CDCl_3} 4.3. When the β-oxyaminoacid bearing H_7 is β-hydroxyleucine, the coupling constant to H_6 of only 1.5 Hz in the semi rigid ring requires the *threo* configuration for this acid in agreement with the isolation of *threo*-β-hydroxyleucine from hydrolysis of several alkaloids (e. g. scutianine, frangulanine).

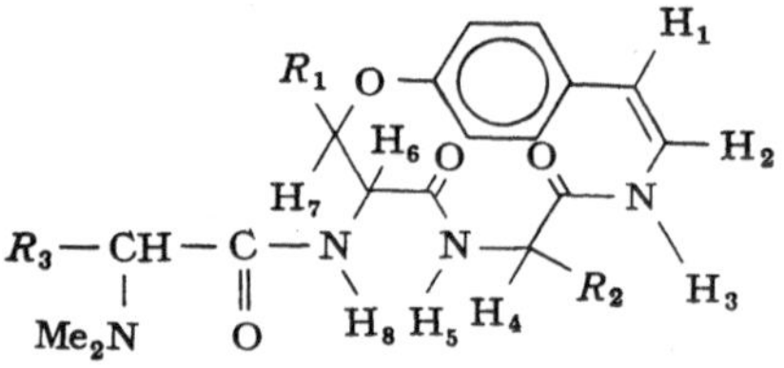

(94)

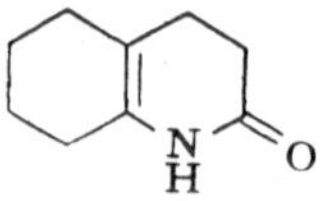

(95)

References, pp. 201—203

Protons H_3, H_4, H_5 and H_6 on the macrocyclic ring are in the shielding zone of the aryloxy ring and thus exhibit up field shifts. Methine protons H_4 and H_6 on the α-carbon atoms of the amino acids in the macrocyclic ring appear at δ^{CDCl_3} 4.0–5.0 in an overlapping pattern separated from other absorption and up field about 0.5–1.0 ppm from their usual position in linear peptides (56). Of the three amide NH protons, H_3 and H_5 experience the shielding effect of the aryloxy ring current and appear in the vicinity of δ^{CDCl_3} 6.4 overlapping the enamide hydrogens H_1 and H_2, while H_8 is less affected and appears in the expected position near δ^{CDCl_3} 7.7. Two of these amide NH protons (H_5 and H_8) are exchanged immediately on shaking the deuteriochloroform solution with deuterium oxide, but the third is not noticeably affected even after an hour at 37° (56). Rapid exchange of the enamide proton H_3 occurs only on addition of acid. The model compound (95) required $\sim$ 15 minutes for 50% exchange of the NH group under the same conditions (56).

In four of the alkaloids (adouétine-Y and -Z, integerrenine and integerressine) where R_1 is phenyl, the basic N,N-dimethylamino group appears as two well separated ($\Delta = 0.4$–0.7 ppm) methyl signals in trifluoroacetic acid and even in carbon tetrachloride. In each case one of the methyl groups is appreciably more shielded than is normal; the shielding is ascribed to the R_1 phenyl group (29, 52).

VII. Mass Spectra of Aryloxy Macrocyclic Peptide Alkaloids

With the earlier investigations of the mass spectra of peptides (16, 17, 45) and the applications of low resolution mass spectroscopy in the early peptide alkaloid work as guides, FEHLHABER has used high resolution mass spectroscopy to formulate the general breakdown pattern of the aryloxymacrocyclic peptide alkaloids (9). The reaction sequence is given in Chart 2. Determination of the basic amino acid units and the way in which they are joined by means of this scheme allows a complete structure to be written except for the distinction between the isomers leucine and isoleucine and the substitution pattern on the aryloxy ring. Leucine and isoleucine on the N_{basic}-terminal amino acid can be distinguished by the secondary decomposition of R_3 in the base peak. The isoleucine ion (96) loses an ethyl radical while the leucyl ion (97) loses propene.

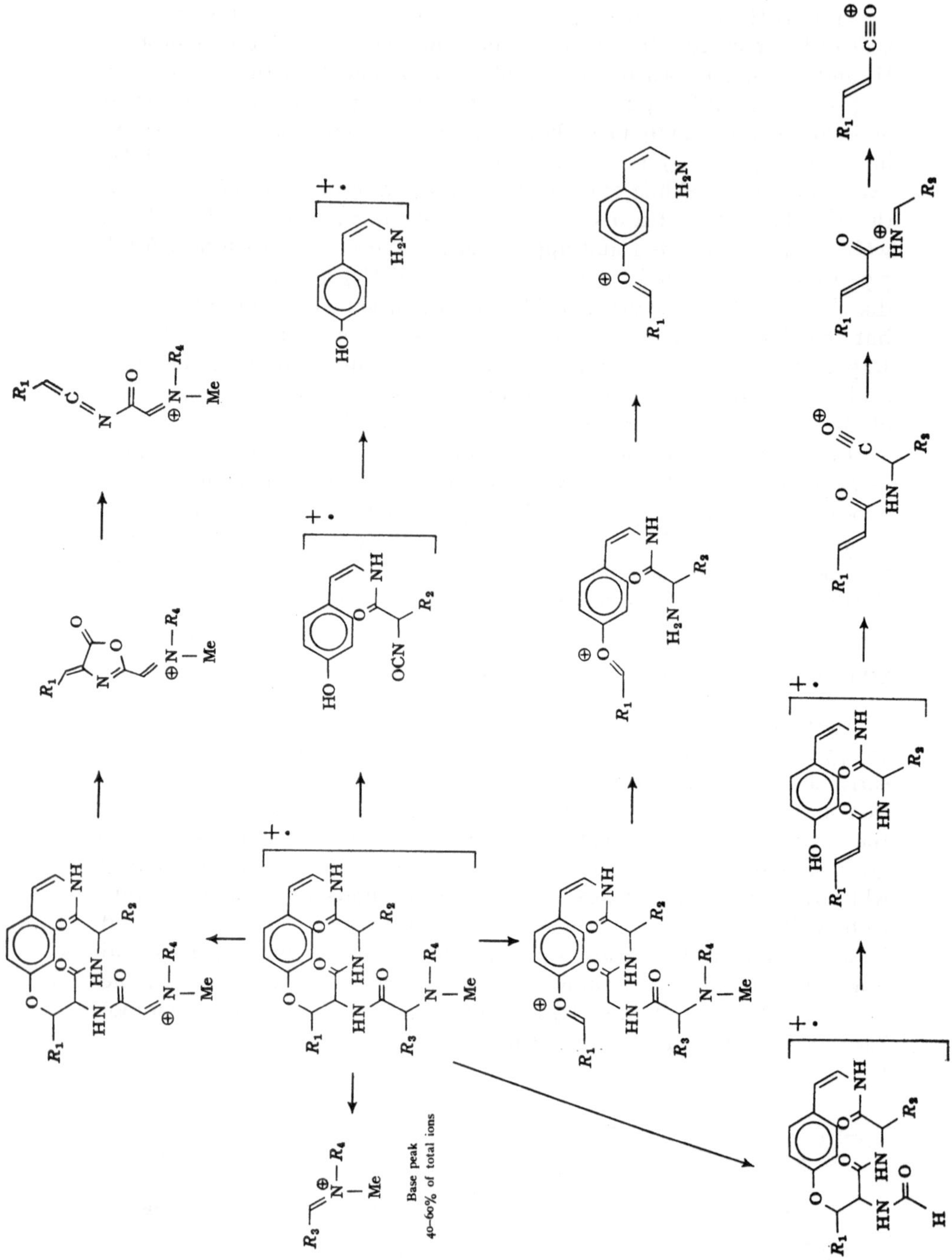

Chart 2. General Mass Spectroscopic Pattern for Aryloxymacrocyclic Peptide Alkaloids

VIII. Pharmacological Properties of Peptide Alkaloids

Many of the plants in which peptide alkaloids have been found have been used in folk medicine throughout the world (*4, 6, 26, 50, 53*). The coagulant properties of an alcoholic extract of the root bark of *Ceanothus americanus* have found use in a commercial oral hemostatic agent (*13*). However, only three pharmacological studies of the alkaloids themselves have been reported (*4, 23, 41*). The aryloxy macrocyclic alkaloids were not found to be particularly toxic. A *Ceanothus americanus* alkaloid mixture had LD_{50} 96 mg/kg when administered intravenously in rats (*23*); adouétine-Z had LD_{50} 52 mg/kg on intravenous administration to mice (*4*). Although a tea prepared from *Ceanothus americanus* had been found to exert a hypotensive effect on oral administration to rats (*57*), a mixture of alkaloids from the same plant and from *Ceanothus velutinus* was not found to have significant hypotensive action (*23, 41*). A more complete study of a single pure alkaloid, adouétine-Z, has not revealed any particularly promising medical application (*4*).

IX. Tables

Table 1. *Basicity of Peptide Alkaloids*

Alkaloid	pKa	Solvent	References
Adouétine-Z	5.38	Methylcellosolve (MCS)	(*28*)
Ceanothine-B	5.85	95% MeOH–5% H_2O	(*55*)
Pandamine	6.02	MCS	(*27*)
O-Acetylpandamine	5.94	MCS	(*27*)
Scutianine	5.94	None stated	(*53*)
Zizyphine	5.77	80% MCS — 20% H_2O	(*26*)
	6.23	80% MCS — 20% H_2O	
Zizyphinine	6.55	80% MCS — 20% H_2O	(*26*)

Table :

No.	Name (Synonyms) Molecular Formula	Structure
1	Adouétine-X (Ceanothamine-B) $C_{28}H_{44}N_4O_4$	$Me_2Leu \rightarrow 3Hyleu \rightarrow Ileu \rightarrow NH-CH=CH-\bigcirc$
2	Adouétine-Y $C_{34}H_{40}N_4O_4$	$Me_2Phe \rightarrow 3Hyphe \rightarrow Ileu \rightarrow NH-CH=CH-\bigcirc$
3	Adouétine-Y′ (Myrianthine-B) $C_{31}H_{42}N_4O_4$	$Me_2Phe \rightarrow 3Hyleu \rightarrow Ileu \rightarrow NH-CH=CH-\bigcirc$
4	Adouétine-Z $C_{42}H_{45}N_5O_5$	$Me_2Phe \rightarrow Pro \rightarrow 3Hyphe \rightarrow Phe \rightarrow NH-CH=CH-\bigcirc$
5	Americine $C_{31}H_{39}N_5O_4$	$Me_2Val \rightarrow 3Hyleu \rightarrow Try \rightarrow NH-CH=CH-\bigcirc$
6	Aralionine $C_{34}H_{38}N_4O_5$	$Me_2Ileu \rightarrow 3Hyphe \rightarrow NH-CH-CO-NH-CH=CH-\bigcirc$ $\bigcirc-C=O$
	Ceanothamine-A Ceanothamine-B	See Frangulanine See Adouétine-X
7	Ceanothine-A $C_{30}H_{40}N_4O_4$	$MePhe \rightarrow 3Hyleu \rightarrow \begin{Bmatrix} Leu \\ Ileu \end{Bmatrix} \rightarrow NH-CH=CH-\bigcirc$
8	Ceanothine-B $C_{29}H_{36}N_4O_4$	$MePro \rightarrow 3Hyleu \rightarrow Phe \rightarrow NH-CH=CH-\bigcirc$
9	Ceanothine-C $C_{26}H_{38}N_4O_4$	$MePro \rightarrow 3Hyleu \rightarrow \begin{Bmatrix} Leu \\ Ileu \end{Bmatrix} \rightarrow NH-CH=CH-\bigcirc$
10	Ceanothine-D $C_{26}H_{38}N_4O_4$	$MePro \rightarrow 3Hyileu \rightarrow Leu \rightarrow NH-CH=CH-\bigcirc$
11	Ceanothine-E $C_{34}H_{40}N_4O_4$	$Me_2Phe \rightarrow 3Hyphe \rightarrow Leu \rightarrow NH-CH=CH-\bigcirc$

Peptide Alkaloids

Melting Point	$[\alpha]_D$ (Solvent)	Species isolated from	References
277–279°	− 316° (CHCl₃)	*Waltheria americana* L.	*(28, 29)*
278–280°	− 338° (CHCl₃)	*Ceanothus americanus* L.	*(43)*
279–280.5°	− 370° (CHCl₃)	*Ceanothus americanus* L.	*(55)*
292°	− 230° (CHCl₃: MeOH 9 : 1)	*Waltheria americana* L.	*(28, 29)*
287–289°	− 213° (CHCl₃)	*Ceanothus americanus* L.	*(43)*
289–290.5°	− 305° (CHCl₃)	*Waltheria americana* L.	*(28, 29)*
302° dec.	− 294° (CHCl₃)	*Myrianthus arboreus* P. Beauv. *Melochia corchorifolia*	*(24)* *(51)*
140–145°	− 184° (CHCl₃)	*Waltheria americana* L.	*(28, 29)*
135.5–137° and 142–182°	− 198° (MeOH)	*Ceanothus americanus* L.	*(20)*
165–167°	+ 82° (MeOH)	*Araliorhamnus vaginatus* Perrier	*(47)*
256–259°	− 256° (CHCl₃)	*Ceanothus americanus* L.	*(55)*
238.5–240.5°	− 293° (CHCl₃)	*Ceanothus americanus* L.	*(21, 42, 54, 55)*
223–229°	− 368° (CHCl₃)	*Ceanothus americanus* L.	*(9, 43, 55)*
227–229°	− 347° (CHCl₃)	*Ceanothus americanus* L.	*(43)*
238–239°	− 285° (CHCl₃)	*Ceanothus americanus* L.	*(43)*

(Table 2, continued)

No.	Name (Synonyms) Molecular Formula	Structure
12	Franganine $C_{28}H_{44}N_4O_4$	Me$_2$Leu $\rightarrow$ 3Hyleu $\rightarrow$ Leu $\rightarrow$ NH $-$ CH $=$ CH $-$ ⬡
13	Frangufoline $C_{31}H_{42}N_4O_4$	Me$_2$Phe $\rightarrow$ 3Hyleu $\rightarrow$ Leu $\rightarrow$ NH $-$ CH $=$ CH $-$ ⬡
14	Frangulanine (Ceanothamine-A)* $C_{28}H_{44}N_4O_4$	Me$_2$Ileu $\rightarrow$ 3Hyleu $\rightarrow$ Leu $\rightarrow$ NH $-$ CH $=$ CH $-$ ⬡
15	Homaline $C_{30}H_{42}N_4O_2$	(68 β) or (69 β)
16	Hymenocardine $C_{37}H_{50}N_6O_6$	Me$_2$Ileu $\rightarrow$ Val $\rightarrow$ 3Hyval $\rightarrow$ Try $\rightarrow$ NH $-$ CH$_2$ $-$ C(=O) $-$ ⬡
17	Integerrenine $C_{31}H_{42}N_4O_4$	Me$_2$Ileu $\rightarrow$ 3Hyphe $\rightarrow$ Leu $\rightarrow$ NH $-$ CH $=$ CH $-$ ⬡
18	Integerrine $C_{35}H_{39}N_5O_4$	Me$_2$Val $\rightarrow$ 3Hyphe $\rightarrow$ Try $\rightarrow$ NH $-$ CH $=$ CH $-$ ⬡
19	Integerressine $C_{33}H_{38}N_4O_4$	Me$_2$Val $\rightarrow$ 3Hyphe $\rightarrow$ Phe $\rightarrow$ NH $-$ CH $=$ CH $-$ ⬡
20	LBX $C_{26}H_{31}N_3O_4$	(87) or (88) ?
21	LBY $C_{25}H_{33}N_3O_4$	4-Epilunarinol
22	LBZ $C_{26}H_{33}N_3O_4$	Alcohol corresponding to (87) or (88) ?
23	Lasiodine-A $C_{39}H_{49}N_5O_7$	Me$_2$ $-$ 3Hyphe $-$ NH $-$ C $-$ C(=O) $-$ 3Hyphe $\rightarrow$ NH $-$ CH $=$ CH $-$ ⬡ $-$ OH (with MeVal)

*Although the name ceanothamine-A was in the chemical literature two years before that of frangulanine, the latter name has been retained by its originators (43).

Melting Point	$[\alpha]_D$ (Solvent)	Species isolated from	References
248°	− 302° (CHCl₃)	*Rhamnus frangula* L.	(*49*)
		Melochia corchorifola	(*51*)
244°	− 299° (CHCl₃)	*Rhamnus frangula* L.	(*49*)
276–279°	− 293° (CHCl₃)	*Ceanothus americanus* L.	(*55*)
275–276°	− 288° (CHCl₃)	*Rhamnus frangula* L.	(*50*)
134°	− 34° (CHCl₃)	*Homalium* sp. from Africa	(*33, 34*)
261°	− 124° (CHCl₃)	*Hymenocardia acida* Tul.	(*30, 31*)
278°	− 228° (CHCl₃)	*Ceanothus integerrimus* Hook and Arn.	(*52*)
258°	not reported	*Ceanothus integerrimus* Hook and Arn.	(*48*)
285° dec.	− 164° (CHCl₃)	*Ceanothus integerrimus* Hook and Arn.	(*52*)
250° dec.	+ 201° (CHCl₃)	*Lunaria biennis* Moench	(*39*)
268–273°	+ 108° (EtOH)	*Lunaria biennis* Moench	(*39*)
amorphous	not reported	*Lunaria biennis* Moench	(*39*)
195°	+ 38° (CHCl₃)	*Lasiodiscus marmoratus* C. H. Wright	(*25*)

Therefore, to avoid proliferation of trivial names the name ceanothamine-A is withdrawn.

(Table 2, continued)

No.	Name (Synonyms) Molecular Formula	Structure
24	Lasiodine-B $C_{35}H_{47}N_5O_5$	MePhe → Pro → 3Hyleu → Leu → NH − CH = CH − (ring)
25	Lunariamine $C_{24}H_{33}N_3O_4$	Not yet known
26	Lunaridine $C_{25}H_{31}N_3O_4$	Stereoisomer of Lunarine?
27	Lunarine $C_{25}H_{31}N_3O_4$	(61)
28	Myrianthine-A $C_{31}H_{42}N_4O_4$	Me₂Leu → 3Hyphe → Ileu → NH − CH = CH − (ring)
	Myrianthine-B	See Adouétine-Y′
29	Myrianthine-C $C_{27}H_{42}N_4O_4$	Me₂Leu → 3Hyleu → Val → NH − CH = CH − (ring)
30	Numismine $C_{25}H_{33}N_3O_4$	Not yet known
31	Pandamine $C_{31}H_{44}N_4O_5$	Me₂Ileu → 3Hyleu → Phe → NH − CH₂ − CH − (ring) / OH
32	Pandaminine $C_{30}H_{42}N_4O_5$	Me₂Val → 3Hyleu → Phe → NH − CH₂ − CH − (ring) / OH
33	Scutianine $C_{39}H_{47}N_5O_5$	Me₂Phe → Pro → 3Hyleu → Phe → NH − CH = CH − (ring)
34	Zizyphine $C_{33}H_{49}N_5O_6$	Me₂Ileu → Ileu → NH − CH = CH (ring structure) MeO −
35	Zizyphinine	Perhaps zizyphine with Val in place of one of the Ileu moieties?

Melting Point	[α]D (Solvent)	Species isolated from	References
221°	— 301° CHCl₃: MeOH 1 : 1)	*Lasiodiscus marmoratus* C. H. Wright	(*25*)
290° (block) 248° (evac. cap.)	no rotation in organic solvents	*Lunaria biennis* Moench	(*19*)
253–254° dec.	0° (HOAc)	*Lunaria rediviva* L.	(*18*)
255° (evac. cap.) 248–250° dec.	+ 233° (CHCl₃?) + 212° (CHCl₃)	*Lunaria biennis* Moench	(*19*) (*5*)
227–231° 224–226° 225–240° (evac. cap.)	+ 300° (CHCl₃) + 298° (CHCl₃)	*Lunaria biennis* Monech	(*5, 44, 46*) (*15*) (*19*)
232–235°	+ 291° (CHCl₃)	*Lunaria rediviva* L.	(*18*)
286°	— 263° (CHCl₃)	*Myrianthus arboreus* P. Beauv.	(*24*)
294°	— 228° (CHCl₃)	*Myrianthus arboreus* P. Beauv.	(*24*)
278–280°	no observable [α]D	*Lunaria biennis* Moench	(*36*)
256°	— 103° (CHCl₃)	*Panda oleosa* Pierre	(*27, 32*)
272°	— 117° (CHCl₃)	*Panda oleosa* Pierre	(*27*)
186–187°	— 399° (CHCl₃)	*Scutia buxifolia* Reiss	(*53*)
121° dec.	— 464° (CHCl₃)	*Zizyphus oenoplia* Mill.	(*26, 61*)
amorphous	— 457° (CHCl₃)	*Zizyphus oenoplia* Mill.	(*26*)

Table 3. *Sources of Peptide Alkaloids*

Plant	Family	Part	% Yield of Crude Bases	No. of Alkaloids Separated	Isolation References
Araliorhamnus vaginatus Perrier	Rhamnaceae	Leaves	0.17	I [a]	(47)
Ceanothus americanus L.	Rhamnaceae	Root bark	0.23 0.83	8 [b]	(55) (20)
Ceanothus integerrimus Hook and Arn.	Rhamnaceae	Roots	0.068	3	(52)
Homalium sp.	Flacourtiaceae	Leaves	not given	I	(33) [e]
Hymenocardia acida Tul.	Hymenocardiaceae	Root bark	0.026	I	(31)
Lasiodiscus marmoratus C. H. Wright	Rhamnaceae	Leaves	0.17	2	(25)
Lunaria biennis Moench	Brassicaceae	Seeds	0.83 0.95	7	(5) (35)
Lunaria rediviva L.	Brassicaceae	Seeds	0.9	2	(18)
Melochia corchorifolia	Sterculiaceae	Leaves and woody part	not given	3	(51) [e]
Myrianthus arboreus P. Beauv.	Urticaceae	Leaves	0.006	3	(24)
Panda oleosa Pierre	Pandaceae	Root bark	0.32	2	(27)
Rhamnus frangula L.	Rhamnaceae	Bark[f]	0.056	3 [c]	(49)
Scutia buxifolia Reiss	Rhamnaceae	Bark	0.17	I [d]	(53)
Waltheria americana L.	Sterculiaceae	Whole plant	0.049	4	(28)
Zizyphus oenoplia Mill.	Rhamnaceae	Root bark	0.86	2	(26)

[a] At least 3 more alkaloids are present; [b] Many more alkaloids are present; [c] At least 4 more alkaloids are present; [d] At least 11 more alkaloids are present; [e] Isolation precedure not given; [f] The same mixture of alkaloids is present in the leaves in smaller amount.

References

1. BERTHO, A. und W. S. LIANG: Notiz über ein Alkaloid aus *Ceanothus americanus.* Arch. Pharmaz. **271**, 273 (1933).

2. BLADON, P., M. CHAIGNEAU, M.-M. JANOT, J. LE MEN, P. POTIER et A. MELERA: Isolement de Dérivés Diphényliques par Fusion Alcaline de la Lunarine. Tetrahedron Letters **1961**, 321.

3. BLADON, P., R. IKAN, F.-S. SPRING and A. D. TAIT: The Chemistry of Lunaria Alkaloids. I. Tetrahedron Letters **1959**, No. 9, 18.

4. BLANPIN, O., M. PAÏS et M. A. QUEVAUVILLER: Étude Pharmacodynamique d'Adouétine Z, Alcaloïde du *Waltheria americana.* Ann. pharm. franç. **21**, 147 (1963).

5. BOIT, H.-G.: Über die Alkaloide von *Lunaria biennis.* Chem. Ber. **87**, 1082 (1954).

6. CLARK, A. H.: The Alkaloids of *Ceanothus americanus.* Amer. J. Pharm. **98**, 147 (1926).

7. — The Alkaloids of Ceanothus americanus. II. Extraction of the Alkaloid. Amer. J. Pharm. **100**, 240 (1928).

8. CLINCH, J. H. M.: Analysis of the Leaves of *Ceanothus americanus* L. Amer. J. Pharm. **56**, 131 (1884).

9. FEHLHABER, H.-W.: Massenspektrometrische Strukturermittlung von Peptid-Alkaloiden. Z. analyt. Chem. **235**, 91 (1968).

10. GERLACH, F. C.: Ceanothus. Amer. J. Pharm. **63**, 332 (1891).

11. GORDIN, H. M.: On the Alkaloids of *Ceanothus americanus.* Pharm. Rev. **18**, 266 (1900).

12. GOUTAREL, R.: Peptides Basiques, Un Type Nouveau d'Alcaloïde. Colloq. Int. Centre Nat. Rech. Sci. No. 144, 159 (1966).

13. GROOT, J. T.: Pharmacology of *Ceanothus americanus.* I. Preliminary Studies: Hemodynamics and the Effects of Coagulation. J. Pharmacol. Exp. Therapeut. **30**, 275 (1927).

14. HAIRS, E.: Un Alcaloïde des Graines du *Lunaria biennis.* Bull. Acad. roy. Belgique **1909**, 1042 [Chem. Abstr. **4**, 1892 (1910)].

15. HANSEN, O. R.: Lunarine, an Alkaloid from *Lunaria biennis.* Acta Chem. Scand. **1**, 656 (1947).

16. HEYNS, K. und H.-F. GRÜTZMACHER: Massenspektren von Freien und N-Acetylierten Aminosäuren. Liebigs Ann. Chem. **667**, 194 (1963).

17. — — Massenspektrometrische Untersuchungen von acylierten Peptiden. Tetrahedron Letters **1963**, 1761.

18. HUNECK, S.: Über die Alkaloide von *Lunaria rediviva* L. Naturwiss. **49**, 233 (1962).

19. JANOT, M.-M. et J. LE MEN: Sur les Alcaloïdes de *Lunaria biennis* Moench. (Crucifères). Bull. soc. chim. France **1956**, 1840.

20. KLEIN, F. K. and H. RAPOPORT: Ceanothus Alkaloids. Americine. J. Amer. Chem. Soc. **90**, 2398 (1968).

21. — — The Structure of Ceanothine-B. J. Amer. Chem. Soc. **90**, 3576 (1968).

22. MA, J. C. N. and E. W. WARNHOFF: On the Use of Nuclear Magnetic Resonance for the Detection, Estimation, and Characterization of N-Methyl Groups. Canad. J. Chem. **43**, 1849 (1965).

23. MANIAN, A. A.: A Study of a Purified Alkaloidal Principle of *Ceanothus americanus.* Ph. D. Thesis. Purdue University, 1954.

24. MARCHAND, J., X. MONSEUR et M. PAÏS: Alcaloïdes Peptidiques — VII (5). Myrianthines A, B et C, Alcaloïdes du *Myrianthus arboreus* P. Beauv. (Urticacées). Ann. pharm. franç. **26**, 771 (1968).

25. Marchand, J., M. Païs, X. Monseur et F.-X. Jarreau: Alcaloïdes Peptidiques — VII. Les Lasiodines A et B, Alcaloïdes du *Lasiodiscus marmoratus* C. H. Wright (Rhamnacées). Tetrahedron **25**, 937 (1969).

26. Ménard, E. L., J. M. Müller, A. F. Thomas, S. S. Bhatnagar und N. J. Dastoor: Über die Inhaltsstoffe von *Zizyphus oenoplia* Mill. 1. Mitteilung: Isolierung der Inhaltsstoffe. Helv. Chim. Acta **46**, 1801 (1963).

27. Païs, M., F.-X. Jarreau, X. Lusinchi et R. Goutarel: Alcaloïdes Peptidiques, III (1). Pandamine et Pandaminine, Alcaloïdes du *Panda oleosa* Pierre (Pandacées). Ann. chimie **1966**, 83.

28. Païs, M., J. Mainil et R. Goutarel: Les Adouétines X, Y et Z, Alcaloïdes du *Waltheria americana* L. (Sterculiacées). Ann. pharm. franç. **21**, 139 (1963).

29. Païs, M., J. Marchand, F.-X. Jarreau et R. Goutarel: Alcaloïdes Peptidiques. V. Structures des Adouétines X, Y, Y′ et Z, Alcaloïdes du *Waltheria americana* L. (Sterculiacées). Bull. soc. chim. France **1968**, 1145.

30. Païs, M., J. Marchand, X. Monseur, F.-X. Jarreau et R. Goutarel: Alcaloïdes Peptidiques. IV. Structure de l'Hymenocardine, Alcaloïde de *l'Hymenocardia acida* Tul. (Euphorbiacées). C. R. hebd. séances Acad. Sci., Sér. C, **264**, 1409 (1967).

31. Païs, M., J. Marchand, G. Ratle et F.-X. Jarreau: Alcaloïdes Peptidiques. (VI) (1). l'Hymenocardine, Alcaloide de *l'Hymenocardia acida* Tul. (Hymenocardiacées). Bull. soc. chim. France **1968**, 2979.

32. Païs, M., X. Monseur, X. Lusinchi et R. Goutarel: Alcaloïdes Peptidiques. II. Structure de la Pandamine, Alcaloïde du *Panda oleosa* Pierre (Pandacées). Bull. soc. chim. France **1964**, 817.

33. Païs, M., G. Ratle, R. Sarfati et F.-X. Jarreau: L'homaline, Nouveau Type d'Alcaloïde Isolé d'un *Homalium* Sp. Africain (Flacourtiacées). C. R. hebd. séances Acad. Sci., Ser. C, **266**, 37 (1968).

34. — — — — A Propos de la Structure de l'Homaline. C. R. hebd. séances Acad. Sci., Ser. C, **267**, 82 (1968).

35. Potier, P. et J. Le Men: Étude Chimique de la Lunarine, Alcaloïde du *Lunaria biennis* Moench. (Crucifères). Bull. soc. chim. France **1959**, 456.

36. Potier, P., J. Le Men et M.-M. Janot: Sur la Numismine, Nouvel Alcaloïde Isolé des Graines du *Lunaria biennis*. Bull. soc. chim. France **1959**, 201.

37. Potier, P., J. Le Men, M.-M. Janot and P. Bladon: The Isolation of Spermidine by Degradation of Lunarine. Tetrahedron Letters **1960**, No. 18, 36.

38. Potier, P., J. Le Men, M.-M. Janot, P. Bladon, A. G. Brown and C. S. Wilson: The Structure of Lunarine. Tetrahedron Letters **1963**, 293.

39. Poupat, C., B. Rodriguez, H.-P. Husson, P. Potier et M.-M. Janot: Nouveaux Alcaloïdes des Graines de la Monnaie du Pape, *Lunaria biennis* Moench (Crucifères). C. R. hebd. séances Acad. Sci., Sér. C, **269**, 335 (1969).

40. Richards, L. W. and E. V. Lynn: An Examination of *Ceanothus velutinus*. J. Amer. Pharm. Assoc. **23**, 332 (1934).

41. Roscoe, C. W. and N. A. Hall: A Preliminary Study of the Alkaloidal Principles of *Ceanothus americanus* and *Ceanothus velutinus*. J. Amer. Pharm. Assoc. Sci. Ed. **49**, 108 (1960).

42. Servis, R. E. and A. I. Kosak: A Revised Structure for Ceanothine-B. J. Amer. Chem. Soc. **90**, 4179 (1968).

43. Servis, R. E., A. I. Kosak, R. Tschesche, E. Frohberg and H. W. Fehlhaber: Peptide Alkaloids from *Ceanothus americanus* L. (Rhamnaceae). J. Amer. Chem. Soc. **91**, 5619 (1969).

44. Steinegger, E. and T. Reichstein: Lunarin, das kristallisierte Alkaloid aus *Lunaria biennis*. Pharmac. Acta. Helv. **22**, 258 (1947).

45. SVEC, H. J. and G. A. JUNK: The Mass Spectra of Dipeptides. J. Amer. Chem. Soc. 86, 2278 (1964).

46. TAMURA, C., G. A. SIM, J. A. D. JEFFREYS, P. BLADON and G. FERGUSON: Lunarine. Chem. Commun. 1965, 485.

47. TSCHESCHE, R., L. BEHRENDT und H.-W. FEHLHABER: Alkaloide aus Rhamnaceen, VI. Aralionin, ein Peptid-Alkaloid aus *Araliorhammus vaginatus* Perrier. Chem. Ber. 102, 50 (1969).

48. TSCHESCHE, R., E. FROHBERG und H.-W. FEHLHABER: Alkaloide aus Rhamnaceen, IV. Integerrin, ein weiteres Peptid-Alkaloid aus *Ceanothus integerrimus* Hook und Arn. Tetrahedron Letters 1968, 1311.

49. TSCHESCHE, R. und H. LAST: Alkaloide aus Rhamnaceen, V. Franganin und Frangufolin. Zwei weitere Peptid-Alkaloide aus *Rhamnus frangula* L. Tetrahedron Letters 1968, 2993.

50. TSCHESCHE, R., H. LAST und H.-W. FEHLHABER: Alkaloide aus Rhamnaceen, III. Frangulanin, ein Peptid-Alkaloid aus *Rhamnus frangula* L. Chem. Ber. 100, 3937 (1967).

51. TSCHESCHE, R. und I. REUTEL: Alkaloïde aus Sterculiaceen, I. Über Peptidalkaloide aus *Melochia corchorifolia*. Tetrahedron Letters 1968, 3817.

52. TSCHESCHE, R., J. RHEINGANS, H.-W. FEHLHABER und G. LEGLER: Integerrissin und Integerrenin, zwei Peptid-Alkaloide aus *Ceanothus integerrimus* Hook und Arn. Chem. Ber. 100, 3924 (1967).

53. TSCHESCHE, R., R. WELTERS und H.-W. FEHLHABER: Alkaloide aus Rhamnaceen, I. Scutianin, ein cyclisches Peptid-Alkaloid aus *Scutia buxifolia* Reiss. Chem. Ber. 100, 323 (1967).

54. WARNHOFF, E. W., J. C. N. MA and P. REYNOLDS-WARNHOFF: Ceanothine-B, a Naturally Occurring Oxazacylononadiene. J. Amer. Chem. Soc. 87, 4198 (1965).

55. WARNHOFF, E. W., S. K. PRADHAN and J. C. N. MA: Ceanothus Alkaloids I. Isolation, Separation and Characterization. Canad. J. Chem. 43, 2594 (1965).

56. — unpublished results.

57. WASTL, H.: Influence of Tea from Leaves of *Ceanothus americanus* on Blood Pressure of Hypertensive Rats. Federat. Proc. (Amer. Soc. Exp. Biol.) 7, 131 (1948).

58. WIELAND, TH.: The Toxic Peptides of *Amanita Phalloides*. Fortschr. Chem. organ. Naturstoffe 25, 214 (1967).

59. WIESNER, K., D. M. MacDONALD, C. BANKIEWICZ and D. E. ORR: Structure of Pithecolobine. II. Canad. J. Chem. 46, 1881 (1968).

60. WIESNER, K., Z. VALENTA, D. E. ORR, V. LIEDE and G. KOHAN: Structure of Pithecolobine. III. The Synthesis of the 1,5- and 1,3-Desoxypithecolobines. Canad. J. Chem. 46, 3617 (1968).

61. ZBIRAL, E., E. L. MÉNARD und J. M. MÜLLER: Über die Inhaltsstoffe von *Zizyphus oenoplia* Mill. 2. Mitteilung. Zur Konstitutionsermittlung des Zizyphins. Helv. Chim. Acta. 48, 404 (1965).

(Received, February 16, 1970)

Insektensexuallockstoffe

Von **K. Eiter**, Köln

I. Einleitung

Unter Insektensexuallockstoffen versteht man Substanzen, die von einem copulationsbereiten Insekt aus bestimmten Organen, den sacculi lateralis, abgesondert werden und zur Anlockung des Partners dienen; hierbei kann es sich um hochartspezifische Einzelsubstanzen als auch um Substanzgemische handeln, wobei vielleicht Synergismen eine bestimmte Rolle spielen könnten. Vielfach beobachtete Antagonismen, die von der teilweisen Maskierung bis zur vollständigen Aufhebung des Lockeffektes von Sexuallockstoffen etwa durch das Vorhandensein geometrisch isomerer Verbindungen verursacht werden sollen, sind vielfach beschrieben worden (6, 23, 44, 45, 76, 109), bedürfen jedoch eingehender wissenschaftlicher Bestätigung. In Erweiterung des Begriffes der Insektensexuallockstoffe hat P. Karlson (61) den Terminus „Pheromone" geprägt, der über das vorher Gesagte alle jene im Insektenreich vorkommenden Substanzen umfassen soll, die über die Lockwirkung und sexuelle Erregung hinaus eine Korrelation innerhalb von Insektenpopulationen bewirken, etwa im Sinne der „Queen-Substance" bei der Honigbiene *Apis mellifera* (*20*) oder den Weg-Markierungssubstanzen zur Erreichung günstiger Holzquellen bei den Termiten (*95, 99*).

Synthetische Substanzen mit gewissen Lockwirkungen auf Insekten sollen hier nicht näher beschrieben werden, da diese „Lockstoffattrappen" sich in ihren Aktivitäten meistens um mehrere Zehnerpotenzen, ausgedrückt in Lockstoffeinheiten (LE) von den artspezifischen natürlichen Lockstoffen unterscheiden. Unter Lockstoffeinheit versteht man jene Konzentration des Lockstoffes in μg/ml Lösungsmittel (Petroläther Kp. 30—50°), die imstande ist, bei 50% der Versuchstiere durch Vorhalten eines mit der Lösung benetzten Glasstabes in ~ 1 cm Abstand vor die Antennen eine eindeutige Schwirreaktion zu verursachen; auch sogenannte Fraßlockstoffe sowie Substanzen, die über gewisse Lichteffekte (auch Luminiszenzen) zur Anlockung von Insekten führen können, sollen in diesem Rahmen nicht erwähnt werden.

Auch die Insektenhormone, die bei der Metamorphose eine wichtige Rolle spielen und die neben den Insektensexuallockstoffen theoretisch für die Entwicklung eines neuzeitlichen Pflanzenschutzes von hervorragender Bedeutung sind, sollen hier kurz erwähnt werden, da es sich hierbei um gleichfalls hochwirksame Pheromone handelt, die chemisch von den mehrfach ungesättigten aliphatischen Alkoholen oder Estern bis zu den Terpenen und Steroiden reichen, wobei mittlerweile synthetisierte Kunstprodukte, z. B. auf dem Gebiet der Sterilisantien gleichfalls unberücksichtigt bleiben sollen.

Über die Insektensexuallockstoffe sind schon eine Reihe von Zusammenfassungen erschienen wie z. B. die Monographie „Insect Sex Attractants" von M. Jacobson (*46*), „Chemistry of Sex Attractants of Insects" von M. Ohno (*73*), „Insect Sex Attractants" von Y. und B. Pyatnova, L. L. Ivanov, and A. S. Kyskina (*77*) sowie „Insekten-Sexuallockstoffe" von K. Eiter in R. Wegler: „Chemie der Pflanzenschutz- und Schädlingsbekämpfungsmittel" (*110*). Diese Übersicht soll besonders auf die Chemie der Insektensexuallockstoffe eingehen, wobei versucht werden wird, eine kritische Sichtung des vorliegenden Materials vorzunehmen, da seit der epochalen Isolierung und Aufklärung des Bombykols (10-cis-12-trans-Hexadecadienols-(1) durch A. Butenandt und Mitarb. (*1, 23*) sowie die sterisch einheitliche Darstellung dieses Pheromons durch E. Truscheit und K. Eiter (*15, 16, 102*) sowie A. Butenandt und Mitarb. (*17*) die Isolierung und Synthese einer Reihe von Sexuallockstoffen beschrieben wurde, deren Nacharbeitung zu biologisch unwirksamen Verbindungen führte; soweit nicht chemisch eindeutige Synthesen falsch ermittelte Konstitutionen von Insektensexuallockstoffen grundsätzlich abklären konnten (etwa beim Lockstoff von *Periplaneta americana*, der amerikanischen Küchenschabe) wird von den Bearbeitern von Insektensexuallockstoffen volkswirtschaftlich wichtiger Schadinsekten, etwa des Schwammspinners (gipsy moth) oder der Baumwollkapselraupe (*Pectinophora gossybiella* S.) darauf hingewiesen, daß geringste Beimengungen von geometrisch isomeren Verbindungen die Lockstoffwirkung völlig blockieren könnten, ein Umstand, der z. B. an dem von E. TRUSCHEIT und K. EITER sowie A. BUTENANDT und seiner Schule bearbeiteten klassischen Beispiel des Bombykols niemals beobachtet wurde! Da der Spruch „*natura non facit saltus*" gerade in der Biologie seine tiefe Bedeutung besitzt, müssen die bislang bekanntgewordenen chemischen Fakten mit Akribie geprüft und daraus Schlußfolgerungen gezogen werden.

II. Insektensexuallockstoffe

a) Olefinische Verbindungen

1. Sexuallockstoff der Wasserwanze *(Lethocerus indicus, Belostoma indica):*

Das Männchen der vornehmlich in Kambodscha beheimateten Wasserwanze besitzt am Abdomen zwei Röhrchen mit einer durchsichtigen citronenartig riechenden Flüssigkeit, von der sich die Entomologen noch nicht sicher sind, ob sie die weibliche Wasserwanze anlockt oder zur Abwehr dient. Da die Substanz als 2-*trans*-Hexenolacetat-(1) von A. BUTENANDT und N. D. TAM (*19*) erkannt wurde, soll sie hier unter den Sexuallockstoffen aufgeführt werden, da die höheren Homologen von (1), das 2,4-Hexadienol-(1) (2), das 2,4,6-Octatrienol-(1) (3) mit zunehmender Zahl an konjugierten Doppelbindungen auch eine erhöhte Lockwirkung zeigt.

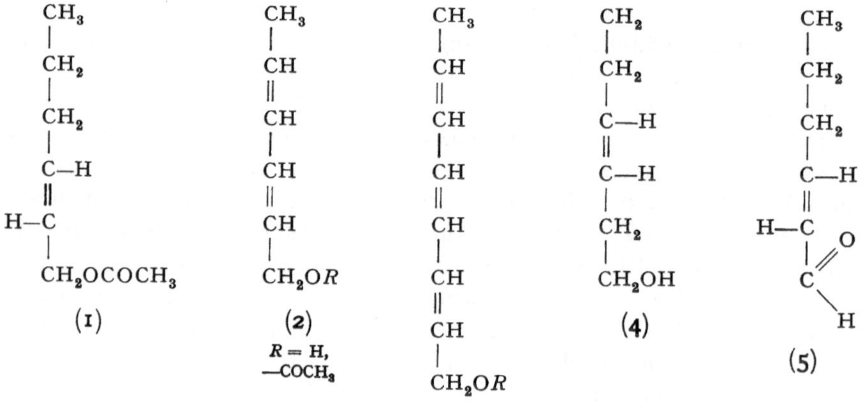

(1) (2) (3) (4) (5)

$R = H, -COCH_3$

Literaturverzeichnis: SS. 250—255

Die Konstitution von (1) konnte durch folgende Synthese bestätigt werden:

$$CH_3CH_2CH_2CHO + CH_2\begin{array}{l}COOH\\ \\COOC_2H_5\end{array} \xrightarrow[\text{Piperidin}]{\text{Pyr.}} CH_3-CH_2-CH_2-\overset{\overset{\textstyle H}{|}}{C}=\overset{\overset{\textstyle H}{|}}{C}-COOC_2H_5$$

$$(6)$$

$$\downarrow \text{LAH}$$

$$CH_3COOCH_2-\overset{\overset{\textstyle H}{|}}{C}=\overset{\overset{\textstyle}{|}}{\underset{\underset{\textstyle H}{|}}{C}}-CH_2CH_2CH_3 \xleftarrow{CH_3COCl} CH_3-CH_2-CH_2-\overset{\overset{\textstyle H}{|}}{C}=\overset{\overset{\textstyle}{|}}{\underset{\underset{\textstyle H}{|}}{C}}-CH_2OH$$

$$(8) \qquad\qquad (7)$$

Das 3-*cis*-Hexenol-(1) (4), auch Blätteralkohol bezeichnet, ist ebenso wie der Blätteraldehyd (5) das 2-*trans*-Hexenal-(1) in der Natur weit verbreitet und hauptsächlich in frischen grünen Blättern enthalten; vermutlich besitzen diese beiden Substanzen eine Fraßlockstoffwirkung, da sich Larven und Würmer bevorzugt von grünen Blättern ernähren.

Blätteralkohol und Blätteraldehyd werden übrigens als Faktoren angesehen, die für die Bildung der Schutzfarben der Nymphen von *Papilio xuthens* (*71*) verantwortlich sein sollen. Der Blätteraldehyd (5) wird unter anderem als Geruchsstoff von Schaben ausgeschieden und zwar von *Eurycotis floridana* (Walker) (*85*), der Ameise *Crematogaster (Atropogyne) africana* (Mayer) (*11*) und *Pentatomidae Coreidae* (*106*). Bei diesen Spezies scheint der Duft bevorzugt als Wegmarkierung sowie als Abwehr- und Schutzgeruch für Feinde zu dienen. Auch als Stimulans für die Abgabe des Sexuallockstoffes von *Antheraea polyphemus* wurde der Blätteraldehyd erkannt (*78*).

2-trans-Hexenal-(1) ist synthetisch sehr gut zugänglich aus n-Butyraldehyd und Vinyläthyläther in Gegenwart von BF_3-Ätherat (*4*)

$$n\text{-}CH_2CH_2CH_2CHO + CH_2=CH-OC_2H_5 \xrightarrow[\text{H}^+]{BF_3\text{-Ätherat}} CH_3CH_2CH_2-\overset{\overset{\textstyle H}{|}}{C}=\overset{\overset{\textstyle}{|}}{\underset{\underset{\textstyle H}{|}}{C}}-C\overset{\textstyle O}{\diagup\!\!\!\diagdown}H$$

$$(9)$$

2. Sexuallockstoff des Seidenspinners *(Bombyx mori L.)*:

A. Butenandt, R. Beckmann, D. Stamm und E. Hecker (*1, 23*) gelang es, den Sexuallockstoff des Seidenspinners aus den Hinterleibsegmenten weiblicher Spinner zu isolieren und über das p-Nitroazobenzoat chromatographisch zu reinigen, so daß eine Menge von etwa 15 mg Sexuallockstoff aus 500.000 Abdominalsegmenten gewonnen, zur Konstitutionsermittlung zur Verfügung stand. Auf Grund IR-spektroskopischer Befunde und chemischer Abbaureaktionen wurde festgestellt, daß es sich bei dem Lockstoff des Seidenspinners um ein cis-trans konjugiertes Hexadecadien-(10,12)-ol-(1) handeln muß, wobei nur durch sterisch einheitliche Totalsynthesen und Vergleich der biologischen Aktivität mit dem Naturprodukt festzustellen war, ob es sich um das 10-*cis*-12-*trans*-(13) oder das 10-*trans*-12-*cis*-Hexadecadienol-(1) (12) handelte.

$$CH_3-(CH_2)_2-\overset{\overset{\displaystyle H}{|}}{C}=\overset{\overset{\displaystyle H}{|}}{C}-\underset{\underset{\displaystyle H}{|}}{C}=\underset{\underset{\displaystyle H}{|}}{C}-(CH_2)_8-CH_2OH$$

(10)

$$CH_3-(CH_2)_2-\overset{\overset{\displaystyle H}{|}}{C}=\underset{\underset{\displaystyle H}{|}}{C}-\overset{\overset{\displaystyle H}{|}}{C}=\underset{\underset{\displaystyle CH_2OH}{|}}{C}-(CH_2)_8$$

(11)

$$CH_3-(CH_2)_2-\overset{\overset{\displaystyle H}{|}}{C}=\overset{\overset{\displaystyle H}{|}}{C}-\underset{\underset{\displaystyle H}{|}}{C}=\overset{\overset{\displaystyle H}{|}}{C}-(CH_2)_8-CH_2OH$$

(12)

$$CH_3-(CH_2)_2-\overset{\overset{\displaystyle H}{|}}{C}=\underset{\underset{\displaystyle H}{|}}{C}-\overset{\overset{\displaystyle H}{|}}{C}=\overset{\overset{\displaystyle H}{|}}{C}-(CH_2)_8-CH_2OH$$ \; CH_2OH

(13)

E. Truscheit und K. Eiter (*15, 16, 102, 103, 104, 105*) hatten nach dem Aufbauschema

$$C_6 + C_{10} = C_{16}$$

$$CH_3-(CH_2)_2-CH=CH-CH=P(C_6H_5)_3 + OHC-(CH_2)_8-R \xrightarrow{-(C_6H_5)_3PO}$$

(14) \qquad\qquad\qquad (15)

$$\rightarrow CH_3-(CH_2)_2-CH=CH-CH=CH-(CH_2)_8-R$$

(16) a: R = COOCH₃, b: R = CH₂OCOCH₃, c: R = COOH, d: R = CH₂OH

zuerst ein Hexadecadienol erhalten, das ein Gemisch von 10-*trans*-12-*trans*- (11) und 10-*cis*-12-*trans*-Hexadecadienol (13) darstellte, in seinen physikalischen Eigenschaften jedoch praktisch identisch mit dem von Butenandt und Mitarb. isolierten Naturstoff war. Da in der Synthese die Wittig-Reaktion mit dem Phosphorylen aus 2-*trans*-Hexenylbromid-(1) (20) und Decanalester durchgeführt worden war und die bei der Olefinierung gebildete neue Doppelbindung in ihrer *cis*- und *trans*-Form anfällt, konnte es sich im Syntheseprodukt nur um ein Gemisch aus 10-*trans*-12-*trans* (11) und 10-*cis*-12-*trans*-Hexadecadienol-(1) (13) handeln. Die Prüfung der biologischen Aktivität ergab für die LE 10⁻² μg während das Naturprodukt 10⁻¹⁰ μg besaß. Die Differenz von acht Zehner-

potenzen in der Wirksamkeit beim Seidenspinner war beweisend, daß
es sich beim natürlichen Lockstoff nur um das 10-*trans*-12-*cis*-Hexa-
decadienol-(1) (*12*) handeln konnte.

Im weiteren Verlauf der Untersuchungen wurden ebenso wie von
A. BUTENANDT und Mitarb. (*17*) alle nur theoretisch möglichen Hexadecadienole sterisch einheitlich dargestellt und ihre biologischen Aktivitäten ermittelt.

Aus Blätteraldehyd (2-*trans*-Hexenal) erhält man durch Reduktion
mit LAH

$$CH_3-(CH_2)_2-\overset{\overset{\textstyle H}{|}}{C}=\overset{\overset{\textstyle H}{|}}{C}-CHO \xrightarrow{\text{LAH}} CH_3-(CH_2)_2-\overset{\overset{\textstyle H}{|}}{C}=\overset{\overset{\textstyle H}{|}}{C}-CH_2OH$$

(5)   (17) $\downarrow$ PBr$_3$

$$CH_3-(CH_2)_2-\overset{\overset{\textstyle H}{|}}{C}=\overset{\overset{\textstyle H}{|}}{C}-CH_2\overset{\overset{\textstyle \oplus}{}}{P}\underset{(C_6H_5)_3}{}Br^{\ominus} \xleftarrow{(C_6H_5)_3P} CH_3-(CH_2)_2-\overset{\overset{\textstyle H}{|}}{C}=\overset{\overset{\textstyle H}{|}}{C}-CH_2Br$$

$\downarrow$ C$_4$H$_9$Li (19)   (18)

$$CH_3-(CH_2)_2-\overset{\overset{\textstyle H}{|}}{C}=\overset{\overset{\textstyle H}{|}}{C}-CH-P(C_6H_5)_3 \xrightarrow[\text{säureester(10)}]{\text{+ Decanal-(1)-}} CH_3-(CH_2)_2-\overset{\overset{\textstyle H}{|}}{C}=\overset{\overset{\textstyle H}{|}}{C}-\overset{\overset{\textstyle H}{|}}{C}=\overset{\overset{\textstyle (CH_2)_8-COO R}{\overset{\textstyle H}{|}}}{C}$$

(20)   (21)

+

$$CH_3-(CH_2)_2-\overset{\overset{\textstyle H}{|}}{C}=\overset{\overset{\textstyle H}{|}}{C}-\overset{\overset{\textstyle H}{|}}{C}=\overset{\overset{\textstyle H}{|}}{\underset{\underset{\textstyle (CH_2)_8-COO R}{|}}{C}}$$

(22)

(Lithiumalanat) das 2-*trans*-Hexenol-(1) (*17*), das mit PBr$_3$ in das Bromid
(18) übergeführt wurde, daraus war mit Triphenylphosphin leicht das
Phosphoniumsalz (19) herzustellen, das mit Butyllithium zum Ylen (20)
umgesetzt und mit Decanalester zu einem Gemisch von 10-*trans*-12-
trans-(21) und 10-*cis*-12-*trans*-Hexadecadiensäure-(1)-methylester (22)
olefiniert werden konnte. Die alkalische Verseifung des Estergemisches
lieferte zwei isomere Säuren, die kristallisiert erhalten wurden. Die
trans-trans konfigurierte Säure schmilzt bei 54—55°, während die 10-*cis*-

12-*trans*-Säure den Schmelzp. 26—27° zeigte. Darüber hinaus konnte die trans-trans-Säure durch Clathratbildung mit Harnstoff noch in Spuren vom 10-*cis*-12-*trans* Isomeren abgetrennt werden, was die Reindarstellung weiters erleichterte.

Die Reduktion sowohl der reinen 10-*trans*-12-*trans*-Hexadecadiensäure-(1) als auch der 10-*cis*-12-*trans*-Hexadecadiensäure-(1) mit Lithiumalanat (LAH) lieferte das sterisch einheitliche 10-*trans*-12-*trans*-Hexadecadienol-(1) (16a) bzw. das 10-*cis*-12-*trans*-Hexadecadienol-(1) (13).

Durch Verwendung von 2-*cis*-Hexenylbromid, hergestellt aus Hexin-(2)-ol-(1) durch Reduktion mit einem modifizierten Lindlar-Katalysator und Bromierung des sterisch einheitlichen 2-*cis*-Hexenol-(1) erhielt man in ähnlicher Reaktionsfolge ein Gemisch isomerer 10,12-Hexadecadiensäuren-(1), die durch fraktionierte Kristallisation völlig rein erhalten werden konnten; 10-*cis*-12-*cis*-Hexadecadiensäure-(1) schmilzt bei 34 bis 35°, das 10-*trans*-12-*cis*-Isomere bei 25—26°. Reduktion mit LAH lieferte das 10-*cis*-12-*cis*-Hexadecadienol-(1) (10) bzw. das 10-*trans*-12-*cis*-Hexadecadienol-(1) (12) (Bombykol)

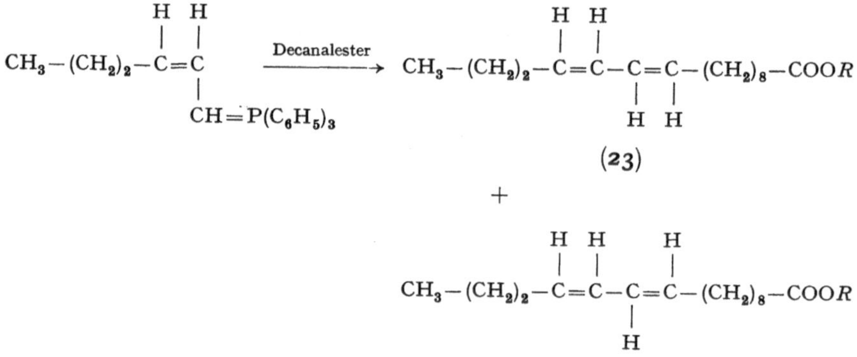

(23)

+

(24) a: $R = CH_3$, b: $R = H$

Aufbauschema

$$C_4 + C_{12} = C_{16}$$

Die Aufbaumethode $C_4 + C_{12} = C_{16}$ ermöglicht die Fixierung der Doppelbindung in 12-Stellung in der trans-Konfiguration, da sie in diesem Falle in der C_{12}-Aldehydkomponente bereits vorgebildet ist

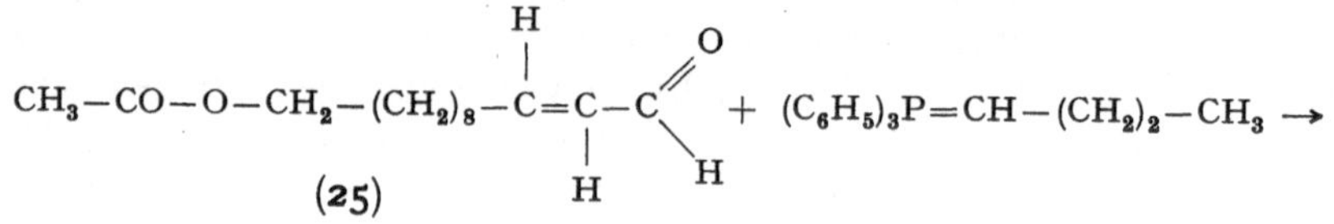

(25)

Literaturverzeichnis: SS. 250—255

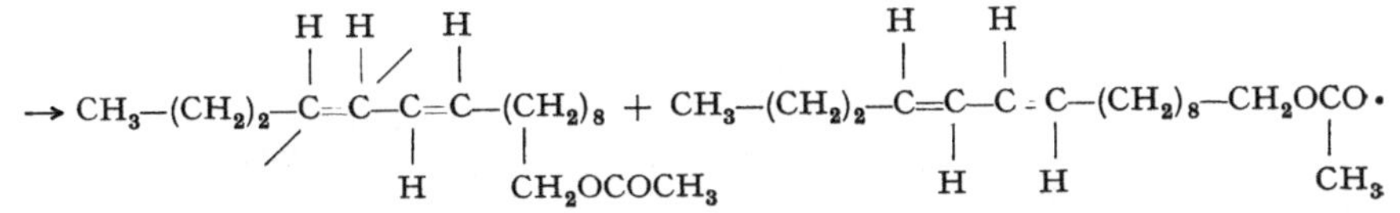

$$\rightarrow CH_3-(CH_2)_2-\overset{\overset{H}{|}}{C}=\overset{\overset{H}{|}}{\underset{\underset{H}{|}}{C}}-\overset{\overset{H}{|}}{C}=\overset{H}{\underset{\underset{CH_2OCOCH_3}{|}}{C}}-(CH_2)_8 + CH_3-(CH_2)_2-\overset{\overset{H}{|}}{C}=\overset{\overset{H}{|}}{\underset{\underset{H}{|}}{C}}-\overset{}{C}=\overset{}{\underset{\underset{CH_3}{|}}{C}}-(CH_2)_8-CH_2OCO\cdot$$

(26) (27)

Die C_{12}-Aldehydkomponente war aus Traumatinsäuredimethylester gut zugänglich, da nach der Verseifung der bei der Olefinierung erhaltenen Ester die 10-*trans*-12-*cis*-Verbindung (24 b) in Hauptmenge und die 10-*trans*-12-*trans*-Verbindung nur in Spuren und damit als Harnstoff-Einschlußverbindung leicht abtrennbar, anfiel, war diese Methode besonders zur Darstellung von Bombykol geeignet.

Aufbauschema

$$C_3 + C_{13} = C_{16}$$

Nach dieser Aufbaumethode können theoretisch alle vier isomeren Hexadecadienole synthetisiert werden, besonders in präparativer Hinsicht zeigte sich dieses Verfahren für die Herstellung von 10-*trans*-12-*cis*-Hexadecadienol-(I) (Bombykol) sowie des 10-*cis*-12-*cis*-Isomeren sehr geeignet.

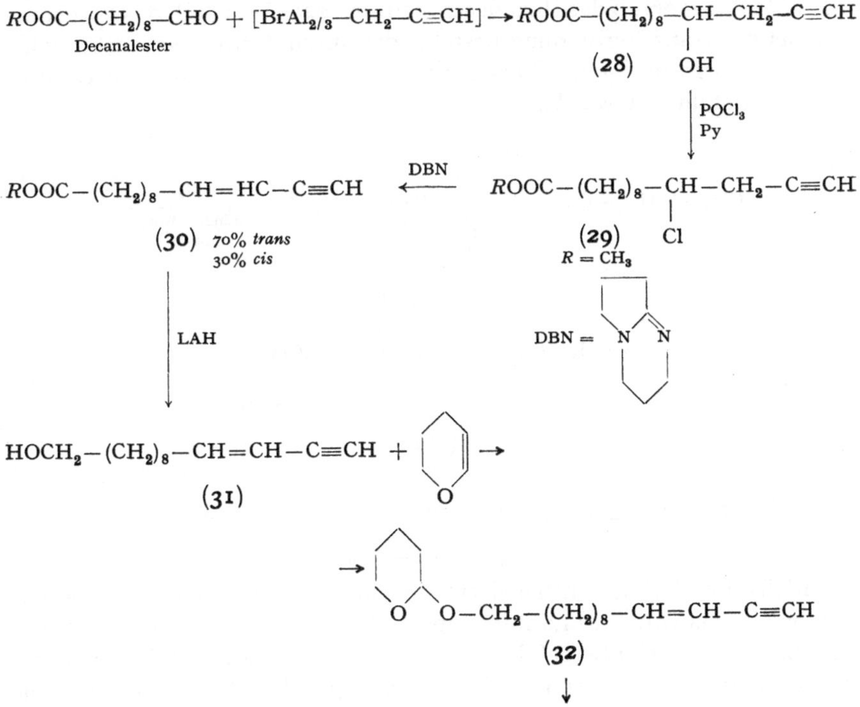

$$ROOC-(CH_2)_8-CHO + [BrAl_{2/3}-CH_2-C\equiv CH] \rightarrow ROOC-(CH_2)_8-\overset{}{\underset{\underset{OH}{|}}{CH}}-CH_2-C\equiv CH$$

Decanalester (28)

$$\Big\downarrow \; \begin{array}{l}POCl_3\\ Py\end{array}$$

$$ROOC-(CH_2)_8-CH=HC-C\equiv CH \overset{DBN}{\longleftarrow} ROOC-(CH_2)_8-\overset{}{\underset{\underset{Cl}{|}}{CH}}-CH_2-C\equiv CH$$

(30) 70% *trans* (29)
30% *cis* R = CH$_3$

$$\Big\downarrow \; LAH \qquad\qquad DBN = \; \text{(structure)}$$

$$HOCH_2-(CH_2)_8-CH=CH-C\equiv CH + \;\text{(structure)} \rightarrow$$

(31)

$$\rightarrow \;\text{(structure)}\; O-CH_2-(CH_2)_8-CH=CH-C\equiv CH$$

(32)

$$\downarrow$$

(**32**)

$H^{\oplus}$

1. LiNH$_2$/fl. NH$_3$
2. CH$_3$CH$_2$CH$_2$Br

$\text{O} \quad \text{O}-\text{CH}_2-(\text{CH}_2)_8-\text{CH}=\text{CH}-\text{C}\equiv\text{C}-(\text{CH}_2)_2-\text{CH}_3$

(**33**)

$\text{HOCH}_2-(\text{CH}_2)_8-\text{CH}=\text{CH}-\text{C}\equiv\text{C}-(\text{CH}_2)_2-\text{CH}_3 \leftarrow$

(**34**)

Die aluminiumorganische Verbindung des Propargylbromids konnte sehr glatt mit dem Decanalester zum Acetylencarbinol (**28**) kondensiert werden, das in das Chlorid (**29**) übergeführt und mit DBN zur En-in-Verbindung (**30**) dehydrohalogeniert werden konnte; bei dieser durch Dehydrohalogenierung erfolgten Einführung der Doppelbindung entsteht ein Gemisch von vorwiegend *trans*-Verbindung neben etwas *cis*-Isomerem. Reduktion dieses Gemisches mit Lithiumalanat ergibt ein Alkoholgemisch (**31**), das in den Pyranyläther (**32**) übergeführt wurde und mit Lithiumamid in fl. NH$_3$ mit n-Propylbromid an der Acetylengruppierung alkyliert werden konnte. Hydrolyse des 1-[Tetrahydropyranyloxy]-hexadien-(10)-in-(12) (**32**) ergab das Hexadecen -(10)-in-(12)-ol-(1) (**34**), das vornehmlich aus der trans-Verbindung besteht, die durch Tieftemperaturkristallisation bei − 15 bis − 20° C bis zur Erreichung einer konstanten Extinktion umkristallisiert wurde.

$$\text{CH}_3-(\text{CH}_2)_2-\text{C}\equiv\text{C}-\overset{\overset{\displaystyle H}{|}}{\underset{\underset{\displaystyle H}{|}}{\text{C}}}=\text{C}-(\text{CH}_2)_8-\text{CH}_2\text{OH} \xrightarrow[\substack{\text{Lindlar Kat.}\\ \text{modif.}}]{\text{H}_2}$$

$$\rightarrow \text{CH}_3-(\text{CH}_2)_2-\overset{\overset{\displaystyle H}{|}}{\text{C}}=\overset{\overset{\displaystyle H}{|}}{\underset{\underset{\displaystyle H}{|}}{\text{C}}}-\overset{\overset{\displaystyle H}{|}}{\underset{\underset{\displaystyle CH_2OH}{|}}{\text{C}}}=\text{C}-(\text{CH}_2)_8$$

(**12**)

A. Butenandt und Mitarb. (*17*) haben nach dem Aufbauschema

$$C_6 + C_{10} = C_{16}$$

gleichfalls Bombykol synthetisiert; sie gingen dabei vom 2-Hexinol-(1) aus, das in das Bromid und mit Triphenylphosphin in das Phosphoniumsalz bzw. Ylen (**35**) übergeführt wurde. Kondensation mit Decanalester ergab ein cis-trans-Isomerengemisch etwa zu gleichen Teilen, das über die

Literaturverzeichnis: SS. 250—255

Harnstoffeinschlußverbindungen aufgetrennt worden war. Das 10-*trans*-en-12-in-(37) konnte mit Lindlar-Katalysator partiell zum 10-*trans*-12-*cis*-Hexadecadiensäureester-(1) partiell hydriert werden, eine Reduktion mit LAH lieferte Bombykol (12)

$$CH_3-(CH_2)_2-C\equiv C-CH=P(C_6H_5)_3 + OHC-(CH_2)_8-COOC_2H_5 \rightarrow$$

$$(35)$$

$$\rightarrow CH_3-(CH_2)_2-C\equiv C-CH=CH-(CH_2)_8-COOC_2H_5 \xrightarrow{\text{Harnstofftrennung}}$$

$$(38)$$

$$\rightarrow CH_3-(CH_2)_2-C\equiv C-\overset{\displaystyle H}{\underset{\displaystyle H}{\overset{|}{C}}}=\overset{}{\underset{\displaystyle COOC_2H_5}{\overset{|}{C}}}-(CH_2)_8 \quad \xrightarrow[\text{2. LAH}]{\text{1. Lindlar}}$$

$$(37)$$

$$\rightarrow CH_3-(CH_2)_2-\overset{\displaystyle H}{\overset{|}{C}}=\overset{\displaystyle H}{\overset{|}{C}}-\overset{\displaystyle H}{\overset{|}{C}}=\overset{}{\underset{\displaystyle H}{\overset{|}{C}}}-(CH_2)_8-CH_2OH$$

$$(12)$$

Auf dem Wege

$$C_5 + C_{11} = C_{16}$$

konnte auch 10-*trans*-12-*cis*-Hexadecadienol erhalten werden.

$$CH_3(CH_2)_2-C\equiv CLi + CH_2-CH-(CH_2)_8-CH_2-O- \overset{\triangle}{\underset{O}{\bigcirc}} \rightarrow$$

$$(38) \qquad\qquad (39)$$

$$\rightarrow CH_3-(CH_2)_2-C\equiv C-CH_2-\underset{\displaystyle OH}{\overset{|}{C}H}-(CH_2)_8-CH_2-O-\overset{\triangle}{\underset{O}{\bigcirc}} \xrightarrow[\substack{\text{2. } H^{\oplus} \\ \text{3. } H_2/Pd/C}]{\text{1. Tos.-Cl, NaOH}}$$

$$(40)$$

$$\rightarrow CH_3(CH_2)_2-\overset{\displaystyle H}{\overset{|}{C}}=\overset{\displaystyle H}{\overset{|}{C}}-\overset{}{\underset{\displaystyle H}{\overset{|}{C}}}=\overset{\displaystyle H}{\overset{|}{C}}-(CH_2)_8-CH_2OH$$

$$(12)$$

Bei der Dehydratisierung von (40) fiel die entstehende en-in-Verbindung als *cis-trans*-Isomerengemisch an, das aufgetrennt und nur die 10-*trans*-Komponente zur Partialreduktion verwendet wurde.

Nach dem Aufbauprinzip

$$C_7 + C_9 = C_{16}$$

konnte gleichfalls von A. Butenandt und Mitarb. eine Synthese von Hexadecadienolen durchgeführt werden; hierbei ließ man die Natriumverbindung des 3-*trans-cis*-Heptenin-(1)-Gemisches (41) mit dem Tetrahydropyranyläther des 9-Bromnonanols (42) reagieren, wobei nach Reduktion der Dreifachverbindung mit Wasserstoff und Lindlar-Katalysator ein Gemisch von 10-*cis*-12-*trans*- und 10-*cis*-12-*cis*-Hexadecadienol-(1) (44) anfiel.

$$CH_3(CH_2)_2CH=CH-C\equiv CNa \; + \; Br(CH_2)_9O-\text{THP} \xrightarrow{\text{fl. NH}_3}$$

$$(41) \qquad\qquad (42)$$

$$\rightarrow CH_3(CH_2)_2-CH=CH-C\equiv C-(CH_2)_9-O-\text{THP} \xrightarrow[\text{2. H}_2/\text{Pd/C}]{\text{1. H}^{\oplus}}$$

$$(43)$$

$$\rightarrow CH_3-(CH_2)_2-CH=CH-\overset{\overset{\text{H}}{|}}{C}=\overset{\overset{\text{H}}{|}}{\underset{\underset{\underset{CH_2OH}{|}}{(CH_2)_8}}{C}}$$

$$(44)$$

Das 10-*trans*-12-*trans*-Hexadecadienol wurde aus dem 10-*cis*-12-*trans*-Hexadecadienol-(1) durch Photoisomerisierung in Gegenwart von Jod gewonnen.

Tabelle 1. *Aktivitäten der geometrisch isomeren 10,12-Hexadecadienole-(1) in Lockstoffeinheiten*

Verbindung	LE μg/ml n. Butenandt et al.	LE μg/ml n. Truscheit et al.
Bombykol	10^{-10}	—
10-cis-12-cis-Hexadecadienol-(1)	1	—
10-cis-12-trans-Hexadecadienol-(1)	10^{-3}	10^{-5}
10-trans-12-cis-Hexadecadienol-(1)	10^{-12}	10^{-13}
10-trans-12-trans-Hexadecadienol-(1)	10	10

Literaturverzeichnis: SS. 250—255

Die Aktivitäten der isomeren Hexadecadienole am männlichen Seidenspinner waren gut reproduzierbar festzustellen und zeigten zwischen den von E. TRUSCHEIT und K. EITER sowie A. BUTENANDT, STAMM und HECKER hergestellten Verbindungen nur geringfügige Unterschiede. Wie aus dem bereits dargelegten hervorgeht, konnten weder an Gemischen geometrisch isomerer Verbindungen noch an Gemischen der Reinsubstanzen mit von der Synthese herrührenden Verunreinigungen Maskierungen oder gar völliges Verschwinden der biologischen Aktivitäten beobachtet werden. Immer wurden die der Konzentration der Reinsubstanz, d. h. der Verdünnung entsprechenden biologischen Aktivitäten gefunden, die Hexadecadienole zeigten keine antagonistischen Effekte, wie dies bei einer Reihe später isolierter und geprüfter Insektensexuallockstoffe beobachtet wurde.

3. Sexuallockstoff des Schwammspinners *(Porthetria Lymanthria dispar)*, gipsy moth:

Der Schwammspinner (gipsy moth) ist ein in Europa heimisches und nach Amerika eingeschlepptes Insekt, das wegen seiner Schäden in den Waldbeständen von großer wirtschaftlicher Bedeutung ist. Es ist schon lange bekannt, daß Extrakte der Hinterleibsegmente weiblicher Schwammspinner Schwammspinnermännchen anlocken können (*41*).

M. JACOBSON, M. BEROZA und W. A. JONES (*51*) haben den Sexuallockstoff aus den Tieren isoliert und für ihn als Struktur (+)10-Acetoxyhexadecen-(7-*cis*)-ol-(1) (*45*) angegeben. $[\alpha]_D^{23} + 7,9$ (c = 1,0 u. CHCl$_3$) LE $10^{-12}\,\mu$g/ml.

$$\underset{\underset{\displaystyle OCOCH_3}{|}}{CH_3(CH_2)_5-CH}-C_2H-\overset{\overset{\displaystyle H}{|}}{C}=\overset{\overset{\displaystyle H}{|}}{C}-(CH_2)_5-CH_2OH$$

(45)

Die amerikanischen Autoren haben den Lockstoff als Racemat synthetisiert und festgestellt, daß die Racemform in den biologischen Eigenschaften völlig identisch mit dem natürlichen Lockstoff, auch „Gyptol" benannt, ist. Ausgehend vom Heptaldehyd wird Propargylbromid nach Reformatsky mit Zink zum Acetylencarbinol (47) umgesetzt,

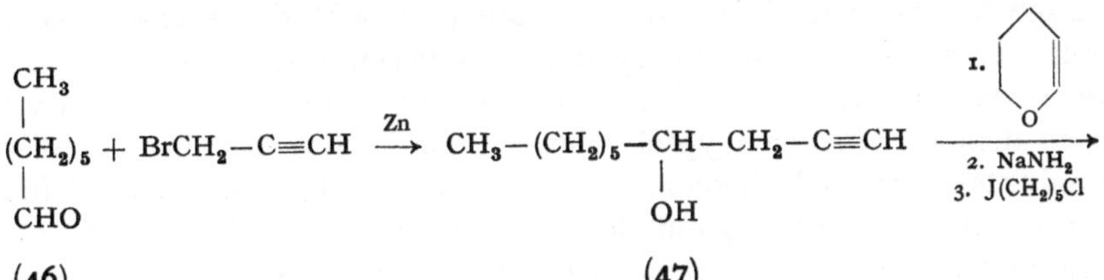

$$\underset{\underset{\displaystyle CHO}{|}}{\overset{\overset{\displaystyle CH_3}{|}}{(CH_2)_5}} + BrCH_2-C\equiv CH \xrightarrow{Zn} \underset{\underset{\displaystyle OH}{|}}{CH_3-(CH_2)_5-CH}-CH_2-C\equiv CH \xrightarrow[\substack{2.\ NaNH_2 \\ 3.\ J(CH_2)_5Cl}]{1.}$$

(46)   (47)

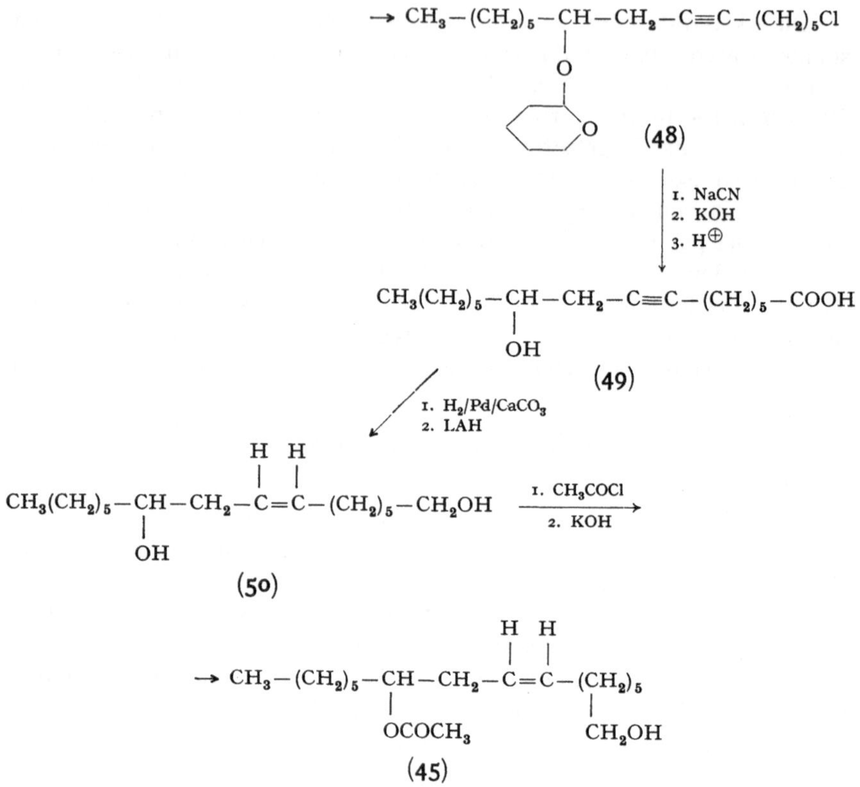

$$\rightarrow CH_3-(CH_2)_5-CH-CH_2-C\equiv C-(CH_2)_5Cl$$

(48)

1. NaCN
2. KOH
3. H$^\oplus$

$$CH_3(CH_2)_5-CH-CH_2-C\equiv C-(CH_2)_5-COOH$$
$$\quad\quad\quad\quad | \quad\quad\quad\quad\quad\quad\quad\quad\quad\quad\quad OH$$

(49)

1. H$_2$/Pd/CaCO$_3$
2. LAH

$$CH_3(CH_2)_5-CH-CH_2-\overset{H}{\underset{}{C}}=\overset{H}{\underset{}{C}}-(CH_2)_5-CH_2OH$$
$$\quad\quad\quad\quad\quad | \quad\quad\quad\quad\quad\quad\quad\quad OH$$

(50)

1. CH$_3$COCl
2. KOH

$$\rightarrow CH_3-(CH_2)_5-CH-CH_2-\overset{H}{\underset{}{C}}=\overset{H}{\underset{}{C}}-(CH_2)_5$$
$$\quad\quad\quad\quad\quad\quad\quad | \quad\quad\quad\quad\quad\quad\quad\quad\quad\quad\quad |$$
$$\quad\quad\quad\quad\quad\quad OCOCH_3 \quad\quad\quad\quad\quad\quad CH_2OH$$

(45)

dessen Hydroxylfunktion durch Bildung des Hexahydropyranyläthers verschlossen wird; Umsetzung dieses Produktes mit Na in flüssigem Ammoniak und 1-Chlor-5-Jodpentan führt zum Chloracetylenäther (48) der mit NaCN das Nitril liefert. Verseifung des Nitrils ergibt schließlich eine Hydroxyacetylencarbonsäure (49), an der die Acetylengruppe partiell zur *cis*-Doppelbindung hydriert wurde. Reduktion mit Lithiumalanat und Acetylierung liefert ein Diazetat, das nach partieller Verseifung Gyptol ergab, das in allen Eigenschaften identisch mit dem natürlichen Lockstoff war.

M. JACOBSON und Mitarb. (*47, 53, 109*) konnten Analoge des Gyptols ausgehend von natürlicher Ricinolsäure (+)-14-Hydroxy-11-*cis*-eicosensäure-(1) sowie (+)-9-Hydroxy-12-*cis*-octadecensäure-(1) synthetisieren, wobei das Produkt aus Octadecensäure die stärkste biologische Aktivität nach dem natürlichen Gyptol zeigte. Beim technischen Einsatz dieses Produktes wurde jedoch keine biologische Aktivität gefunden, ein Befund, den die Autoren (*57, 58*) mit dem Vorhandensein von *trans*-Verbindung oder anderen chemischen Substanzen begründeten. So konnte festgestellt werden, daß in Gegenwart von etwa 20% trans-Isomerem die gesamte Aktivität verloren ging bzw. ein Zusatz von geringen Mengen z. B. des Ricinoleylalkohols genügt, um völlig inaktive Substanzen zu erhalten. Sowohl D. STEFANOVIC, B. GRUJIČ und D. MIĆIĆ (96) als auch unabhängig davon K. EITER, E. TRUSCHEIT und M. BONESS (33) haben Gyptol synthetisiert und für völlig inaktiv befunden.

Ausgehend von der gut kristallisierenden 10-Hydroxyhexadecin-(7)-säure-(1) konnten die deutschen Autoren D,L-10-Acetoxyhexadecen-(7-cis)-ol-(1) (Gyptol) herstellen.

$$Cl-(CH_2)_5-C\equiv CLi + CH_2-CH-C_6H_{13} \rightarrow C_6H_{13}-CH-CH_2-C\equiv C-(CH_2)_5-R$$

$$\underset{O}{\diagdown\diagup}$$

$$\underset{OH}{|}$$

(51)	(52)	(53)

(a) $R = Cl$
(b) $R = CN$
(c) $R = COOH$

Die kristallisierte 10-Hydroxyhexadecin-(7)-säure-(1) (53c) wurde mit vergiftetem Lindlar-Katalysator an der Dreifachbindung partiell zur 10-Hydroxy-(7-cis)-hexadecensäure-(1) hydriert, die über Hexadecen-(7-cis)-diol-(1,10) und das entsprechende Diazetat in das D,L-10-Acetoxy-hexadecen-(7-cis)-ol-(1) (45) übergeführt wurde. Dieses Syntheseprodukt enthielt auf Grund des IR- und NMR-Spektrums wenn überhaupt sicherlich weniger als 3% an *trans*-Verbindung, so daß dies unter keinen Umständen der Grund für die völlige Inaktivität des Syntheseproduktes im biologischen Test sein kann.

$$CH_3-(CH_2)_5-CH-CH_2-C=C-(CH_2)_7-CH_2OH$$

$$\underset{OCOCH_3}{|} \qquad \underset{H}{|}\ \underset{H}{|}$$

(54)

Das höhere Homologe des Gyptols, das 12-Acetoxyoctadecen-(9-cis)-ol-(1) (Gyplure) (54) haben die deutschen Autoren gleichfalls aus natürlicher Ricinolsäure dargestellt und im biologischen Test keinerlei Aktivität gefunden. Nachdem D. SCHNEIDER und I. BOECKH (88) im Elektro-antennogramm (EAG) beim Schwammspinner mit dem synthetischen Gyptol und Gyplure keinerlei Reaktion feststellen konnten, die mit aktiven Duftdrüsen von *Porthetria-dispar*-Weibchen gut beobachtet wurden, konnten sie auch die Einwendungen von M. JACOBSON (47, 53, 109) widerlegen, nach denen Beimengungen, ganz gleich welcher Art im Syntheseprodukt völlige Inaktivierung verursachten; Syntheseprodukt im Verhältnis 9 : 1 mit von den amerikanischen Autoren übersandten wirksamen D,L-10-Acetoxyhexadecen-(7-cis)-ol-(1) gemischt, zeigte im

EAG ein einwandfreies Reaktionspotential. Demnach scheint die Maskierungstheorie widerlegt zu sein, wobei erst eine Neubearbeitung des Sexuallockstoffes von *Porthetria dispar* eine Klärung des Problems bringen kann.

4. Sexuallockstoff der Amerikanischen Küchenschabe *(Periplaneta americana)*:

Aus weiblichen Tieren von *Periplaneta americana* konnten Wharton und Mitarb. *(112)* einen Sexuallockstoff präparieren, der bei männlichen Tieren typisches Kopulationsverhalten hervorruft. M. Jacobson und Mitarb. *(52)* berichteten danach, den Sexuallockstoff isoliert und in seiner Konstitution aufgeklärt zu haben. Für die Struktur des Küchenschabenlockstoffes wurde aus den analytischen und spektroskopischen Daten die Konstitution eines Cyclopropansystems *(55)* vorgeschlagen, obwohl

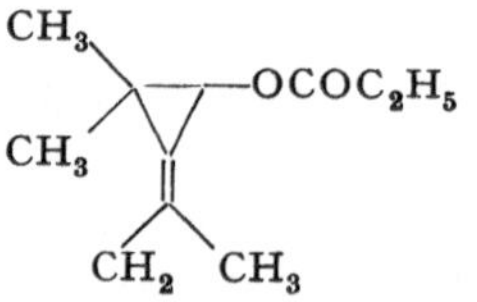

(55) 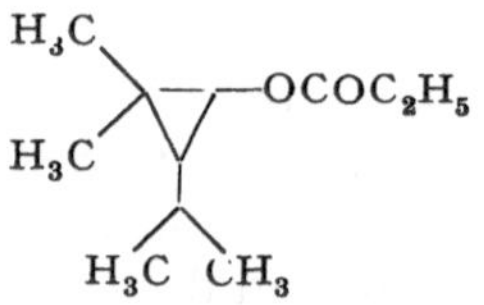   (56)

Wharton darauf hinwies *(49, 111)*, daß auf Grund der Retentionszeiten im Gaschromatogramm der von M. Jacobson isolierte Stoff nicht mit dem Lockstoff von Wharton identisch sein konnte. M. Jacobson und Mitarb. konnten den angeblichen Sexuallockstoff *(55)* zum Dihydroprodukt *(56)* hydrieren, das mit einem Syntheseprodukt in allen chemischen und physikalischen Eigenschaften identisch war; nachdem vor allem durch die Arbeit von Day und Whiting sowie anderer Forscher *(25, 30, 69, 108)* die Totalsynthese des Körpers *(55)* gelungen war und die erhaltene Substanz nicht die geringste biologische Aktivität zeigte, publizierten M. Jacobson und Mitarb. *(50)*, daß ein Irrtum bei der Konstitutionsermittlung und im natürlichen Lockstoff offensichtlich noch ein Gemisch von Substanzen vorgelegen hatte.

Auch Wakabayashi *(25, 30, 69, 108)* glückte eine Synthese des von Jacobson vorgeschlagenen Cyclopropanderivates auf folgendem Wege:

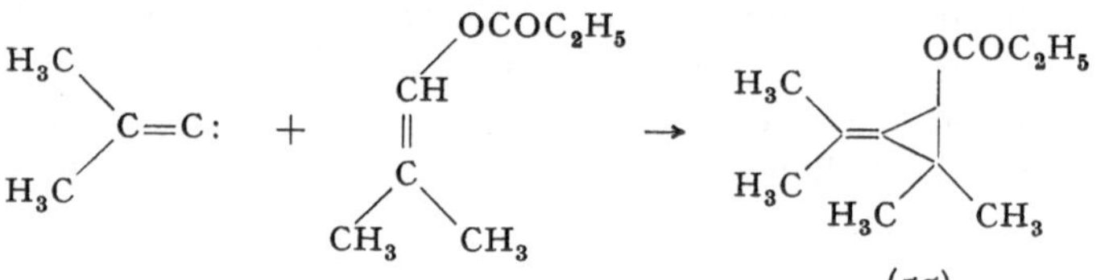

(55)

Das Produkt war mit der von WHITING und DAY erhaltenen Substanz identisch und zeigte natürlich keine biologische Aktivität; demnach ist die Struktur des Sexuallockstoffes der amerikanischen Küchenschabe unbekannt und bedarf einer neuerlichen Bearbeitung.

5. Sexuallockstoff der roten Baumwollkapselraupe (*Pectinophora gossybiella* Saunders), pink bollworm moth:

Nachdem ONYE (*74*) zeigte, daß männliche Tiere durch Methylenchloridextrakt kopulierender Insekten angelockt wurden und BERGER et al. (*7*) beobachteten, daß Männchen der roten Baumwollkapselraupe in charakteristischer Weise durch Extrakte weiblicher Abdominalspitzen erregt werden, konnten JONES, JACOBSON und MARTIN (*60*) den Sexuallockstoff isolieren und durch Konstitutionsermittlung und Synthese zeigen, daß es sich bei dem Naturstoff um das Acetat des 10-Propyltridecadien-(5-*trans*-9)-ol-(1) (57) handelt:

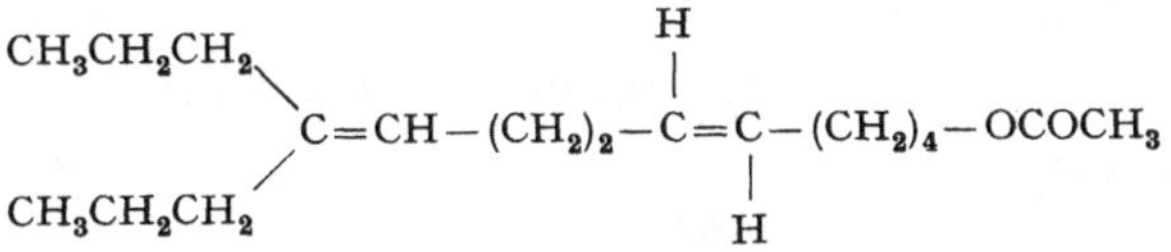

$$CH_3CH_2CH_2\diagdown \atop CH_3CH_2CH_2\diagup C=CH-(CH_2)_2-\overset{H}{\underset{H}{\overset{|}{C}}}=\overset{|}{C}-(CH_2)_4-OCOCH_3$$

(57)

Die sterisch einheitliche Synthese des Sexuallockstoffes, der die Bezeichnung „Propylure" erhielt, vollzogen JACOBSON und Mitarb. nach dem Aufbauprinzip

$$C_7 + C_2 = C_9; \quad C_9 + C_1 = C_{10}; \quad C_{10} + C_6 = C_{16}$$

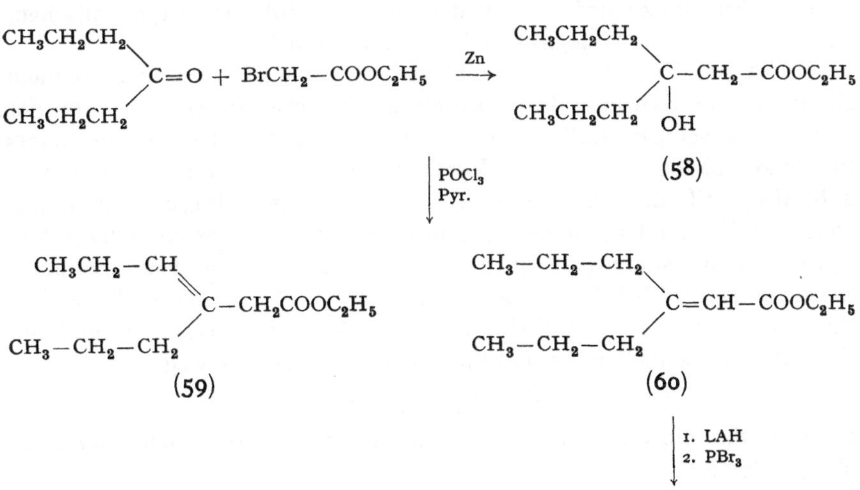

$$CH_3CH_2CH_2\diagdown \atop CH_3CH_2CH_2\diagup C=O + BrCH_2-COOC_2H_5 \xrightarrow{Zn} \begin{array}{c} CH_3CH_2CH_2\diagdown \\ \\ CH_3CH_2CH_2\diagup \end{array} \underset{OH}{\overset{|}{C}}-CH_2-COOC_2H_5$$

(58)

$$\Bigg\downarrow \begin{array}{l} POCl_3 \\ Pyr. \end{array}$$

$$CH_3CH_2-CH\diagdown \atop CH_3-CH_2-CH_2\diagup C-CH_2COOC_2H_5$$

(59)

$$CH_3-CH_2-CH_2\diagdown \atop CH_3-CH_2-CH_2\diagup C=CH-COOC_2H_5$$

(60)

$$\Bigg\downarrow \begin{array}{l} 1.\ LAH \\ 2.\ PBr_3 \end{array}$$

(60)

↓

$$CH_3CH_2CH_2 \diagdown$$
$$ C=CH-CH_2-COOH$$
$$CH_3CH_2CH_2 \diagup$$

(62)

$$\xleftarrow[\text{2. OH}^-]{\text{1. NaCN}}$$

$$CH_3CH_2CH_2 \diagdown$$
$$ C=CH-CH_2Br$$
$$CH_3CH_2CH_2 \diagup$$

(61)

$$\Big\downarrow \begin{array}{l} \text{1. LAH} \\ \text{2. PBr}_3 \end{array}$$

$$CH_3-CH_2-CH_2 \diagdown$$
$$ C=CH$$
$$CH_3-CH_2-CH_2 \diagup \quad |$$
$$ CH_2$$
$$ |$$
$$ CH_2Br$$

(63)

$$NaC\equiv C-(CH_2)_3CH_2O- \longrightarrow$$

→

$$CH_3-CH_2-CH_2 \diagdown$$
$$ C=CH-CH_2-CH_2-C\equiv C-(CH_2)_3-CH_2O-$$
$$CH_3-CH_2-CH_2 \diagup$$

(64)

$$\xrightarrow[\begin{array}{l}\text{2. H}^{\oplus} \\ \text{3. CH}_3\text{COCl}\end{array}]{\text{1. Na/fl. NH}_3}$$

→

$$CH_3-CH_2-CH_2 \diagdown \qquad\qquad H$$
$$ \qquad\qquad\quad |$$
$$ C=CH-CH_2-CH_2-C=C-(CH_2)_3-CH_2OCOCH_3$$
$$CH_2-CH_2-CH_2 \diagup \qquad\qquad |$$
$$ \qquad\qquad\quad H$$

(57)

Diese Substanz wurde als mit dem Naturprodukt in allen chemischen, physikalischen und biologischen Daten identisch befunden.

K. Eiter, E. Truscheit und M. Boness (33) hatten bald danach gleichfalls eine Synthese des „Propylure" beschrieben, wobei sie von der Annahme ausgingen, daß ähnlich wie beim Lockstoff des Seidenspinners Bombykol biologisch ähnliche Voraussetzungen vorhanden sein sollten, d. h. also, daß ein Syntheseprodukt, das ein oder mehrere Isomere enthielt, bei Vorhandensein des „natürlichen" Lockstoffes unbedingt biologisch nachweisbare Aktivitäten gemäß seiner vorhandenen Konzentration zeigen sollte! Es wurde deshalb auf die sterisch einheitliche Synthese des 5-*trans*-10-propyltridecadien-(9)-ol-(1) (57) verzichtet und aus Gründen der schnellen Durchführbarkeit der Syntheseweg

$$C_7 + C_4 = C_{11}; \quad C_{11} + C_5 = C_{16}$$

gewählt, der einwandfrei nachweisbar 10-Propyltridecadien-(5-*trans*-9)-ol-(1) lieferte.

Literaturverzeichnis: SS. 250—255

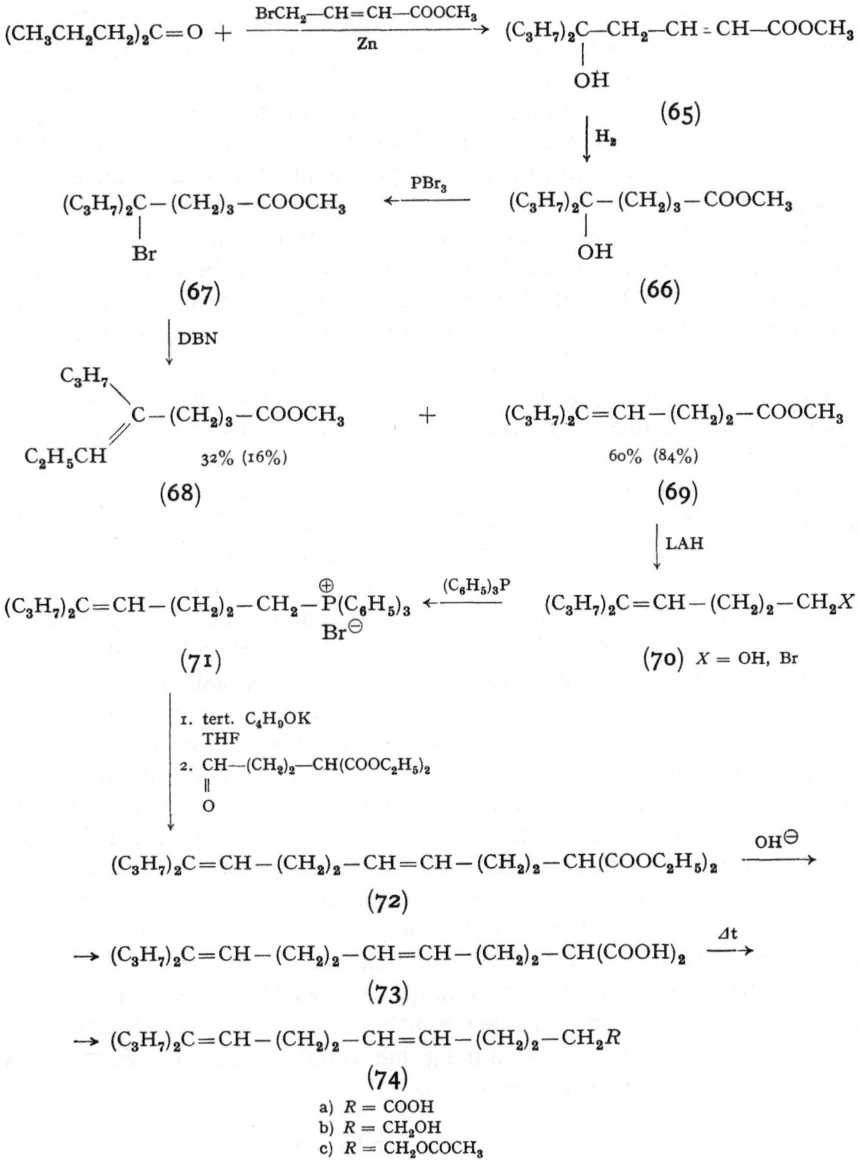

$$(CH_3CH_2CH_2)_2C=O \ + \ \xrightarrow[Zn]{BrCH_2-CH=CH-COOCH_3} \ (C_3H_7)_2C-CH_2-CH=CH-COOCH_3$$

$$\underset{OH}{|}$$

(65)

$$\downarrow H_2$$

$$(C_3H_7)_2C-(CH_2)_3-COOCH_3 \ \xleftarrow{PBr_3} \ (C_3H_7)_2C-(CH_2)_3-COOCH_3$$

$$\underset{Br}{|} \qquad\qquad\qquad \underset{OH}{|}$$

(67)   (66)

$$\downarrow DBN$$

$$\underset{C_2H_5CH}{\overset{C_3H_7}{\diagdown}}C-(CH_2)_3-COOCH_3 \qquad + \qquad (C_3H_7)_2C=CH-(CH_2)_2-COOCH_3$$

32% (16%)   60% (84%)

(68)   (69)

$$\downarrow LAH$$

$$(C_3H_7)_2C=CH-(CH_2)_2-CH_2-\overset{\oplus}{\underset{Br^{\ominus}}{P(C_6H_5)_3}} \ \xleftarrow{(C_6H_5)_3P} \ (C_3H_7)_2C=CH-(CH_2)_2-CH_2X$$

(71)   (70) $X = $ OH, Br

$$\downarrow \begin{array}{l} \text{1. tert. } C_4H_9OK \\ \quad THF \\ \text{2. } CH-(CH_2)_2-CH(COOC_2H_5)_2 \\ \quad \overset{\|}{O} \end{array}$$

$$(C_3H_7)_2C=CH-(CH_2)_2-CH=CH-(CH_2)_2-CH(COOC_2H_5)_2 \ \xrightarrow{OH^{\ominus}}$$

(72)

$$\rightarrow (C_3H_7)_2C=CH-(CH_2)_2-CH=CH-(CH_2)_2-CH(COOH)_2 \ \xrightarrow{\Delta t}$$

(73)

$$\rightarrow (C_3H_7)_2C=CH-(CH_2)_2-CH=CH-(CH_2)_2-CH_2R$$

(74)

a) $R = $ COOH
b) $R = $ CH$_2$OH
c) $R = $ CH$_2$OCOCH$_3$

Bei diesem Syntheseverfahren fiel in der Stufe der Dehydrohalogenierung von (67) ein Substanzgemisch isomerer ungesättigter Ester an, in denen die Doppelbindung nur in den Positionen Δ^4 oder Δ^5 sein konnte. Sowohl durch Ozonisierung wie auch durch gaschromatographische Analyse konnte festgestellt werden, daß das Substanzgemisch in Hauptmenge als gewünschtes Δ^4-Produkt vorlag, so daß im Endprodukt -bei den weiteren überschaubaren Synthesestufen auch tatsächlich genügend 5-*trans*-Propylure vorhanden sein mußte, was auch der Fall war. Im biologischen Test hatte dieses Gemisch bei *Pectinophora* nicht die geringste Aktivität gezeigt, so daß Eiter, Truscheit und Boness den Schluß zogen, daß die Konstitution des Propylure nicht richtig ermittelt sein konnte.

Mittlerweile hat M. Jacobson (*48*) Propylure nach K. Eiter, E. Truscheit und M. Boness synthetisiert, das Syntheseprodukt durch Kombination von Destillationsmethoden und präparativer Dünnschicht-chromatographie an Silbernitrat dotiertem Silicagel aufgetrennt und das reine *trans*-Produkt als biologisch voll aktiv befunden; es wurde weiters nachgewiesen, daß 15% an *cis*-Isomerem „unverständlicher Weise" zur völligen Maskierung und damit Inaktivierung der *trans*-Verbindung führten.

An dieser Stelle muß darauf hingewiesen werden, daß die Verwirrung über Maskierungs- und Inaktivierungseffekte von Insektensexuallock-stoffen durch Beimengung von geometrischen Isomeren oder anderen Fremdstoffen vollständig ist, wenn N. Green, M. Jacobson und J. C. Keller (*39*) berichten, daß 7-*cis*-Hexadecenolacetat-(1) („Hexalure") weitaus wirksamer bei *Pectinophora gossypiella* Saunders ist als das natürliche *trans*-Propylure, das im Feldversuch überhaupt erst wirksam wird, wenn Aktivatoren vorhanden sind (*59*)! Diese Fülle von uner-klärlichen Beobachtungen steht im völligen Gegensatz zu den Erfah-rungen, die beim synthetischen und damit natürlichen Bombykol ebenso wie bei Extrakten vom Schwammspinner *(Porthetria dispar)* gemacht worden sind. Hier wirken in Verfolg naturwissenschaftlicher Prinzipien die Sexuallockstoffe in der Relation Dosis-Wirkung, was auch verständlich ist, da in der Natur neben den Sexuallockstoffen immer eine Flut olfak-torisch wirksamer Moleküle vorhanden sein wird, die jedoch niemals maskierend oder desaktivierend sein können.

Da die rote Baumwollkapselraupe ein wichtiges Schadinsekt ist, haben auch andere Laboratorien Synthesen des „Propylure" und zwar als sterisch einheitliche Verbindung hergestellt. Nach G. Pattenden (*75*) wird nach dem Syntheseprinzip

$$C_3 + C_6 = C_9; \ C_9 + C_7 = C_{16}$$

vorgegangen.

$$(CH_3O)_2CH-CH_2-CH_2Br \ + \ HC{\equiv}C-(CH_2)_3-CH_2-O-\!\!\!\triangleleft\!\!\!\underset{O}{\triangleright} \ \rightarrow$$

(75)   (76)

Literaturverzeichnis: SS. 250—255

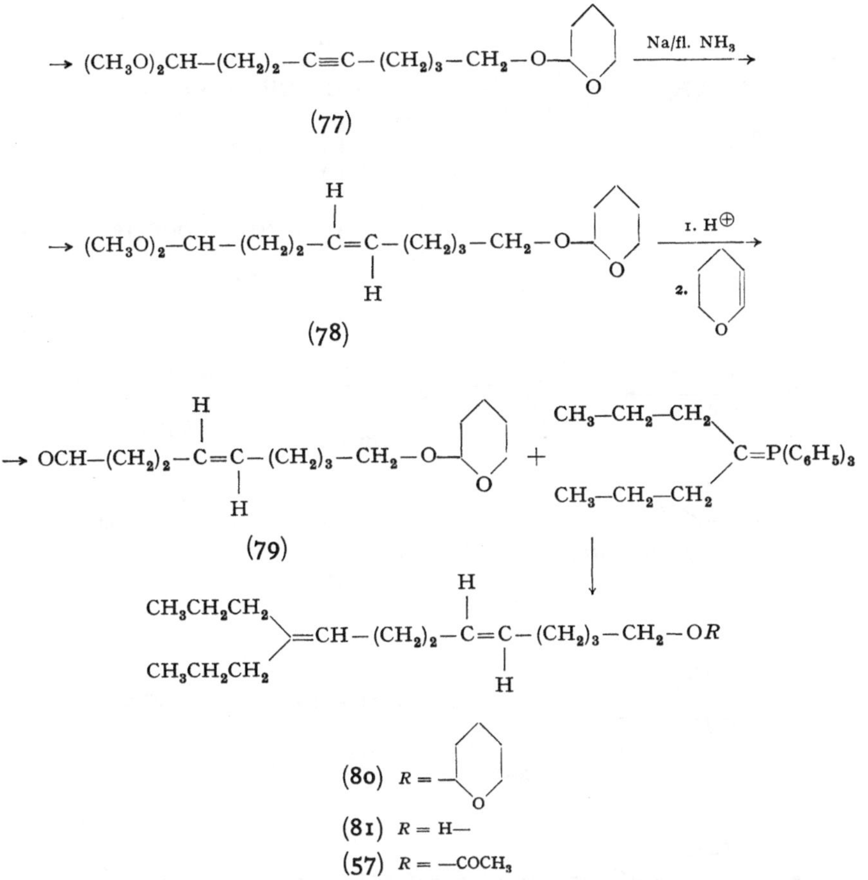

$$\rightarrow (CH_3O)_2CH-(CH_2)_2-C\equiv C-(CH_2)_3-CH_2-O-\underset{O}{\bigotimes} \quad \xrightarrow{Na/fl.\ NH_3}$$

$$(77)$$

$$\rightarrow (CH_3O)_2-CH-(CH_2)_2-\underset{\underset{H}{|}}{\overset{\overset{H}{|}}{C}}=C-(CH_2)_3-CH_2-O-\underset{O}{\bigotimes} \quad \xrightarrow[\ 2.\ ]{1.\ H^\oplus}$$

$$(78)$$

$$\rightarrow OCH-(CH_2)_2-\underset{\underset{H}{|}}{\overset{\overset{H}{|}}{C}}=C-(CH_2)_3-CH_2-O-\underset{O}{\bigotimes} \quad + \quad \underset{CH_3-CH_2-CH_2}{\overset{CH_3-CH_2-CH_2}{>}}C=P(C_6H_5)_3$$

$$(79)$$

$$\underset{CH_3CH_2CH_2}{\overset{CH_3CH_2CH_2}{>}}=CH-(CH_2)_2-\underset{\underset{H}{|}}{\overset{\overset{H}{|}}{C}}=C-(CH_2)_3-CH_2-OR$$

(80) $R = -\underset{O}{\bigotimes}$

(81) $R = H-$

(57) $R = -COCH_3$

β-Brompropionaldehyddimethylacetal (75) wird mit der Natriumverbindung des Pyranyläthers von Hexin-(1)-ol-(6) (76) in fl. NH₃ umgesetzt und anschließend die Dreifachbindung mit Natrium in flüssigem
Ammoniak zur *trans*-Doppelbindung (78) reduziert. Verseifung zum
Nonen-(4)-al-(1)-ol-(9) ergibt den Hydroxyaldehyd, der nochmals in den
Pyranyläther (79) übergeführt werden mußte um mit dem Phosphorylen
zum Propylure kondensiert zu werden.

Eine besonders elegante Synthese des *trans*-Propylure beschrieben
M. STOLL und I. FLAMENT (*98*). Cyclopenten-(1)-carbonsäureäthylester
(*82*) diente als Ausgangsmaterial für die Synthese, die nach dem Prinzip

$$C_6 + C_2 = C_8;\ C_8 + C_9 = C_{16} + C_1$$

durchgeführt wurde.

$$\text{(82)} \xrightarrow[\text{(iC}_3\text{H}_7)_2\text{NMgBr}]{\text{CH}_3\text{COOC}_2\text{H}_5} \text{(83)} \rightleftharpoons \text{(84)} \longrightarrow$$

Structures:

(82) cyclopentene–$COOC_2H_5$

(83) cyclopentene–$\overset{OH}{C}=CH-COOC_2H_5$

(84) cyclopentene–$CO-CH_2-COOC_2H_5$

$$\xrightarrow[\text{(iC}_3\text{H}_7\text{O})_3\text{Al}]{(C_3H_7)_2\overset{OH}{C}-CH=CH_2}$$

(85) cyclopentene–$CO-CH_2-CH_2-CH=C\big<\genfrac{}{}{0pt}{}{C_3H_7}{C_3H_7}$ $\xrightarrow{H_2O_2}$

(86) $\xrightarrow{CH_3-\text{C}_6\text{H}_4-SO_2NHNH_2}$

(87) $\genfrac{}{}{0pt}{}{C_3H_7}{C_3H_7}\!\!>\!\!=CH-CH_2-CH_2-C\equiv C-CH_2$; $OHC-CH_2-CH_2$ $\xrightarrow{LAH}$

(88) $\genfrac{}{}{0pt}{}{C_3H_7}{C_3H_7}\!\!>\!\!=CH-CH_2-CH_2-C\equiv C-CH_2$; $HOCH_2-CH_2-CH_2$ $\xrightarrow{Na/fl.NH_3}$

(89) $\genfrac{}{}{0pt}{}{C_3H_7}{C_3H_7}\!\!>\!\!=CH-CH_2-CH_2-\overset{H}{C}=C-CH_2$; H ; $CH_2-CH_2-CH_2OH$ $\xrightarrow{(CH_3CO)_2O/\text{Pyr.}}$

(57) $\genfrac{}{}{0pt}{}{C_3H_7}{C_3H_7}\!\!>\!\!=CH-CH_2-CH_2-\overset{H}{\underset{H}{C}}=C-CH_2-CH_2-CH_2-CH_2-OCOCH_3$

Literaturverzeichnis: SS. 250—255

Das von G. Pattenden als auch von den schweizer Autoren herge-
stellte Propylure wurde von M. Jacobson in USA getestet und als biologisch aktiv befunden.

6. Sexuallockstoff des cabbage looper *(Trichoplusia ni)*:

Aus den Abdominalspitzen weiblicher Cabbage Looper hat R. B.
Berger (5) ein unreines aber biologisch hoch aktives Material isolieren
können, aus dem als Wirkprinzip Dodecen-(7-cis)-ol-acetat-(1) (90)
isoliert wurde.

$$\begin{array}{cc} H & H \\ | & | \\ CH_3-(CH_2)_3-C=C-(CH_2)_5-CH_2OCOCH_3 \end{array}$$

(90)

Die Konstitution konnte durch Totalsynthese nach dem Schema

$$C_6 + C_5 = C_{11}; \quad C_{11} + C_1 = C_{12}$$

bestätigt werden.

$$CH_3-(CH_2)_3-C\equiv CH + JCH_2(CH_2)_3-CH_2Cl \xrightarrow[\text{fl. NH}_3]{\text{NaNH}_2}$$

(91)   (92)

$$\rightarrow CH_3-(CH_2)_3-C\equiv C-CH_2-(CH_2)_3-CH_2Cl \xrightarrow[\text{2. OH}^{\ominus}]{\text{1. NaCN}} CH_3-(CH_2)_3C\equiv C-(CH_2)_5COOH$$

(93)   (94)

$$\left.\begin{array}{l}\text{1. Pd(CaCO}_3)\\ \text{2. LAH}\end{array}\right\downarrow$$

$$\begin{array}{cc} H & H \\ | & | \\ CH_3-(CH_2)_3-C=C-(CH_2)_5-CH_2OH \end{array} \xrightarrow{CH_3COCl}$$

(95)

$$\rightarrow \begin{array}{cc} H & H \\ | & | \\ CH_3-(CH_2)_3-C=C-(CH_2)_5-CH_2OCOCH_3 \end{array}$$

(90)

Das hierbei angewandte Reaktionsschema zeigte keine Besonder-
heiten.

N. Green, M. Jacobson, T. J. Henneberry und A. N. Kishaba (38)
haben nach dem Aufbauprinzip

$$C_6 + C_2 = C_8; \quad C_8 + C_4 = C_{12}$$

gleichfalls sterisch einheitliches *cis*-Dodecen-(7)-olacetat-(1) herstellen
können, das in seiner biologischen Aktivität wirksamer als das Natur-

produkt war. Auch Toba et al. (*100*) stellten für das *7-cis*-Dodecenol-acetat (90) sowohl im Labor als auch im Freilandversuch gleich hohe Aktivitäten fest.

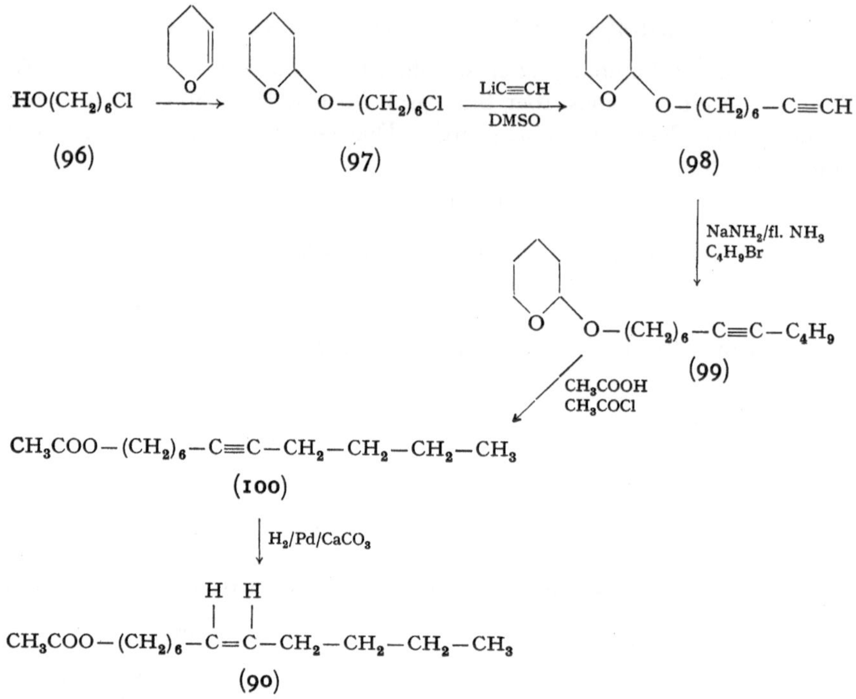

$$HO(CH_2)_6Cl \longrightarrow \quad O-(CH_2)_6Cl \quad \xrightarrow[DMSO]{LiC\equiv CH} \quad O-(CH_2)_6-C\equiv CH$$

$$(96) \qquad\qquad (97) \qquad\qquad\qquad (98)$$

$$\Big\downarrow \begin{array}{l} NaNH_2/fl.\ NH_3 \\ C_4H_9Br \end{array}$$

$$O-(CH_2)_6-C\equiv C-C_4H_9$$

$$(99)$$

$$\swarrow \begin{array}{l} CH_3COOH \\ CH_3COCl \end{array}$$

$$CH_3COO-(CH_2)_6-C\equiv C-CH_2-CH_2-CH_2-CH_3$$

$$(\mathbf{100})$$

$$\Big\downarrow H_2/Pd/CaCO_3$$

$$CH_3COO-(CH_2)_6-\overset{H}{\underset{|}{C}}=\overset{H}{\underset{|}{C}}-CH_2-CH_2-CH_2-CH_3$$

$$(90)$$

N. Green et al. konnten das *reine 7-trans*-Dodecenolacetat-(1) (*107*) auf einem sehr interessanten Syntheseweg herstellen, wobei sie diese trans-Verbindung nach der Methode

$$C_5 + C_4 = C_9$$
$$C_9 + C_3 = C_{11} + C_1$$
$$C_{11} + C_1 = C_{12}$$

aufbauten.

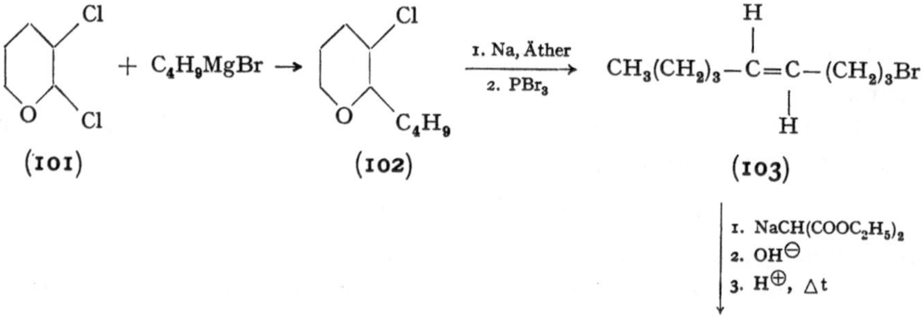

$$\begin{array}{c} Cl \\ \\ O \quad Cl \end{array} + C_4H_9MgBr \longrightarrow \begin{array}{c} Cl \\ \\ O \quad C_4H_9 \end{array} \xrightarrow[2.\ PBr_3]{1.\ Na,\ Äther} \quad CH_3(CH_2)_3-\overset{H}{\underset{\underset{H}{|}}{C}}=C-(CH_2)_3Br$$

$$(\mathbf{101}) \qquad\qquad (\mathbf{102}) \qquad\qquad\qquad (\mathbf{103})$$

$$\Big\downarrow \begin{array}{l} 1.\ NaCH(COOC_2H_5)_2 \\ 2.\ OH^{\ominus} \\ 3.\ H^{\oplus},\ \Delta t \end{array}$$

(103)
↓

$$CH_3(CH_2)_3-\overset{\overset{\displaystyle H}{|}}{C}=\overset{\overset{\displaystyle}{|}}{\underset{\underset{\displaystyle H}{|}}{C}}-(CH_2)_4-COOH \xrightarrow[\text{2. PBr}_3]{\text{1. LAH}} CH_3(CH_2)_3-\overset{\overset{\displaystyle H}{|}}{C}=\overset{\overset{\displaystyle}{|}}{\underset{\underset{\displaystyle H}{|}}{C}}-(CH_2)_4-CH_2Br$$

(104)   (105)

$$\downarrow \begin{array}{l}\text{1. Mg}\\\text{2. CO}_2\end{array}$$

$$CH_3-(CH_2)_3\overset{\overset{\displaystyle H}{|}}{C}=\overset{\underset{\displaystyle H}{|}}{C}-(CH_2)_6OCOCH_3 \xleftarrow[\text{2. CH}_3\text{COCl}]{\text{1. LAH}} CH_3-(CH_2)_3-\overset{\overset{\displaystyle H}{|}}{C}=\overset{\underset{\displaystyle H}{|}}{C}-(CH_2)_5COOH$$

(107)   (106)

7. Sexuallockstoff des red banded leaf roller *(Argyrotaenia velutinana)*:

W. L. ROELOFS et al. *(80)* konnten aus Weibchen von Argyrotaenia velutinana einen Sexuallockstoff isolieren, der durch Synthese und Vergleich mit dem Naturprodukt als 11-*cis*-Tetradecenylacetat (108) erkannt wurde.

$$CH_3-CH_2-\overset{\overset{\displaystyle H}{|}}{C}=\overset{\overset{\displaystyle H}{|}}{C}-(CH_2)_9-CH_2OCOCH_3 \quad (\text{,,Riblure``})$$

(108)

Während 0,1 μg dieser Verbindung am red banded leaf roller eindeutige Lockwirkung zeigten, erwies sich das entsprechende trans-Isomere als völlig unwirksam.

Die Synthese erfolgte nach dem Aufbauprinzip

$$C_{10} + C_4 = C_{14}$$

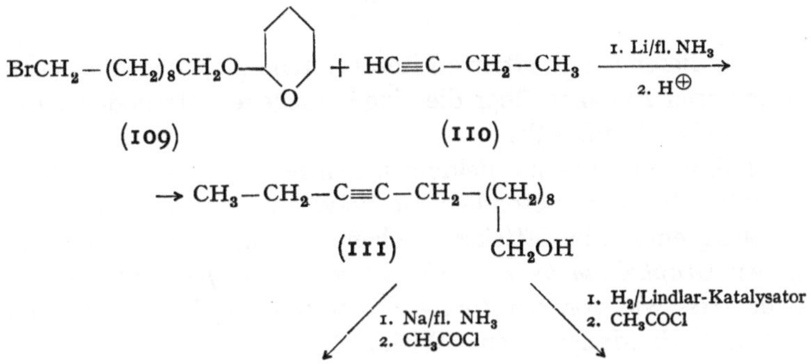

$$BrCH_2-(CH_2)_8CH_2O-\underset{O}{\overset{\bigcirc}{\bigcirc}} \quad + \quad HC\equiv C-CH_2-CH_3 \xrightarrow[\text{2. H}^\oplus]{\text{1. Li/fl. NH}_3}$$

(109)   (110)

$$\rightarrow CH_3-CH_2-C\equiv C-CH_2-(CH_2)_8$$

(111) CH_2OH

↙ 1. Na/fl. NH$_3$ ↘ 1. H$_2$/Lindlar-Katalysator
   2. CH$_3$COCl   2. CH$_3$COCl

$$(\text{III})$$

$$\underset{\underset{\text{trans }(\text{II2})}{}}{CH_3COOCH_2-(CH_2)_8-CH_2-\overset{\overset{\text{H}}{|}}{C}=\overset{\overset{}{}}{C}-CH_2CH_3} \qquad \underset{\underset{\text{cis }(108)}{}}{CH_3-CH_2-\overset{\overset{}{|}}{C}=\overset{\overset{}{|}}{C}-CH_2-(CH_2)_8}$$

während eine einstufige weitere Synthese nach dem Aufbauschema

$$C_{11} + C_3 = C_{14}$$

folgenden Verlauf nahm und zu einem Gemisch von geometrisch isomeren Tetradecen-(11)-ylacetaten-(1) führt.

$$CH_3COO-(CH_2)_{11}-Br + P(C_6H_5)_3 \rightarrow CH_3COO-(CH_2)_{11}-P^{\oplus}(C_6H_5)_3 \xrightarrow[\text{DMF}]{CH_3ONa}$$
$$(113) \hspace{6cm} (114) \hspace{0.3cm} Br^{\ominus}$$

$$\rightarrow CH_3COO-(CH_2)_{10}-CH=P(C_6H_5)_3 + OHC-C_2H_5 \rightarrow$$
$$(116)$$

$$\rightarrow CH_3COO-(CH_2)_{10}-CH=CH-C_2H_5$$
$$(108) + (112) \hspace{0.3cm} cis/trans$$

8. Sexuallockstoff des Pfirsichtriebbohrers *(Grapholitha molesta)*:

W. L. Roelofs et al. *(83)* konnten aus den Weibchen des Pfirsich-triebbohrers (oriental fruit moth) Dodecen-(8-cis)-ylacetat-(1) **(132)** isolieren und auf zwei voneinander unabhängigen Wegen synthetisieren.

$$CH_3-(CH_2)_2-\overset{\overset{}{|}}{C}=\overset{\overset{}{|}}{C}-(CH_2)_7-OCOCH_3$$
$$\hspace{2.5cm} \overset{}{H} \hspace{0.3cm} \overset{}{H}$$
$$(132)$$

Die in dieser Arbeit zitierte Literaturstelle *(80)* über die Synthesewege gibt keinerlei Angaben über die eingeschlagenen Methoden zum Aufbau dieses Lockstoffmoleküls.

In Laboratoriums- und Feldversuchen zeigte die synthetische Substanz eine gute Wirkung; besonders an Polyäthylen absorbiert konnten mit 10—200 μg eindeutige Effekte gesehen werden. Nahe verwandte Schäd-linge wie *Grapholitha prunivora* und *Grapholitha puchardi* wurden sowohl von der *cis-* wie von der *trans-*Verbindung angelockt, während andere getestete Verbindungen keine Wirkung zeigten.

Literaturverzeichnis: SS. 250—255

9. Sexuallockstoff der Männchen von *Danaus plexippus* (monarch butterfly):

J. Meinwald, A. M. Chalmers, T. E. Pliske und T. Eisner (67) berichteten über die Isolierung eines Pheromons aus den Haarstiften von männlichen Danaus plexippus-Tieren, wobei die Substanz (116) festgestellt werden konnte.

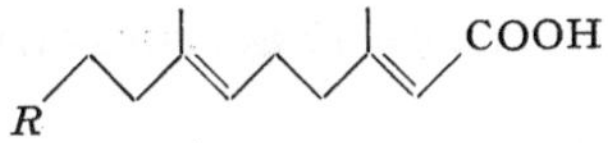

$$R \qquad COOH \qquad \text{a) } R = COOH$$
$$\text{b) } R = CH_2OH$$
(116)

Durch spektroskopische Methoden konnte die Konstitution als *trans, trans*-3,7-Dimethyl-2,6-decadien-1,10-disäure (116a) fixiert und durch Synthese bestätigt werden. Durch Oxydation der Säure (116a) nach Cornforth (27) sowie durch Synthese aus dem Diol (117)

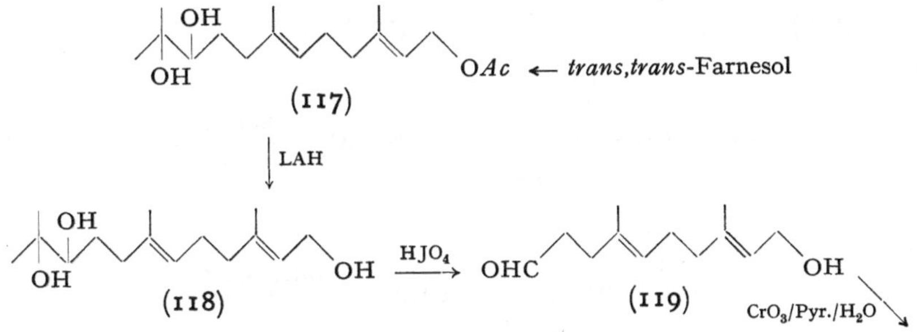

(117) $\qquad OAc \leftarrow$ *trans,trans*-Farnesol

$\downarrow$ LAH

(118) $\qquad$ OH $\xrightarrow{HJO_4}$ OHC $\qquad$ OH (119) $\qquad$ CrO$_3$/Pyr./H$_2$O

(116a)

erhält man die Dicarbonsäure (116a), die mit dem natürlichen Pheromon in allen Eigenschaften identisch war.

10. Sexuallockstoff von *Trogoderma inclusum* Le Conte:

J. O. Rodin et al. (79) konnten aus 250000 Weibchen zwei der insgesamt vier Komponenten des Sexuallockstoffes der zu den Dermestiden gehörigen Vorratsschädlinge *Trogoderma inclusum* isolieren, identifizieren und synthetisieren. Es handelt sich um das (—)14-Methyl-8-cis-hexadecenol-(1)

$$CH_3-CH_2-CH-(CH_2)_4-C=C-(CH_2)_6-CH_2OH$$
$$\quad\quad\quad | \quad\quad\quad\quad\quad\quad | \; |$$
$$\quad\quad\quad CH_3 \quad\quad\quad\quad\quad H \; H$$
(120)

sowie um (—)14-Methyl-8-cis-hexadecensäuremethylester-(1)

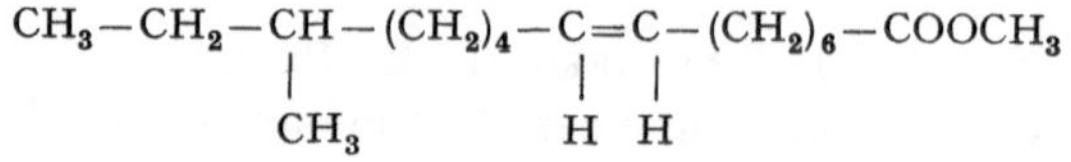

$$CH_3-CH_2-CH-(CH_2)_4-C=C-(CH_2)_6-COOCH_3$$
$$\underset{CH_3}{\mid} \qquad \underset{H}{\mid} \underset{H}{\mid}$$

$$(121)$$

10^{-3} bis $10^{-2}\,\mu g$ dieser Substanzen übten noch eine anziehende Wirkung auf die Männchen aus, aber auch andere *Trogoderma*-Arten sprechen auf den Wirkstoff an.

Die Synthese erfolgte nach dem Schema

$$C_4 + C_3 = C_7$$
$$C_7 + C_2 = C_9$$
$$C_9 + C_8 = C_{17}$$

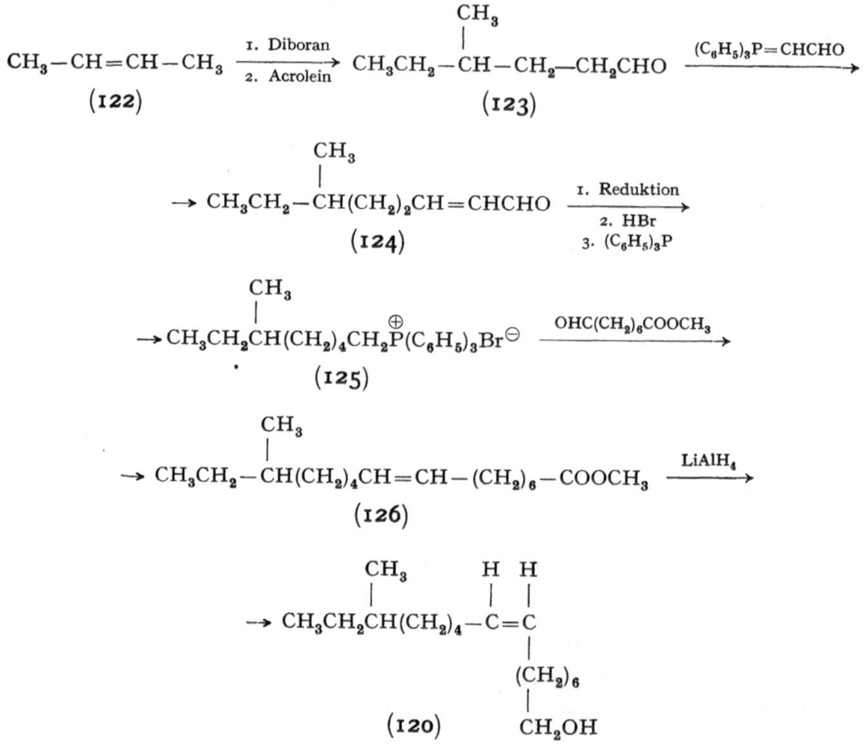

$$CH_3-CH=CH-CH_3 \xrightarrow[\text{2. Acrolein}]{\text{1. Diboran}} CH_3CH_2-\overset{CH_3}{\underset{\mid}{CH}}-CH_2-CH_2CHO \xrightarrow{(C_6H_5)_3P=CHCHO}$$

$$(122) \hspace{6cm} (123)$$

$$\rightarrow CH_3CH_2-\overset{CH_3}{\underset{\mid}{CH}}(CH_2)_2CH=CHCHO \xrightarrow[\substack{\text{2. HBr} \\ \text{3. }(C_6H_5)_3P}]{\text{1. Reduktion}}$$

$$(124)$$

$$\rightarrow CH_3CH_2\overset{CH_3}{\underset{\mid}{CH}}(CH_2)_4CH_2\overset{\oplus}{P}(C_6H_5)_3Br^{\ominus} \xrightarrow{OHC(CH_2)_6COOCH_3}$$

$$(125)$$

$$\rightarrow CH_3CH_2-\overset{CH_3}{\underset{\mid}{CH}}(CH_2)_4CH=CH-(CH_2)_6-COOCH_3 \xrightarrow{LiAlH_4}$$

$$(126)$$

$$\rightarrow CH_3CH_2CH(CH_2)_4-\overset{CH_3}{\underset{\mid}{}}\;\overset{H\;\;H}{\underset{\mid\;\;\mid}{C}}=C$$
$$\underset{\underset{\underset{CH_2OH}{\mid}}{(CH_2)_6}}{\mid}$$

$$(120)$$

Buten-(2) (**122**) konnte nach Umsetzung mit Diboran und Acrolein in das 4-Methylhexanal-(1) (**123**) umgewandelt und mit dem Phosphorylen des halbseitigen Glyoxals zum 6-Methylocten-(2)-al-(1) (**124**) kondensiert werden. Reduktion des α,β-ungesättigten Aldehyds zum gesättigten

Alkohol führte nach Behandlung mit HBr und Triphenylphosphin zum Phosphoniumsalz (125), das nach WITTIG mit dem Octanal-(1)-säure methylester-(8) zu (126) olefiniert werden konnte. *Cis*- und *trans*-Form trennte man durch Dünnschichtchromatographie an Silicagel und Silbernitrat, um nach Isolierung der *cis*-Verbindung diese mittels Lithiumalanat in 14-Methyl-8-*cis*-hexadecenol-(1) (120) zu überführen.

11. Sexuallockstoff des fall army worm *(Spodoptera frugiperda)*:

Der Sexuallockstoff von *Spodoptera frugiperda*, einer Lepidopteren-Art, die in den USA große Schäden bei Mais und Sorghum verursacht, konnte von A. A. SEKUL und A. N. SPARKS (89) isoliert, in seiner Konstitution aufgeklärt und auch synthetisiert werden; es handelt sich bei dem Sexuallockstoff um das Tetradecen-(9-*cis*)-ylacetat-(1)

$$\begin{array}{cc} H & H \\ | & | \\ CH_3-(CH_2)_3-C=C-(CH_2)_8-OCOCH_3 \end{array}$$

$$(127)$$

das in Konzentrationen von $3 \cdot 10^{-12}\,\mu g$ einwandfreie biologische Aktivität zeigte.

Nach dem Syntheseprinzip

$$C_8 + C_2 = C_{10}$$
$$C_{10} + C_4 = C_{14}$$

$$RO-(CH_2)_8Cl \xrightarrow[\text{DMSO}]{\text{LiC}\equiv\text{CH}} RO-(CH_2)_8-C\equiv CH \xrightarrow[\text{LiNH}_2]{\text{C}_4\text{H}_9\text{Br}}$$

$$(128) \qquad\qquad (129)$$

$$\longrightarrow RO-(CH_2)_8-C\equiv C-(CH_2)_3CH_3 \xrightarrow{\text{Pd/CaCO}_3} RO-(CH_2)_8-\overset{\displaystyle H}{\underset{\displaystyle }{C}}=\overset{\displaystyle H}{\underset{\displaystyle }{C}}-(CH_2)_3CH_3$$

$$(130) \qquad\qquad (131)$$

$$\Big\downarrow \begin{array}{l} \text{AlCl}_3 \\ \text{CH}_3\text{COCl} \end{array}$$

$$\begin{array}{cc} H & H \\ | & | \\ CH_3COO-(CH_2)_8-C=C-(CH_2)_3CH_3 \end{array}$$

$$(127)$$

war die Substanz sterisch einheitlich herstellbar.

W. L. ROELOFS und A. COMEAU (81) haben mit 9-*cis*- und 9-*trans*-Tetradecenylacetat in jeweils Reinsubstanz und als Gemisch Freilandversuche mit *Argyrotaenia velutinana* (red banded leaf roller) gemacht und festgestellt, daß die graue Form dieser Insektenspezies spezifisch von 9-*cis*-Tetradecenylacetat und die gelbe Form nur von 9-*trans*-Tetradecenylacetat angelockt wird. Eine Mischung der beiden Komponenten ergab für beide Insektenspezies völlige Blockierung der Anlockwirkung, während die Zugabe von z. B. 7-*trans*-Tetradecenylacetat zu den aktiven C_{14}-Körpern keine Inhibierung der Aktivität verursachten; solche Befunde werden solange schlecht in die Wirkungsweise von Insektensexuallockstoffen einzuordnen sein, solange nicht an den verschiedensten Objekten von völlig verschiedenen Beobachtern ähnliche Befunde erhoben werden (*82*).

12. Sexuallockstoffe des schwarzen Teppichkäfers [*Attagenus megatoma* (Fabricius) oder *Attagenus piceus* (Olivier)], Black carpet Beetle:

W. E. BURKHOLDER und R. J. DICKE (*14*) gelang 1966 der Nachweis, daß beim black carpet beetle zwischen weiblichen und männlichen Tieren ein Sexuallockstoff wirksam ist, der von R. M. SILVERSTEIN et al. (*92*) isoliert und identifiziert wurde und dessen Konstitution durch Synthese erhärtet werden konnte. Beim Lockstoff des Teppichkäfers handelt es sich um die Tetradecadien-(3,5)-säure-(1) (*132*) die als *cis-trans* konjugiertes System erkannt worden war. Die durchgeführte Synthese der beiden geometrisch isomeren Säuren zeigte, daß es sich beim Naturprodukt um die Tetradecadien-(3-*trans*,5-*cis*)-säure-(1) (*132*) handeln mußte.

$$C_8H_{17}-\overset{\overset{\displaystyle H}{|}}{C}=\overset{\overset{\displaystyle H}{|}}{\underset{\underset{\displaystyle H}{|}}{C}}-\overset{\overset{\displaystyle H}{|}}{C}=\overset{\overset{\displaystyle H}{|}}{C}-CH_2-COOH$$

(132)

Während die 3-*trans*-5-*cis*-Tetradecadiensäure-(1) nach der beim Aufbau von 3-trans-5-cis-Tridecadiensäure-(1) (*24*) angegebenen Methode durchgeführt wurde, konnte die 3-*cis*-5-*trans*-Tetradecadiensäure-(1) nach dem Aufbauprinzip

$$C_9 + C_3 = C_{12}$$
$$C_{12} + C_2 = C_{14}$$

hergestellt werden.

$$C_8H_{17}CHO + BrCH_2C{\equiv}CH \xrightarrow{\text{Zn}} C_8H_{17}\underset{\underset{\displaystyle OH}{|}}{CH}-CH_2-C{\equiv}CH \xrightarrow[\text{OH}^{\ominus}]{\text{Tosylchlorid}}$$

(133)   (134)

Literaturverzeichnis: SS. 250—255

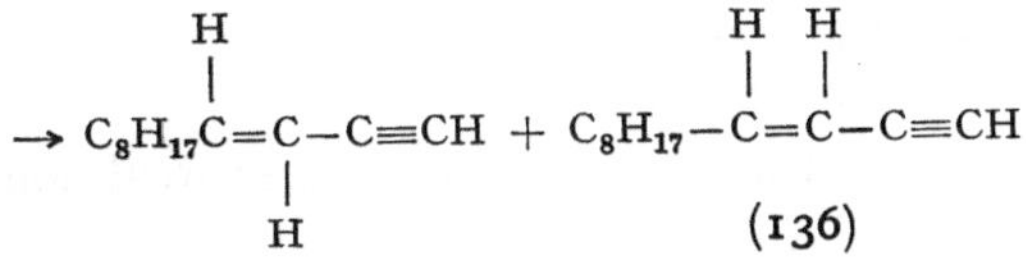

$$\rightarrow\ C_8H_{17}\overset{\overset{\textstyle H}{|}}{C}=C-C\equiv CH\ +\ C_8H_{17}-\overset{\overset{\textstyle H}{|}}{C}=\overset{\overset{\textstyle H}{|}}{C}-C\equiv CH$$

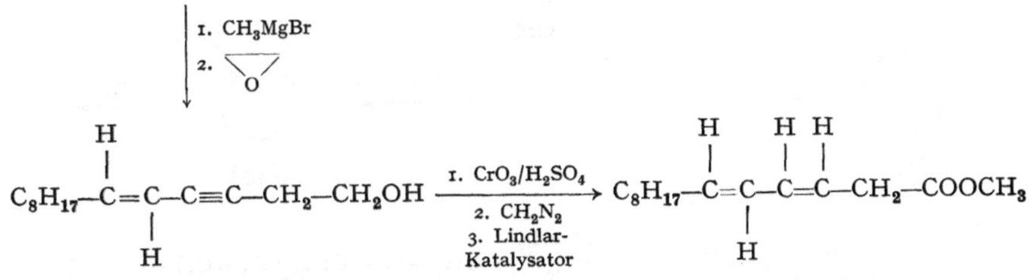

$$\underset{(135)}{\phantom{C_8H_{17}}}\qquad\qquad\qquad (136)$$

$$\Bigg| \begin{array}{l} \text{1. } CH_3MgBr \\[4pt] \text{2. } \triangleleft_O \end{array}$$

$$\underset{(137)}{C_8H_{17}-\overset{\overset{\textstyle H}{|}}{C}=C-C\equiv C-CH_2-CH_2OH}\ \xrightarrow[\substack{\text{2. } CH_2N_2 \\ \text{3. Lindlar-} \\ \text{Katalysator}}]{\text{1. } CrO_3/H_2SO_4}\ \underset{(138)}{C_8H_{17}-\overset{\overset{\textstyle H}{|}}{C}=C-\overset{\overset{\textstyle H}{|}}{C}=\overset{\overset{\textstyle H}{|}}{C}-CH_2-COOCH_3}$$

13. Königinnensubstanz (queen-substance):

Aus den Mandibulardrüsen der Bienenkönigin *(Apis mellifera)* wird die sogenannte Königinnensubstanz (Queen-Substance) als Pheromon abgegeben, die verhindert, daß von den Arbeiterinnen bei lebender Königin eine sogenannte Weiselzelle zur Aufzucht einer neuen Königin gebaut wird.

R. K. CALLOW und N. C. JOHNSTON *(22)* konnten die Königinnensubstanz isolieren und als 2-trans-Decenon-(9)-säure-(1) (139) identifizieren.

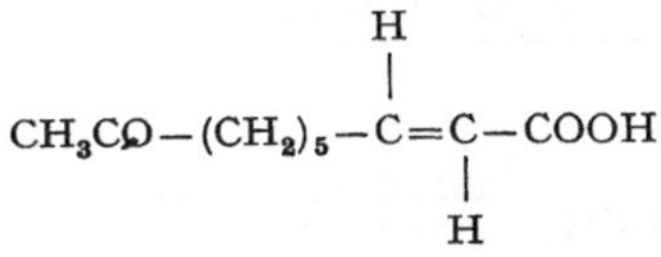

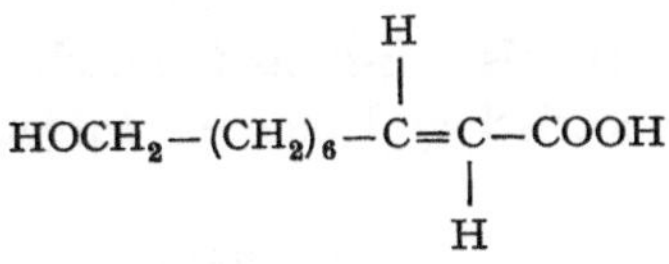

$$\underset{(139)}{CH_3CO-(CH_2)_5-\overset{\overset{\textstyle H}{|}}{C}=C-COOH}\qquad\qquad \underset{(140)}{CH_3CO-(CH_2)_5-\overset{\overset{\textstyle H}{|}}{C}=\overset{\overset{\textstyle H}{|}}{C}-COOH}$$

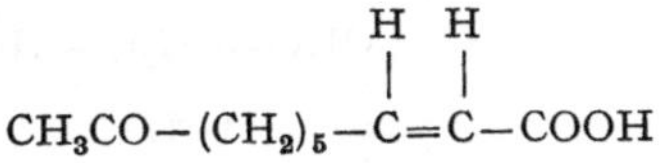

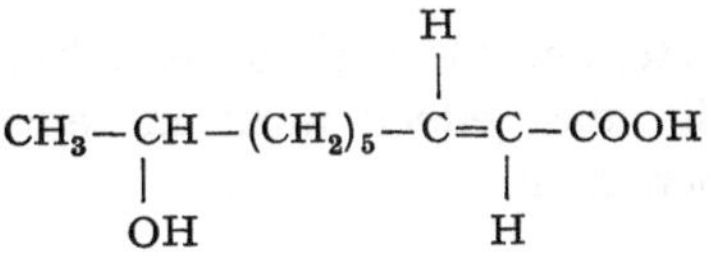

$$\underset{(141)}{HOCH_2-(CH_2)_6-\overset{\overset{\textstyle H}{|}}{C}=C-COOH}\qquad\qquad \underset{(142)}{CH_3-\underset{\underset{\textstyle OH}{|}}{CH}-(CH_2)_5-\overset{\overset{\textstyle H}{|}}{C}=C-COOH}$$

Nach Abklärung der Konstitution der Queen-Substance wurden eine Fülle von Synthesen für diesen interessanten Naturstoff bekannt *(2, 10, 21, 35, 55, 63)* wobei hier nur 2 Synthesen beschrieben werden sollen:

M. BARBIER, E. LEDERER und T. NOMURA *(3)* sowie M. BARBIER und M. F. HÜGEL *(1)* führten die Synthese nach dem Schema

$$C_7 + C_1 = C_8$$
$$C_8 + C_3 = C_{10} + C_1$$

durch und gingen vom Cycloheptanon aus, um in vierstufiger Reaktion Queen-Substance zu erhalten:

$$\text{(143)} \qquad \text{(144)} \qquad \text{(145)}$$

$$CH_3-\underset{\underset{O}{\|}}{C}-(CH_2)_5-CHO \xrightarrow[C_5H_5N]{CH_2(COOH)_2} CH_3-\underset{\underset{O}{\|}}{C}-(CH_2)_5CH=CH-COOH$$

$$\text{(146)} \qquad\qquad\qquad\qquad \text{(139)}$$

K. Eiter (31, 32) konnte das Pheromon im 50—100-g-Maßstab nach dem Aufbauschema

$$C_5 + C_2 = C_7$$
$$C_7 + C_3 = C_{10}$$

herstellen, wobei in interessanter Reaktion Glutaraldehyd (147) mit Triphenylphosphincarbomethoxymethylen nur in halbseitiger Reaktion umgesetzt wird, so daß

$$OHC-(CH_2)_3-CHO \xrightarrow{(C_6H_5)_3P=CH-COOCH_3}$$
$$\text{(147)}$$

$$\rightarrow OCH(CH_2)_3-CH=CH-COOCH_3 \xrightarrow{[HC\equiv C-CH_2Al_{2/3}Br]}$$
$$\text{(148)}$$

ein Hepten-(2)-al-(7)-säuremethylester-(1) als *cis-trans*-Gemisch (148) anfällt, das jedoch nicht in seine geometrischen Isomeren aufgetrennt werden muß.

$$HC\equiv C-CH_2-\underset{\underset{OH}{|}}{CH}-(CH_2)_3-CH=CH-COOCH_3 \xrightarrow[HgSO_4\cdot HgO]{H_2SO_4}$$
$$\text{(149)}$$

$$\rightarrow CH_3-CO-CH=CH-(CH_2)_3-CH=CH-COOCH_3 \xrightarrow[\text{2. Na}_2CO_3]{\text{1. H}_2/\text{mod. Lindlar-Katalysator}}$$
$$\text{(150)}$$

$$\rightarrow \text{2-trans + cis-Decenon-(9)-säure-(1)}$$

Literaturverzeichnis: SS. 250—255

Umsatz dieses Heptenalesters (148) mit Zink nach REFORMATSKY oder vorteilhafter Weise mit der aluminiumorganischen Verbindung des Propargylbromids führt zu einem allenfreien Acetylencarbinol (149), das unter Hydratisierung-Dehydratisierung die α,β-ungesättigte Carbonylverbindung (150) liefert, die mit einem mit Rohchinolin vergifteten Lindlar-Katalysator selektiv an dieser Doppelbindung hydriert werden konnte.

In einem zweiten Verfahren, das jedoch nur mit schlechten Ausbeuten durchführbar war, konnte der 2-Heptenal-(7)-säureester-(1) (148) nach Knövenagel mit Aceton kondensiert werden, wobei die Verbindung (150) isolierbar war.

$$OHC-(CH_2)_3-CH=CH-COOCH_3 + CH_3COCH_3 \xrightarrow[\text{Eisessig}]{\text{Piperidin}}$$
$$(148)$$

$$\rightarrow CH_3-CO-CH=CH-(CH_2)_3$$
$$|$$
$$CH=CH$$
$$|$$
$$(150) \quad COOCH_3$$

Aus der Königinnensubstanz konnte durch Reduktion mit Natriumboranat 9-Hydroxy-2-trans-decensäure-(1) (142) bereitet werden, während die 10-Hydroxy-2-trans-decensäure-(1) (141) als Bestandteil des Weiselfuttersaftes (Royal jelly) aufgefunden und u. a. von E. TRUSCHEIT (101) synthetisiert worden ist.

Durch gaschromatographische Trennung des Weiselfuttersaftes konnten der 9-Hydroxydecensäuremethylester, p-Hydroxybenzoesäuremethylester, Nonansäure, Decansäure, 2-Decensäure, 9-Oxadecansäure, 9-Hydroxydecansäure und p-Methoxybenzoesäure neben 18 noch nicht identifizierten Substanzen nachgewiesen werden.

14. Pheromon der männlichen Hummel *(Bombus terrestris* L.*)*:

G. STEIN (97) konnte aus den Mandibulardrüsen der Männchen Farnesol (157) isolieren, womit die Tiere in den Monaten Juli und August bei Schwarmflügen Duftpunkte setzen, so daß in den so markierten Arealen die Weibchen angelockt und begattet werden können.

$$(157)$$

15. Wegweiserpheromone bei Termiten:

1961 haben Esenther et al. (*34*) gefunden, daß Holz, das von den Pilzen der Gattung *Lenzites trabea* Pers. *ex* Fr. befallen war, eine Substanz enthielt, die für die Termiten *Reticuliterma flavipes* sowie *R. virginicus* Lockwirkung besaß und als Wegweisersubstanz für diese Insekten wirkte. F. Matsamura, H. C. Coppel und A. Tai (*66*) haben sowohl aus pilzbefallenem Holz als auch aus Termiten *R. virginicus* das Pheromon isolieren können und als 3-*cis*, 6-*cis*-8-*trans*-Dodecatrienol-(1)

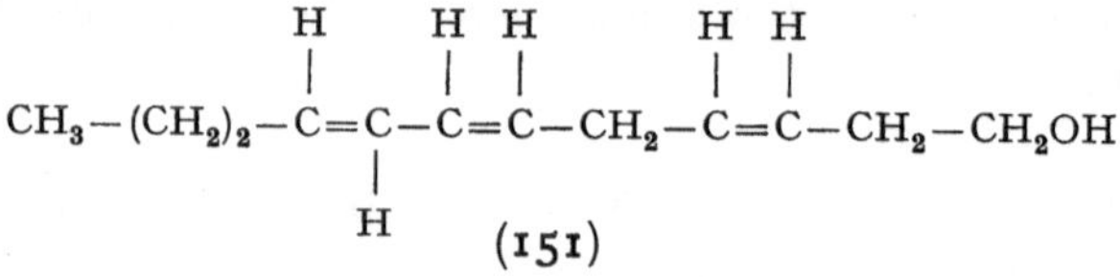

$$CH_3-(CH_2)_2-\overset{\overset{\displaystyle H}{|}}{C}=\overset{\overset{\displaystyle H}{|}}{\underset{\underset{\displaystyle H}{|}}{C}}-\overset{\overset{\displaystyle H}{|}}{C}=\overset{\overset{\displaystyle H}{|}}{C}-CH_2-\overset{\overset{\displaystyle H}{|}}{C}=\overset{\overset{\displaystyle H}{|}}{C}-CH_2-CH_2OH$$

$$(\mathbf{151})$$

identifiziert.

Die Synthese der Substanz gelang nach dem Aufbauschema

$$C_4 + C_3 = C_7$$
$$C_7 + C_3 = C_{10}$$
$$C_{10} + C_2 = C_{12}$$

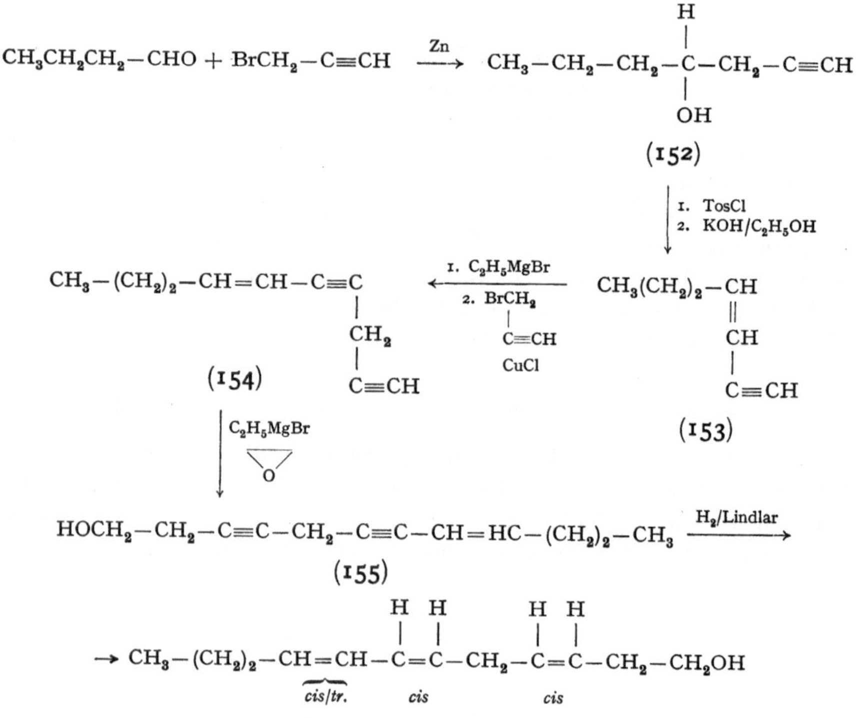

$$CH_3CH_2CH_2-CHO + BrCH_2-C\equiv CH \xrightarrow{Zn} CH_3-CH_2-CH_2-\overset{\overset{\displaystyle H}{|}}{\underset{\underset{\displaystyle OH}{|}}{C}}-CH_2-C\equiv CH$$

$$(\mathbf{152})$$

1. TosCl
2. KOH/C_2H_5OH

$$CH_3-(CH_2)_2-CH=CH-C\equiv C \xleftarrow[\text{2. BrCH}_2]{\text{1. } C_2H_5MgBr} \quad CH_3(CH_2)_2-CH$$

(154) — CH_2 — $C\equiv CH$ — CuCl

(153) — $CH_3(CH_2)_2-CH \overset{||}{} CH$ — $C\equiv CH$

C_2H_5MgBr / O

$$HOCH_2-CH_2-C\equiv C-CH_2-C\equiv C-CH=HC-(CH_2)_2-CH_3 \xrightarrow{H_2/Lindlar}$$

$$(\mathbf{155})$$

$$\rightarrow CH_3-(CH_2)_2-CH=CH-\overset{\overset{\displaystyle H}{|}}{C}=\overset{\overset{\displaystyle H}{|}}{C}-CH_2-\overset{\overset{\displaystyle H}{|}}{C}=\overset{\overset{\displaystyle H}{|}}{C}-CH_2-CH_2OH$$

cis/tr.   *cis*   *cis*

$$(\mathbf{156})$$

Literaturverzeichnis: SS. 250—255

wobei n-Butyraldehyd mit Propargylbromid nach REFORMATZKY umgesetzt das Carbinol (152) ergab, das nach Dehydratisierung eine zur Acetylenbindung konjugierte Äthylenbindung in *cis*- und *trans*-Konfiguration besaß. Durch metallorganische Synthese dieses C_7-Bausteines (153) mit Propargylbromid und Cu(I)-Chlorid entstand das C_{10}-Diinen (154), das in nochmaliger metallorganischer Reaktion mit Äthylenoxyd schließlich das C_{12}-Diinenol (155) lieferte, das zur 3,6-Di*cis*-verbindung selektiv hydriert werden konnte. Synthetisches 3-*cis*-6-*cis*-8-*trans*-Dodecatrienol-(1) besaß bei Termitenarbeiterinnen mit 0,1 μg in 10 ml n-Hexan die gleiche Wegweiserwirkung wie das natürliche Pheromon.

b) Aliphatische Verbindungen

1. Sexuallockstoff des Zuckerrübendrahtwurmes (sugar beet wireworm):

Beim kalifornischen Zuckerrübendrahtwurm (*Limonius californicus* Mannerheim) konnten M. JACOBSON und C. E. LILLY (54) einen Lockstoff isolieren und seine Konstitution identifizieren; die Locksubstanz, die bei weiblichen Tieren in verhältnismäßig hoher Konzentration vorkommt, ist nach IR-Spektrum, R_F-Wert sowie Gaschromatographie des Methylesters die Valeriansäure

$$CH_3CH_2CH_2CH_2—COOH$$

Rohe Naturextrakte stoßen die Männchen ab, während verdünnte Lösungen eine Anlockwirkung mit sexueller Erregbarkeit zeigen.

c) Terpenartige Verbindungen

1. Pheromon des kalifornischen Borkenkäfers *(Ips confusus)*, bark beetle:

In den Exkrementen der Männchen von *Ips confusus* wird ein Substanzgemisch abgesondert, das für die Weibchen Lockwirkung besitzt (94). R. M. SILVERSTEIN, I. O. RODIN und D. L. WOOD (93) konnten zeigen, daß dieses Gemisch aus (—)2-Methyl-6-methylen-7-octen-4-ol (157), (+)*cis*-Verbenol (158) und (+)-2-Methyl-6-methylenoctadien-(2,7)-ol-(4) (159) besteht.

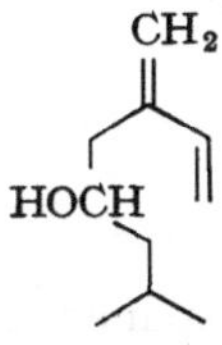

(157) 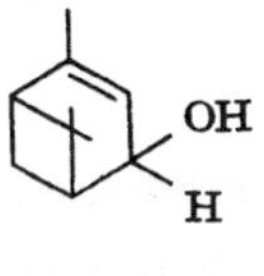   (158) 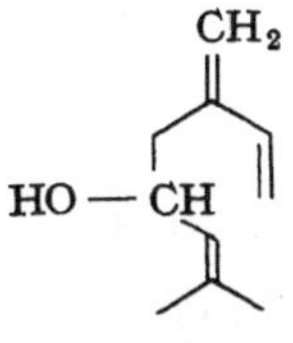   (159)

Die aus dem Naturprodukt isolierten Einzelkomponenten waren wie die Syntheseprodukte allein biologisch inaktiv; das Gemisch der Substanzen (157) und (158) war im Laborversuch bereits wirksam, während im Feldversuch nur dann eine Wirkung beobachtet wurde, wenn alle drei Komponenten im Gemisch vorhanden waren.

2. Lockstoff der männlichen Florfliege *Chrysopa septempunctata* Wesmale (Lace Wing):

S. ISHIN (*43*) und T. SAKAN (*86*) beobachteten, daß männliche Florfliegen durch einen Extrakt der Pflanze *Actinidia polygoma* Mig angelockt werden. Das in dieser Pflanze aufgefundene Gemisch von Lactonen wurde auch in den Ausscheidungen argentinischer und australischer Ameisen des genus *Iridomyrmex* beobachtet; es handelt sich um die Substanzen Iridomyrmecin (160) und Isoiridomyrmecin (161) sowie Irododiol (162)

(160)   (161)   (162)

während 5-Hydroxy-(163), 7-Hydroxymetatabiäther (164), Metatabiol (165), Allometatabiol (166) und Neometatabiol (167) in Pflanzen aufgefunden wurden, wobei die Verbindungen (162), (163) und (167) männliche Florfliegen sexuell stimulieren

(163)   (164)   (165)

(166)   (167)

3. Lockstoff des *Dendroctonus brevicomis*:

R. M. SILVERSTEIN et al. (*91*) haben aus den Nagespänen des Käfers *Dendroctonus brevicomis* einen Lockstoff Brevicomin (168) isoliert und durch Synthese bewiesen, daß es sich bei dem Pheromon um 5-Methyl-7-äthyl-6,8-dioxabicyclo[3.2.1]-octan handelt.

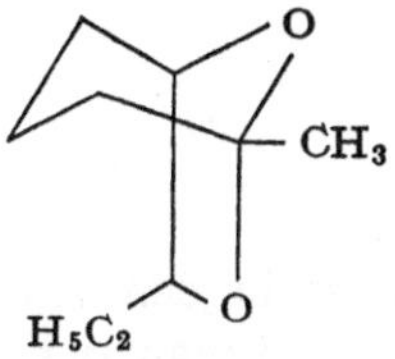

(168)

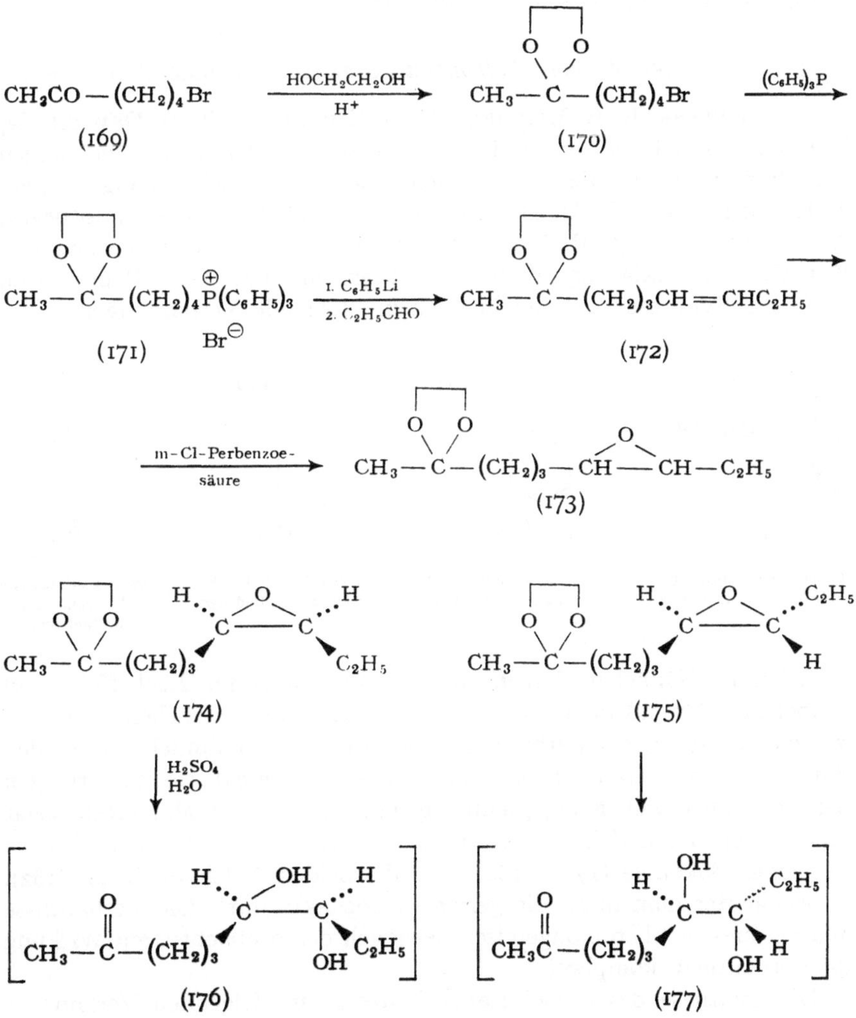

$$CH_3CO-(CH_2)_4Br \xrightarrow[H^+]{HOCH_2CH_2OH} CH_3-\overset{O \quad O}{\underset{\diagdown \diagup}{C}}-(CH_2)_4Br \xrightarrow{(C_6H_5)_3P}$$

(169)   (170)

$$CH_3-\overset{O \quad O}{\underset{\diagdown \diagup}{C}}-(CH_2)_4\overset{\oplus}{P}(C_6H_5)_3 \xrightarrow[2.\ C_2H_5CHO]{1.\ C_6H_5Li} CH_3-\overset{O \quad O}{\underset{\diagdown \diagup}{C}}-(CH_2)_3CH=CHC_2H_5 \longrightarrow$$

(171)   Br$^{\ominus}$ (172)

$$\xrightarrow[säure]{m-Cl-Perbenzoe-} CH_3-\overset{O \quad O}{\underset{\diagdown \diagup}{C}}-(CH_2)_3-\overset{O}{\overset{\diagdown \diagup}{CH-CH}}-C_2H_5$$

(173)

(174)   (175)

$\Bigg\downarrow \begin{smallmatrix} H_2SO_4 \\ H_2O \end{smallmatrix}$

(176)   (177)

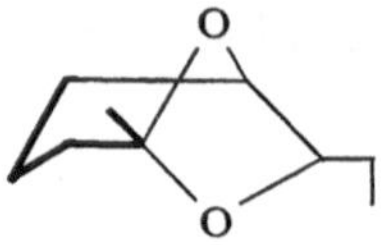

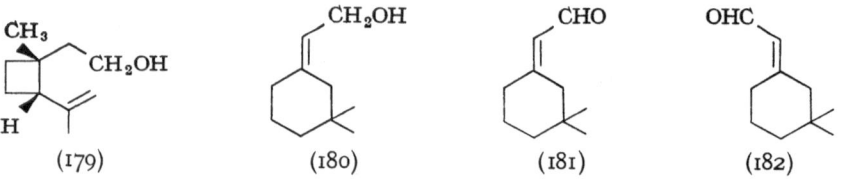

(168) (178)
exo endo

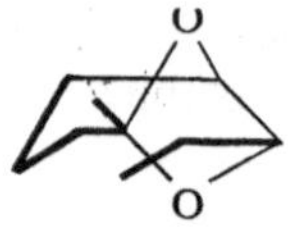

Dendroctonus brevicomis ist in den Kiefernwäldern Kaliforniens ein beachtliches Schadinsekt, so daß man sich durch Anlockung des Schädlings mit dem synthetischen Pheromon und seine Vernichtung einen Bekämpfungserfolg für die Zukunft verspricht.

4. Sexuallockstoff von *Anthonomus grandis* Boheman (boll weevil):

J. C. KELLER, E. B. MITCHELL, G. McKIBBEN und T. B. DEWICH (*62*) konnten aus männlichen Boll Weevil bzw. aus deren Ausscheidungen Duftstoffe isolieren, die auf weibliche Tiere eine Lockwirkung zeigten. J. H. TUMLINSON, D. D. HARDER, R. C. GUELDNER, A. C. THOMPSON, A. A. HEDIN und J. P. MINYARD (*107*) gelang nun die Isolierung und Konstitutionsaufklärung dieses Substanzgemisches von Pheromonen sowie die Synthese. Es handelt sich um folgende Verbindungen

(179) (180) (181) (182)

(+)*cis*-2-Isopropenyl-1-methylcyclobutanäthanol

cis-3,3-dimethyl-$\Delta^{1\alpha}$-cyclohexanäthanol

cis-3,3-dimethyl-$\Delta^{1\alpha}$-cyclohexanacetaldehyd

trans-3,3-dimethyl-$\Delta^{1\alpha}$-cyclohexanacetaldehyd

Lebende Männchen haben nur im Laboratorium Lockeffekte auf Weibchen, während im Feldversuch Männchen und Weibchen angelockt werden; die synthetisch erhaltenen Lockstoffe zeigen ähnlich wie bei den Männchen von *Ips confusus* Le Conte synergistische Effekte; so vermögen Gemische aller vier Komponenten genau wie lebende Männchen sogar noch bessere Lockeffekte zu bewirken. Fehlen jedoch im Gemisch z. B. die beiden Alkohole (179), (180) oder die beiden Aldehyde (181), (182), so beobachtet man nicht die geringste Lockaktivität! Die Verhältnisse sind bei diesem Pheromongemisch bezüglich der beobachtbaren Wirkung außerordentlich komplex!

Die Synthese des Cyclobutanderivates nahm folgenden Verlauf:

Literaturverzeichnis: SS. 250—255

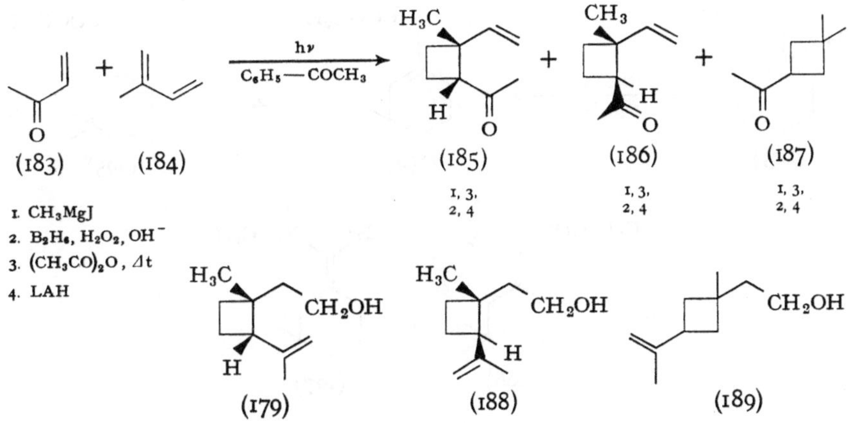

(183) (184) (185) (186) (187)

1. CH₃MgJ

2. B₂H₆, H₂O₂, OH⁻

3. (CH₃CO)₂O, Δt

4. LAH

(179) (188) (189)

Photokondensation von Methylvinylketon und Isopren führte zu einem komplexen Substanzgemisch, aus dem die Cyclobutanderivate (185), (186), (187) gaschromatographisch in geringen Ausbeuten isoliert werden konnten; in den angegebenen Reaktionsfolgen gelangte man zu dem Cyclobutanderivat (179), das eine der natürlichen Pheromonkomponenten darstellt.

Die Cyclohexanverbindungen (180), (181), (182) waren auf folgendem Weg zugänglich:

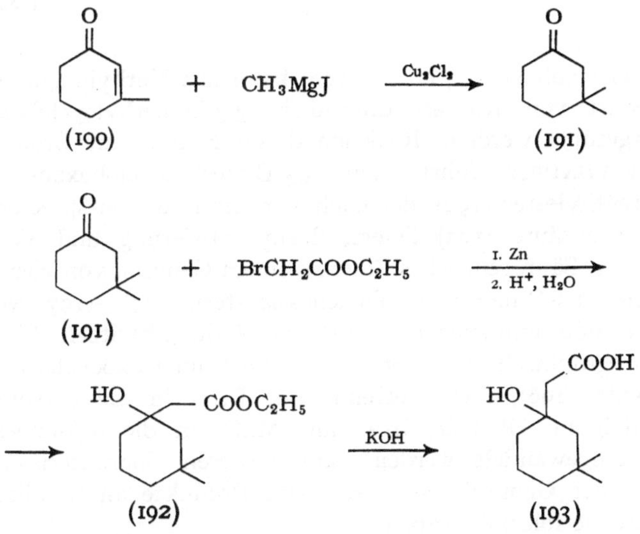

(190) + CH₃MgJ $\xrightarrow{Cu_2Cl_2}$ (191)

(191) + BrCH₂COOC₂H₅ $\xrightarrow[\text{2. H}^+, \text{H}_2\text{O}]{\text{1. Zn}}$

(192) $\xrightarrow{KOH}$ (193)

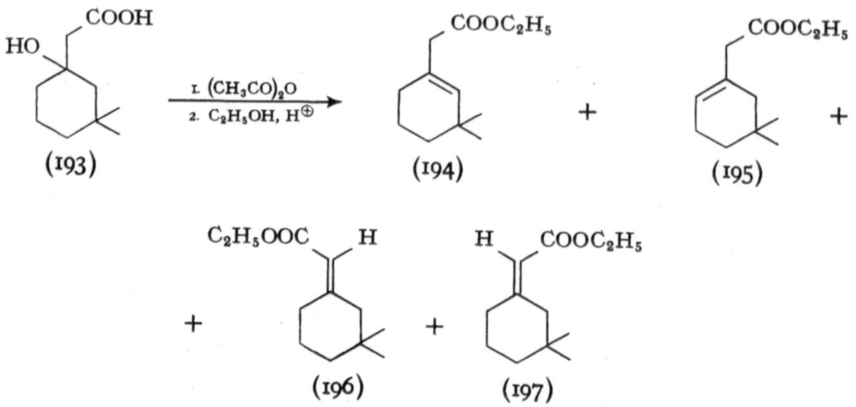

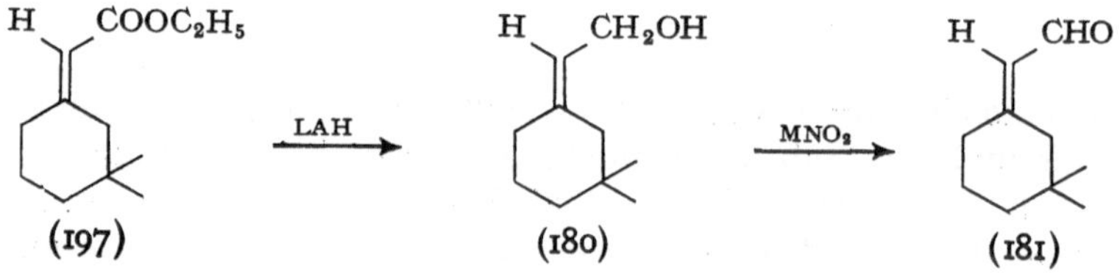

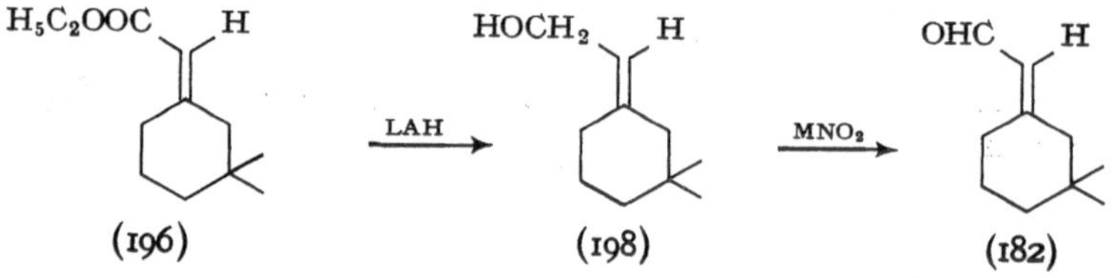

COOH
HO

(193)

1. (CH₃CO)₂O
2. C₂H₅OH, H⊕

COOC₂H₅

(194)   +

COOC₂H₅

(195)   +

C₂H₅OOC   H

+

(196)

H   COOC₂H₅

+

(197)

H   COOC₂H₅

(197)

LAH →

H   CH₂OH

(180)

MNO₂ →

H   CHO

(181)

H₅C₂OOC   H

(196)

LAH →

HOCH₂   H

(198)

MNO₂ →

OHC   H

(182)

3-Methylcyclohexen-(2)-on-(1) (190) kann mit Methylmagnesiumjodid in Gegenwart von Kupfer-I-Chlorid in 3,3-Dimethylcyclohexanon-(1) (191) übergeführt werden. Reaktion dieses Ketons mit Bromessigester nach REFORMATSKY führt zum 3,3-Dimethylcyclohexan-1-hydroxy-1-essigsäureäthylester (192), der nach Verseifung die entsprechende freie Hydroxycarbonsäure (193) liefert. Dehydratisierung und Veresterung des erhaltenen Säuregemisches führt zu einem Gemisch von vier isomeren ungesättigten 3,3-Dimethylcyclohexenessigestern (194)—(197), von denen je zwei eine endo- und exocyclische Doppelbindung besitzen. Sie konnten gaschromatographisch aufgetrennt, und die beiden exocyclischen Cyclohexenderivate (196), (197) getrennt mit LAH in die entsprechenden α,β-ungesättigten Alkohole, bzw. mit MnO₂ in die α,β-ungesättigten Aldehyde umgewandelt werden. Zur besseren Charakterisierung der Aktivitäten der Gemische synthetisierter Produkte an weiblichen Boll weevil dient die folgende Tabelle.

Literaturverzeichnis: SS. 250—255

Tabelle 2. *Aktivität von synthetischen I, II, III und IV*

Probe	Menge in µg				Durchschnitt T/S
I	0,09				0,11
II	0,07				0,06
III	0,12				0,00
IV	0,12				0,11
I, II	0,09,	0,07			0,15
I, III	0,09,	0,12			0,06
I, IV	0,09,	0,12			0,12
II, III	0,07,	0,12			0,24
II, IV	0,07,	0,12			0,12
III, IV	0,12,	0,12			0,16
I, II, III	0,09,	0,07,	0,12		1,00
I, II, IV	0,09,	0,07,	0,12		0,84
I, III, IV	0,09,	0,12,	0,12		0,07
II, III, IV	0,07,	0,12,	0,12		0,18
I, II, III, IV	0,09,	0,07,	0,12,	0,12	1,26

Aktivitäten gemessen an weiblichen Boll weevil im Laboratoriumstest bei optimalen Konzentrationen an I, II, III und IV. Der Quotient T/S bedeutet die Reaktionen mit der Testprobe dividiert durch die Reaktionen, die mit einem Standard (lebendes Männchen oder Wasserdampfdestillat von männlichen Tieren) ermittelt werden.

Für die Darstellung des Cyclobutanderivates als *cis*-Verbindung haben die Bearbeiter der USDA (*114*) einen Weg aufgefunden, der vom Mevalolacton ausgehend zum Racemat der cis-Cyclobutanverbindung führt, das noch in seine optischen Antipoden gespalten werden muß.

J. Sidall, R. Zurflüh, L. L. Dunham und V. L. Spain von der
Zoecon (*114*) haben 3-Methylcyclohexen-(2)-on-(1) und Äthylen UV-
bestrahlt und so einen Vierring-Sechsringbicyclus (**205**) erhalten, von dem
aus sie gleichfalls zum racemischen cis-Cyclobutanpheromon (**179**) kamen

Das racemische cis-2-isopropenyl-1-methylcyclobutanäthanol erhielt
die Bezeichnung „Grandisol", während das Gemisch aller vier syn-
thetisierten Pheromonsubstanzen die Bezeichnung „Grandlure" erhielt;
es bleibt abzuwarten, welche Erfolge im Freiland mit solchen Synthese-
produkten erzielt werden können.

d) Heterocyclische Verbindungen

J. Meinwald und Y. C. Meinwald (*67 a*) hatten aus den männlichen
Tieren von *Lycorea ceres* Cramer nach Entfernung der Haarstifte einen
Extrakt erhalten, der ein Gemisch von Pheromonen darstellte. Aus ihm
konnten Cetylacetat

$$CH_3(CH_2)_{14}-CH_2OCOCH_3$$
$$(211)$$

sowie ein weiterer Ester isoliert werden, der auf Grund der erhobenen
analytischen Befunde nur Octadecen-(11-*cis*)-ylacetat-(1) (**212**) sein

$$CH_3-(CH_2)_5-C=C-(CH_2)_9-CH_2OCOCH_3$$
$$| \quad |$$
$$H \quad H$$

(212)

konnte. Diese Substanz zeigte sich identisch mit einem Syntheseprodukt, das ausgehend von Octadecen-(11-*cis*)-säuremethylester-(1) durch Reduktion mit LAH und Acetylierung erhalten wurde.

$$CH_3(CH_2)_5-C=C-(CH_2)_9-COOCH_3 \xrightarrow{\text{LAH}} CH_3(CH_2)_5-C=C-(CH_2)_9-CH_2OH \underset{(Ac)_2O}{\searrow}$$

(213)   **(214)**   **(212)**

Neben diesen beiden Estern wurde auch noch eine „Ketonkomponente" rein dargestellt, der nach Synthese die Konstitution eines 7-Methyl-2,3-dihydro-1 H-pyrrolizinon-(1) zukam.

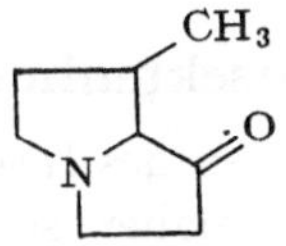

(215)

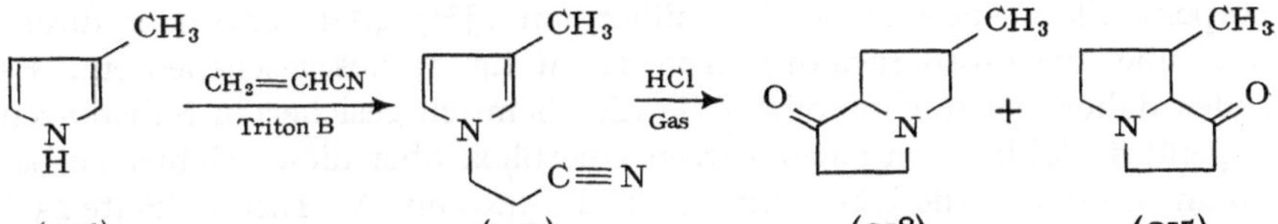

(216)   **(217)**   **(218)**   **(215)**

Aus der Sekretion der Haarstifte von *Danaus gilippus berenice* (queen butterfly) konnten in weiteren Untersuchungen von J. MEINWALD, Y. S. MEINWALD und P. H. MAZZOCCHI (68) neben dem schon oben beschriebenen 7-Methyl-2,3-dihydro-1/H/-pyrrolizinon-(1) **(215)** weitere ungesättigte Verbindungen isoliert und in ihrer Konstitution aufgeklärt werden. Die Hauptkomponente dürfte 3,7-Dimethyldecadien-(2-*trans*, 6-*trans*)-1,10-diol sein.

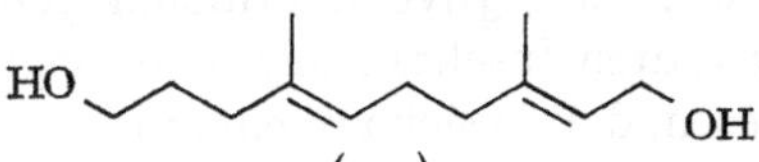

(219)

das auch synthetisch aus Farnesol *(trans, trans)* (220) auf folgendem Weg
erhalten werden konnte

III. Weitere Pheromone

Sexuallockstoff von *Heliothis virescens* (F.), tobacco budworm:

C. R. Gentry, F. R. Lawson und J. D. Hoffmann (*37*) haben aus den
Abdominalsegmenten weiblicher Tiere der Gattung *Heliothis virescens* (F.)
einen Sexuallockstoff isolieren können. Die Konstitution des Pheromons
ist noch nicht aufgeklärt worden.

IV. Insektenhormone

Die neurosekretorischen Zellen des Insektengehirns geben Substanzen
ab, unter denen ein noch nicht näher isoliertes Brain-Hormon BH (vermutlich Peptidhormon) als prothoracotropes Hormon die Ausschüttung
von Häutungshormon Moultinghormon (MH) bewirkt. Das aus den
corpora allata stammende Juvenilhormon (JH) greift entweder direkt
oder über die Prothoracaldrüsen in das Metamorphosegeschehen ein. Es
sollen daher hier nur unvollständig die chemisch gesicherten Substanzen
angeführt werden, um einen kurzen Überblick über dieses Gebiet zu be-
kommen (*8*); auf die Übersicht von H. Hikino und Y. Hikino (Seite 256)
dieses Bandes wird auch verwiesen.

a) Juvenilhormon

Nachdem P. Schmialek (*87*) 1961 zeigte, daß aus den Exkreten des
Mehlwurms *Tenebrio* Farnesol und Farnesal gewonnen werden konnte,
welche Juvenilhormonaktivitäten besaßen, fand er bald danach, daß
der Methyläther des Farnesols eine wesentlich stärkere Juvenilhormonaktivität zeigte. 1965 hatten Stämer und Williams (*113*) gefunden, daß
Papier, das aus „Balsamtannenholzzellulose" gewonnen wurde, einen
„paper-factor" enthielt, der Juvenilhormonaktivität jedoch nur bei
Pyrrhocoris und verwandten Insekten zeigte; Bowers, Fales, Thompson
und Uebel (*13*) bewiesen, daß es sich um ein monocyclisches Sesquiterpen,
den Methylester der Todomatusäure (Juvabion), handelt.

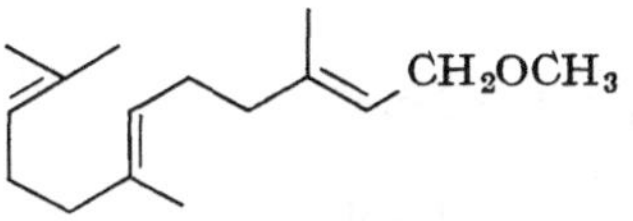

CH$_2$OCH$_3$

Farnesylmethyläther

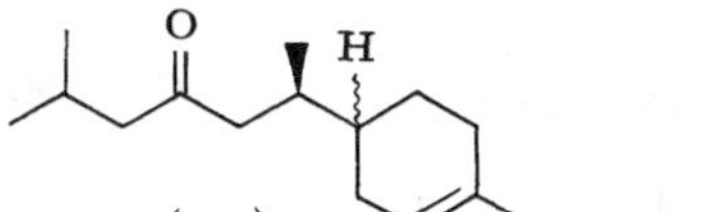

O H

(222) COOCH$_3$

(+) Juvabion

Schließlich konnten K. H. DAHM, B. M. TROST und H. RÖLLER (*28, 29*) zeigen, daß das natürliche Juvenilhormon des Seidenspinners Hyalophora cecropia *trans,trans,cis*-10-Epoxy-7-äthyl-3,11-dimethyl-2,6-trisdecadiensäuremethylester-(1) **(223)** ist.

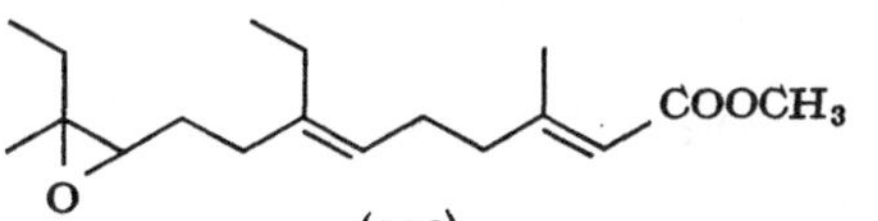

COOCH$_3$

O

(223)

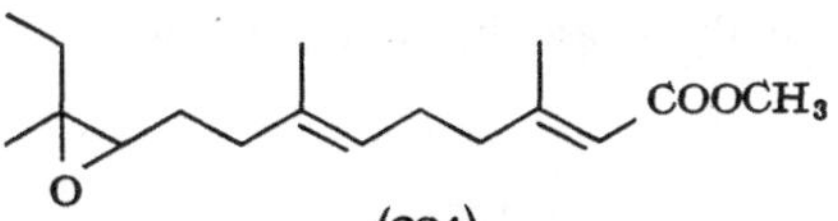

COOCH$_3$

O

(224)

A. S. MEYER und H. A. SCHNEIDERMAN (*70*) konnten eine zweite Komponente mit Juvenilhormonaktivität aus *Hyalophora cecropia* isolieren und zwar *trans,trans*-3,7,11-Trimethyl-10-epoxy-2,6-tridecadiensäuremethylester-(1) **(224)** in dem gegenüber dem zuerst aufgefundenen Hormon an C$_7$ eine Methyl- statt einer Äthylgruppe steht. Das ursprünglich von H. RÖLLER und Mitarb. aufgefundene Juvenilhormon wurde mehrfach synthetisch hergestellt (*26, 40, 115*); stellvertretend soll hier die hoch stereoselektive Synthese racemischen Juvenilhormons von W. S. JOHNSON, T. LI, D. J. FAULKNER und S. F. CAMPBELL (*56*) näher beschrieben werden

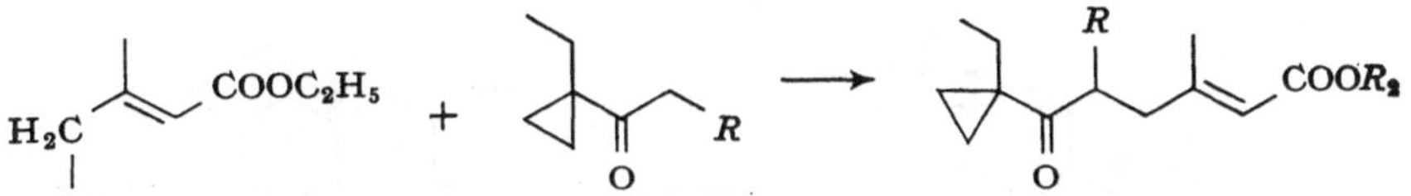

H$_2$C COOC$_2$H$_5$
|
Br **(225)**

+

R
O
(226)

⟶

R
R
COOR$_2$
O
(227)

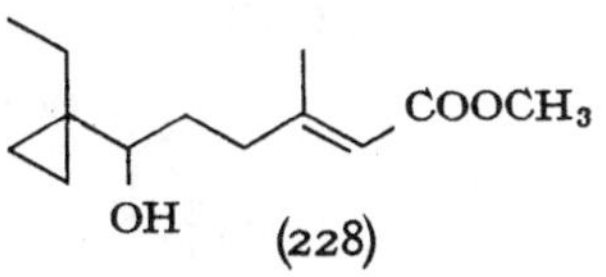

COOCH$_3$
OH **(228)**

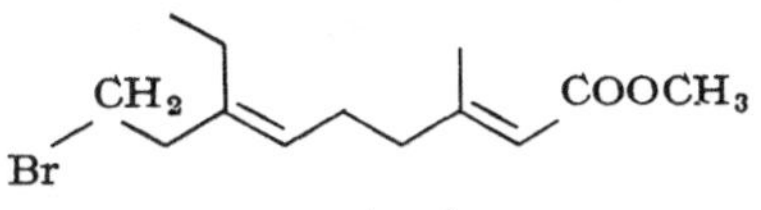

CH$_2$
Br
COOCH$_3$
(229)

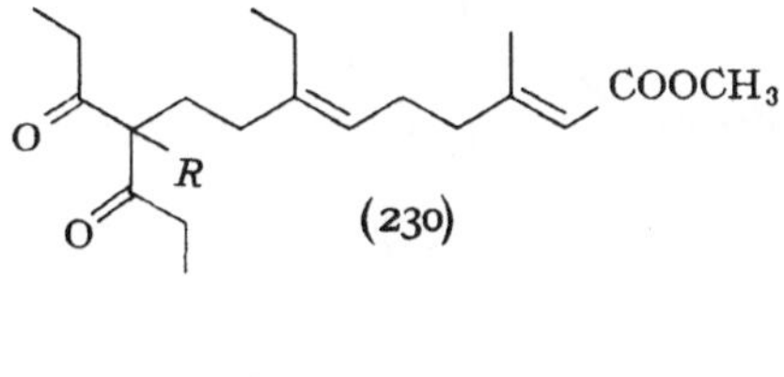

COOCH$_3$
O
R
O
(230)

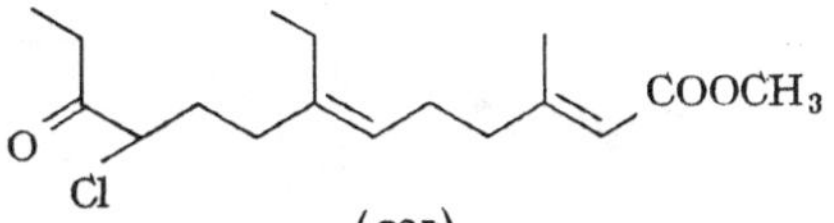

COOCH$_3$
O
Cl
(231)

(232)   (223) JH

b) Häutungshormon (MH = moulting-Hormon)

Das aus den Prothoracaldrüsen stammende Häutungshormon (Ecdyson) konnten BUTENANDT und KARLSON (*18*) 1954 isolieren, worauf in einer Reihe von Arbeiten die Chemie des Ecdysons untersucht wurde. Die endgültige Konstitution konnte durch Röntgen-Kristallstruktur-analyse (*42*) bewiesen werden.

(233)

In der Folge haben zwei Arbeitsgruppen und zwar Syntex (*90*) in den USA und Schering AG—Hofmann-La Roche (*36, 64*) durch unabhängige Synthesen Konstitution und Stereochemie des Ecdysons bewiesen.

(234)   (235)

(236)   (237)

Literaturverzeichnis: SS. 250—255

(233)
Ecdyson

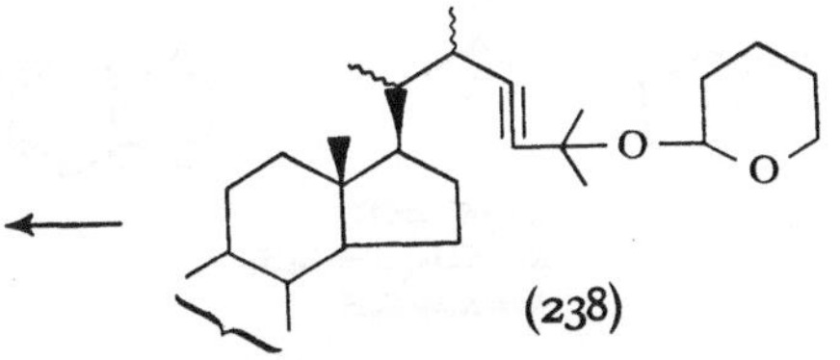

←

(238)

c) Sterilisantien (Juvenilhormonartige Substanzen)

M. ROMANEK, K. SLAMA und F. SORM (*84*) konnten beim Einleiten von Chlorwasserstoff in eine methanolische Lösung von Farnesylsäuremethylester ein Dichlorid (239) isolieren, das überraschend um 5 Zehnerpotenzen bei der Feuerwanze *Pyrrhocoris apterus* wirksamer war als Farnesol. SLAMA et al. (*65*) konnten zeigen, daß weniger als $1\,\mu$g dieses Dichlorids (239) eine dauernde Sterilität bei *Pyrrhocoris apteris* zu erzeugen imstande ist, ein im Hinblick auf die Entwicklung moderner Schädlingsbekämpfung sensationeller Befund.

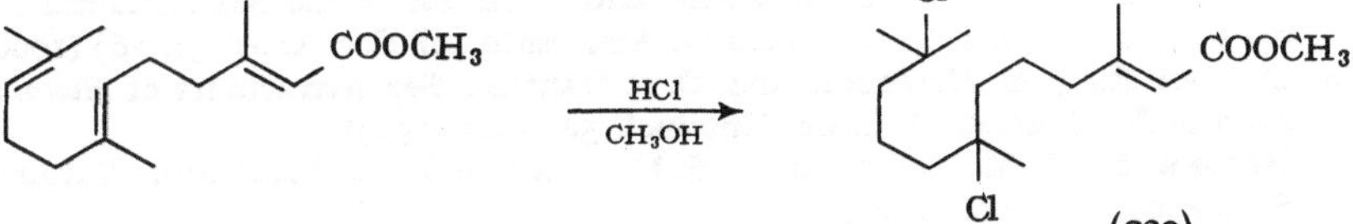

COOCH₃ → $\xrightarrow[CH_3OH]{HCl}$ COOCH₃

(239)

d) Synthetische Produkte

Im Holz und Papier der Balsamtanne, *Abies balsamea*, konnte eine Substanz aufgefunden werden („paper-factor"), die JH-Aktivität zeigte. BOWERS et al. (*13*) haben die Verbindung isoliert und bewiesen, daß es sich um (+)-Juvabion, um den Methylester der Todomatusäure (240) handelt.

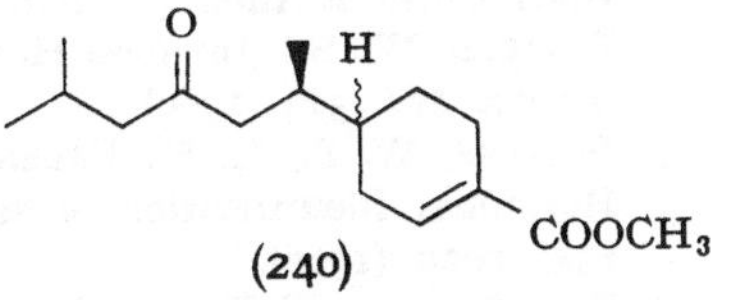

MORI und MATSUI haben (±)-Juvabion synthetisiert, während Juvabionanaloge mit völlig aromatisiertem 6-Ring und hoher biologischer Aktivität aufgefunden wurden (*72*).

(240)
Todomatusäuremethylester
(+)-Juvabion

COOCH₃

In diesem Zusammenhang sollte noch erwähnt werden, daß W. S. BOWERS (*12*) synthetische Verbindungen herstellte, die die wirksame Methylendioxyphenyl-Komponente der Insektizid-Synergisten mit der Sesquiterpenstruktur der Juvenilhormone in sich vereinigen. Der Austausch des 10,11-Epoxyfarnesols durch 6,7-Epoxygeraniol und seiner äthylverzweigten Homologen führte zu Verbindungen, deren biologische Aktivität die der *Cecropia*-Juvenilhormone gewaltig übertrafen.

$$R=R'=CH_3$$
$$R=CH_3, R'-C_2H_5$$
$$R=R'=C_2H_5$$

Literaturverzeichnis

 1. Barbier, M. und M. F. Hügel: Synthèse de l'acide céto-9-dècene-2-cis oique, isomère cis de la „Substance Royale". Bull. Soc. Chim. France 951, 1324 (1961).
 2. Barbier, M., E. Lederer, T. Reichstein und O. Schindler: Auftrennung der sauren Anteile von Extrakten aus Bienenköniginnen (*Apis melliferra* L.) Isolierung des als Königinnensubstanz bezeichneten Pheromons. Helv. Chim. Acta 43, 1682 (1960).
 3. Barbier, M., E. Lederer und T. Nomura: Synthèse de l'acide céto-9-decène-2-trans-oique („substance royale") et de l'acide céto-8 nonène -2-trans- oique. Compt. Rend. Hebd. Science, Acad. Sci. 251, 1133 (1960).
 4. Bayer, O.: Houben-Weyl, Methoden der Organ. Chemie Bd. 711, 120 (1954), Stuttgart: Georg Thieme.
 5. Berger, R.: Isolation, Identification and Synthesis of the Sex Attractant of the Cabbage Looper, *Trichoplusia* ni. Ann. Entomol. Soc. Amer. 59, 767 (1966).
 6. Berger, R., J. M. McGough und D. F. Martin: Sex Attractants of *Heliothis* zea and *H. virescens*. J. Econ. Entomol. 58, 1023 (1965).
 7. Berger, R., J. M. McGough, D. F. Martin und L. R. Ball: Ann. Entomol. Soc. Amer. 57, 606 (1964).
 8. Berkoff, C. E.: The Chemistry and Biochemistry of Insect Hormones Quart. Reviews 23, 372—391 (1961).
 9. Beroza, M.: Nonpersistant Inhibitor of the Gipsy Moth Sex Attractant in Extracts of the Insect. J. Econ. Entomol. 60, 875 (1967).
10. Bestmann, H. J., R. Kunstmann und H. Schulz: Eine Synthese der „Königinnensubstanz" und der *trans*-10-Hydroxydecen-(2)-säure-(1) (royal jelly acid). Liebigs Ann. Chem. 699, 33 (1966).
11. Bevan, C. W. L., A. J. Birch und H. Caswell: An Insect Repellent from Black Cocktail Ants. J. Chem. Soc. (London) 1961, 488.
12. Bowers, W. S.: Juveline Hormone: Activity of Aromatic Terpenoid Ethers. Science 164, 323 (1969).
13. Bowers, W. S., H. M. Fales, M. J. Thompson und E. C. Uebel: Juvenile Hormone: Identification of an Active Compound from Balsam Fir. Science 154, 1020 (1966).
14. Burkholder, W. E. und R. J. Dicke: Evidence of Sex Pheromones in Females of Several Species of Dermestidae. J. Econ. Entomol. 59, 540 (1966).
15. Butenandt, A., R. Beckmann und E. Hecker: Über den Sexuallockstoff des Seidenspinners, I: Der biologische Test und die Isolierung des reinen Sexuallockstoffes Bombykol. Hoppe-Seyler's Z. physiol. Chem. 324, 71 (1961).
16. Butenandt, A., R. Beckmann, D. Stamm und E. Hecker: Über den Sexuallockstoff des Seidenspinners Bombyx mori, Reinherstellung und Konstitution. Z. Naturforsch. 14 b, 283 (1959).
17. Butenandt, A., E. Hecker, M. Hopp und W. Koch: Über den Sexuallockstoff des Seidenspinners, IV: Die Synthese des Bombykols und der cis-trans-Isomeren Hexadecadien-(10,12)-ole-(1). Liebigs Ann. Chem. 658, 39 (1962).

18. BUTENANDT, A. und P. KARLSON: Über die Isolierung eines Metamorphose-Hormons der Insekten in kristallisierter Form. Z. Naturforsch. **9 b**, 389 (1954).

19. BUTENANDT, A. und NGUYEN-DANG TAM: Über einen geschlechtsspezifischen Duftstoff der Wasserwanze *Belostoma indica* Vitalis (*Lethocerus indicus* Lep.). Hoppe-Seyler's Z. physiol. Chem. **308**, 277 (1957).

20. BUTLER, C. G., R. K. CALLOW und I. R. CHAPMAN: 9-Hydroxydec-trans-2-enoic Acid, A Pheromone Stabilizing Honeybee Swarms. Nature **201**, 733 (1964).

21. BUTLER, C. G., R. K. CALLOW und N. C. JOHNSTON: Proc. Roy. Soc. (London) Ser. B. **155**, 417 (1962).

22. CALLOW, R. K. und N. C. JOHNSTON: Bee World **91**, 152 (1960).

23. CASIDA, J. E., H. C. COPPEL und T. WATANABE: Purification and Potency of the Sex Attractant from the Introduced Pine Sawfly, *Diprion similis*. J. Econ. Entomol. **56**, 18 (1963).

24. CELMER, W. D. und I. A. SOLOMONS: Mycomycin, IV. Stereoisomeric 3,5-Diene Fatty Acid Esthers. J. Amer. Chem. Soc. **75**, 3430 (1953).

25. CHAPMAN, I. R.: Photolysis of the Tetrahydropyranyl Ether of 3-Hydroxy-2,2,4,4-Tetramethylcyclobutanone Tosylhydrazone. Tetrahedron Letters 1966, 113.

26. COREY, E. J., J. A. KATZENELLENBOGEN, N. W. GILMAN, S. A. ROMAN und B. W. ERICKSON: Stereospecific Total Synthesis of the dl-C_{18} Cecropia Juvenile Hormone. J. Amer. Chem. Soc. **90**, 5618 (1968).

27. CORNFORTH, R. H., J. W. CORNFORTH und G. POPJAK: Preparation of R-And-S-Mevalonolactones. Tetrahedron **18**, 1351 (1962).

28. DAHM, K. H., H. RÖLLER und B. M. TROST: Life Science **7**, 129 (1968).

29. DAHM, K. H., B. M. TROST und H. RÖLLER: The Juvenile Hormone V. Synthesis of the Racemic Juvenile Hormone. J. Amer. Chem. Soc. **89**, 5292 (1967).

30. DAY, A. C. und M. C. WHITING: The Structure of the Sex Attractant of the American Cockroach. Proc. Chem. Soc. **1964**, 368; Proc. Chem. Soc. **1966**, 64.

31. EITER, K.: Neue Synthesen der Königinnensubstanz. Angew. Chemie **73**, 619 (1961).

32. EITER, K.: Neue Synthesen der Königinnensubstanz und der 9-Hydroxy-2-trans-decensäure-(1). Liebigs Ann. Chem. **658**, 91 (1962).

33. EITER, K., E. TRUSCHEIT und M. BONESS: Synthesen von D,L-10-Acetoxyhexa-decen-(7-cis)-ol-(1), 12-Acetoxy-octadecen-(9-cis)-ol-(1) (,,Gyplure") und 1-Acet-oxy-10-propyltridecadien-(5-trans-9). Liebigs Ann. Chem. **709**, 29—45 (1967).

34. ESENTHER, G. R., T. C. ALLEN, J. E. CASIDA und R. D. SHENEFELT: Termite Attractant from Fungus-Infected Wood. Science **134**, 50 (1961).

35. FRAY, I. G. und R. ROBINSON: Synthesis of *cis*-10-Hydroxydec-2-Enoic Acid. Tetrahedron Letters **13**, 34 (1966).

36. FURLENMEIER, A., A. FÜRST, A. LANGEMANN, G. WALDVOGEL, P. HOCKS, U. KERB und R. WIECHERT: Zur Synthese des Ecdysons. Helv. Chimica Acta **50**, 2387 (1967).

37. GENTRY, C. R., F. R. LAWSON und J. D. HOFFMANN: A Sex Attractant in the Tobacco Budworm. J. Econ. Entomol. **57**, 819 (1964).

38. GREEN, N., M. JACOBSON, T. J. HENNEBERRY und A. N. KISHABA: Insect Sex Attractants. VI. 7-Dodecen-1-ol Acetates and Congeners. J. Med. Chem. **10**, 533 (1967).

39. GREEN, N., M. JACOBSON und J. C. KELLER: Hexalure, an Insect Sex Attractant Discovered by Empirical Screening. Experientia **25**, 682 (1969).

40. HOFFMANN, W., H. PASEDACH und H. POMMER: Reaktionen von Allylalkoholen mit aktiven Methin- und Methylenverbindungen. Liebigs Ann. Chem. **729**, 52—63 (1969).

41. Holbrook, R. F., M. Beroza und E. D. Burgess: Gypsy moth (Porthetria dispar) Detection With the Natural female Sex Lure. J. Econ. Entomol. **53**, 751 (1960).

42. Huber, R. und W. Hoppe: Die Kristall- und Molekülstrukturanalyse des Insektenverpuppungshormons Ecdyson mit der automatisierten Faltmolekül-methode. Chem. Ber. **98**, 2403 (1965).

43. Ishii, S.: Bull. Japanese Sc. Appl. Entomol. Zoology **8**, 334 (1964).

44. Jacobson, M.: Sex Pheromone of the Pink Bollworm Moth: Biological Masking by its Geometrical Isomer. Science **163**, 190 (1969).

45. Jacobson, M.: Recent progress in the Chemistry of Insect Sex Attractants, Advances in Chem. **41**, 1 (1965).

46. Jacobson, M.: Insect Sex Attractants, New York: J. Wiley and Sons Inc. 1965.

47. Jacobson, M.: Synthesis of a Highly Potent Gypsy Moth Sex Attractant. J. org. Chem. **25**, 2074 (1960).

48. Jacobson, M.: Sex Pheromone of the Pink Bollworm Moth: Biological Masking by its Geometrical Isomer. Science **163**, 190 (1969).

49. Jacobson, M. und M. Beroza: Sex Attractants of the American Cockroach. Science **142**, 1258 (1963).

50. Jacobson, M. und M. Beroza: American Cockroach Sex Attractant. Science **147**, 748 (1965).

51. Jacobson, M., M. Beroza und W. A. Jones: Isolation, Identification, and Synthesis of the Sex Attractant of Gypsy Moth. Science **132**, 1011 (1960); J. Amer. Chem. Soc. **83**, 4819 (1961).

52. Jacobson, M., M. Beroza und R. T. Yamamoto: Isolation and Identification of the Sex Attractant of the American Cockroach. Science **139**, 48 (1963).

53. Jacobson, M. und W. A. Jones: Insect sex attractants, II: The Synthesis of a Highly Potent Gypsy Moth Sex Attractant and Some Related Compounds. J. Org. Chem. **27**, 2523 (1962).

54. Jacobson, M. und C. E. Lilly: Sex Attractants of Sugar Beet Wireworm: Identification and Biological Activity. Science **159**, 208 (1968).

55. Jaegger, R. H. und R. Robinson: A Simple Synthesis of ,,Queen Substance". Tetrahedron **14**, 320 (1961).

56. Johnson, W. S., T. Li, D. J. Faulkner und S. F. Campbell: A Highly Stereo-selective Synthesis of the Racemic Juvenile Hormone. J. Amer. Chem. Soc. **90**, 6225 (1968).

57. Jones, W. A. und M. Jacobson: Insect Sex Attractants, V: The Synthesis of Some Additional Compounds Related to Gyplure. J. Med. Chem. **7**, 373 (1964).

58. Jones, W. A. und M. Jacobson: Insect Sex Attractants, IV: The Determina-tion of Gyplure in its Mixtures by Adsorption and Gas Chromatography. J. Med. Chem. **14**, 22 (1964).

59. Jones, W. A. und M. Jacobson: Isolation of N,N-Diethyl-m-Toluamide (Deet) from Female Pink Bollworm Moths. Science **159**, 99 (1968).

60. Jones, W. A., M. Jacobson und D. F. Martin: Sex Attractants of the Pink Bollworm Moth: Isolation, Identification and Synthesis. Science **152**, 1516 (1966).

61. Karlson, P. und A. Butenandt: Pheromones (ecto-hormones) in Insects. Ann. Rev. Entomol. **4**, 39 (1959).

62. Keller, J. C., E. B. Mitchell, G. McKibben und T. B. Dewich: A Sex Attractant for Female Boll Weevils from Males. J. Econ. Entomol. **57**, 609 (1964).

63. Kennedy, J., N. J. McCorkindale und R. A. Raphael: A New Synthesis of Queen Substance. J. Chem. Soc. **1961**, 3813.

64. KERB, U., G. SCHULTZ, P. HOCKS, R. WIECHERT, A. FURLENMEIER, A. FÜRST, A. LANGEMANN und G. WALDVOGEL: Zur Synthese des natürlichen Häutungshormons. Helv. Chim. Acta **49**, 1601 (1966).

65. MASNER, P., K. SLÁMA und V. LAUDA: Sexually Spread Insect Sterility Induced by the Analogues of Juvenile Hormone. Nature **219**, 395 (1968).

66. MATSUMURA, F., H. C. COPPEL, A. TAI: Isolation and Identification of Termite Trail-Following Pheromone. Nature **219**, 963 (1968).

67. MEINWALD, J., A. M. CHALMERS, T. E. PLISKE und T. EISNER: Identification and Synthesis of *trans,trans*-3,7-Dimethyl-2,6-decadien-1,10-dioic Acid, a Component of the Pheromonal Secretion of the Male Monarch Butterfly. Chemical Communications Nr. 3 **86** (1969).

67a. MEINWALD, J. und Y. C. MEINWALD: J. Am. Soc. **88**, 1305 (1966). Structure and Synthesis of the Major Components in the Hair pencil Secretion of a Male Butterfly, *Lycorea ceres ceres* (Cramer).

68. MEINWALD, J., Y. C. MEINWALD und P. H. MAZZOCCHI: Sex Pheromone of the Queen Butterfly. Science **164**, 1174 (1969).

69. MEINWALD, J., J. W. WHEELER, A. A. NIMETZ und J. S. LIU: Synthesis of Some 1-Substituted 2,2-Dimethyl-3-isopropylidencyclopropanes. J. Org. Chem. **30**, 1038 (1965).

70. MEYER, A. S. und H. A. SCHNEIDERMAN: Fed. Proc. **27**, 393 (1968).

71. MITAKA, J.: (Kagaku) **27**, 93 (1957).

72. MORI, K. und M. MATSUI: Synthesis of Compounds with Juvenile Hormone Activity I Juvabione (Methyl Todomatuate). Tetrahedron **24**, 3127 (1968).

73. OHNO, M.: Chemistry of Sex Attractants of Insects. Bull. Inst. Chem. Research. Kyoto Univers. **45**, 207—228 (1967).

74. OUYE, M. T. und B. A. BUTT: J. Econ. Entomol. **55**, 419 (1962).

75. PATTENDEN, G.: A Synthesis of Propylure, Sex Attractant of the Pink Bollworm Moth. J. Chem. Soc. (London) **18**, 2385 (1968).

76. PRÜFFER, J.: Zool. Pol. **2**, 43 (1937).

77. PYATNOVA, Y., B. L. L. IVANOV und A. S. KYSKINA: Insect Sex Attractants. Russian Chemical Reviews **38**, (2) 126 (1969).

78. RIDDIFORD, L. M.: Stimulans für Sexuallockstoffabgabe bei Weibchen von *Antherea polyphema*. Science **158**, 139 (1967).

79. RODIN, J. O., R. M. SILVERSTEIN, W. E. BURKHOLDER und J. E. GORMAN: Sex Attractant of Female Dermestid Beetle. Science **165**, 904 (1969).

80. ROELOFS, W. L. und H. ARN: Sex Attractant of the Red Banded Leaf Roll Moth *Argyrothenia velutinana*. Nature **219**, 513 (1968).

81. ROELOFS, W. L. und A. COMEAU: Sex Pheromone Spezificity: Taxonomic and Evolutionary Aspects in Lepidoptera. Science **165**, 398 (1969).

82. ROELOFS, W. L. und A. COMEAU: Sex Pheromone Perception. Nature **220**, 600 (1968).

83. ROELOFS, W. L., A. COMEAU und R. SELLE: Pheromone of the Oriental Fruit Moth. Nature **224**, 723 (1969).

84. ROMANUK, M., K. SLAMA und F. SORM: Proc. Nat. Acad. Sci. USA **57**, 349 (1967).

85. ROTH, L. M., W. D. NIEGISCH und W. H. STAHL: Occurrence of 2-Hexenal in the Cockroach *Eurycotis floridana*. Science **123**, 670 (1956).

86. SAKAN, T.: Attractants. 8. Sympos. Chemistry of Nat. Products 189 (1964); Nippon Kagaku Zasshi **81**, 1320 (1960).

87. SCHMIALEK, P.: Die Identifizierung zweier im Tenebriokot und in Hefe vorkommender Substanzen mit Juvenilhormonwirkung. Z. Naturforsch. **16 b**, 461 (1961).

88. Schneider, D. und J. Boeckh: Chemical Sense Communication in Insects. Symposia Soc. Exp. Biol. **20**, 273 (1965).

89. Sekul, A. A. und A. N. Sparks: The Chemical Sex Lure of the Female Fall Army Worm Has Been Synthesized. Chem. Engn. News **45**, 345 (1967) J. Econ. Entomol. **60**, 1270 (1967).

90. Sidall, J. B., A.. D. Cross und J. H. Fried: Synthetic Studies on Insect Homones. II: The Synthetic of Ecdysone. J. Amer. Chem. Soc. **88**, 862 (1966).

91. Silverstein, R. M., R. S. Brownlee, E. T. Bellos, D. L. Wood und L. E. Browne: Brevicomin: Principal Sex Attractant in the Frass of the Female Western Pine Beetle. Chem. Engn. News **45**, 21 (1967); Science **159**, 889 (1968).

92. Silverstein, R. M., J. O. Rodin, W. E. Burkholder und J. E. Gorman: Sex Attractant of the Black Carpet Beetle. Science **157**, 85 (1967).

93. Silverstein, R. M., J. O. Rodin und D. L. Wood: Sex Attractants in Frass Produced by Male *Ips confusus* in Ponderosa Pine. Science **154**, 509 (1966).

94. Silverstein, R. M., J. O. Rodin, D. L. Wood und L. E. Browne: Identification of two new Terpene Alkohols from Frass produced by *Ips confusus* in Ponderosa Pine. Tetrahedron **22**, 1929 (1966).

95. Smythe, R. V. und H. C. Coppel: Ann. Entomol. Soc. Amer. **59**, 1008 (1966).

96. Stefanovič, D., B. Grujič und F. Mićić: Faštita Bilja **73**, 235 (1963).

97. Stein, G.: Untersuchungen über den Sexuallockstoff des Hummelmännchens. Biol. Zttl. **82**, 343 (1963).

98. Stoll, M. und I. Flament: Synthèse du „Propylure", Pheromone sexuelle de *Pectinophora gossypiella* Saunders. Helv. Chim. Acta **52**, 1996 (1969).

99. Stuart, A.M.: Mechanism of Trail-laying in Two Species of Termites. Nature **189**, 419 (1961).

100. Toba, M. M., A. N. Kishaba und W. W. Wolf: Bioassay of the Synthetic Female Sex Pheromone of the Cabbage Looper. J. Econ. Entomol. **61**, 812 (1968).

101. Truscheit, E.: Tätigkeitsbericht Nr. 9 vom 1. Sept. 1963. Wiss. Hauptlaboratorium der Farbenfabriken Bayer AG.

102. Truscheit, E. und K. Eiter: Synthese der vier isomeren Hexadecadien-(10,12)-ole-(1). Liebigs Ann. Chem. **658**.

103. Truscheit, E. und K. Eiter: Ungesättigte aliphatische Alkohole. DBP 1.163.313 (20. Februar 1964).

104. Truscheit, E., K. Eiter, A. Butenandt und E. Hecker: Verfahren zur Herstellung sterisch einheitlicher Dien-(10,12)-ole-(1). Österr. Patent 223.182 (10. September 1962).

105. Truscheit, E., K. Eiter, A. Butenandt und E. Hecker: Verfahren zur Herstellung von 10-*trans*, 12-*cis* oder 10-*cis*, 12-*cis*-Dien-(10,12)-olen-(1), DBP 1.138.037 (18. Oktober 1962).

106. Tsuyuki, T., Y. Ogatha und I. Yamamoto: Schimik, Agrar. Biol. Chem. Jap. **29**, 419 (1965).

107. Tumlinson, J. H., D. D. Hardee, R. C. Gueldner, A. C. Thompson, P. A. Hedin und J. P. Minyard: Sex Pheromones Produced by Male Boll Weevil: Isolation, Identification and Synthesis. Science **166**, 1010 (1969), Chem. Engn. News **47**, 36 (1969).

108. Wakabayashi, N.: A new Synthesis of 2,2-Dimethyl-3-isopropylidencyclopropyl-Propionate. J. org. Chemistry **32**, 489 (1967).

109. Waters, R. M. und M. Jacobson: Attractiveness of gyplure Masked by Impurities. J. Econ. Entomol. **58**, 370 (1965).

110. WEGLER, R.: Chemie der Pflanzenschutz- und Schädlingsbekämpfungsmittel. Bd. 1, S. 497. Berlin-Heidelberg-New York: Springer 1970.

111. WHARTON, D. R. A., E. D. BLACK und C. MERRITT jr.: Sex Attractant of the American Cockroach. Science **142**, 1257 (1963).

112. WHARTON, D. R. A., E. D. BLACK, C. MERRITT jr., M. L. WHARTON, M. BAZINET und J. T. WALSH: Isolation of the Sex attractant of the American Cockroach. Science **137**, 1062 (1962).

113. WILLIAMS, C. M. und K. SLAMA: Biol. Bull. **130**, 247 (1966).

114. Zoecon Corp. and USDA-Teams Achieve Synthesis of Last Terpenoid Component of Boll Weevil Attractant. Chem. Engn. News. Nr. 4, **48**, 40—43 (1970).

115. ZURFLÜH, R., E. N. WALL, J. B. SIDALL und J. A. EDWARDS: Synthetic Studies on Insect Hormones VII. An Approach to Stereospecific Synthesis of Juvenile Hormones. J. Amer. Chem. Soc. **90**, 6224 (1968).

(Eingelaufen am 6. Juli 1970)

Arthropod Molting Hormones

By HIROSHI HIKINO and YASUKO HIKINO, Sendai

With 1 Figure

Contents

I. Introduction

While much is known about the hormones of vertebrates, knowledge of the hormones of invertebrates is far less complete. However, the chemistry of the molting hormones and the juvenile hormones of insects has made surprisingly rapid advances during the past few years and has now become a subject of research which is attracting the interest of both chemists and biologists. This review article is an attempt to summarize recent developments in our knowledge regarding the chemistry, synthesis and metabolism of the arthropod molting hormones, but will in the main exclude consideration of biological properties which have been frequently the object of excellent reviews (*71*).

References are cited from published papers up to the middle of 1969, while some important papers which appeared thereafter are also included.

For the presentation of this article, thanks are owed to earlier articles on the subject of the physiology of insects and crustaceans.

II. Hormonal Regulation of Growth
of Insects and Crustaceans (*23, 51*)

The growth and development of an insect from an immature larva directly or indirectly (*via* a pupa) into a reproducing imago of very different form, structure and habit of life, is separated by periodical moltings. Although there are differences among insect groups, the basic molting pattern may be best exemplified by a silkmoth whose life cycle consists of four larval molts, one larval-pupal molt and one pupal-adult molt, the larval-pupal-adult transformation being called metamorphosis. The challenge to clarify the origin of these phenomena has fascinated scientists since olden times. The knowledge so far accumulated demonstrates that the postembryonic development is under control by chemical substances.

The cyclical molting is initiated by the brain hormone which is secreted in the neurosecretory cells of the brain and then triggers glands in the prothorax, the prothoracic glands. The activated prothoracic glands respond to this stimulus by releasing directly and/or indirectly the molting hormone which in turn reacts on various cells and starts them to grow. During the larval stages, glands known as the corpora allata concurrently excrete the juvenile hormone which favors larval synthesis. Thus, in response to the molting hormone, larval epidermic cells in the presence of a high concentration of the juvenile hormone deposit larval cuticle and the molt results in the retention of the larval characters. When the larva reaches the mature stage, the corpora allata cease secreting. At the next molt, the same epidermal cells respond to the molting hormone and to the low concentration of the juvenile hormone by secreting pupal cuticle, and under these hormonal conditions other tissues either break down or transform into pupal structures. During the pupal-adult transformation, since no juvenile hormone remains, the same cells under the sole action of the molting hormone deposit imaginal cuticle and the other pupal tissues either break down or develop into adult structures.

The growth of crustaceans is physiologically very similar to that of insects. Thus, a crustacean also periodically casts off its hard exoskeleton for increasing size during the course of its growth. The process participating in this molting and cuticle regeneration is regulated by chemical substances.

Extirpation and implantation experiments have shown that the Y-organ, a glandular mass in the thoracic region, produces the molting hormones. It is believed that the Y-organ is under the control of an antagonist, the molt inhibiting hormone, which is secreted by the neurosecretory cells of the X-organ in the eyestalk and inhibits molting when liberated in the general circulation. When release of the molt inhibiting hormone stops, the reins of inhibition from the Y-organ are removed. This results in secretion of the molting hormone which leads to the premolt and finally to ecdysis (molting).

III. Occurrence in Animals

With the advance of endocrinological konwledge concerning the molting of insects, attempts to isolate the pure hormones were initiated. Pioneering work on the purification of the hormone responsible for molting was performed by Becker (2, 1), who obtained relatively potent extracts from the blowfly, Calliphora erythrocephala, by 1941. This work was continued by Butenandt and Karlson (5) in 1943. Following preliminary experiments first on the pupae of Calliphora erythrocephala and, after 1947, on the pupae of the commercial silkmoth, Bombyx mori, these workers started, in 1953, to extract a large amount of male pupae of Bombyx mori, the female pupae being employed for the isolation of the sex attractant. Painstaking purification procedures with the aid of the Calliphora bioassay (p. 263) led in 1954 at last to the isolation in crystalline form of a minute amount of the hormone which was named ecdysone from the fact that it induces the ecdysis. A further decade was to be required for establishment of the structure. The hormone was later isolated from the same source in improved yield utilizing a modified purification procedure (74).

References, pp. 305—312

Shortly after the first isolation of ecdysone, KARLSON (70) isolated from the extracts a second active substance which could only be characterized spectroscopically, since the amount obtained was very small. The name ecdysone was therefore changed to α-ecdysone and the new compound was called β-ecdysone. Further investigation demonstrated that five active fractions were present in Bombyx mori (4). Following the structure elucidation of α-ecdysone in 1965, HOCKS and WIECHERT (41) synthesized it. In the course of attempts to obtain α-ecdysone from Bombyx mori pupae for comparison purposes, these workers isolated another active substance in crystalline form and identified it as 20-hydroxyecdysone. Apparently the same substance was independently isolated by HOFFMEISTER et al. (42, 43) from the same source, Bombyx mori pupae, almost at the same time and named ecdysterone.

Meanwhile, STAMM (113) reported the isolation of α- and β-ecdysone from the adult Moroccan locust, Dociostaurus maroccanus, where β-ecdysone was the predominant hormone. Further, HORN et al. (54) fractionated the extract of the pupae of the Saturniid oak-silk moth, Antherea pernyi, and isolated, together with a smaller amount of α-ecdysone, an active substance whose mass spectrum and behavior on thin-layer chromatography was identical with those of crustecdysone which had been isolated from a crustacean, Jasus lalandei (vide infra). Another new molting hormone from insects was reported by KAPLANIS et al. (69, 140) who isolated from pupae of the tobacco hornworm, Manduca sexta, four substances. Two of these were identified as α-ecdysone and a 20-hydroxyecdysone. The third, tobacco hornworm ecdysone III, was proposed to be a 20,26-dihydroxyecdysone and the fourth was assumed to be a 5α-isomer of the third. Recently, GALBRAITH et al. (22), in an attempt to isolate molting hormones from pupae of the blowflies, Calliphora stygia and C. vicina (= C. erythrocephala), obtained β-ecdysone. Surprisingly enough no α-ecdysone was detected.

Studies dealing with molting hormones of crustaceans must be mentioned here. After preliminary examinations by KARLSON of extracts from whole crustacea, Cragon vulgaris, Emertia talpodia and Libinia spp., which were shown to cause molting in insects, KARLSON and SKINNER (76) examined in search of the molting hormone the Y-organ as well as the carcasses of the green crab, Carcinus maenas, during the intermolt period. Bioassays indicated that the Y-organs stored no measurable quantity of the hormone, thus demonstrating that this gland does not function as a storage organ, while the carcasses contained a fairly large amount. In these experiments, KARLSON was unable to obtain the hormones in a form sufficiently pure to permit characterization, but recognized that the chemical properties of the crustacean molting hormone are closely related to those of the insect hormone α-ecdysone and that the former is more polar than the latter. Further, the positive response of *Calliphora* to the crustacean extracts indicated the interchangeability of insect and crustacean molting hormones.

These attempts were continued by HORN *et al.* (*24, 54, 52*) who have since 1962 investigated the sea crayfish, *Jasus lalandei*. After a long series of purification procedures, two molting hormones were isolated in substantially pure but non-crystalline form. The first was called crustecdysone and postulated to be a 20-hydroxyecdysone (*24*). The second was first thought to be α-ecdysone (*54*) but was later assigned a 2-deoxy-crustecdysone structure (*15*).

Quite recently, there appeared an interesting report by FAUX *et al.* (*6*) who examined extracts of the female marine crab, *Callinectes sapidus*. In the early premolt stage, callinecdysone A (either or both of the epimers of inokosterone) was the only hormone which could be detected. At the later premolt stage, callinecdysone A was accompanied by a small amount of β-ecdysone. Finally, after molting, β-ecdysone was the major hormone present along with a smaller amount of callinecdysone B (makisterone A or its C-24 epimer). This is the first report of the discovery in the animal kingdom of active substances originally discovered in the plant kingdom (pp. 261–262). Discovery of callinecdysone B suggests that C_{28} analogs also play the part of molting hormones.

Thus seven kinds of arthropod molting hormones have so far been discovered (Table I).

Mention must be made of the nomenclature of the molting hormones which has given rise to much confusion. Although the names α- and β-ecdysone proposed by KARLSON were later abandoned by the author himself, they have enjoyed wide circulation among entomologists. On the other hand, as reviewed in the previous section, several arthropod molting hormones having similar properties were isolated from various sources and given different names, *i. e.*, β-ecdysone, crustecdysone, 20-hydroxyecdysone and ecdysterone, the same structure being proposed for the last three. Partial clarification came with the demonstration that β-ecdysone and 20-hydroxyecdysone were identical (*40*). Subsequently comparison of 20-hydroxyecdysone, ecdysterone and crustecdysone of plant origin (p. 261) proved their identity (*14*). However, no conclusive evidence for the identity of the crustecdysone from the arthropods with crustecdysone from the plant was presented, though the properties of these substances were described as identical in many respects (*14*). Even after the identification of the 20-hydroxyecdysones with β-ecdysone, the names have been used interchangeably by different authors. Although the original β-ecdysone was apparently obtained in a fairly impure form, priority must be given to this name. In this review article the names α- and β-ecdysone will be adopted.

The subject has been developed under the generalized heading of the molting hormones or the ecdysones. However, with the discovery of related active substances from the plant kingdom, the general term molting hormone has seemed inappropriate. There has been a tendency to use the terms zooecdysone and phytoecdysone for the ecdysone analogs of animal and plant origin, respectively. In this article, ecdysterol is also adopted as the more general term.

References, pp. 305—312

IV. Occurrence in Plants

In contradistinction to the isolation of ecdysterols from animals which were achieved as the results of planned investigations, isolation of ecdysterols from plants were accomplished purely by chance. NAKA-NISHI *et al.* (*96*) studied chemical constituents of a Formosan folk drug prepared from the leaves of *Podocarpus nakaii* which reputedly possessed anti-cancer activity. As frequently happens, however, no anti-cancer substance was obtained, although four compounds were isolated in 1964. In 1966, these workers elucidated the constitution of the main constituent and realized that it was an ecdysone analog. Since bioassay revealed that these substances had strong molting hormone activity, the main constituent was named ponasterone A and the three minor ones ponasterone B, C and D. This constitutes the first recognition of ecdysterols in the plant kingdom.

Meanwhile, TAKEMOTO *et al.* (*130, 131*) studied a crude drug prepared from the roots of *Achyranthes fauriei* which is utilized as a diuretic in Oriental medicine. In an attempt to isolate a principle having the diuretic activity, these workers obtained, in 1965, a polar material which was later shown to be a mixture of two related isomers. Subsequent to structure elucidation of the 20-hydroxyecdysones in 1966, it was realized that the two isomeric compounds from *Achyranthes fauriei* were very similar to the zooecdysones. They were named inokosterone and iso-inokosterone, the latter being identified as β-ecdysone and the former concluded to be a position isomer.

Meanwhile, GALBRAITH and HORN (*11, 12*) reported the isolation of β-ecdysone from *Podocarpus elatus*. However, as their report clearly stated, the idea that *Podocarpus* plants may contain ecdysone analogs was derived from NAKANISHI's announcement of the isolation and structure elucidation of the phytoecdysone, ponasterone A. Therefore, in all fairness the isolation of β-ecdysone by the Australian workers is not the first isolation of a phytoecdysone, though it is the first report bibliographically.

Following the recognition that ecdysterols occurred in the plant kingdom, it was noticed that some substances which had already been isolated from vegetable sources were also ecdysone analogs. Thus, a constituent of *Polypodium vulgare*, polypodine A which may have been obtained in crude form as early as 1933, was found to be identical with β-ecdysone (*65*). Further, a substance which had been isolated from *Vitex megapotamica* in 1965 was also identified as β-ecdysone (*102*).

The discovery of ecdysterols in a number of plant sources which have no close taxonomic relationship to each other suggested that active substances might be widely distributed in the plant kingdom. Consequently, it seemed natural to survey plants in order to seek good sources of ecdysterols and new ones. Accordingly very extensive screening tests with the aid of the bioassay have been carried out by a few groups (*133, 112, 60, 63*). Some eighty families which contain species showing positive molting hormone activity have so far been discovered (Table 2).

It appears that so far as ecdysterols are concerned, the accepted chemotaxonomic theory that related plants contain related compounds does not neccesarily seem to hold. However, the screening operations used a bioassay which has a limit of detectability, reportedly 1 ppm even in the sensitive *Musca* test (p. 264) (*112*). Further, since the results of screening tests on certain plants are contradictory, variation of contents depending upon the season and location may be significant (*65, 27, 60, 124, 17*). Furthermore, co-occurrence of substances possessing anti-molting hormone activity is not improbable. This may interfere with evaluation of the bioassay. In fact, an anti-ecdysterol, ajugalactone (16) was recently found in a plant, *Ajuga decumbens* (*85*). It appears, therefore, that the data obtained to date are insufficient to draw definite chemotaxonomic conclusions.

Thus far, a number of ecdysterols has been isolated from those plant extracts which showed molting hormone activity in the screening tests (Table 3). Furthermore, structures of some 30 new phytoecdysones have been determined (Table 4). Undoubtedly the list of new active substances will continued to expand.

The following are the first reported ecdysterols having the significant features indicated, apart from differences in the positions of hydroxyls in the side-chains. Cyasterone (32) (C_{29} stigmastane skeleton, γ-lactone ring), polypodine B (36) (5β-hydroxyl group), ponasterone B (17) ($2\alpha,3\alpha$-dihydroxyl groups), rubrosterone (40) (C_{19} etiocholane skeleton*, 17-oxo group), makisterone A (46) (C_{28} ergostane skeleton), ponasteroside A (41) (glycoside), capitasterone (35) (δ-lactone ring), shidasterone (43) (abnormal stereochemistry in side-chain), ajugasterone B (58) (double bond in side-chain), viticosterone E (52) (acetoxyl group), podecdysone B (56) (additional double bond in nucleus), ajugasterone C (44) (11α-hydroxyl group), stachysterone A (54) (rearranged nucleus) and stachysterone D (57) (ether linkage).

Some plants are far superior to animals as a source of ecdysterols because of their much higher ecdysterol content and, therefore, because of the greater simplicity of extraction. This will undoubtedly facilitate studies of the mode of action of ecdysterols in higher animals as well as in arthropods. Such studies have so far been almost completely frustrated by the minute supply of material available from animals.

The discovery of ecdysterols in the plant kingdom raises the question whether they have any beneficial or adverse effects on the plants themselves or on phytophagous animals in their natural habitat. No definite answers to these questions

---

* This is also the first substance possessing this carbon skeleton isolated from the plant kingdom.

have been presented so far and experimental clarification is required. In this connection, the view has been expressed that the presence of ecdysterols in plants makes them resistant to insect attack (*11, 68*). However, this cannot always be true, since the larvae of *Bombyx mori* and *Milionia basalis pryeri*, for instance, live on the leaves of *Morus* sp. and *Podocarpus macrophyllus*, respectively, which are now known to contain considerable quantities of ecdysterols.

V. Isolation and Assay

Since ecdysterols, polyhydroxy sterols, are soluble in polar solvents, but sparingly soluble in non-polar solvents, animal and plant materials are extracted with polar solvent. Purification of the extracts is performed by a combination of solvent extraction, partition between various solvents, liquid chromatography, preparative thin-layer chromatography and crystallization. Counter-current distribution has also been utilized in some cases. A difficultly separable mixture of ecdysterols has been acetylated to give a mixture of acetates whose separation is sometimes much easier. This affords pure acetates from which the free ecdysterols can be regenerated by hydrolysis (*59, 127, 37*).

Characterization of ecdysterols is most commonly carried out by silica gel thin-layer chromatography. Although ecdysterols can be located as dark spots on silica gel plates containing a fluorescent agent under UV light, characteristic color reactions with sulfuric acid or the vanillin-sulfuric acid spray reagent on the plates have proven to be very useful preliminary identification. Hori (*50*) developed a liquid chromatographic method which utilizes an Amberlite XAD-2 column and elution with linear gradient 20–70% aqueous ethanol, eluates being monitored by the absorption at 254 nm. Identification and estimation are carried out by measuring the elution time and the area of the peak for each ecdysterol on a chromatogram. This technique can also be used for preparative purposes by collecting the eluates. Takemoto *et al.* (*129*) described a method for spectrographic determination of β-ecdysone and inokosterone which absorb at 250 nm, the latter being estimated by the fluorescence at 520 nm which is produced by treatment with conc. sulfuric acid and then with ammonia.

Bioassay of the molting hormones has been an extremely useful tool in purifying these biologically active substances.

The assay procedure must be productive, sensitive and dependable. In this sense, one of the most commonly used methods is the puparium assay represented by the *Calliphora* test which was devised by Fraenkel (*7*) and developed to a more quantitative degree by Karlson and Hauser (*72*). Mature larvae of the blowfly, *Calliphora erythrocephala*, are ligated on the anterior third. The front parts of the animals, which are pupated only in front of the ligature after 24 hr, are cut off and the isolated abdomens are injected with test solution. After 24 hr of the injection, the larval segments are scored for pupation. A number of other test animals (*e. g.*, the house fly, *Musca domestica* (67), the fleshfly, *Sarcophaga peregrina* (97), etc) and modified assay procedures (112) have been proposed. A unique modification of this assay, the so-called dipping test, has been devised by Sato et al. (104). Larvae of the rice stem borer, *Chilo supressalis*, are ligated at the metathoraces and dipped for 5–10 sec into methanol solutions of test materials. The presence of active substances is evaluated by pupation of the abdomens within 48 hr. This assay is claimed to have an advantage for the easy detection of ecdysterols in crude plant extracts.

For evaluation of molting hormone activity, another assay is also performed on isolated abdomens or brainless pupae of the silkmoth, *Bombyx mori* or *Samia cynthia* (144). Isolated abdomens which are prepared by ligation of unpigmented pupae or pupae which are deprived of their brains, survive for extended periods without further development. Activity is signalled by the observation that test objects which received active substances undergo adult development in 14–20 days following the injection.

VI. Structure Determination and Chemistry

1. α-Ecdysone (Ecdysone) (1)

(1)
α-Ecdysone

The structure elucidation of α-ecdysone was a history of painstaking struggles because of lack of material and underdevelopment of physico-chemical techniques (5, 74, 48, 47, 46, 103, 75). The first step was the determination of the molecular formula. Analytical data for the first sample gave an elemental composition $C_{4.4}H_{7.3}O_1$. Since the initial molecular weight determination afforded an erroneous value of *ca* 310, the empirical formula was first reported to be $C_{18}H_{30}O_4$ and was quoted as such for several years. Later, X-ray analysis and mass spectrometric determination resulted in revision of the molecular weight to 464 which led to the correct formula $C_{27}H_{44}O_6$.

Of the six double-bond equivalents indicated by the molecular formula, two were accounted for by one double bond and one carbonyl group. It was, therefore, concluded that α-ecdysone contained four rings, which led to the suspicion that it might be a steroid. This was supported by the facts that it was biosynthesized from cholesterol (73) and that it had five methyl groups. The steroid nature was substantiated when it was found that dehydrogenation gave cyclopentanophenanthrene derivatives.

The UV, IR and NMR properties (λ_{max} 244 nm, ν_{max} 1657 cm^{-1} and δ 6.18 ppm (1 H)) indicated that the double bond and carbonyl group constituted a β,β-disubstituted, α,β-unsaturated ketone system in a six- or larger-membered ring. Originally, the chromophore was ascribed to the presence of a 9(11)-en-12-one because of the chemical shift of the 18-methyl signal (0.75 ppm) in the NMR spectrum, but in disregard of the contribution of the hydroxyl group now known to be at 14α (ca − 0.23 ppm). Later, the presence of a 7-en-6-one was postulated because of UV and IR spectral data.

Arrangement of the five hydroxyl groups was also a very difficult problem. Since the NMR spectrum showed two somewhat deshielded methyl singlets, the presence of a hydroxyl at C-25 was suggested. Acid treatment of α-ecdysone furnished two products, an unstable conjugated ketone (λ_{max} 293 nm, ν_{max} 1666 cm^{-1}) and a more stable isolated ketone (λ_{max} 248 nm, ν_{max} 1710 cm^{-1}). This could only be rationalized in terms of a 14-hydroxy-7-en-6-one chromophore for α-ecdysone which underwent transformation as depicted in Chart 1. The presence of at least

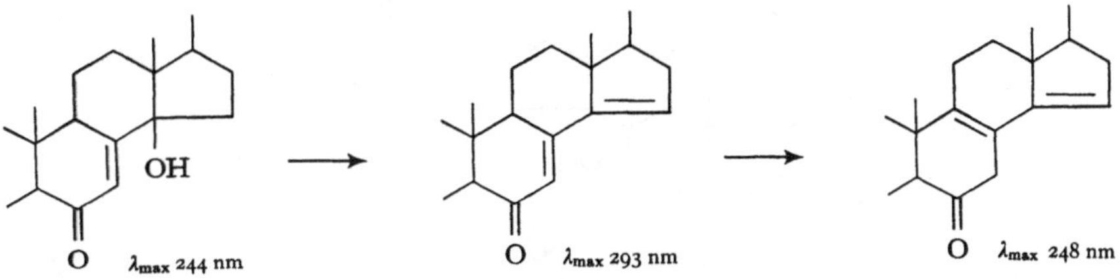

Chart 1. Acid Treatment of α-Ecdysone

one 1,2-glycol function in the α-ecdysone nucleus was inferred from the results of periodate oxidation. On the assumption that α-ecdysone possessed a cholestane skeleton, attachment of the glycol system to either the 2,3- or 3,4-position was considered likely. Major peaks in the mass spectrum (MS) at m/e 99 and 81 were attributed to side-chain fragments and major peaks at m/e 346 and 328 were associated with nuclear fragments and concomitant hydrogen transfer, generated by fission between C-20 and C-22 due to the presence of a 22-hydroxyl group. Chemical investigation thus led to the conclusion that α-ecdysone might be represented by (2).

Final establishment of the full structure including stereochemistry had to await the X-ray work of HUBER and HOPPE (55). For this purpose, preparation of α-ecdysone derivatives with heavy atoms was first tried;

however, none of the derivatives prepared crystallized sufficiently well for an X-ray analysis. Finally, X-ray crystallography of α-ecdysone itself utilizing the bent molecule method resulted in successful solution of the structure and absolute configuration as (1). This unambiguous determination of the stereostructure of α-ecdysone permitted it to serve as a reference substance for the structure determinations of its analogs.

2. β-Ecdysone (Crustecdysone, 20-Hydroxyecdysone, Ecdysterone) (3)

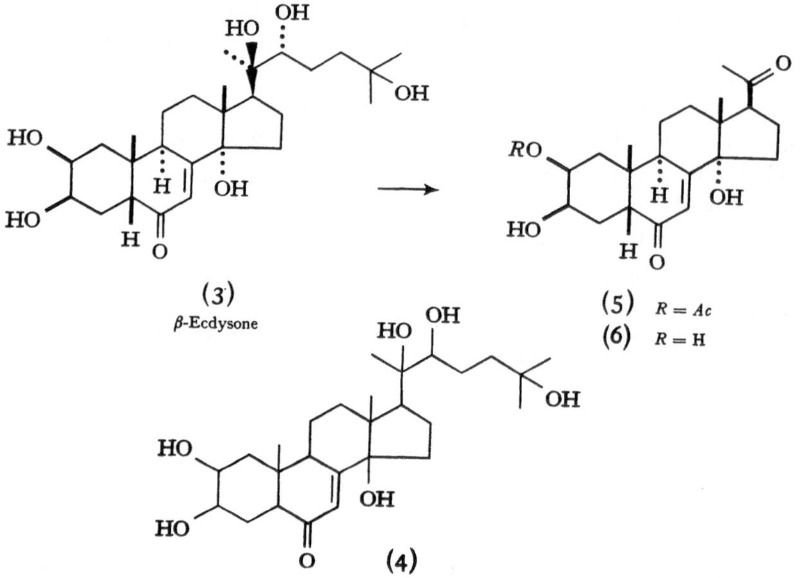

(3) β-Ecdysone

(4)

(5) $R = Ac$
(6) $R = H$

Chart 2. Some Reactions of β-Ecdysone

Following the discovery of α-ecdysone (1), a number of analogs, β-ecdysone, crustecdysone, 20-hydroxyecdysone and ecdysterone, were isolated from various sources.

Crustecdysone, the first crustacean molting hormone, had the molecular formula $C_{27}H_{44}O_7$, one oxygen more than α-ecdysone, and showed very similar chemical and physico-chemical properties. However, the line positions and splitting patterns of two NMR methyl signals of crustecdysone differed markedly from those of the C-18 and C-21 methyl signals in the spectrum of α-ecdysone, the difference being interpreted as arising from the presence of an additional hydroxyl group at C-20 in crustecdysone. This assignment was supported by the MS of crustecdysone which showed the same side-chain fragment peaks at m/e 99 and 81 as α-ecdysone, but had nuclear fragment peaks at m/e 363 and 345,

References, pp. 305—312

one mass unit less than the corresponding ions in the MS of α-ecdysone. This was taken to indicate that the side-chain C-20 : C-22 bond cleavage took place without hydrogen rearrangement which is expected of a vicinal diol. On the basis of the above evidence, structure (4) was proposed for crustecdysone (24).

20-Hydroxyecdysone and ecdysterone, the second insect molting hormone, were deduced to have the same structure (4) on the basis of similar evidence (41, 42, 43) and later shown to be identical with β-ecdysone (14). The stereochemistry was assumed to be the same as that of α-ecdysone and, in particular, the configuration at C-20 was assumed to be the same as that of cholesterol (41). The A/B cis ring fusion was confirmed by the ORD curve which exhibited a positive Cotton effect of $a + 73$ (130). The stereochemistry of the tetracyclic nucleus was finally established by conversion of β-ecdysone into the known methyl ketone (6) (110, 120). If β-ecdysone were directly biosynthesized from α-ecdysone, the configurations at C-20 and C-22 must be both R, since biological hydroxylation is known to proceed with the retention of the configuration. In fact, it has recently been proved that α-ecdysone is rapidly transformed to β-ecdysone in animals (80, 85). In the course of a subsequent synthesis of β-ecdysone it was shown that Grignard reaction of a pregnan-20-one resulted in the predominant formation of a C-20 epimer possessing the same configuration as β-ecdysone (79, 86), the 20(R)-configuration being deduced on the basis of Cram's rule. The accumulated evidence showed that β-ecdysone is represented by stereoformula (3).

3. Tobacco Hornworm Ecdysone III (20,26-Dihydroxyecdysone) (7)

The next insect molting hormone, tobacco hornworm ecdysone III, was obtained in such small amounts that no physical data were recorded except the NMR and mass spectra. Comparison of its chemical and spectroscopic properties with those of α-ecdysone (1) suggested that both had a steroid nucleus with similarly oriented substituents. While α-

Tabacco Hornwarm Ecdysone III

ecdysone had five methyl singlets, the zooecdysone exhibited four methyl singlets. Among them, the C-18, C-19 and C-21 methyl protons resonated at similar fields. The remaining methyl signal which must be ascribable to the terminal carbon atom of the side-chain, appeared at 1.48 ppm as did the C-27 methyl signal of a model steroid having the 25,26-dihydroxy structure. The MS fragmentation pattern also placed the additional hydroxyl group in the side-chain and not in the steroid nucleus. The intense peak at m/e 31 (CH_2OH^+) agreed with the presence of a 25,26-dihydroxy structure. All the above data demonstrated that tobacco hornworm ecdysone III was a 20,26-dihydroxy-α-ecdysone (140).

4. Deoxycrustecdysone (8)

Structural investigation of deoxycrustecdysone, the second crustacean molting hormone, was also very defficult. The amount available was so small that not even the IR spectrum was recorded. The characteristic UV absorption suggested the presence of the 7-en-6-one chromophore. The NMR spectrum showed five NMR methyl singlets whose chemical shifts were closely similar to those of β-ecdysone (3), indicating that deoxycrustecdysone had a structure very similar to that of β-ecdysone and that hydroxyls were present at C-14, C-20 and C-25. The MS revealed that the side-chains of deoxycrustecdysone and β-ecdysone were the same and that the tetracyclic nucleus of the former had one oxygen less than the latter. The remaining unassigned hydroxyl group was attached to C-3β by comparing the rate of acetylation of the hydroxyl at the A ring of deoxycrustecdysone with rates of acetylation of several model compounds (15).

(8)
Deoxycrustecdysone

5. Ponasterone A (9)

Ponasterone A was the first reported phytoecdysone and also the first compound which could be investigated chemically in detail, since it was obtained in greater quantity. Hence, the structure elucidation of ponasterone A will be described here in some detail. Since determination of the molecular formula presented difficulties, ponasterone A was initially considered to be a C_{29} nortriterpenoid, and it was only after all partial structures had been derived that MS measurements of the acetonides established the molecular formula $C_{27}H_{44}O_6$, indicating it to be a steroid.

Spectral properties indicated the presence of a β,β-disubstituted, α,β-unsaturated ketone moiety. The presence of four hydroxyl groups was established by the formation of the diacetonide (11) which, because of its IR and NMR spectra, was revealed to have a fifth tertiary hydroxyl group. Therefore, the six oxygen atoms were represented by five hydroxyls and one carbonyl group. Since ponasterone A on acetylation gave the triacetate (**12**) whose NMR spectrum indicated the presence of three sencondary acetoxyl groups, ponasterone A contained three secondary hydroxyls and two tertiary hydroxyls, four of which constituted two α-glycol systems in view of the consumption of periodate. Introduction of two acetyls upon acetylation of the monoacetonide (**10**) showed that a secondary and a tertiary hydroxyl group participated in formation of the first acetonide, the remaining two secondary hydroxyls being then involved in formation of the second acetonide. Acid treatment of ponasterone A gave two isomeric dienones having UV maxima at 294 and

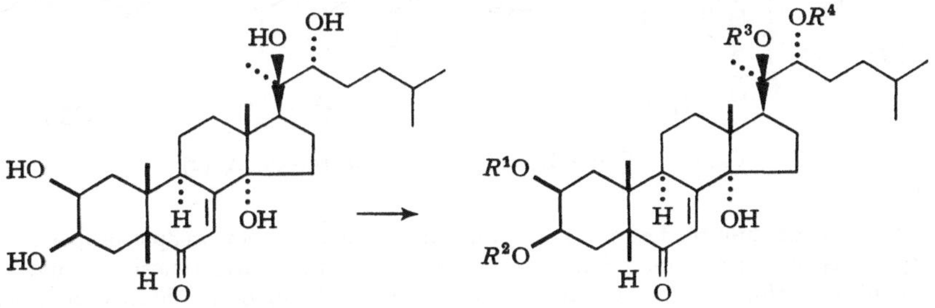

(**9**)

Ponasterone A

(**10**) $R^1 = R^2 = H$, R^3, $R^4 = C(CH_3)_2$

(**11**) R^1, $R^2 = R^3$, $R^4 = C(CH_3)_2$

(**12**) $R^1 = R^2 = R^4 = Ac$, $R^3 = H$

Chart 3. Some Reactions of Ponasterone A (1)

241 nm, indicating that ponasterone A had a steroid nucleus containing the 14-hydroxy-7-en-6-one system. The side-chain structure was established by periodate oxidation to give isohexanal. In the NMR spectrum of the acetate (**12**), the three carbinyl (2-H, 3-H, 22-H), olefinic (7-H) and allylic (9-H) proton signals were well separated and decoupling measurements deduced the partial structures in ring A and ring B, C. The chemical shift of the C-19 methyl protons was in complete accord with that of α-ecdysone (**1**) and the ORD curve showed a positive Cotton effect ($a + 68$) almost superimposable on that of 22-isoecdysone, indicating a $2\beta,3\beta$-dihydroxy-5β(H)-structure for ponasterone A. The C-18 methyl chemical shift showed the C-14 hydroxyl to have the α-configuration (96). The nuclear structure of ponasterone A was later established by its conversion into the known methyl ketone (**6**) (94). It was clear that the absolute configurations at C-20 and C-22 in ponasterone A und β-ecdysone (**3**) were identical because of the chemical shifts of the C-18 and C-21 methyl signals of ponasterone A, β-ecdysone and their acetates and because of the positions and patterns of the C-22 hydrogen signals of the acetates (95). This was later substantiated by synthetic

evidence (57) (p. 287). The 22(R)-configuration was confirmed by converting ponasterone A *via* the dehydro-derivative (13) into the acetoxymethyl ketone (14) which showed a negative Cotton effect opposite in sign to that of an analog (15) derived from L-leucine (85).

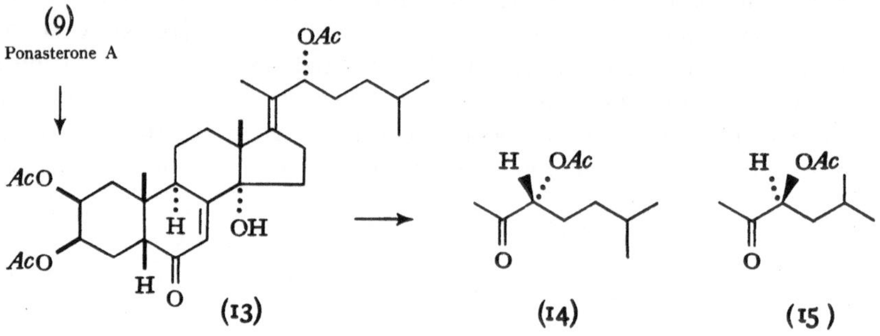

Chart 4. Some Reactions of Ponasterone A (2)

It is worthwhile to describe here in outline form the spectroscopic properties of ecdysterols, because it may fairly be said that structural investigations could not have been performed successfully without the use of modern physical methods.

Since the high number of hydroxyl groups in free ecdysterols renders them highly polar, NMR spectra of ecdysterols have customarily been measured in pyridine for solubility reasons, though pyridine is not necessarily an ideal solvent for purposes of comparison due to its strong solvating power and anisotropic effect. After sufficient data had been accumulated, however, resonance frequencies of the methyl protons provided valuable information about the structures, since they are sensitive to both the stereochemistry of and substituents on the steroid skeleton (Table 5). However, the carbinyl proton signals are usually difficult to analyze due to overlapping. On the other hand, acetates are much less polar and readily soluble in less polar solvents such as chloroform. Further information with respect to the nature and orientation of substituent groups may be obtained not only from the chemical shifts of methyl signals but also from analysis of signals with the aid of double resonance experiments. The intramolecular nuclear Overhauser effect between the C-3 and C-9 hydrogens provides rigid proof for the A-ring conformation (114, 29, 85).

MS's also provide a powerful tool for elucidating the structures of ecdysterols. The principal fragmentations upon electron impact which lead to characteristic peaks can be classified as follows: 1) the molecular ion, 2) ion due to the nuclear fragment and 3) ion due to the side-chain fragment, loss of water and/or methyl being involved in formation of most of the ions, sometimes to such an extent that the molecular ion is hardly detectable. All compounds except, for instance, deoxycrustecdysone (8) and polypodine B (36), give the same nuclear fragments. The fragments of the exceptions are shifted by 16 mass units, because the net difference amounts to the absence or presence of one oxygen atom. The side-chain fragments vary depending upon the structure of the side-chain. Thus, displacement of peaks arising from nuclear or side-chain fragments is of considerable value in deducing the structure (Fig. 1). In addition, the presence of a substituent at C-24 facilitates the C-23 : C-24 bond fission, a fact which provides further information about the side-chain structure. Fragmentations of the acetates and the acetonides of ecdysterols upon electron impact have also been examined (83).

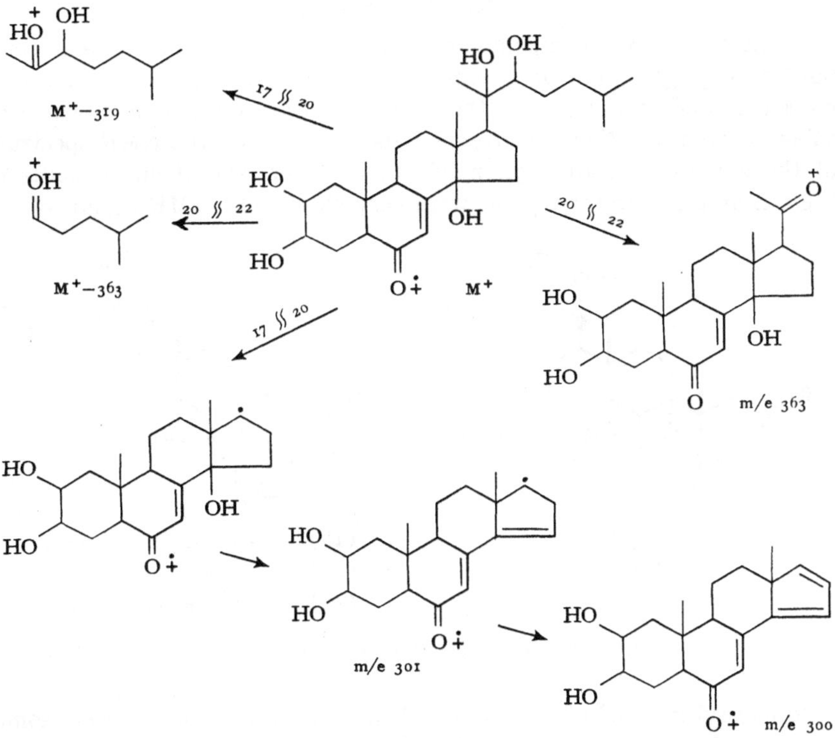

Fig. 1. Principal mass fragmentation of an ecdysterol. Each ion further loses water(s) and/or methyl(s)

ORD and CD curves of the common ecdysterols exhibit positive and negative Cotton effects at *ca* 240 nm ($\pi - \pi^*$) and *ca* 340 nm ($n - \pi^*$), respectively (Table 6). Comparison of the data with those of certain reference substances indicates that ecdysterol A/B and C/D ring junctions all belong to the *cis* and *trans* series, respectively, the A/B *cis*, C/D *cis*-system as in ajugalactone (16) being revealed by the positive Cotton effects of both $\pi - \pi^*$ and $n - \pi^*$ transitions (85). Alterations in the neighborhood of the enone system as, for example, in ponasterone B (17), polypodine B (36) and ponasteroside A (41) are reflected in the characteristic changes in the ORD and CD curves (95, 66, 114, 29).

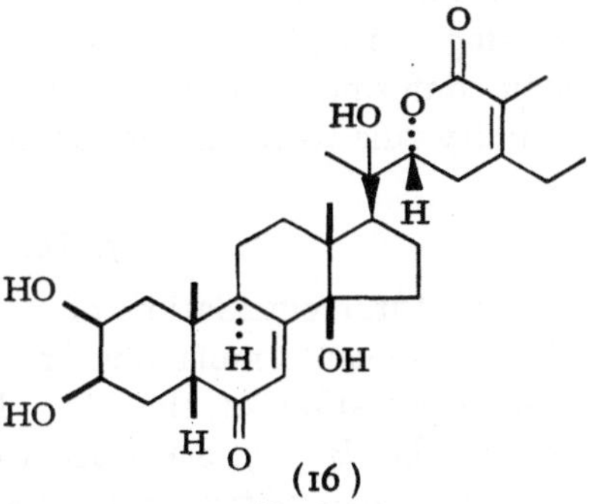

(16)

Ajugalactone

6. Ponasterone B (17)

Ponasterone B, a companion of ponasterone A (9), had the same composition $C_{27}H_{44}O_6$. The great similarity of the MS's of both 20,22-acetonides (10) and (18) suggested that they might be stereoisomers. This was substantiated by other spectroscopic data, especially the NMR spectrum of the 2,3,22-triacetate (19) in which, however, the chemical shifts of the C-2 and C-3 proton signals were very close. The NMR signals of the

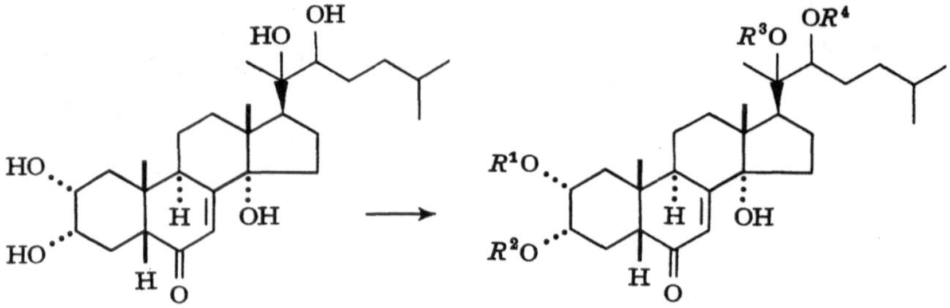

(17)
Ponasterone B

(18) $R^1 = R^2 = H$, R^3, $R^4 = C(CH_3)_2$

(19) $R^1 = R^2 = R^4 = Ac$, $R^3 = H$

(20) $R^1 = R^2 = Bz$, R^3, $R^4 = C(CH_3)_2$

Chart 5. Some Reactions of Ponasterone B

C-19 methyl protons of ponasterone B and its triacetate (19) were somewhat displaced as compared with those of ponasterone A and its triacetate (12). These data suggested that ponasterone B differed from ponasterone A in the configuration(s) at C-2 and/or C-3. The presence of an intramolecular hydrogen-bond in the IR spectrum of the acetonide (18) and the observation that only one of the C-2 or C-3 proton of the triacetate (19) exhibited axial-axial coupling to a neighboring proton could only be rationalized by the $2\alpha,3\alpha$-hydroxy arrangement whether the ring A adopted the chair or twist-boat conformation (95). Later, the chair conformation of the A ring was deduced by applying the dibenzoate chirality rule to the monoacetonide dibenzoate (20) (85).

7. Ponasterone C (21)

Structural elucidation of ponasterone C was somewhat circuitous. The molecular formula was first deduced incorrectly as $C_{27}H_{44}O_7$ due to misinterpretation of the MS. Different from ponasterone A (9) and ponasterone B (17), acetylation of ponasterone C proceeded stepwise to give the 2,22,24-triacetate and 2,3,22,24-tetraacetate. Nevertheless, the NMR data of the C-2, C-3 and C-19 protons of ponasterone C and its tetraacetate coincided remarkably well with those of ponasterone B and its triacetate, respectively. These spectral data, even in the absence of knowledge of the correct composition of the derivatives, led to the conclusion that the stereostructure of the A-ring of ponasterone C was identical with that of ponasterone B which possessed the 5β(H)-$2\alpha,3\alpha$-dihydroxy structure. Chemical and physico-chemical study of ponasterone C and its acetates, in particular, the formation of the isohexenal on periodate oxidation established the side-chain structure. The combined evidence was, therefore, initially interpreted in terms of constitution (22) for ponasterone C. The ORD and CD data, which should be

OH OH
HO

HO

HO

HO
HO O
H OH

(21)

Ponasterone C

OH OH
HO

HO

HO

H
O
H OH

(22)

but, in fact, were not consistent with those of ponasterone B remained ambiguous (95). However, reexamination of the MS of ponasterone C led to the realization that the tetracyclic nucleus contained an extra oxygen atom and that the correct molecular formula was $C_{27}H_{44}O_8$. Immediately after the revision of the molecular formula, it was shown that ponasterone C had all the features of a $2\beta,3\beta,5\beta$-trihydroxy A-ring structure, as are observed, for instance, in polypodine B (36) (85).

8. Inokosterone (23)

The structure of inokosterone, $C_{27}H_{44}O_7$, was examined by comparing the substance with its companion, β-ecdysone (3). The NMR spectra of inokosterone and its tetraacetate showed that the difference between these two ecdysterols lay in their side-chains. Periodate oxidation of inokosterone cleaved the 2,3- and 20,22-diol systems to give nuclear and side-chain fragments. The nuclear fragment (24) was converted by silica gel treatment into the enal (26) which was identical with the enal prepared from β-ecdysone by the same method. The side-chain fragment (25) on permanganate oxidation afforded $(\pm)$-α-methylglutaric acid (27) (130, 132). In order to eliminate the possibility that the asymmetric center at C-25 had been racemized during oxidation in alkali medium, inokosterone was subjected to partial acetylation to yield the 2,26-diacetate (28) whose periodate oxidation furnished the methyl ketone (5) and the acetoxy-aldehyde (29). The latter was transformed to the di-acetate (31) which was again found to be optically inactive, a fact which indicates inokosterone to be a mixture of C-25 epimers (120).

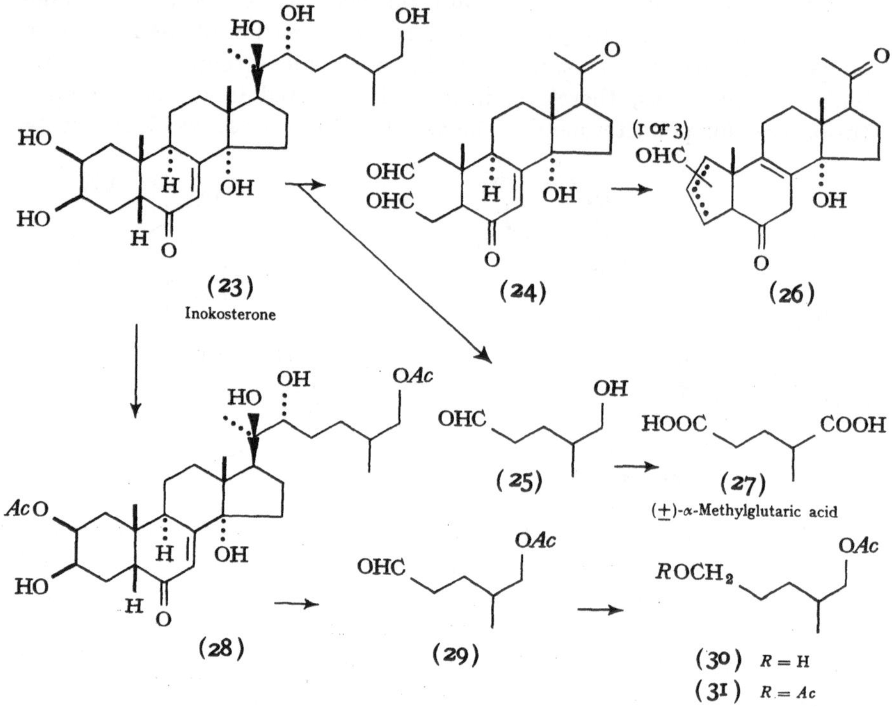

Chart 6. Some Reactions of Inokosterone

9. Cyasterone (32), Isocyasterone (33), Sengosterone (34) and Capitasterone (35)

Cyasterone, $C_{29}H_{44}O_8$, was the first ecdysterol shown to possess the stigmastane skeleton. Its special feature was the presence of a γ-lactone ring, as evidenced by an IR band at 1752 cm^{-1}. After the substituents in the steroidal tetracycle were shown to be the same as in β-ecdysone (3), the side-chain structure was elucidated by the complete analysis of the NMR spectrum with the aid of double resonance experiments of the side-chain fragment, the aldehyde-lactone, formed by periodate oxidation of cyasterone (125, 30). Later isocyasterone, a stereoisomer of cyasterone, was found in the same plant source. Absolute configurations at C-28 of cyasterone and isocyasterone was established to be both R by appli-cation of Hudson-Klyne lactone rule (37). Sengosterone, an immediate

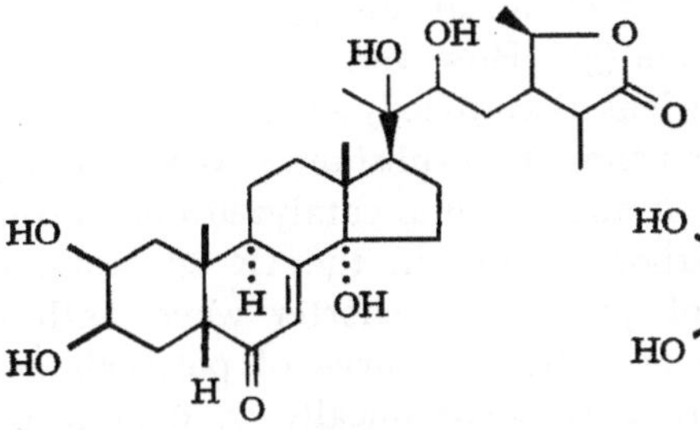

(32)
Cyasterone

(34)
Sengosterone

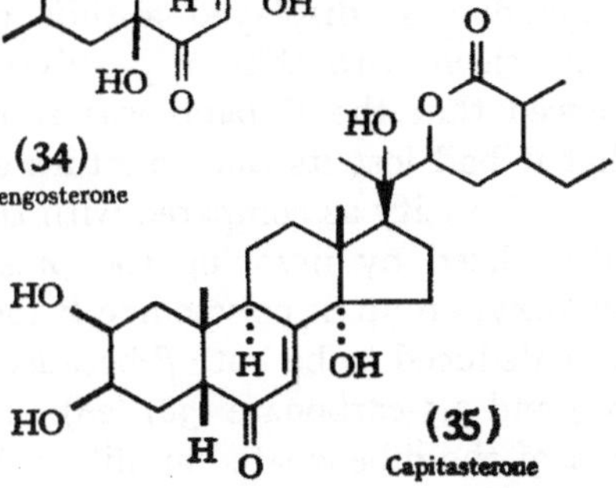

(33)
Isocyasterone

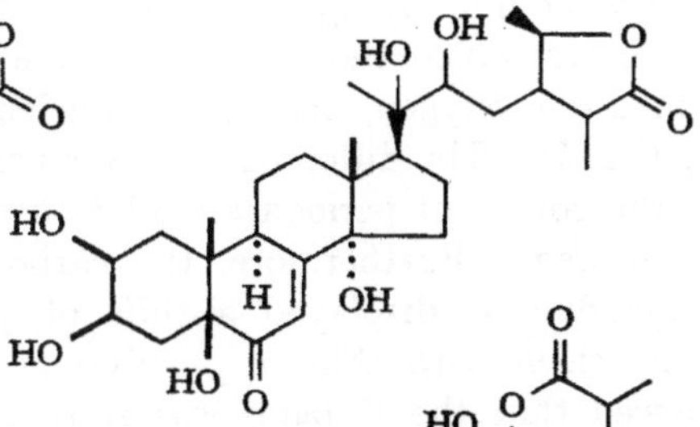

(35)
Capitasterone

metabolite of cyasterone, was revealed by physico-chemical data to have the same structural features as cyasterone except that it possessed an additional 5β-hydroxyl group (36). In elucidation of the capitasterone structure, IR, MS and NMR considerations played a key role (128).

10. Polypodine B (36)

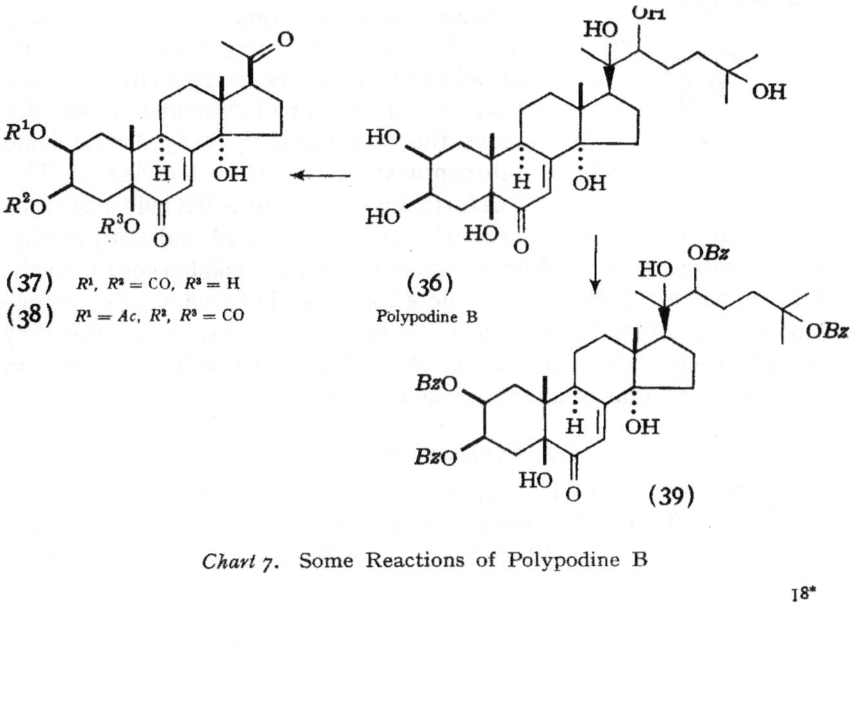

(37) R^1, R^2 = CO, R^3 = H
(38) R^1 = Ac, R^2, R^3 = CO

(36)
Polypodine B

(39)

Chart 7. Some Reactions of Polypodine B

Polypodine B, $C_{27}H_{44}O_8$, furnished chemical and spectroscopic data which resembled those of β-ecdysone (3). However, polypodine B contained one oxygen more in the nucleus than β-ecdysone, as pointed out by the MS. The differences between these two substances were observed in the course of periodate oxidation and alkaline catalyzed deuteration experiment. Furthermore, the carbonyl band in the IR spectrum of polypodine B displayed a shift of 32 cm^{-1} to shorter wavelength in comparison with that of β-ecdysone. The CD curve of polypodine B showed that the R band was shifted hypsochromically by 9.5 nm and that it had lost its fine structure and exhibited an increase in the molecular elipticity as compared with that of β-ecdysone. All these data were rationalized by invoking the presence of an extra oxygen as a C-5β hydroxyl group in polypodine B (66). The configurations at C-2 and C-3 were deduced to be both β because of transformations to a 2,3-carbonate (37) and 3,5-carbonate (38) (28). This was further confirmed by application of the dibenzoate chirality rule to the 2,3-dibenzoate (39) (85).

11. Rubrosterone (40)

Rubrosterone had a unique composition $C_{19}H_{26}O_5$. Besides the usual absorption bands in the IR spectrum, the significant feature was a band at 1741 cm^{-1} characteristic of a cyclopentanone. Since chemical and

(40)
Rubrosterone

physico-chemical study pointed to the presence of a $2\beta,3\beta,14\alpha$-trihydroxy-7-en-6-one system in a 5β-steroidal nucleus, indicating it to possess the etiocholane skeleton, the remaining problem was the location of the ketone group in a five-membered ring (the D-ring). Its attachment to C-17 was deduced by observing that acid treatment of rubrosterone did not move the resulting C-14 : C-15 double bond into conjugation with the carbonyl group. This was further confirmed by NMR evidence which showed that the observed chemical shifts of the C-18 and C-19 methyl signals of rubrosterone and its 2,3-diacetate were in good accord with the calculated values. It is worthy of note that the CD curve of rubrosterone exhibited a positive Cotton effect for the $n-\pi^*$ transition of the C-17 carbonyl group, whose maximum showed a violet shift of 16 nm in comparison with that of androsterone (121, 122).

12. Ponasteroside A (41)

Ponasteroside A, $C_{33}H_{54}O_{11}$, was a glycoside, since on enzymatic hydrolysis it afforded ponasterone A (9) and glucose (42). In the NMR spectrum the chemical shift of the C-19 methyl signal was displaced as

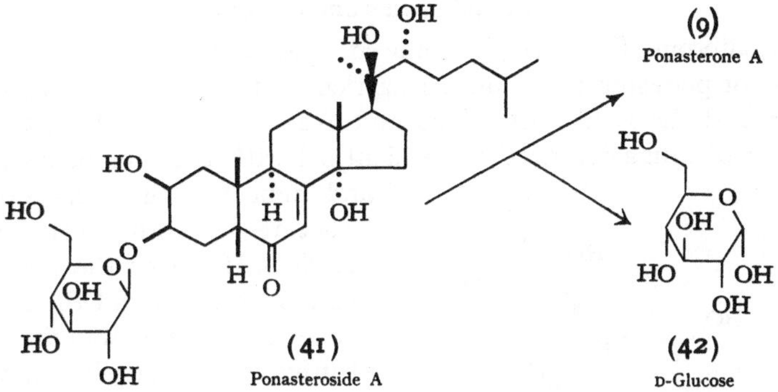

Chart 8. Some Reaction of Ponasteroside A

compared with that of the aglycone ponasterone A, suggesting that the glucose unit was linked with ponasterone A through ring A. The NMR spectrum of the hexaacetate showed that an axial hydroxyl group participated in the glycoside linkage. On the other hand, that ring A of ponasteroside A had the chair conformation was deduced by the ORD curve, the splitting pattern of the 5β-hydrogen NMR signal, and the observation of an intramolecular nuclear Overhauser effect between the hydrogen on the ethereal oxygen-bearing carbon and the C-9 allylic hydrogen. The combined evidence indicated that the C-3 hydroxyl was joined to glucose. The stereochemistry of the glucose residue as the β-D-glucopyranosyl moiety was deduced by analysis of the NMR data and applying the method of molecular rotation differences (114, 29).

13. Shidasterone (43)

A unique phytoecdysone is shidasterone, $C_{27}H_{44}O_7$. Physicochemical data of shidasterone indicated it to be a stereoisomer of β-ecdysone (3) with respect to some point(s) in the side-chain. The most significant difference in the chemical properties of the two substances was that shidasterone gave a 2,3-diacetate, while β-ecdysone afforded a 2,3,22-triacetate under the same conditions. On the assumption that shidasterone is biosynthesized from cholesterol, it was first suggested that shidasterone was probably the C-22 epimer of β-ecdysone. Since this was shown not to be so, shidasterone is now considered to be the C-20 epimer (126).

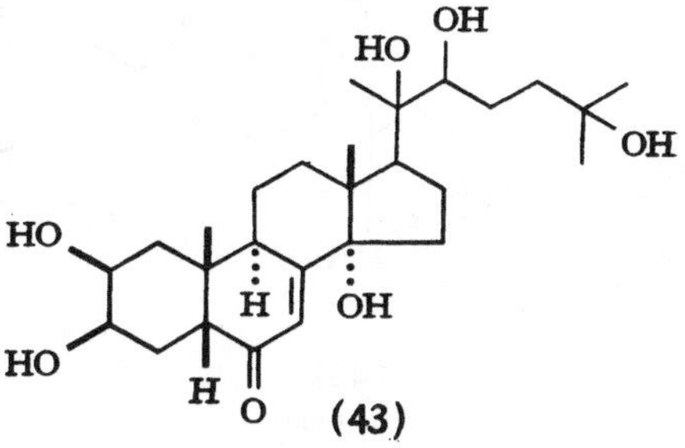

14. Ajugasterone C (44)

Ajugasterone C, $C_{27}H_{44}O_7$, showed spectroscopic properties similar to those of ponasterone A (9). In particular, the side-chain fragments in the MS and the C-21, C-22, C-26 and C-27 proton signals in the NMR spectrum of ajugasterone C were identical with those of ponasterone A and indicated that the side-chain structures of both substances were identical. However, the nuclear fragments in the MS and the formation of a tetraacetate showed the presence of an extra secondary hydroxyl group in the nuclear part of ajugasterone C as compared with ponasterone A. Differences between the two substances were observed in the chemical shifts of the NMR angular methyl signals, especially that of C-19, and the line positions and splitting patterns of the C-9 proton signals (coupled to the C-7 and C-11β protons in ajugasterone C and to the C-7, C-11α and C-11β protons in ponasterone A). These results showed that ajugasterone C was 11α-hydroxyponasterone A (62). The spatial disposition of the 2β,3β,11α-trihydroxyls was corroborated by applying the dibenzoate chirality rule (84).

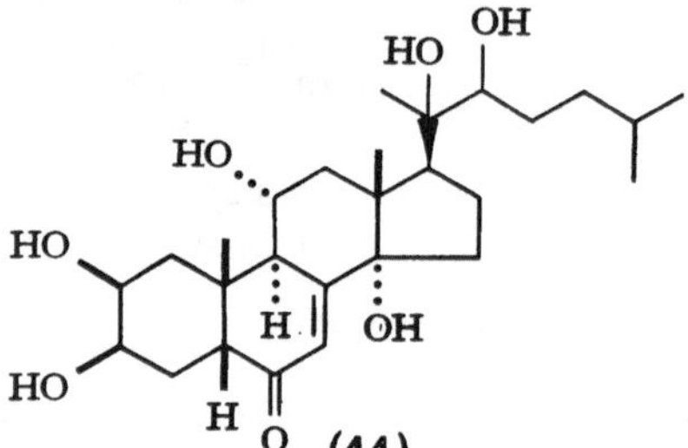

(44)
Ajugasterone C

15. Pterosterone (45), Makisterone A (46), Makisterone B (47), Lemmasterone (Makisterone C, Podecdysone A) (48), Amarasterone A (49), Makisterone D (50) and Amarasterone B (51)

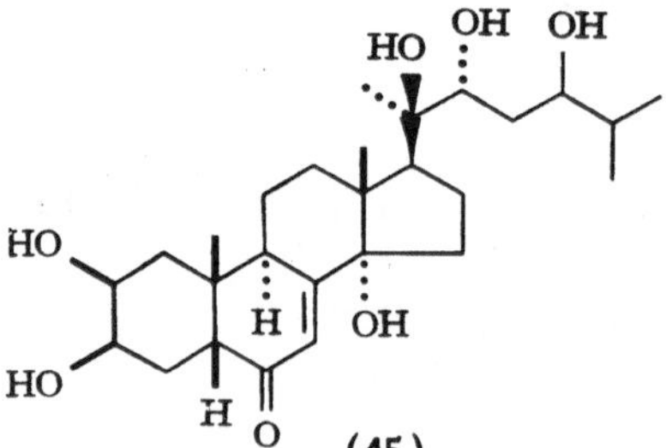

(45)
Pterosterone

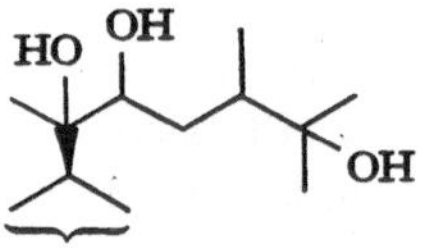

(46)
Makisterone A

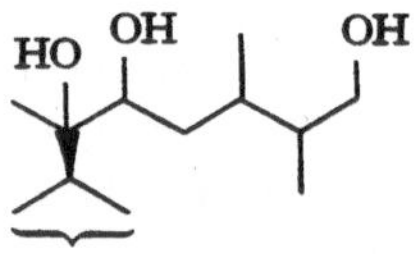

(47)
Makisterone B

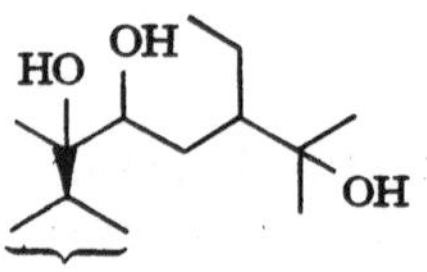

(48)
Lemmasterone

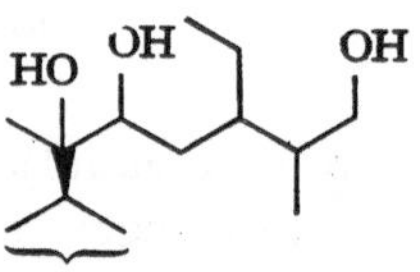

(49)
Amarasterone A

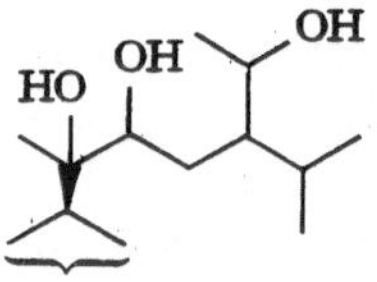

(50)
Makisterone D

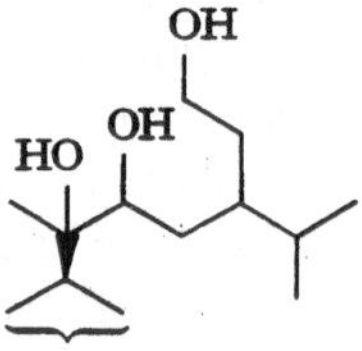

(51)
Amarasterone B

These compounds may result biogenetically from hydroxylation of ponasterone A (9) and its C-24 alkylated congeners. Structures were elucidated by physico-chemical studies of substances (45)—(51) and their acetates (*115, 61, 59, 117, 21, 127*).

16. Viticosterone E (52), Stachysterone C (53), Stachysterone A (54), Stachysterone B (55), Podecdysone B (56) and Stachysterone D (57)

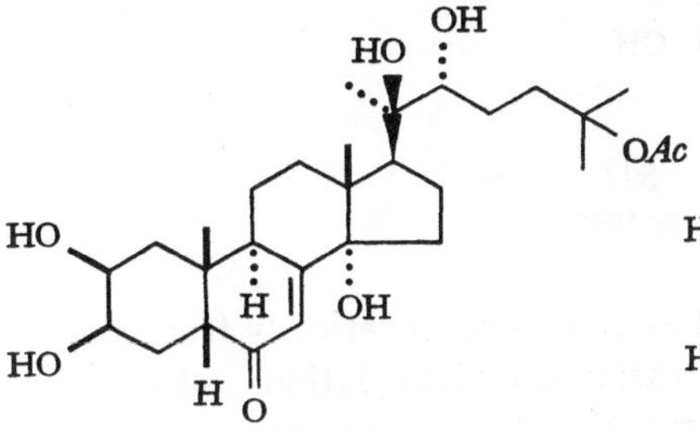

(52)
Viticosterone E

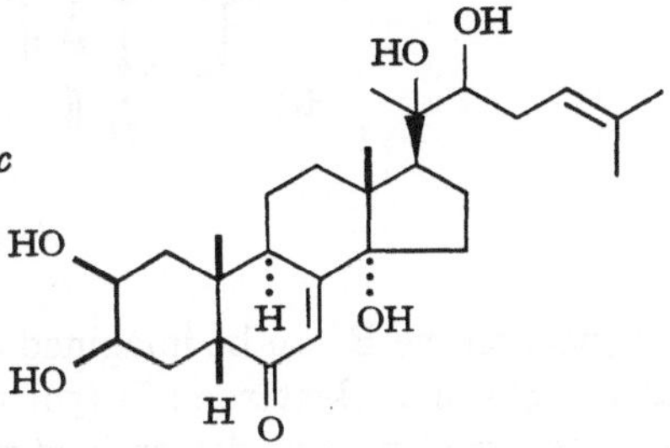

(53)
Stachysterone C

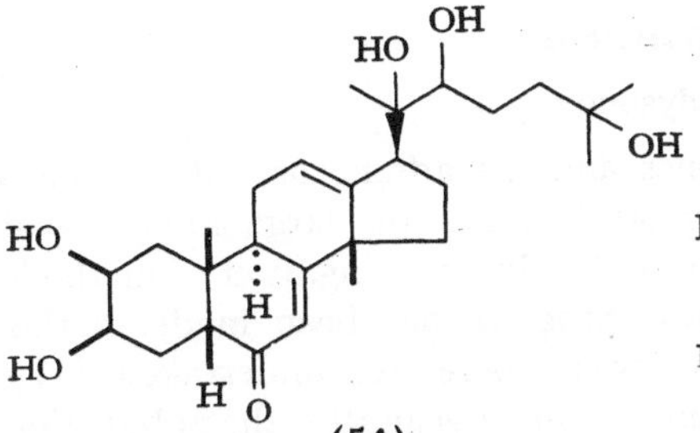

(54)
Stachysterone A

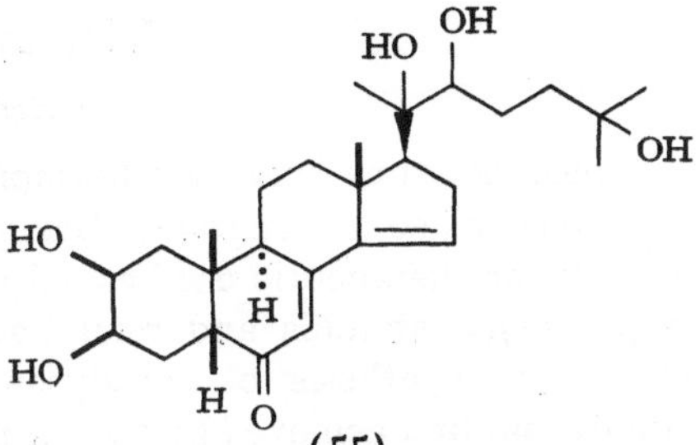

(55)
Stachysterone B

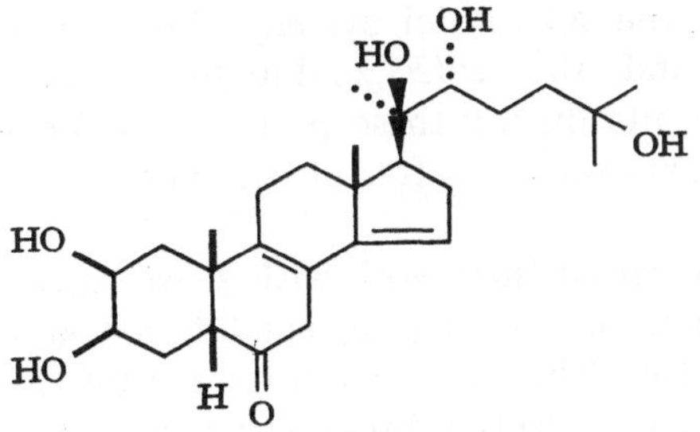

(56)
Podecdysone B

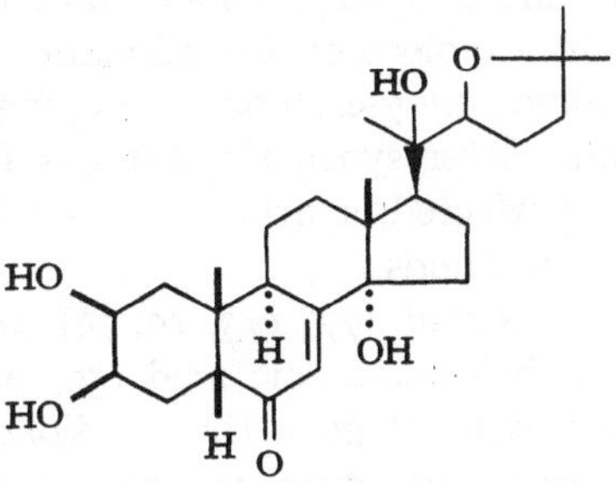

(57)
Stachysterone D

These appear to be formed biogenetically from β-ecdysone (3) by acetylation or dehydration. Structures were deduced by interpretation of their spectroscopic properties (*101, 85, 17*).

17. Ajugasterone B (58)

(58)
Ajugasterone B

Ajugasterone B can be imagined as arising biogenetically from lemmasterone (48) or makisterone D (50) through hydroxylation and dehydration. Its structure was determined mainly on spectroscopic grounds (58). Ajugasterone D is thought to be a position isomer of ajugasterone B (85).

VII. Synthesis

1. α-Ecdysone (1)

Since the time when sex hormones and the adrenocortical hormones were recognized as steroids, the great demand for large amounts of steroids for therapeutic use has led to extensive investigation of methods of producing steroids and remarkable progress has been made in this area. The synthesis of α-ecdysone was, therefore, undertaken very rapidly and first achieved by two groups within a year after the publication of the structure determination. A synthetic route must involve the following key steps which fulfil the stereochemical requirements of the finished molecule: Introduction of the $2\beta,3\beta$-diol system, the $5\beta(\mathrm{H})$-14α-hydroxy-7-en-6-one grouping and the $22(R),25$-dihydroxy side-chain. After synthetic methods for introducing these partial structures were perfected, synthesis of α-ecdysone was a matter of combination of these methods.

Kerb et al. (77, 143, 10, 78) commenced their work with 3-oxo-23,24-bisnorchol-4-en-22-oic acid (59) which on ketalization, epoxidation and acid treatment gave the 3,6-dione (60). The construction of the $2\beta,3\beta$-dihydroxy structure was carried out by selective bromination at C-2α, stereospecific and selective reduction of the C-3 carbonyl and acetylation of the C-3 hydroxyl followed by silver acetate treatment. This led to a 1:1 mixture of the C-5 epimers of the $2\beta,3\beta$-diacetoxy-6-one (61) and (62). Bromination of the ketones (61) and (62) yielded the same $5\alpha(\mathrm{H})$-7α-bromo-6-one which on dehydrobromination gave the key intermediate

(63). Partial hydrolysis of the carbomethoxy group led, with partial inversion at C-5 and C-20, to a mixture of the four possible isomeric acids from which one acid was isolated. Selective reduction of the carboxyl group gave the aldehyde which was coupled with a five-carbon Grignard reagent. Of the resulting C-22 epimers, one (64) was converted into α-ecdysone (1) by hydrogenation of the triple bond. This was followed by hydrolysis of the acetoxyl group and selenium dioxide hydroxylation at C-14 from the less hindered α-side with coincident removal of the protecting group. The other C-22 epimer (65) was converted by similar procedures into the inactive 22-epi-α-ecdysone, 22-isoecdysone (66). During the above work, three deoxy-derivatives of α-ecdysone were prepared (38).

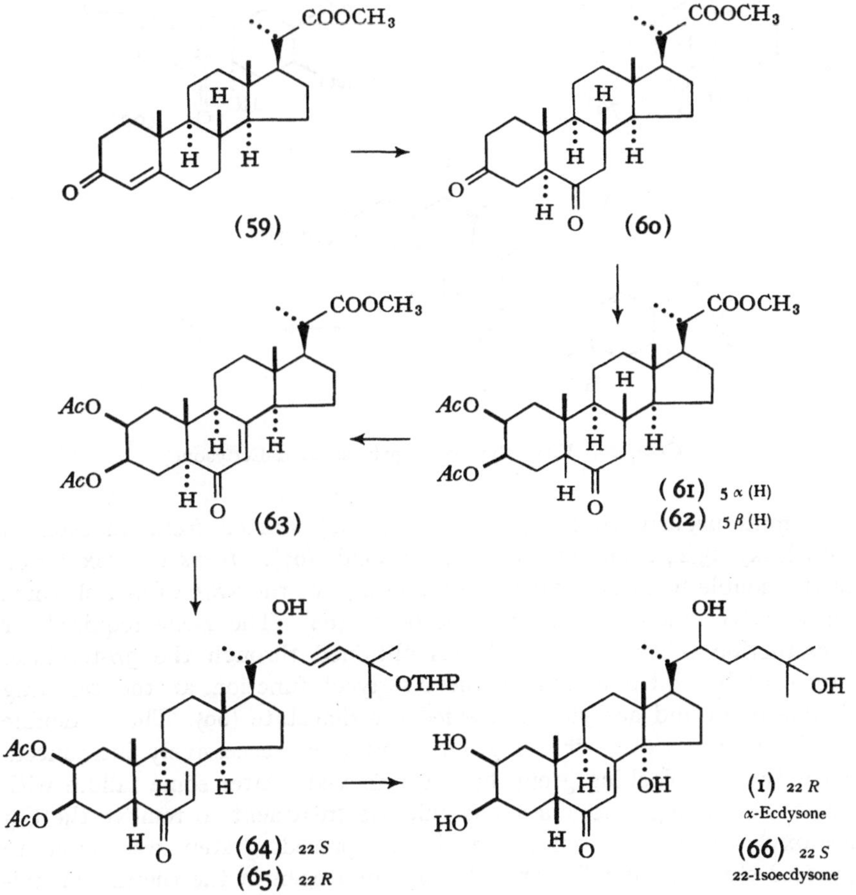

Chart 9. First Schering-Roche Synthesis of α-Ecdysone

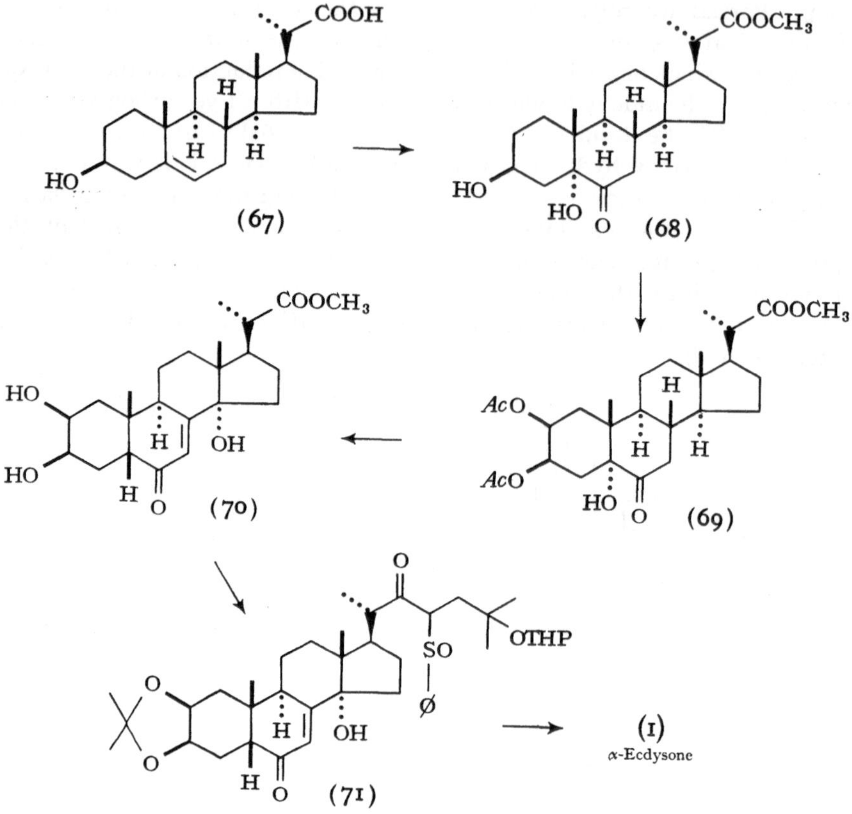

Chart 10. First Syntex Synthesis of α-Ecdysone

Simultaneously SIDDALL et al. (*111, 109*) started from an ester of 3-hydroxy-23,24-bisnorchol-5-en-22-oic acid (67). *trans*-Hydroxylation of the double bond and partial hydrolysis gave the 3β,5α,6β-triol which on selective oxidation afforded the 6-one (68). The 2-ene required for construction of the 2β,3β-diol was obtained through the 3β-tosylate. Stereospecific introduction of the *cis*-glycol function at the resulting double bond and acetylation yielded the diacetate (69). The *Δ*⁷-double bond and the 14α-hydroxyl group were then successively introduced. Then the C-5 blocking group had to removed. After some failure with agents containing zinc, chromous chloride treatment to remove the C-5 acetoxyl which had been formed in the preceding step, was found to proceed stereospecifically, and placed a hydrogen in the thermodynamically less stable α-configuration. Inversion at C-5 with alkali furnished the key intermediate (70) together with its 5-epimer in the ratio 2 : 1.

The sequence of the last steps involved construction of the side-chain. Initial attempts to use the Grignard reaction caused degradation of the nucleus. After the glycol and the keto group were protected, a sulfinyl-stabilized secondary carbanion was coupled with the ester (70) to give the sulfoxyphenyl ketone (71) from which the phenylsulfinyl group was removed by hydrogenolysis. Hydride reduction of the C-22 carbonyl group yielded a roughly 1 : 1 mixture of the C-22 epimers which by selective oxidation of the allylic C-6 hydroxy group followed by removal of the protecting groups furnished a mixture of the four possible C-20 and 22 stereoisomers from which α-ecdysone (1) was isolated.

The completion of these syntheses was, of course, highly significant. However, there were considerable disadvantages in that both involved multistep routes from readily available steroids and that certain steps proceeded without stereospecific control.

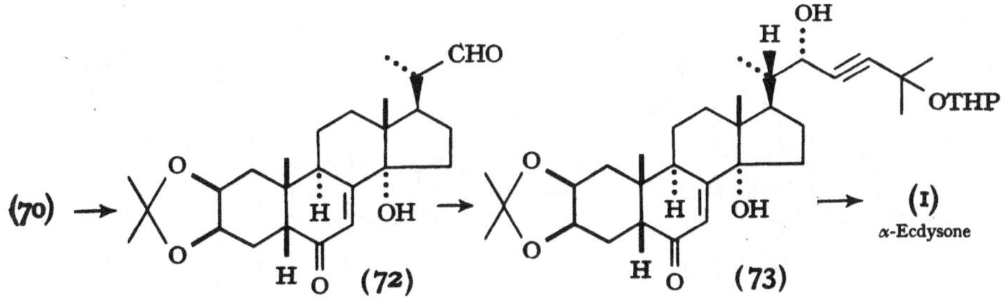

Chart 11. Second Syntex Synthesis of α-Ecdysone

HARRISON *et al*. (*25*) soon announced an alternative synthesis of α-ecdysone which proceeded without equilibration at C-20 and, therefore, afforded only α-ecdysone (1) and 22-isoecdysone (66). The point of departure was the acetonide of the previous intermediate (70) which on stepwise reduction and oxidation furnished the aldehyde (72). The side-chain structure was completed by procedures similar to those which had been utilized by KERB *et al*. (77).

FURLENMEIER *et al*. (8, 9) also reported a second synthesis of α-ecdysone which involved fewer steps. The known enone (74), prepared from ergosterol acetate, was transformed *via* Δ^2-olefin (75) to the diacetate (76) by procedures similar to those used by SIDDALL *et al*. (*111*). The ergosterol side-chain was removed by ozonolysis to give the aldehyde (77). The ecdysone side-chain was attached by condensation with the Grignard reagent used previously, subsequent reduction of the triple bond afforded a mixture of diastereomers in which the major isomer was the desired 22(S)-compound (78). After the 14α-hydroxyl group was inserted, equilibration with alkali gave the 5β-analog which on removal of the protecting group furnished α-ecdysone (1). By interchange of the reaction sequence, *i. e.* by first submitting the aldehyde (77) to hydroxylation at C-14, 22-isoecdysone (66) was obtained stereospecifically.

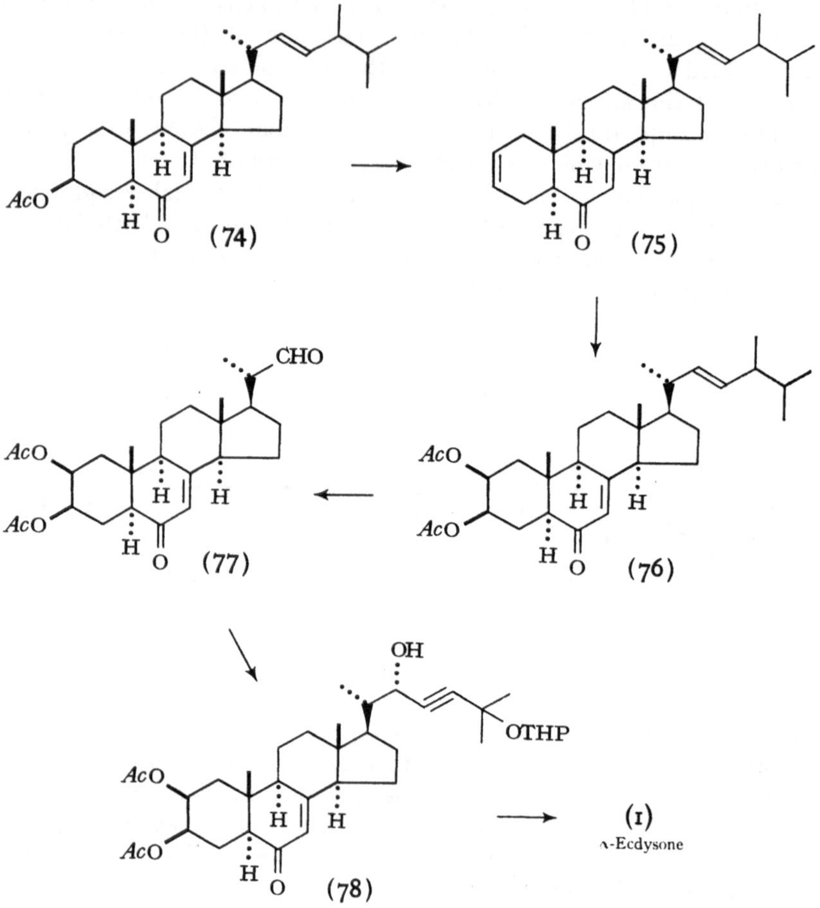

Chart 12. Second Schering-Roche Synthesis of α-Ecdysone

Thus, these syntheses were to make this interesting steroid available in greater quantities without requiring the extraction of a large amount of animal material. In fact, the Schering group was interested in its biological activity and, consequently, prepared some 50 g of α-ecdysone from a vast amount of ergosterol at a great expense. Meanwhile, the discovery of high concentrations of ecdysterols in plant sources made the chemical synthesis for biological investigation almost obsolete, though it is still valuable for structure proof and for the preparation of some analogs difficultly obtainable from nature.

References, pp. 305—312

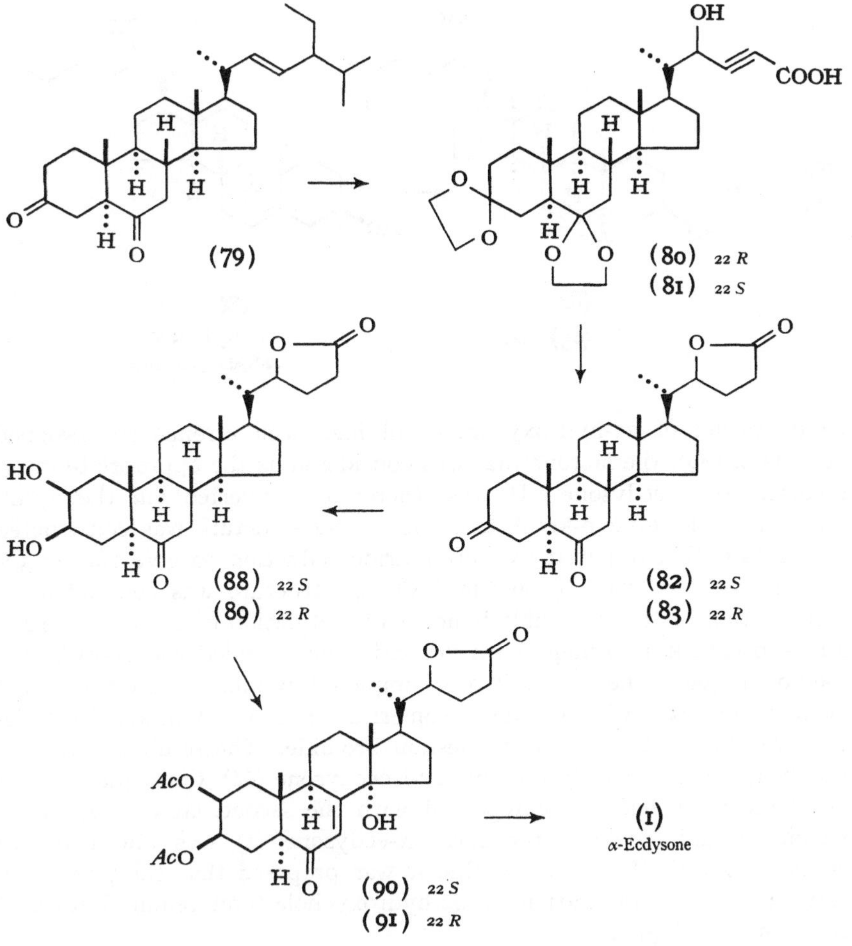

Chart 13. Teikoku Synthesis of α-Ecdysone

In the meantime, MORI *et al.* (*88, 87, 93, 89, 91, 90*) had also been engaged in the synthesis of α-ecdysone since 1965. The starting material, stigmast-22-ene-3,6-dione (79), on ketalization, ozonolysis and treatment with ethynyl magnesium bromide gave the ethynyl derivative which was converted into a mixture of unsaturated acids (80) and (81). Hydrogenation of the triple bond followed by acid hydrolysis of the ketal linkages yielded the two isomeric lactones (82) and (83). To determine the C-22 configurations, both lactones (82) and (83) were transformed in eight steps to the isomeric diols (84) and (85), respectively, the latter being also derived from natural 22-hydroxycholesterol. Since the C-22

(82)
(83) →

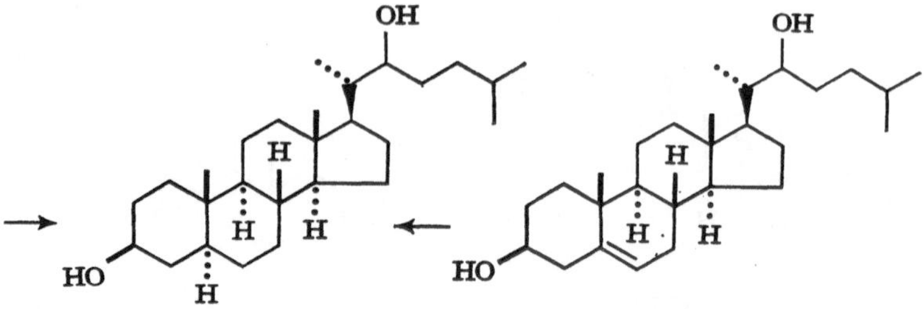

(84) 22 S
(85) 22 R

(86) 22 S
(87) 22 R
22-Hydroxycholesterol

configuration of 22-hydroxycholesterol had been erroneously assigned as S as in (86), the lactone (82) was considered as the appropriate intermediate for α-ecdysone. It was, therefore, converted *via* the 3β-ol-6-one into the 3-one-6-ketal, and the A-ring structure was constructed by autoxidation followed by borohydride reduction to give the 2β,3β-dihydroxy-6-one (88). Then the B,C-ring structure was completed by introduction of the Δ⁷-double bond, enol-acetylation of the enone group and subsequent treatment with peracid which yielded the 14-hydroxy-7-en-6-one (90). The synthesis was completed by alkali hydrolysis of the acetoxyl groups with coincident isomerization at C-5 followed by Grignard reaction with methyl magnesium bromide. The resulting product, however, was not α-ecdysone but 22-isoecdysone (66). Consequently, the synthetic work had to be repeated with the second lactone (83) from which through similar procedures α-ecdysone (1) was obtained. In connection with the above results, it was proposed that the previously assigned C-22 configuration of 22-hydroxycholesterol required reversal to R as in (87) (92).

2. β-Ecdysone (3) and Ponasterone A (9)

After completion of the α-ecdysone synthesis, the next efforts were directed towards the synthesis of its subsequently discovered analogs. The first announcement in this area were from Hüppi and Siddall (56, 57) who completed a synthesis of β-ecdysone and ponasterone A. This was a matter of preparing a suitable 20-hydroxy-20-formyl intermediate (94), since their previous synthesis of α-ecdysone had solved the problems of elaborating the nucleus. The starting material was the intermediate (69) which was converted by four steps to the acetonide (92) which on subsequent reduction followed by stepwise oxidation gave the aldehyde (93). Hydroxylation at C-20 by selective enol acetylation of the aldehyde function and epoxidation of the resulting $\Delta^{20(22)}$ double bond followed by alkali hydrolysis afforded the desired intermediate (94) together with its C-20 epimer (5 : 3). After protection of the C-20 hydroxyl and hydroxylation at C-14α, the side-chain was constructed by alkylation with a Grignard reagent. Inversion of the configuration at C-5 with alkali led to the

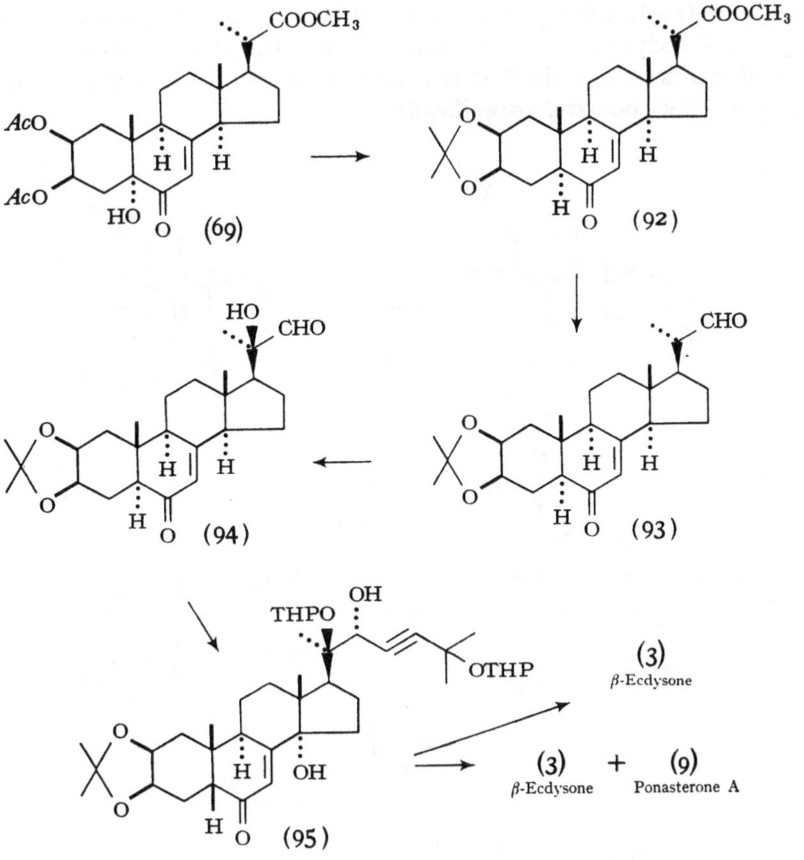

Chart 14. Syntex Synthesis of β-Ecdysone and Ponasterone A

intermediate (95). When the triple bond was first hydrogenated and the protecting groups were then removed, β-ecdysone (3) was obtained. On the other hand, prior removal of the protecting groups in the intermediate (95) gave the free hexaol whose triple bond was hydrogenated to give the hydrogenolysis product ponasterone A (9) along with β-ecdysone. A disadvantage of this synthesis is the method of introducing oxygen at C-20 which was effected without steric control.

Another β-ecdysone synthesis developed by Kerb *et al.* (*79*) employed as starting material progesterone (*96*). The key intermediate (*97*) obtained by four steps was modified by a similar route to that employed in the α-ecdysone synthesis to give the methyl ketone (*98*). The side-chain was then introduced by a Grignard reaction. This yielded the acetylene derivative (*99*) which on hydration of the triple bond gave the dehydro-β-ecdysone (*100*). Finally, hydride reduction of the C-22 carbonyl group furnished β-ecdysone and its C-22 epimer. The disadvantage of this series of synthesis is again that the Grignard reaction and hydride reduction proceeded non-stereospecifically.

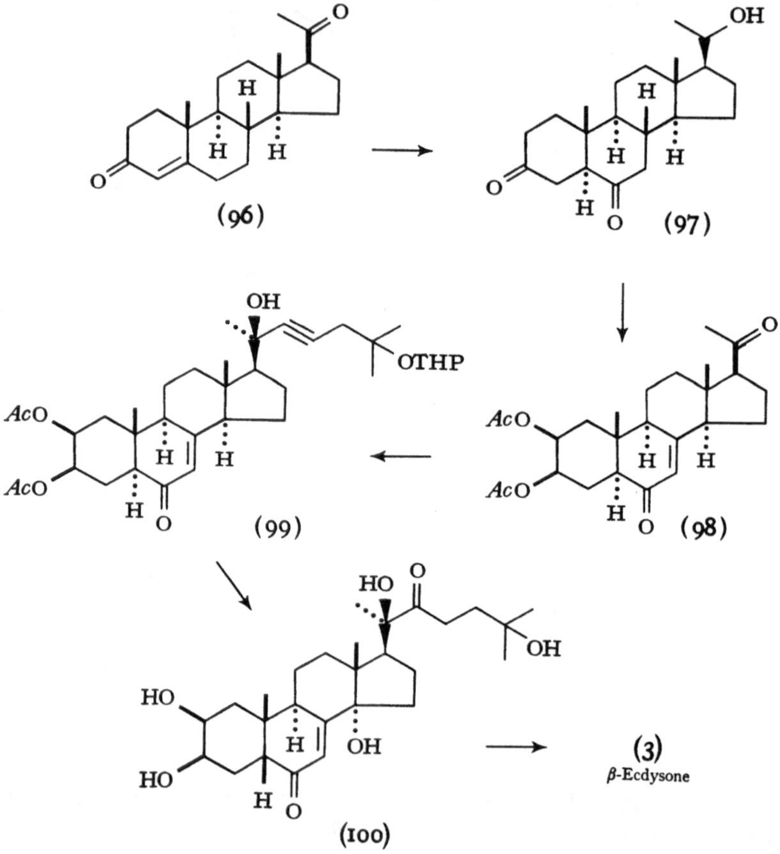

Chart 15. Schering-Roche Synthesis of β-Ecdysone

A new synthesis of β-ecdysone by Mori and Shibata (*86*) was stereoselective in every step. Thus pregn-5-ene-3β,20(R)-diol (*101*) was transformed *via* a route similar to that employed in the α-ecdysone synthesis to the methyl ketone (6) whose Grignard reaction with vinyl magnesium bromide afforded the vinyl derivative (102) with high stereospecificity. After selective ozonolysis of the vinyl group, the resulting aldehyde (103) was selectively reacted with a Grignard reagent to give only one acetylenic intermediate (104) which was converted to β-ecdysone.

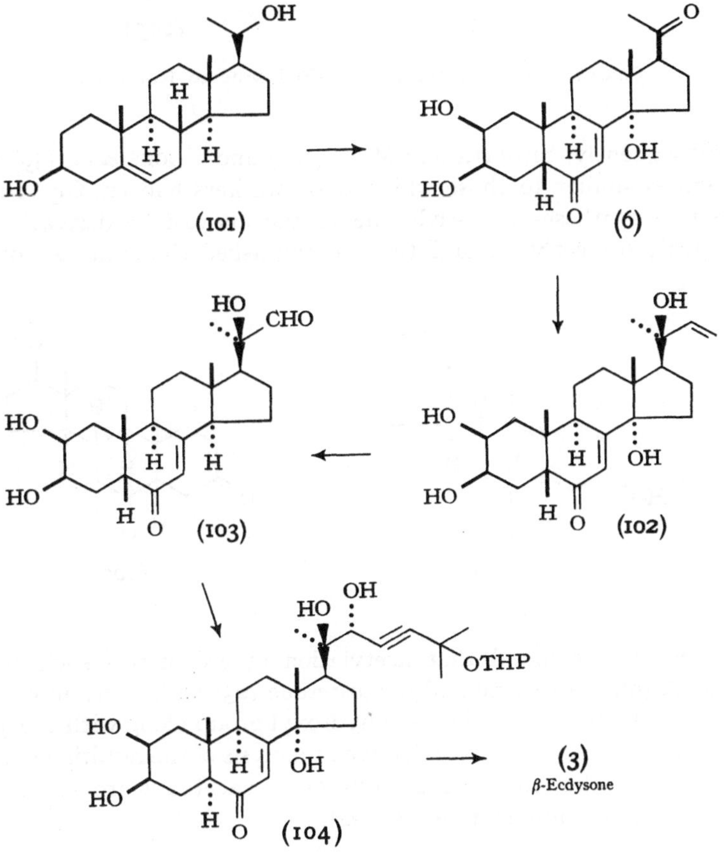

Chart 16. Teikoku Synthesis of β-Ecdysone

3. Rubrosterone (40)

Immediately after the discovery of rubrosterone, HIKINO *et al.* (*31, 32*) confirmed its stereostructure by the transformation of β-ecdysone (3), already synthesized, to rubrosterone (40) during which the selective oxidation of the methyl ketone (6) to the acetate (105) was the key step.

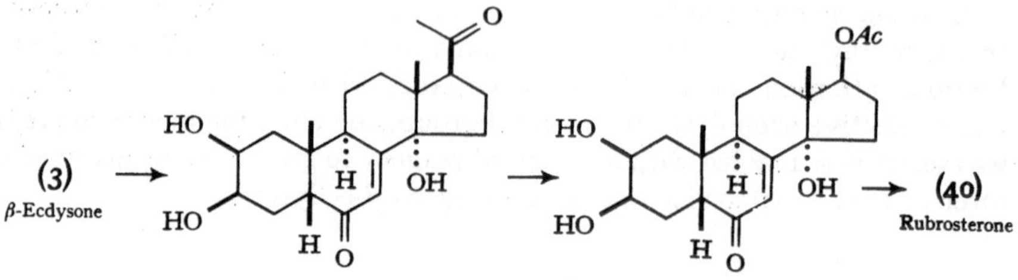

(6) (105)

Chart 17. Transformation of β-Ecdysone to Rubrosterone

Subsequently, Shibata and Mori (*108*) and Hocks *et al.* (*39*) applied procedures similar to those which these workers had employed in their respective syntheses of α-ecdysone to the androstane derivatives (106) and (107), respectively, and thus accomplished the synthesis of rubrosterone.

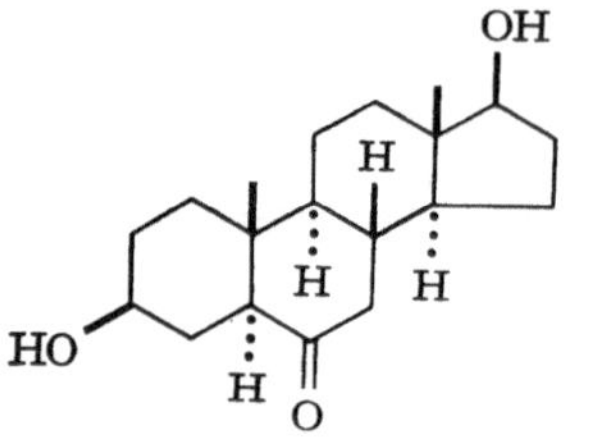

(106)

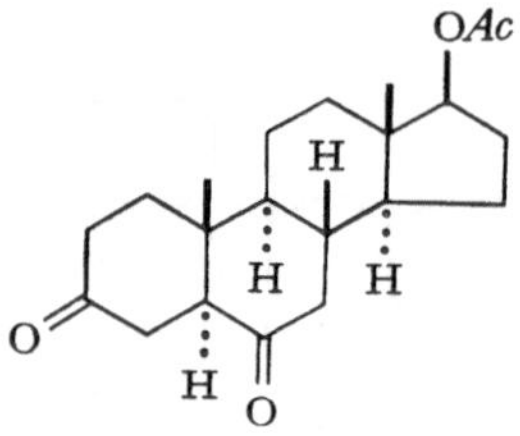

(107)

Later, it was found that acetylation of ecdysterols such as ponasterone A (9), β-ecdysone and inokosterone (23) under certain conditions led to the elimination of the C-20 hydroxyl group along with the acetylation of the hydroxyl groups. Selective ozonolysis of the resulting C-17:C-20 double bond followed by hydrolysis of the acetoxyl groups completed the synthesis of rubrosterone (*85, 32*).

4. Miscellaneous

Another position isomer of α-ecdysone, 20(*S*)-hydroxy-22-deoxy-α-ecdysone was synthesized by Galbraith *et al.* (*16, 19*) from the known methyl ketone (*6*) by using a trimethyl silyl ether as a protecting group for sterically hindered C-14α hydroxyl group. 2-Deoxy-3-*epi*-β-ecdysone was also synthesized (*18*).

Šorm *et al.* (*49, 142*) synthesized a number of α-ecdysone analogs and found that ergostane, cholestane and pregnane derivatives containing 3β-hydroxyl and 6-keto groups inhibited the postecdysial hardening and sclerotization of the cuticle in *Pyrrhocoris apterus* larvae.

References, pp. 305—312

VIII. Biosynthesis and Catabolism

After discovery of the ecdysterols in nature, the metabolism of these substances has naturally become an interesting subject of research.

Plants with the possible exception of certain micro-organisms are capable of synthesizing steroids which are generally called phytosterols. Higher animals, while unable to utilize C_{28} and C_{29} phytosterols in their diet, are capable of synthesizing C_{27} steroids by themselves. Insects, however, are incapable of synthesizing steroids essential to life and, therefore, depend on exogenous sources. Thus, insects dealkylate physiologically inactive C_{28} and C_{29} phytosterols to essential C_{27} steroids.

Zooecdysones are most probably synthesized *via* cholesterol. In fact, at a time when the steroidal nature of α-ecdysone (1) was only suspected and not yet established, the conversion of labelled cholesterol into α-ecdysone in *Calliphora erythrocephala* larvae was demonstrated by KARLSON and HOFFMEISTER (73). Recently MORIYAMA *et al.* (85) also showed transformation of labelled cholesterol to α-ecdysone and β-ecdysone (3) in *Bombyx mori* larvae, α-ecdysone rather than β-ecdysone being predominantly biosynthesized. A similar conversion of α-ecdysone in third instar larvae of *Calliphora stygia* was observed by HORN *et al.* (53).

Although a number of biogenetic hypotheses can be envisaged, the actual pathways of biosynthesis of the zooecdysones from cholesterol remain to be elucidated. In view of the coexistence of α-ecdysone, β-ecdysone and 20,26-dihydroxyecdysone in an insect, *Manduca sexta*, KAPLANIS *et al.* (69) suggested that these substances are all intermediates in a single synthesis-degradation scheme. Actually, KING and SIDDALL (80) recently proved that a young shrimp, *Cragon nigricanda*, a male crab, *Uca pugilator*, and fifth instar larvae of the browfly, *Calliphora vicina*, were all able to convert α-ecdysone into β-ecdysone. In the experiments with *Calliphora*, it was found that a minimum of 25% of the injected tritium still remained as α- and β-ecdysones after 12–24 hr, in spite of the claim by OHTAKI *et al.* (99) that α-ecdysone in *Sarcophaga* was rapidly deactivated by an inactivation mechanism present in the tissues, the half-life time of 1 μg of α-ecdysone being *ca* 1 hr. MORIYAMA *et al.* (85) also showed that in every part of *Bombyx mori* larvae, α-ecdysone was converted into β-ecdysone at a rapid rate, three other metabolites being detected. This observation is inconsistent with their finding that α-ecdysone was biosynthesized in a much larger quantity than β-ecdysone. To explain this discrepancy, it was suggested that the biosynthesized α-ecdysone exists in a form bound loosely to a biomolecule. THOMSON *et al.* (141) further found that labelled 25-deoxy-α-ecdysone (108) was converted in *Calliphora stygia* larvae and prepupae into ponasterone A (9) and inokosterone (23) as well as into β-ecdysone. Surprisingly, labelled α-ecdysone was not detected among the products. As ponasterone A and inokosterone were not detected in extracts of the normal *Calliphora*, it

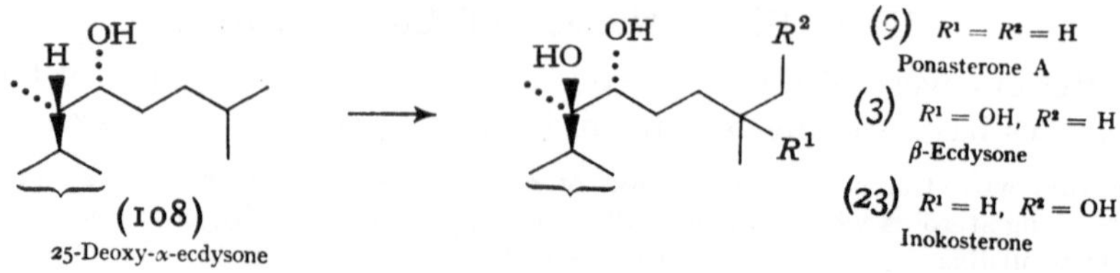

Chart 18. Metabolic Pathway of 25-Deoxy-α-ecdysone in *Calliphora*

was concluded that 25-deoxy-α-ecdysone (108) was not a major natural precursor of β-ecdysone in this insect. When crustecdysone was isolated from the crayfish, *Jasus lalandei*, Horn et al. (54) suggested that the ecdysterol might be metabolized to the methyl ketone (6), but later these workers (110) concluded that this was unlikely since no methyl ketone (6) could not be detected in the crayfish extracts. However, Galbraith et al. (20) recently found that in *Calliphora stygia* β-ecdysone was catabolized partly by the C-20 : C-22 side-chain scission to 4-methyl-4-hydroxy-pentanoic acid. This finding indicates the concurrent formation of the methyl ketone (6) which may undergo further degradation to the less active metabolites such as rubrosterone (40) as depicted in Chart 19.

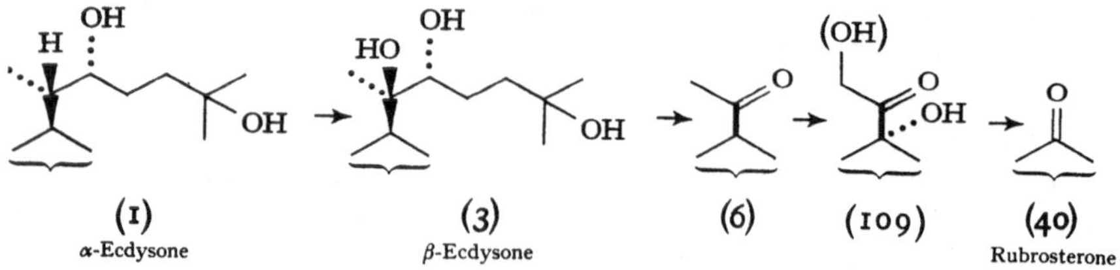

Chart 19. Probable Metabolic Pathway of Ecdysterols

The localization of the hormone producing organs in insects has been the object of protracted controversy (23). Thus, in 1934, the secretion center of the molting hormone was shown to be present within the head. On the basis of experiments on an insect, *Rhodnius*, this center was first considered to be the corpora allata and, in 1940, to be the neurosecretory cells of the brain. It was then assumed that the prothoracic glands took part in the pupation of *Bombyx mori* larvae, this being soon supported by experimental evidence. Since then, it has long been believed that the prothoracic glands are the genuine endocrine organs of the molting hormone. However, doubt has recently raised as to whether the glands produce the hormone directly. In particular, it has been found that transplantation of the living prothoracic glands causes isolated pupal abdomens of *Bombyx mori* to develop into normal adult abdomens, while injection of the prothoracic gland extract has not succeeded in giving a positive response. In fact, Moriyama et al. (85) found that biosynthesis of α- and β-ecdysones took place mainly in the posterior half of *Bombyx mori* larvae lacking the glands, a fact which suggests that the prothoracic glands do not neccessarily produce the molting hormone. Hence, the term prothoracic gland hormone commonly used for the molting hormone, may quite possibly be abandoned.

It was observed that the levels of molting hormones in insects undergo cyclic changes which are synchronized with the life cycles (*3*, *106*, *107*, *69*). However, these levels were determined only by bioassay of the hormone activity in the entire insect body and qualitative changes were ignored. The periodic rise and fall of the molting hormone observed by Faux *et al.* (*6*) in the crab, *Callinectes sapidus* (p. 260,) provides interesting information on the biosynthesis and catabolism of ecdysterols in an animal. Thus, the results seem to suggest that a molting hormone once produced is degraded selectively and at the same time another molting hormone is biosynthesized selectively.

It is very conceivable that phytoecdysones are also biosynthesized from cholesterol, and indeed, Heftmann *et al.* (*26*, *105*) and Hikino *et al.* (*34*, *35*) demonstrated the conversion of cholesterol-^{14}C to radioactive β-ecdysone, and ponasterone A and β-ecdysone by *Podocarpus elatus* and *P. macrophyllus* seedlings, respectively. However, actual biosynthetic pathways from cholesterol to the phytoecdysones again remain unknown.

The common occurrence of ponasterone A, α-ecdysone and β-ecdysone in certain ferns (Table 3) implies that two alternative biosynthetic pathways from the hypothetical precursor, 25-deoxy-α-ecdysone (*108*), to β-ecdysone through α-ecdysone and ponasterone A are operating in plants.

The C_{29} ecdysterols are probably metabolites of β-sitosterol or clinasterol which, of course, may be synthesized *via* cholesterol. *Inter alia,* the simultaneous occurrence of a series of C_{29} phytoecdysones in the same source, *Cyathula capitata*, suggests the presence of the metabolic pathway shown in Chart 20.

Of most interest biosynthetically is the formation of ponasterone B (*17*) and shidasterone (*44*). If ponasterone B were derived from colesterol, inversion of the 3β-hydroxyl group should be involved, since ponasterone B has the 2α,3α-dihydroxy arrangement. This may be analogous to the conversion of 3β-hydroxy-Δ^5-steroids to 3α-hydroxy-5β(H)-steroids operative in the biosynthesis of bile acids from cholesterol by higher animals. Shidasterone is considered to be a stereoisomer of β-ecdysone (most probably the C-20 epimer) whose biosynthesis must involve some unusual mechanism near the junction of the nucleus and the side-chain.

Although nothing is known about the actual catabolism of ecdysterols in plants, it is most likely that rubrosterone (*40*), which has been so far found only in *Achyranthes* spp., is derived from the coexisting ecdysterols, β-ecdysone and inokosterone, through a degradation sequence analogous to that operating in the case of cholesterol (from 20(*R*),22(*R*)-dihydroxycholesterol to dehydroepiandrosterone) as shown in Chart 19.

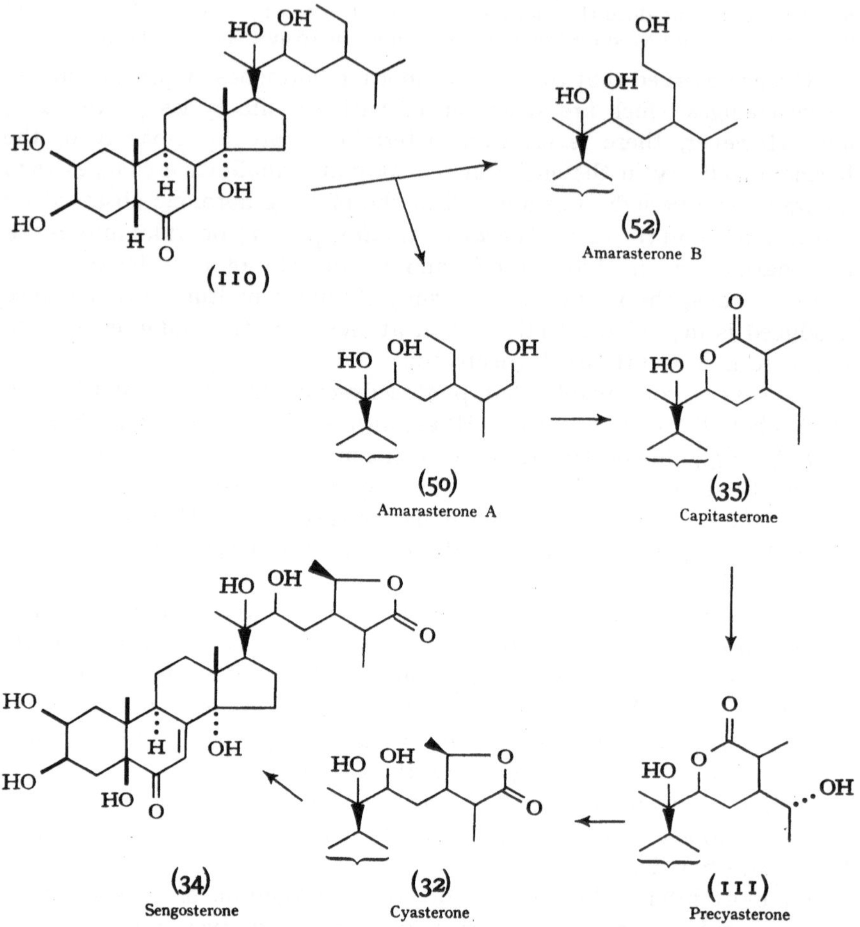

Chart 20. Probable Metabolic Pathway of Ecdysterols in *Cyathula capitata*

These hypotheses concerning the biosynthesis and catabolism of ecdysterols now await experimental verification.

Quite recently, Hikino *et al.* (*33*) showed that callus tissues derived from *Achyranthes* spp. synthesized the same ecdysterols, β-ecdysone and inokosterone, as those produced by the parent plants. This technique will provide a powerful tool for clarification of the biosynthetic-degradative scheme of the phytoecdysones.

IX. Tables

Table 1. *Animals Known to Contain Ecdysterols*

	Source	Ecdysterol	Yield	References
Insects				
Bombyx mori	pupa	α-ecdysone	25 mg/500 kg	(5)
			250 mg/ton	(74)
		β-ecdysone	2.5 mg/500 kg	(70)
			50 mg/2.8 ton (dried)	(41)
			9 mg/ton	(42)
Dociostaurus maroccanus	adult	α-ecdysone $\Big\}$ β-ecdysone	24 mg/10 kg	(113)
Antherea pernyi	pupa	crustecdysone ?	0.2 mg/31 kg	(54)
		α-ecdysone		(54)
Manduca sexta	pupa	α-ecdysone	3.7 mg/12.7 kg	(69)
		β-ecdysone	4.4 mg/12.7 kg	(69)
		20,26-dihydroxy-ecdysone	2 mg/40.2 kg	(140)
		tobacco hornworm ecdysone IV	1 mg/40.2 kg	(140)
Calliphora stygia	larva	β-ecdysone		(22)
Calliphora vicina (= C. erythro-cephala)	larva	β-ecdysone		(22)
Crustacea				
Jasus lalandei	adult (waste)	crustecdysone (= β-ecdysone ?)	2 mg/ton	(24, 52)
		deoxycrustecdy-sone	0.05 mg/ton	(15, 52)
Callinectes sapidus	adult (early premolt)	callinecdysone A (= inoko-sterone ?)	5 μg/kg	(6)
	adult (later premolt)	callinecdysone A (= inoko-sterone ?)	20 μg/kg	(6)
		β-ecdysone	4 μg/kg	(6)
	adult (after molt)	β-ecdysone	280 μg/kg	(6)
		callinecdysone B (= makisterone A ?)	24 μg/kg	(6)

Table 2. *Plant Families Containing Ecdysterols**

Pteridophyta	·	Polygonaceae	·
·		Chenopodiaceae	·
Lycopodiaceae	·	Amaranthaceae	·
Selaginellaceae	·	Nyctaginaceae	·
Psilotaceae	·	Cynocrambaceae	·
·		Phytolaccaceae	·
·	Triuridaceae	Aizoaceae	·
Marattiaceae	Gramineae	Portulacaceae	·
Osmundaceae	Cyperaceae	Basellaceae	·
Schizaeaceae	·	Caryophyllaceae	·
Gleicheniaceae	·	·	
Hymenophyllaceae	·	Ceratophyllaceae	Tiliaceae
Cyatheaceae	·	·	Malvaceae
·	·	·	
·	·	·	Stercuriaceae
Polypodiaceae	·	Ranunculaceae	·
Aspidiaceae	·	Lardizabalaceae	·
·	·	Berberidaceae	·
Blechnaceae	Stemonaceae	·	·
·	Liliaceae	Magnoliaceae	·
·	Amaryllidaceae	·	Tamaricaceae
Plagiogyriaceae	·	·	Cistaceae
Pteridaceae	Dioscoreaceae	·	Violaceae
·	Iridaceae	·	Flacourtiaceae
Parkeriaceae	Musaceae	·	Stachyuraceae
·	Zingiberaceae	Papaveraceae	Passifloraceae
·	·	Capparidaceae	·
·	·	Cruciferae	·
Gymnospermae	·	·	·
Cycadaceae	·	·	·
Ginkgoaceae	**Dicotyledoneae**	·	·
Podocarpaceae	**Archichlamydeae**	·	·
Cephalotaxaceae	·	·	·
Taxaceae	·	·	·
·	·	·	·
·	·	Platanaceae	·
·	·	Rosaceae	·
Cupressaceae	·	Leguminosae	·
·	·	·	
Taxodiaceae	·	Oxalidaceae	
Pinaceae	Ulmaceae	·	**Sympetalae**
Ephedraceae	Moraceae	·	·
·	Urticaceae	·	·
·	·	Zygophyllaceae	·
Angiospermae	·	Rutaceae	·
Monocotyledoneae	·	Simaroubaceae	·
·	·	·	·
·	·	·	·
·	·	·	·

* Mainly based on ref (*63*).

References, pp. 305—312

(Table 2, continued)

	Verbenaceae	.	.
.	Labiatae	.	.
.	Solanaceae	.	Cucurbitaceae
Polemoniaceae	Scrophulariaceae	.	Campanulaceae
.	.	.	.
Boraginaceae	.	.	Compositae

Table 3. *Plants Known to Contain Ecdysterols*

Plant	Source	Ecdysterol	Yield (%)	References
Osumundaceae				
Osumunda asiatica Ohwi	whole	α-ecdysone		(*123*)
		β-ecdysone		(*123*)
		ponasterone A		(*123*)
O. japonica Thunberg	whole	α-ecdysone		(*123*)
		β-ecdysone		(*123*)
		ponasterone A		(*123*)
Gleicheniaceae				
Gleichenia glauca Hooker	leaf	ponasterone A		*
Polypodiaceae				
Cryosinus hastatus Copel	whole	β-ecdysone		*
Lemmaphyllum micro- phyllum Presl	whole	α-ecdysone		(*117*)
		β-ecdysone		(*119*)
		lemmasterone		(*119*)
		pterosterone		(*117*)
Neocheiropteris ensata Ching	whole	α-ecdysone		*
		β-ecdysone		(*119*)
Pleopeltis thunbergiana Kaulf	whole	β-ecdysone		(*119*)
Polypodium japonicum Makino	rhizome	β-ecdysone	0.05	(*64*)
P. vulgare Linné	rhizome	α-ecdysone	0.0018	(*27*)
		β-ecdysone	> 1	(*65*)
			0.07	(*27*)
		polypodine B	0.04	(*66*)
Aspidiaceae				
Athyrium niponicum Hance	whole	β-ecdysone		*
		ponasterone A		*
		pterosterone		*

* Unpublished results in our laboratory.

(Table 3, continued)

Plant	Source	Ecdysterol	Yield (%)	References
A. yokoscense Christ	whole	β-ecdysone		*
Lastrea japonica Copeland	whole	β-ecdysone		*
L. thelypteris Bory	whole	β-ecdysone		*
		ponasterone A		*
		pterosterone		(*118*)
Matteuccia struthio-pteris Todaro	whole	β-ecdysone		(*118*)
		ponasterone A		*
Onoclea sensibilis Linné	whole	β-ecdysone		*
		ponasterone A		(*118*)
		pterosterone		(*118*)
Blechnaceae				
Blechnum amabile Makino	whole	β-ecdysone		(*139*)
		ponasterone A		(*139*)
B. niponicum Makino	whole	β-ecdysone		(*139*)
		ponasterone A		(*139*)
		shidasterone		(*126, 139*)
Pteridaceae				
Pteridium aquilinum Kuhn	leaf	α-ecdysone	0.000045	(*68*)
		β-ecdysone	0.0001	(*68*)
P. aquilinum Kuhn var. *latiusculum* Underwood	whole	ponasterone A		(*116*)
		ponasteroside A		(*114, 116*)
		pterosterone		(*116*)
Podocarpaceae				
Podocarpus elatus R. Brown	wood	β-ecdysone	0.001	(*11, 12*)
	bark	β-ecdysone	0.05	(*21*)
		lemmasterone		(*21*)
		makisterone A		(*17*)
		podecdysone B		(*21*)
		podecdysone C		(*21*)
		podecdysone E		(*17*)
		podecdysone F		(*17*)
P. macrophyllus Lambert	leaf	β-ecdysone	0.0007, 0.01	(*60, 61*)
		lemmasterone	0.0001	(*61*)
		makisterone A	0.001	(*61*)
		makisterone B	0.0001	(*61*)
		makisterone D	0.0001	(*61*)
		ponasterone A	0.038, 0.05	(*60, 61*)

* Unpublished results in our laboratory.

References, pp. 305—312

(Table 3, continued)

Plant	Source	Ecdysterol	Yield (%)	References
P. macrophyllus Lambert var. *maki* Siebold (= *P. chinensis* Sweet)	leaf	ponasterone A	0.03	*(60)*
P. nakaii Hayata	leaf	ponasterone A	0.04, 0.16	*(96, 95)*
		ponasterone B	0.001, 0.007	*(96, 95)*
		ponasterone C	0.01, 0.0017	*(96, 95)*
		ponasterone D	0.0004	*(96)*
P. neriifolius D. Don	leaf	ponasterone A		*(13)*
Taxaceae				
Taxus baccata Linné	leaf	β-ecdysone	0.03	*(44)*
T. cuspidata Siebold et Zuccarini	leaf	β-ecdysone	0.0014	*(60)*
		ponasterone A	0.045	*(60)*
T. cuspidata Siebold et Zuccarini var. *nana* Rehder	leaf	ponasterone A		*(124)*
Liliaceae				
Trillium smallii Maximowicz	rhizome	β-ecdysone	0.008	*(64)*
T. tschonoskii Maximowicz	rhizome	β-ecdysone	0.01	*(64)*
Moraceae				
Morus sp.	leaf	β-ecdysone	0.0001	*(134)*
		inokosterone	0.001	*(134)*
Amaranthaceae				
Achyranthes bidentata Blume	root	β-ecdysone		*(135)*
		inokosterone		*(135)*
A. fauriei Léveillé et Vaniot	root	β-ecdysone	0.01	*(130, 131)*
		inokosterone	0.01	*(130, 131)*
		rubrosterone		*
A. japonica Nakai	whole	β-ecdysone		*(135)*
		inokosterone		*(135)*
A. longifolia Makino	whole	β-ecdysone		*(137)*
		inokosterone		*(135)*

* Unpublished results in our laboratory.

(Table 3, continued)

Plant	Source	Ecdysterol	Yield (%)	References
A. obtusifolia Lamarck	root	β-ecdysone	0.025	(*138*)
		rubrosterone		*
A. rubrofusca Wight	root	β-ecdysone	0.049	(*137*)
		inokosterone		(*135*)
		rubrosterone		(*121, 137*)
Bosea yervamora Linné	root	β-ecdysone	0.02	(*136*)
Cyathula capitata Moquin-Tandon	root	amarasterone A		(*127*)
		amarasterone B		(*127*)
		capitasterone		(*128*)
		cyasterone	0.02	(*125*)
		isocyasterone		(*37*)
		sengosterone		(*36*)
C. prostrata Blume	whole	β-ecdysone		(*135*)

Stachyuraceae

Stachyurus praecox Siebold et Zuccarini	bark	β-ecdysone	0.06	(*64*)
		stachysterone A		(*85*)
		stachysterone B		(*85*)
		stachysterone C		(*85*)
		stachysterone D		(*85*)

Verbenaceae

Vitex megapotamica Moldenke	leaf	β-ecdysone	0.088	(*102*)
		inokosterone		(*100*)
		polypodine B		(*101*)
		pterosterone		(*101*)
		viticosterone E		(*101*)

Labiatae

Ajuga decumbens Thunberg	leaf	ajugasterone C	0.0015	(*62*)
	whole	cyasterone	0.008	(*64*)
		β-ecdysone	0.012	(*64*)
A. incisa Maximowicz	whole	ajugasterone A		(*58*)
		ajugasterone B		(*58*)
		ajugasterone D		(*85*)
		cyasterone	0.008	(*64*)
		β-ecdysone	0.012	(*64*)
A. japonica Miquel	leaf	ajugasterone C	0.01	(*62*)
		cyasterone	0.058	(*62*)
		β-ecdysone	0.2	(*62*)
A. nipponensis Makino	whole	cyasterone	0.08	(*64*)
		β-ecdysone	0.012	(*64*)

* Unpublished results in our laboratory.

References, pp. 305—312

Table 4. *Properties of Ecdysterols*

Name (Synonym)	Physical Properties		References	Molting Hormone Activity
	m. p.	$[\alpha]_D$		
α-Ecdysone (1) (Ecdysone)	235–237°	+ 58.5 ± 2° (E)	(5)	C: 0.0075 µg* (5), M: 0.005 µg* (67), S: 0.05 µg* (98), Y: 5 µg** (145)
β-Ecdysone (3) (20-Hydroxyecdysone,	237.5–239.5°	—	(41)	C: α × 1.25 (38), H: 25 µg* (104),
Ecdysterone,	234°	—	(43)	S: α × 2 (98), Y: α × 1/2 (145)
Isoinokosterone)	242° (dec)	+ 63.0° (M)	(130)	
20,26-Dihydroxy-ecdysone (7)	non-crystal.	—	(140)	M: α × 1/10–1/15 (140)
Deoxycrustecdysone (8)	non-crystal.	—	(15)	C: β × 1 (15)
Ponasterone A (9)	259–260° (dec)	+ 90° (M)	(96)	C: β × 1/4 (45), S: α × 3.2 (98), Y: α × 2 (145)
Ponasterone B (17)	non-crystal.	—	(95)	C: β × 1/30 (45), S: α × 1.5 (98), Y: α × 2 (145)
Ponasterone C (21)	270–272° (dec)	—	(95)	C: β × 1/10 (45), S: α × 1.4 (98), Y: α × 1 (145)
Inokosterone (23)	255° (dec) (SH)	+ 59.4° (M)	(130)	B: β × 1 (82), M: 0.05 µg* (82), S: α × 0.55 (98), Y: α × 1/2 (145)
Cyasterone (32)	164–166° (SH)	+ 64.5° (P)	(125)	S: α × 2.2 (98), Y: α × 20 (145)
Polypodine B (36)	254–257° (MH)	+ 93° (M)	(66)	C: α × 4 (66)
Pterosterone (45)	229–230° (MH)	− 7.4° (M)	(118, 115)	S: active (118)
Lemmasterone (48) (Makisterone C, Podecdysone A)	258–259° 263–265° (dec) 262–264° (dec)	— — —	(119) (59) (21)	S: active (119) H: 0.5–1 µg*** (59) C: β × 1 (21)
Rubrosterone (40)	244–245° (dec)	+ 119° (M)	(121, 122)	S: < β × 1/1000 (122), B: inactive (122)
Makisterone A (46)	263–265° (dec)	—	(61)	H: 0.5–1 µg*** (61)
Makisterone B (47)	172–173° (dec)	—	(59)	H: 0.5–1 µg*** (59)
Makisterone D (50)	non-crystal.	—	(59)	H: 0.5–1 µg*** (59)
Ponasteroside A (41)	278–279.5°	+ 28.5° (P)	(114)	S: active (114)

(Table 4, continued)

Name (Synonym)	Physical Properties		References	Molting Hormone Activity
	m. p.	[α]D		
Capitasterone (35)	234–235°	—	(128)	S: active (128)
Amarasterone A (49)	210–211°	—	(127)	S: active (127)
Amarasterone B (51)	284–285°	—	(127)	S: active (127)
Shidasterone (43)	257–258°	—	(126)	S: active (126)
Ajugasterone B (58)	240° (dec)	—	(58)	H: active (58)
Viticosterone E (52)	198–199°	—	(101)	C: α × 1/14 (101)
Podecdysone B (56)	125–127°	—	(17)	C: β × 1/5 (17)
Ajugasterone C (44)	non-crystal.	—	(62)	H: active (62)
Sengosterone (34)	159–161°	+ 39.6° (P)	(36)	S: active (36)
Stachysterone A (54)	non-crystal.	—	(85)	H: active (85)
Stachysterone B (55)	non-crystal.	—	(85)	H: active (85)
Stachysterone C (53)	235–240° (dec)	—	(85)	H: active (85)
Stachysterone D (57)	245–250° (dec)	—	(85)	H: inactive or slightly active (8
Isocyasterone (33)	non-crystal.	—	(37)	S: active (37)
Ajugalactone (16)	225–230° (dec)	—	(85)	H: anti-pona-sterone A

Abbreviations: SH = semihydrate, MH = monohydrate, M = methanol, E = ethan
P = pyridine, B = *Bombyx* test, C = *Calliphora* test, H = *Chilo* test, M = *Musca* te:
S = *Sarcophaga* test, Y = *Samia* test.

* Critical dose causing puparium formation of 50(–70)% of larval abdomens.
** Critical dose causing 50% of pupae to form normal moth.
*** Critical dose causing puparium formation of 100% of larval abdomens.

Table 5. *NMR Data of some Ecdysterols*

	Methyl Chemical Shifts (C₅D₅N, ppm from internal TMS)					References
	C-18	C-19	C-21	C-26	C-27	
α-Ecdysone (**1**)	0.74	1.07	1.28	1.38	1.38	(*47*)
Ponasterone A (**9**)	1.16	1.03	1.51	0.82	0.82	(*96*)
Ponasterone B (**17**)	1.17	1.11	1.54	0.82	0.82	(*95*)
Deoxycrustecdysone (**8**)	1.21	1.04	1.57	1.35	1.35	(*15*)
Ajugasterone C (**44**)	1.21	1.27	1.51	0.82	0.82	(*62*)
Pterosterone (**45**)	1.18	1.05	1.54	1.00	1.00	(*115*)
β-Ecdysone (**3**)	1.19	1.05	1.55	1.35	1.35	(*41*)
Inokosterone (**23**)	1.19	1.07	1.52	—	1.03	(*130*)
Shidasterone (**43**)	1.06	1.06	1.49	1.22	1.22	(*126*)
Polypodine B (**36**)	1.15	1.09	1.52	1.35	1.35	(*101*)
20,26-Dihydroxyecdysone (**7**)	1.22	1.08	1.58	—	1.48	(*140*)
Viticosterone E (**52**)	1.20	1.07	1.60	1.45	1.52	(*101*)
Stachysterone C (**53**)	1.17	1.05	1.55	1.60	1.66	(*85*)
Stachysterone B (**55**)	1.00/1.33	1.00/1.33	1.49	1.42	1.41	(*85*)
Podecdysone B (**56**)	1.35	1.08	1.55	1.43	1.43	(*17*)
Stachysterone D (**57**)	1.20	1.06	1.37	1.20	1.06	(*85*)
Stachysterone A (**54**)	0.92/1.20	0.92/1.20	1.51	1.41	1.41	(*85*)
Ponasteroside A (**41**)	1.17	0.88	1.54	0.84	0.84	(*114*)

Table 6. *ORD and CD Data of Ecdysterols**

| | ORD | | CD | | | |
| | π — π* | n — π* | π — π* | | n — π* | |
	a	a	nm	[θ]	nm	[θ]
5β(H)-Ecdysterols	−200 ~ −304	+ 42 ~ + 82	243	−10500 ~ −13700	338 ~ 340	+ 3380 ~ + 5300
5β(OH)-Ecdysterols	−240 ~ −268	+ 106 ~ + 119	245 ~ 251	−17900 ~ −18070	324 ~ 328	+ 7490 ~ + 9900
Ponasteroside A (41)	−288 (MeOH)	+ 107 (MeOH)	—	—	339	+ 4500
Ponasterone B (17)	−180	+ 57	—	—	—	—
Rubrosterone (40)	—	—	246	−41100	342	+ 4830
Ajugalactone (16)	—	—	244	+ 38200	340	+ 2840

* Accumulated from many ref's. Determined in dioxane unless specified to the contrary.

References

1. BECKER, E.: Über Versuche zur Anreicherung und physiologischen Charakterisierung des Wirkstoffs der Puparisierung. Biol. Zbl. **61**, 360 (1941).

2. BECKER, E. und E. PLAGGE: Über das die Pupariumbildung auslösende Hormon der Fliegen. Biol. Zbl. **59**, 326 (1939).

3. BURDETTE, W. J.: Changes in Titer of Ecdysone in *Bombyx mori* during Metamorphosis. Science **135**, 432 (1962).

4. BURDETTE, W. J. and M. W. BULLOCK: Ecdysone: Five Biologically Active Fractions from *Bombyx*. Science **140**, 1311 (1963).

5. BUTENANDT, A. und P. KARLSON: Über die Isolierung eines Metamorphose-Hormons der Insekten in kristallisierter Form. Z. Naturforsch. **9b**, 389 (1954).

6. FAUX, A., D. H. S. HORN, E. J. MIDDLETON, H. M. FALES and M. E. LOWE: Moulting Hormones of a Crab during Ecdysis. Chem. Commun. **1969**, 175.

7. FRAENKEL, G.: Hormone Causing Pupation in the Blowfly *Calliphora erythrocephala*. Proc. Roy. Soc. London Ser. B **118**, 1 (1935).

8. FURLENMEIER, A., A. FÜRST, A. LANGEMANN, G. WALDVOGEL, P. HOCKS, U. KERB and R. WIECHERT: The Synthesis of Ecdysone. Experientia **22**, 573 (1966).

9. — — — — — — Zur Synthese des Ecdysons. Helv. Chim. Acta **50**, 2387 (1967).

10. FURLENMEIER, A., A. FÜRST, A. LANGEMANN, G. WALDVOGEL, U. KERB, P. HOCKS und R. WIECHERT: Zur Synthese des Ecdysons. Synthesen von $2\beta,3\beta,14\alpha$-Trihydroxy-6-keto-Δ^7-A/B-*cis*-Steroiden. Helv. Chim. Acta **49**, 1591 (1966).

11. GALBRAITH, M. N. and D. H. S. HORN: An Insect-Moulting Hormone from a Plant. Chem. Commun. **1966**, 905.

12. — — Insect Moulting Hormones: Crustecdysone (20-Hydroxyecdysone) from *Podocarpus elatus*. Austral. J. Chem. **22**, 1045 (1969).

13. — — unpublished results.

14. GALBRAITH, M. N., D. H. S. HORN, P. HOCKS, G. SCHULZ and H. HOFFMEISTER: The Identity of the 20-Hydroxy-Ecdysones from Various Sources. Naturwiss. **54**, 471 (1967).

15. GALBRAITH, M. N., D. H. S. HORN, E. J. MIDDLETON and R. J. HACKNEY: Structure of Deoxycrustecdysone, a Second Crustacean Moulting Hormone. Chem. Commun. **1968**, 83.

16. — — — — A Silyl Ether as a Protecting Group in the Synthesis of an Isomer of Ecdysone, 20*S*-Hydroxy-22-deoxyecdysone. Chem. Commun. **1968**, 466.

17. — — — — The Structure of Podecdysone B, a New Phytoecdysone. Chem. Commun. **1969**, 402.

18. — — — — Moulting Hormones of Insects and Crustaceans: The Synthesis of 2-Deoxy-3-epicrustecdysone. Austral. J. Chem. **22**, 1059 (1969).

19. — — — — Moulting Hormones of Insects and Crustaceans: The Synthesis of 22-Deoxycrustecdysone. Austral. J. Chem. **22**, 1517 (1969).

20. GALBRAITH, M. N., D. H. S. HORN, E. J. MIDDLETON, J. A. THOMSON, J. B. SIDDALL and W. HAFFERL: The Catabolism of Crustecdysone in the Blowfly *Calliphora stygia*. Chem. Commun. **1969**, 1134.

21. GALBRAITH, M. N., D. H. S. HORN, Q. N. PORTER and R. J. HACKNEY: Structure of Podecdysone A, a Steroid with Moulting Hormone Activity from the Bark of *Podocarpus elatus* R. Br. Chem. Commun. **1968**, 971.

22. GALBRAITH, M. N., D. H. S. HORN, J. A. THOMSON, G. J. NEUFELD and R. J. HACKNEY: Insect Moulting Hormones: Crustecdysone in *Calliphora*. J. Insect Physiol. **15**, 1225 (1969).

23. Gilbert, L. I.: Hormones Controlling Reproduction and Molting in Invertebrates. Comparative Endocrinology, Vol. II. p. 1. New York: Academic Press. 1963.

24. Hampshire, F. and D. H. S. Horn: Structure of Crustecdysone, a Crustacean Moulting Hormone. Chem. Commun. 1966, 37.

25. Harrison, I. T., J. B. Siddall and J. H. Fried: Synthetic Studies on Insect Hormones. III. An Alternative Synthesis of Ecdysone and 22-Isoecdysone. Tetrahedron Letters 1966, 3457.

26. Heftmann, E., H. H. Sauer and R. D. Bennett: Biosynthesis of Ecdysterone from Cholesterol by a Plant. Naturwiss. 55, 37 (1968).

27. Heinrich, G. und H. Hoffmeister: Ecdyson als Begleitsubstanz des Ecdysterons in *Polypodium vulgare* L. Experientia 23, 995 (1967).

28. — — 5β-Hydroxyecdysteron, ein Pflanzensteroid mit Häutungshormonaktivität aus *Polypodium vulgare* L. Tetrahedron Letters 1968, 6063; cf. Addendum: *Ibid.* 1969, No. 30, ii.

29. Hikino, H., S. Arihara and T. Takemoto: Ponasteroside A, a Glycoside of Insect Metamorphosing Substance from *Pteridium aquilinum* var. *latiusculum*: Structure and Absolute Configuration. Tetrahedron 25, 3909 (1969).

30. Hikino, H., Y. Hikino, K. Nomoto and T. Takemoto: Cyasterone, an Insect Metamorphosing Substance from *Cyathula capitata*: Structure. Tetrahedron 24, 4895 (1968).

31. Hikino, H., Y. Hikino and T. Takemoto: Synthesis of Rubrosterone, a Metabolite of Insect-Moulting Substances from *Achyranthes rubrofusca*. Tetrahedron Letters 1968, 4255.

32. — — — Rubrosterone, a Metabolite of Insect Metamorphosing Substance from *Achyranthes rubrofusca*: Synthesis. Tetrahedron 25, 3389 (1969).

33. Hikino, H., H. Jin and T. Takemoto: unpublished results.

34. Hikino, H., T. Kohama and T. Takemoto: Biosynthesis of Ponasterone A, an Insect-Moulting Substance from *Podocarpus macrophyllus*. Chem. Pharm. Bull. (Japan) 17, 415 (1969).

35. — — — Biosynthesis of Ponasterone A and Ecdysterone from Cholesterol in *Podocarpus macrophyllus*. Phytochem. 9, 367 (1970).

36. Hikino, H., K. Nomoto and T. Takemoto: Structure of Sengosterone, a Novel C_{29} Insect-Moulting Substance from *Cyathula capitata*. Tetrahedron Letters 1969, 1417.

37. — — — unpublished results.

38. Hocks, P., A. Jäger, U. Kerb, R. Wiechert, A. Furlenmeier, A. Fürst, A. Langemann und G. Waldvogel: Synthetische Steroide mit Häutungshormonaktivität. Angew. Chem. 78, 680 (1966).

39. Hocks, P., U. Kerb, R. Wiechert, A. Furlenmeier und A. Fürst: Die Synthese des Rubrosterons. Tetrahedron Letters 1968, 4281.

40. Hocks, P., G. Schulz und P. Karlson: Die Struktur des β-Ecdysons. Naturwiss. 54, 44 (1967).

41. Hocks, P. und R. Wiechert: 20-Hydroxy-Ecdyson, Isoliert aus Insekten. Tetrahedron Letters 1966, 2989.

42. Hoffmeister, H. und H.-F. Grützmacher: Zur Chemie des Ecdysterons. Tetrahedron Letters 1966, 4017.

43. Hoffmeister, H., H.-F. Grützmacher und K. Dünnebeil: Untersuchungen über die Struktur und biochemische Wirkung von Ecdysteron. Z. Naturforsch. 22b, 66 (1967).

44. Hoffmeister, H., G. Heinrich, G. B. Staal und W. J. van der Burg: Über das Vorkommen von Ecdysteron in Eiben. Naturwiss. 54, 471 (1967).

45. HOFFMEISTER, H., K. NAKANISHI, M. KOREEDA and H. Y. HSU: The Moulting Hormone Activity of Ponasterones in the *Calliphora* Test. J. Insect Physiol. **14**, 53 (1968).

46. HOFFMEISTER, H. und C. RUFER: Zur Chemie des Ecdysons, IV. Farbreaktionen von gesättigten und ungesättigten Steroidketonen. Chem. Ber. **98**, 2376 (1965).

47. HOFFMEISTER, H., C. RUFER, H. H. KELLER, H. SCHAIRER und P. KARLSON: Zur Chemie des Ecdysons, III. Vergleichende spektrometrische Untersuchungen an α,β-ungesättigten Steroidketonen. Chem. Ber. **98**, 2361 (1965).

48. HOPPE, W. und R. HUBER: Zur Chemie des Ecdysons, II. Bestimmung des Sterin-Skeletts und seiner Orientierung mit diffuser Röntgenstreuung in Kristallen von Ecdyson. Chem. Ber. **98**, 2353 (1965).

49. HORA, J., L. LÁBLER, A. KASAL, V. ČERNÝ, F. ŠORM and K. SLÁMA: Molting Deficiencies Produced by some Sterol Derivatives in an Insect (*Pyrrhocoris apterus* L.). Steroids **8**, 887 (1966).

50. HORI, M.: Automatic Column Chromatographic Method for Insect-Moulting Steroids. Steroids **14**, 33 (1969).

51. HORN, D. H. S.: Moulting Hormones of Crustaceans. unpublished work.

52. HORN, D. H. S., S. FABBRI, F. HAMPSHIRE and M. E. LOWE: Isolation of Crustecdysone (20 R-Hydroxyecdysone) from a Crayfish (*Jasus lalandei* H. Milne-Edwards). Biochem. J. **109**, 399 (1968).

53. HORN, D. H. S., E. J. MIDDLETON and J. A. THOMSON: unpublished results.

54. HORN, D. H. S., E. J. MIDDLETON, J. A. WUNDERLICH and F. HAMPSHIRE: Identity of the Moulting Hormones of Insects and Crustaceans. Chem. Commun. **1966**, 339.

55. HUBER, R. und W. HOPPE: Zur Chemie des Ecdysons, VII. Die Kristall- und Molekülstruktur Analyse des Insektenverpuppungshormons Ecdyson mit der automatisierten Faltmolekülmethode. Chem. Ber. **98**, 2403 (1965).

56. HÜPPI, G. and J. B. SIDDALL: Synthetic Studies on Insect Hormones. V. The Synthesis of Crustecdysone (20-Hydroxyecdysone). J. Amer. Chem. Soc. **89**, 6790 (1967).

57. — — Synthetic Studies on Insect Hormones VI. The Synthesis of Ponasterone A and its Stereochemical Identity with Crustecdysone. Tetrahedron Letters **1968**, 1113.

58. IMAI, S., S. FUJIOKA, E. MURATA, K. OTSUKA and K. NAKANISHI: Structure of the Phytoecdysone, Ajugasterone B. Chem. Commun. **1969**, 82.

59. IMAI, S., S. FUJIOKA, E. MURATA, Y. SASAKAWA and K. NAKANISHI: The Structures of Three Additional Phytoecdysones from *Podocarpus macrophyllus*, Makisterone B, C and D. Tetrahedron Letters **1968**, 3887.

60. IMAI, S., S. FUJIOKA, K. NAKANISHI, M. KOREEDA and T. KUROKAWA: Extraction of Ponasterone A and Ecdysterone from Podocarpaceae and Related Plants. Steroids **10**, 557 (1967).

61. IMAI, S., M. HORI, S. FUJIOKA, E. MURATA, M. GOTO and K. NAKANISHI: Isolation of Four New Phytoecdysones, Makisterone A, B, C, D and the Structure of Makisterone A, a C_{28} Steroid. Tetrahedron Letters **1968**, 3883.

62. IMAI, S., E. MURATA, S. FUJIOKA, M. KOREEDA and K. NAKANISHI: Structure of Ajugasterone C, a Phytoecdysone with an 11-Hydroxy-group. Chem. Commun. **1969**, 546.

63. IMAI, S., T. TOYOSATO, M. SAKAI, Y. SATO, S. FUJIOKA, E. MURATA and M. GOTO: Screening Results of Plants for Phytoecdysones. Chem. Pharm. Bull. (Japan) **17**, 335 (1969).

64. — — — — — — — Isolation of Cyasterone and Ecdysterone from Plant Materials. Chem. Pharm. Bull. (Japan) **17**, 340 (1969).

65. Jizba, J., V. Herout and F. Šorm: Isolation of Ecdysterone (Crustecdysone) from *Polypodium vulgare* L. Rhizomes. Tetrahedron Letters **1967**, 1689.

66. — — — Polypodine B — a Novel Ecdysone-like Substance from Plant Material. Tetrahedron Letters **1967**, 5139.

67. Káplanis, J. A., L. A. Tabor, M. J. Thomson, W. E. Robbins and T. J. Shortino: Assay for Ecdysone (Molting Hormone) Activity Using the House Fly, *Musca domestica* L. Steroids **8**, 625 (1966).

68. Kaplanis, J. N., M. J. Thompson, W. E. Robbins and B. M. Bryce: Insect Hormones: Alpha Ecdysone and 20-Hydroxyecdysone in Bracken Fern. Science **157**, 1436 (1967).

69. Kaplanis, J. N., M. J. Thompson, R. T. Yamamoto, W. E. Robbins and S. J. Louloudes: Ecdysones from the Pupa of the Tobacco Hornworm, *Manduca sexta* (Johannson). Steroids **8**, 605 (1966).

70. Karlson, P.: Biochemical Studies on Insect Hormones. Vitamins and Hormones **14**, 227 (1956).

71. e. g., — Ecdyson das Häutungshormon der Insekten. Naturwiss. **53**, 445 (1966).

72. Karlson, P. and G. Hauser: Bildungsort und Erfolgsorgan des Puparisierungshormons der Fliegen. Z. Naturforsch. **8 b**, 91 (1953).

73. Karlson, P. and H. Hoffmeister: Zur Biologenese des Ecdysons, I. Umwandlung von Cholesterin in Ecdyson. Z. Physiol. Chem. **331**, 298 (1963).

74. Karlson, P., H. Hoffmeister, W. Hoppe und R. Huber: Zur Chemie des Ecdysons, I. Liebigs Ann. Chem. **662**, 1 (1963).

75. Karlson, P., H. Hoffmeister, H. Hummel, P. Hocks und G. Spiteller: Zur Chemie des Ecdysons, VI. Reaktionen des Ecdysonmoleküls. Chem. Ber. **98**, 2394 (1965).

76. Karlson, P. and D. M. Skinner: Attempted Extraction of Crustacean Moulting Hormone from Isolated Y-Organs. Nature **185**, 543 (1960) and earlier references cited therein.

77. Kerb, U., P. Hocks, R. Wiechert, A. Furlenmeier, A. Fürst, A. Langemann und G. Waldvogel: Die Synthese des Ecdysons. Tetrahedron Letters **1966**, 1387.

78. Kerb, U., G. Schulz, P. Hocks, R. Wiechert, A. Furlenmeier, A. Fürst, A. Langemann und G. Waldvogel: Zur Synthese des Ecdysons. Die Synthese des natürlichen Häutungshormons. Helv. Chim. Acta **49**, 1601 (1966).

79. Kerb, U., R. Wiechert, A. Furlenmeier und A. Fürst: Über eine Synthese des Crustecdysons (20-Hydroxyecdyson). Tetrahedron Letters **1968**, 4277.

80. King, D. S. and J. B. Siddall: Conversion of α-Ecdysone to β-Ecdysone by Crustaceans and Insects. Nature **221**, 955 (1969).

81. Kobayashi, M., K. Nakanishi and M. Koreeda: The Moulting Hormone Activity of Ponasterones on *Musca domestica* (Diptera) and *Bombyx mori* (Lepidoptera). Steroids **9**, 529 (1967).

82. Kobayashi, M., T. Takemoto, S. Ogawa and N. Nishimoto: The Moulting Hormone Activity of Ecdysterone and Inokosterone Isolated from Achyranthis Radix. J. Insect Physiol. **13**, 1395 (1967).

83. Koreeda, M., K. Nakanishi, S. Imai, T. Tsuchiya and N. Wasada: Mass Spectrometric Studies of Ecdysone Derivatives. Mass Spectroscopy **17**, 669 (1969).

84. Koreeda, M., N. Harada and K. Nakanishi: Interaction between Nonadjacent Benzoate Groups: an Extention of the Dibenzoate Chirality Rule. Chem. Commun. **1969**, 548.

85. KOREEDA, M., H. MORIYAMA, K. NAKANISHI, S. IMAI and T. OKAUCHI: Bioorganic Studies of the Ecdysones. Symposium Papers of the 12th Symposium on the Chemistry of Natural Products, p. 129, Sapporo 1969.

86. MORI, H. and K. SHIBATA: Synthesis of Ecdysterone. Chem. Pharm. Bull. (Japan) 17, 1970 (1969).

87. MORI, H., K. SHIBATA, K. TSUNEDA and M. SAWAI: Synthesis of Ecdysone. I. A Novel Synthesis of 2β,3β-Dihydroxy Steroids. Chem. Pharm. Bull. (Japan) 15, 460 (1967),

88. — — — — Synthesis of Ecdysone. Chem. Pharm. Bull. (Japan) 16, 563 (1968).

89. — — — — Synthesis of Ecdysone. III. A Novel Synthesis of 14α-Hydroxy-7-en-6-oxo Steroids. Chem. Pharm. Bull (Japan) 16, 1593 (1968).

90. — — — — Synthesis of Ecdysone. V. Synthesis of 22-Isoecdysone. Chem. Pharm. Bull. (Japan) 16, 2416 (1968).

91. — — — — Synthesis of Ecdysone. IV. A Novel Synthesis of the Side Chain Structure of Ecdysone. Chem. Pharm. Bull. (Japan) 17, 690 (1969).

92. MORI, H., K. SHIBATA, K. TSUNEDA, M. SAWAI and K. TSUDA: The Configuration of 22-Hydroxycholesterol. Chem. Pharm. Bull. (Japan) 16, 1407 (1968).

93. MORI, H., K. TSUNEDA, K. SHIBATA and M. SAWAI: Synthesis of Ecdysone. II. Synthesis of and Stereochemistry of 2β,3β-Dihydroxycholestan-6-ones and Their Derivatives. Chem. Pharm. Bull. (Japan) 15, 466 (1967).

94. MORIYAMA, H. and K. NAKANISHI: Insect Hormones. VI. Confirmation of the Skeletal Structure of Ponasterone A. Tetrahedron Letters 1968, 1111.

95. NAKANISHI, K., M. KOREEDA, M. L. CHANG and H. Y. HSU: Insect Hormones. V. The Structures of Ponasterones B and C. Tetrahedron Letters 1968, 1105.

96. NAKANISHI, K., M. KOREEDA, S. DASAKI, M. L. CHANG and H. Y. HSU: Insect Hormones. I. The Structure of Ponastereone A, an Insect-Moulting Hormone from the Leaves of Podocarpus nakaii Hay. Chem. Commun. 1966, 915.

97. OHTAKI, T.: On the Delayed Pupation of the Fleshfly, Sarcophaga peregrina. Jap. J. Med. Sci. Biol. 19, 97 (1966).

98. OHTAKI, T., R. D. MILKMAN and C. M. WILLIAMS: Ecdysone and Ecdysone Analogues: Their Assay on the Fleshfly Sarcophaga peregrina. Proc. Nat. Acad. Sci. (USA) 58, 981 (1967).

99. — — — Dynamics of Ecdysone Secretion and Action in the Fleshfly Sarcophaga peregrina. Biol. Bull. 135, 322 (1968).

100. RIMPLER, H.: Steroide mit ecdysonartiger Wirkung aus Vitex megapotamica. Pharmaz. Ztg. 48, 1799 (1967).

101. — Pterosteron, Polypodin B und ein neues ecdysonartiges Steroid (Viticosteron E) aus Vitex megapotamica (Verbenaceae). Tetrahedron Letters 1969, 329.

102. RIMPLER, H. und G. SCHULZ: Vorkommen von 20-Hydroxy-ecdyson in Vitex megapotamica. Tetrahedron Letters 1967, 2033.

103. RUFER, C., H. HOFFMEISTER, H. SCHAIRER und M. TRAUT: Zur Chemie des Ecdysons, V. Versuche zur Darstellung von Steroid-Δ⁷-6-ketonen mit 3.4-Diol-Gruppierung. Chem. Ber. 98, 2383 (1965).

104. SATO, Y., M. SAKAI, S. IMAI and S. FUJIOKA: Ecdysone Activity of Plant-originated Molting Hormones Applied on the Body Surface of Lepidopterous Larvae. App. Ent. Zool. 3, 49 (1968).

105. SAUER, H. H., R. D. BENNETT and E. HEFTMANN: Ecdysterone Biosynthesis in Podocarpus elata. Phytochem. 7, 2027 (1968).

106. SHAAYA, E. und P. KARLSON: Der Ecdysontiter während der Insektenentwicklung II. Die postembryonale Entwicklung der Schmeißfliege Calliphora erythrocephala Meig. J. Insect Physiol. 11, 65 (1965).

107. SHAAYA, E. und P. KARLSON: Der Ecdysontiter während der Insektenentwicklung IV. Die Entwicklung der Lepidopteren *Bombyx mori* L. und *Cerura vinula* L. Developmental Biology 11, 424 (1965) and previous references cited therein.

108. SHIBATA, K. and H. MORI: Synthesis of Rubrosterone. Chem. Pharm. Bull. (Japan) 16, 1404 (1968).

109. SIDDALL, J. B., A. D. CROSS and J. H. FRIED: Synthetic Studies on Insect Hormones. II. The Synthesis of Ecdysone. J. Amer. Chem. Soc. 88, 862 (1966).

110. SIDDALL, J. B., D. H. S. HORN and E. J. MIDDLETON: Synthetic Studies on Insect Hormones. The Synthesis of a Possible Metabolite of Crustecdysone (20-Hydroxyecdysone). Chem. Commun. 1967, 899.

111. SIDDALL, J. B., J. P. MARSHALL, A. BOWERS, A. D. CROSS, J. A. EDWARDS and J. H. FRIED: Synthetic Studies on Insect Hormones. I. Synthesis of the Tetracyclic Nucleus of Ecdysone. J. Amer. Chem. Soc. 88, 379 (1966).

112. STAAL, G. B.: Plants as a Source of Insect Hormones. Proc. Nederl. Akad. Wetenschappen (Amsterdam) Series C 70, 409 (1967).

113. STAMM, M. D.: Estudios sobre Hormonas de Invertebrados II. Aislamiento de Hormonas de la Metamorfosis en el Ortoptero *Dociostaurus maroccanus.* Anales Real. Soc. Espan. Fis. Quim. (Madrid) 55 B, 171 (1959).

114. TAKEMOTO, T., S. ARIHARA and H. HIKINO: Structure of Ponasteroside A, a Novel Glycoside of Insect-Moulting Substance from *Pteridium aquilinum* var. *latiusculum.* Tetrahedron Letters 1968, 4199.

115. TAKEMOTO, T., S. ARIHARA, Y. HIKINO and H. HIKINO: Structure of Pterosterone, a Novel Insect-Moulting Substance from *Lastrea thelypteris* and *Onoclea sensibilis.* Tetrahedron Letters 1968, 375.

116. — — — — Isolation of Insect Moulting Substances from *Pteridium aquilinum* var. *latiusculum.* Chem. Pharm. Bull. (Japan) 16, 762 (1968).

117. TAKEMOTO, T., Y. HIKINO, T. ARAI and H. HIKINO: Structure of Lemmasterone, a Novel C_{29} Insect-Moulting Substance from *Lemmaphyllum microphyllum.* Tetrahedron Letters 1968, 4061.

118. TAKEMOTO, T., Y. HIKINO, T. ARAI, M. KAWAHARA, C. KONNO, S. ARIHARA and H. HIKINO: Isolation of Insect Moulting Substances from *Matteuccia struthiopteris, Lastrea thelypteris,* and *Onoclea sensibilis.* Chem. Pharm. Bull. (Japan) 15, 1816 (1967).

119. TAKEMOTO, T., Y. HIKINO, T. ARAI, C. KONNO, S. NABETANI and H. HIKINO: Isolation of Insect Moulting Substances from *Pleopeltis thunbergiana, Neocheiropteris ensata,* and *Lemmaphyllum microphyllum.* Chem. Pharm. Bull. (Japan) 16, 759 (1968).

120. TAKEMOTO, T., Y. HIKINO, S. ARIHARA, H. HIKINO, S. OGAWA and N. NISHIMOTO: Absolute Configuration of Inokosterone, an Insect-Moulting Substance from *Achyranthes fauriei.* Tetrahedron Letters 1968, 2475.

121. TAKEMOTO, T., Y. HIKINO, H. HIKINO, S. OGAWA and N. NISHIMOTO: Structure of Rubrosterone, a Novel C_{19} Metabolite of Insect-Moulting Substances from *Achyranthes rubrofusca.* Tetrahedron Letters 1968, 3053.

122. — — — — Rubrosterone, a Metabolite of Insect Metamorphosing Substances from *Achyranthes rubrofusca:* Structure and Absolute Configuration. Tetrahedron 25, 1241 (1969).

123. TAKEMOTO, T., Y. HIKINO, H. JIN, T. ARAI and H. HIKINO: Isolation of Insect Moulting Substances from *Osmunda japonica* and *Osmunda asiatica.* Chem. Pharm. Bull. (Japan) 16, 1636 (1968).

124. TAKEMOTO, T., Y. HIKINO, H. JIN and H. HIKINO: Isolation of Ponasterone A from *Taxus cuspidata* var. *nana.* Yakugaku Zasshi 88, 359 (1968).

125. TAKEMOTO, T., Y. HIKINO, K. NOMOTO and H. HIKINO: Structure of Cyasterone, a Novel C_{29} Insect-Moulting Substance from *Cyathula capitata*. Tetrahedron Letters 1967, 3191.

126. TAKEMOTO, T., Y. HIKINO, T. OKUYAMA, S. ARIHARA and H. HIKINO: Structure of Shidasterone, a Novel Insect-Moulting Substance from *Blechnum niponicum*. Tetrahedron Letters 1968, 6095.

127. TAKEMOTO, T., K. NOMOTO and H. HIKINO: Structure of Amarasterone A and B, Novel C_{29} Insect-Moulting Substances from *Cyathula capitata*. Tetrahedron Letters 1968, 4953.

128. TAKEMOTO, T., K. NOMOTO, Y. HIKINO and H. HIKINO: Structure of Capitasterone, a Novel C_{29} Insect-Moulting Substance from *Cyathula capitata*. Tetrahedron Letters 1968, 4929.

129. TAKEMOTO, T., S. OGAWA, M. MORITA, N. NISHIMOTO, K. DOME and K. MORISHIMA: Studies on the Constituents of Achyranthis Radix. VI. Determination of Insect-Moulting Hormones. Yakugaku Zasshi **88**, 39 (1968).

130. TAKEMOTO, T., S. OGAWA and N. NISHIMOTO: Isolation of the Moulting Hormones of Insects from Achyranthis Radix. Yakugaku Zasshi **87**, 325 (1967).

131. — — — Studies on the Constituents of Achyranthis Radix. II. Isolation of the Insect-Moulting Hormones. Yakugaku Zasshi **87**, 1469 (1967).

132. — — — Studies on the Constituents of Achyranthis Radix. III. Structure of Inokosterone. Yakugaku Zasshi **87**, 1474 (1967).

133. TAKEMOTO, T., S. OGAWA, N. NISHIMOTO, S. ARIHARA and K. BUE: Insect Moulting Activity of Crude Drugs and Plants. I. Yakugaku Zasshi **87**, 1414 (1967).

134. TAKEMOTO, T., S. OGAWA, N. NISHIMOTO, H. HIRAYAMA and S. TANIGUCHI: Isolation of the Insect-Moulting Hormones from Mulberry Leaves. Yakugaku Zasshi **87**, 748 (1967).

135. — — — — — Studies on the Constituents of Achyranthis Radix. VII. The Insect-Moulting Substances of *Achyranthes* and *Cyathula*. Supplement. Yakugaku Zasshi **88**, 1293 (1968).

136. TAKEMOTO, T., S. OGAWA, N. NISHIMOTO and H. HOFFMEISTER: Steroide mit Häutungshormon-Aktivität aus Tieren und Pflanzen. Z. Naturforsch. **22 b** 681 (1967).

137. TAKEMOTO, T., S. OGAWA, N. NISHIMOTO and S. TANIGUCHI: Studies on the Constituents of Achyranthis Radix. IV. Isolation of the Insectmoulting Hormones from Formosan *Achyranthes* spp. Yakugaku Zasshi **87**, 1478 (1967).

138. TAKEMOTO, T., S. OGAWA, N. NISHIMOTO, K.-Y. YEN, K. ABE, T. SATO, K. OSAWA and M. TAKAHASHI: The Isolation of Ecdysterone from the Radix of *Achyranthes obtusifolia* Lam. Yakugaku Zasshi **87**, 1521 (1967).

139. TAKEMOTO, T., T. OKUYAMA, S. ARIHARA, Y. HIKINO and H. HIKINO: Isolation of Insect Moulting Substances from *Blechnum amabile* and *Blechnum niponicum*. Chem. Pharm. Bull. (Japan) **17**, 1973 (1969).

140. THOMPSON, M. J., J. N. KAPLANIS, W. E. ROBBINS and R. T. YAMAMOTO: 20,26-Dihydroxyecdysone, a New Steroid with Moulting Hormone Activity from the Tobacco Hornworm, *Manduca sexta* (Johannson). Chem. Commun. **1967** 650.

141. THOMSON, J. A., J. B. SIDDALL, M. N. GALBRAITH, D. H. S. HORN and E. J. MIDDLETON: The Biosynthesis of Ecdysones in the Blowfly *Calliphora stygia*. Chem. Commun. 1969, 669.

142. VELGOVÁ, H., L. LÁBLER, V. ČERNY, F. ŠORM and K. SLÁMA: Some Further Compounds Producing Molting Deficiencies in an Insect. Collect. Czech. Chem. Comm. **33**, 242 (1968).

143. WIECHERT, R., U. KERB, P. HOCKS, A. FURLENMEIER, A. FÜRST, A. LANGE-
MANN, und G. WALDVOGEL: Zur Synthese des Ecdysons. Synthesen von
2β,3β-Dihydroxy-6-keto-A/B-*cis*-Steroiden. Helv. Chim. Acta **49**, 1581 (1966).

144. WILLIAMS, C.: The Juvenile Hormone, I. Endocrine Activity of the Corpora
allata of the Adult Cecropia Silkworm. Biol. Bull. **116**, 323 (1959) and earlier
references cited therein.

145. — Ecdysone and Ecdysone-analogues: Their Assay and Action on Diapausing
Pupae of the Cynthia Silkworm. Biol. Bull. **134**, 344 (1968).

(Received, November 9, 1969)

Total Synthesis of Prostaglandins

By J. E. PIKE, Kalamazoo, Michigan

With 1 Figure

Contents

I. Introduction

The prostaglandins are a family of naturally occurring hydroxy fatty acids found widely distributed in mammalian tissues (*12, 13, 33*). However the tissue levels are very low and the richest source, human seminal plasma, contains only up to 300 μg/ml of several prostaglandins. Other organs typically yield about 1 μg/g of prostaglandins based on wet tissue weight. Large scale biosynthesis from polyunsaturated fatty acids has been used to obtain sufficient amounts of various prostaglandins to study their chemistry and biology, as well as to allow preliminary clinical evaluation (*23, 53*). With the increasing interest of these hormone-like agents in fertility control, labor induction and renal physiology, to name only a few of the areas of interest, the chances of a practical medical use appear promising. Since in the biosynthetic route limitations are apparent on the large scale availability of enzyme, chemical total synthesis offers the best solution, both for larger scale needs and for the synthesis of structural analogs. Also some of the desired analogs may not be available by bioconversion of unsaturated fatty acids because of the reported limitations of the enzyme specificity (26, 45).

Before discussing in detail the several synthetic routes which have been employed, a brief review of the general chemistry will be given.

II. Structure and Chemical Transformations of the Prostaglandins

The structures of the prostaglandins are on firm ground as a result of the pioneering work of BERGSTRÖM and his associates in Sweden (12). The absolute configuration has been established for prostaglandin E_1 (47) and more recently for the 19-hydroxy derivatives (30).

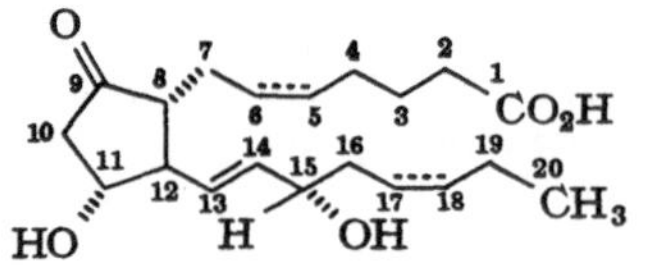

(1) Prostaglandin E_1 (PGE$_1$)
(5,6-dihydro-17,18-dihydro)

(2) Prostaglandin E_2 (PGE$_2$)
(5,6-cis double bond, 17,18-dihydro)

(3) Prostaglandin E_3 (PGE$_3$)
(5,6-cis double bond, 17,18-cis double bond)

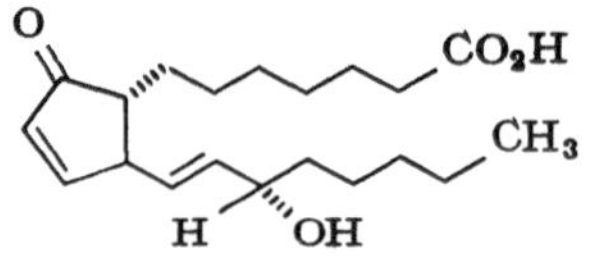

(4) Prostaglandin A_1

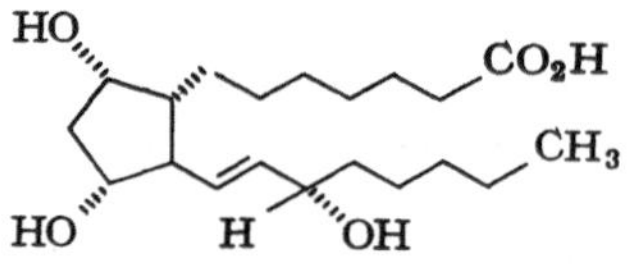

(5) Prostaglandin $F_{1\alpha}$

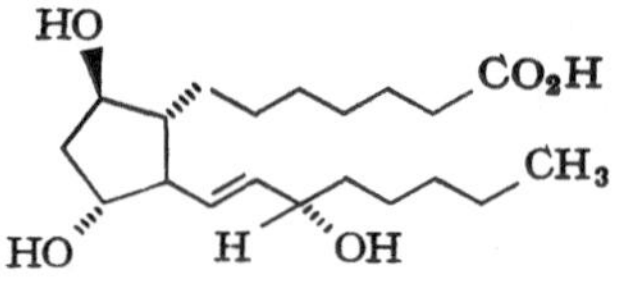

(6) Prostaglandin $F_{1\beta}$

Fig. 1

Structures of the three main classes of prostaglandins are shown in Fig. 1. The numbering system is indicated in the case of the PGE series. Prostaglandin E_1 has one *trans* double bond at the 13,14 position; PGE$_2$ has an additional 5,6-*cis* double bond, and the PGE$_3$ has *cis* double bonds additionally both at the 5,6 and 17,18 positions. The PGA compounds are Δ^{10}-unsaturated ketones obtained by mild acid-catalyzed dehydration of the β-hydroxy-ketone system characteristic of the PGE series (53). The PGF structures are the two hydroxy epimers formed on reduction of the PGE 9-ketone; only the α-isomer is naturally occurring. Stereo-chemistry is designated α or β by analogy with similar steroid practice;

α-substituents are oriented on the same side of the cyclopentane ring as the carboxy side chain (i. e. below in the usual representation) at C-8, and β-substituents are above the plane of the ring and on the same side as the "alkyl" (C_{13}–C_{20}) side chain. The 15-hydroxyl has been established as having the (15-S) stereochemistry in the natural mammalian prostaglandins and this has been also designated as 15α. The other hydroxy epimer at C-15 is therefore (15-R), 15β or 15-*epi*. A more complete discussion of the earlier chemical transformations and identification is given in the review by SAMUELSSON (58) and more details of the somewhat confusing earlier nomenclature are given in the joint Worcester Foundation-Upjohn review (55), which also contains a compilation of analytical data on natural prostaglandins. An earlier review of total synthesis has appeared (3). Also two recent papers have discussed chromatography and dehydration studies of the prostaglandins (1, 2).

Treatment of either the PGE or PGA compounds with base gives the doubly conjugated ketones, with a characteristic U. V. absorption at 278 nm and named PGB derivatives. Thus, PGE_1 (1) with sodium hydroxide yields PGB_1 (7). Epimerization of the C-15-hydroxyl can be effected under acidic conditions (formic acid at 0° C.) to give a mixture of the (15 R) and (15 S) derivatives (52).

Treatment of PGE_1 (1) under very mild basic conditions with potassium acetate in methanol at room temperature causes epimerization at C-8 to give a low, isolable, yield of 8-*iso*-PGE_1 (24). However the equilibrium appears greatly in favor of the natural isomer. Other chemical transformations which have been described are: reduction of the carboxyl to an alcohol with lithium aluminum hydride: selective protection of the 9-ketone in PGE_1 as an oxime, and regeneration of the ketone with nitrous acid (23, 52): esterification of the carboxyl as a trichloroethyl ester which could be selectively removed by treatment with zinc and acetic acid (52).

The instability of the natural prostaglandins both to acidic and basic treatment provides a considerable challenge to the synthetic chemist who must devise approaches in which the final steps are conducted under carefully controlled, mild conditions. To date, no methods have been published for converting either the PGA or PGF compounds back to the parent 11-hydroxy-9-ketones and a total synthesis which is general for

prostaglandins should therefore be directed toward the PGE series.
For purposes of classification the various approaches which started with
the Upjohn and Ayerst publications in 1966 will not be treated chrono-
logically. Instead they are divided into, first, the routes which have led
generally to natural prostaglandins, second those approaches which have
led to prostaglandins in which certain structural features are missing
(e. g. 13,14 double bond, or 15-hydroxyl present as 15-ketone), or where
the approach has been directed toward a simpler objective such as PGB_1
which has only one asymmetric center. Finally, synthetic approaches
will be discussed which have as their initial goal preparation of analogs
(11-deoxy, 7-oxa). A separate discussion will cover resolution and the
significance of the unnatural antipodes.

III. General Approaches to Prostaglandin Synthesis

At present only four of the published approaches appear to qualify
as generally applicable routes which have led to natural prostaglandins
of the E series. These are the three routes carried out in a series of studies
at Harvard by E. J. COREY and his associates, and the bicyclohexane
route devised at McGill University by JUST and SIMONOVITCH and
successfully carried out and further extended at the Upjohn Laboratories.
The nonenzymatic conversion of polyunsaturated fatty acids to PGE_1-
like products described by VAN DORP and co-workers in the Unilever
Laboratories (46) appears general but at present the yields are apparently
too low for this to be a useful route.

A. Initial Corey Synthesis of dl-PGE₁

A Diels-Alder condensation of the diene (8), prepared from 2-bromo-
methyl-1,3-butadiene and 2-lithio-2-n-amyl-1,3-dithiane, and the nitroolefin (9) gave in good yield the adduct (10) (17). Reduction of the nitro
group, followed by formylation of the resulting amine (formic: acetic
anhydride) and ketal exchange in the presence of mercuric chloride gave
the ethylene glycol ketal (11). Hydroxylation of the double bond and
cleavage of the glycol with lead tetraacetate gave the keto-aldehyde (12).
The key ring closure of the 5-membered ring was effected by an aldol
condensation using 0.1 equivalents of 1,5-diazabicyclo[4,3,0]-5-nonene
(48) in methylene chloride at 0° for 24 hrs., and this was followed by
acetylation to give, after chromatography a 45% yield of the cyclopentane derivative (13), together with an epimeric by-product believed
to be the 11β-isomer. Reduction of the ketone (13), followed by ketal
hydrolysis gave the β-hydroxy ketone (14) which was dehydrated using
dicyclohexylcarbodiimide in ether and cupric chloride as catalyst. This
novel method for the dehydration of β-hydroxy carbonyl compounds

under mild non-acidic, non-basic conditions is a general procedure which should find wider application. Selective reduction of the 15-carbonyl function using zinc borohydride gave a mixture of hydroxy epimers; removal of the acetate group with mild base followed by protection of the two hydroxy groups as tetrahydropyranyl ethers and strong base hydrolysis gave the amino acid (17) again as a mixture of the two 15-isomers. N-Bromination of (17) with NBS followed by base catalyzed dehydrobromination gave the 9-imine which was hydrolyzed at pH 2. Chromatography gave dl-PGE$_1$ (18), m. p. 112.8–113.1°, with the correct

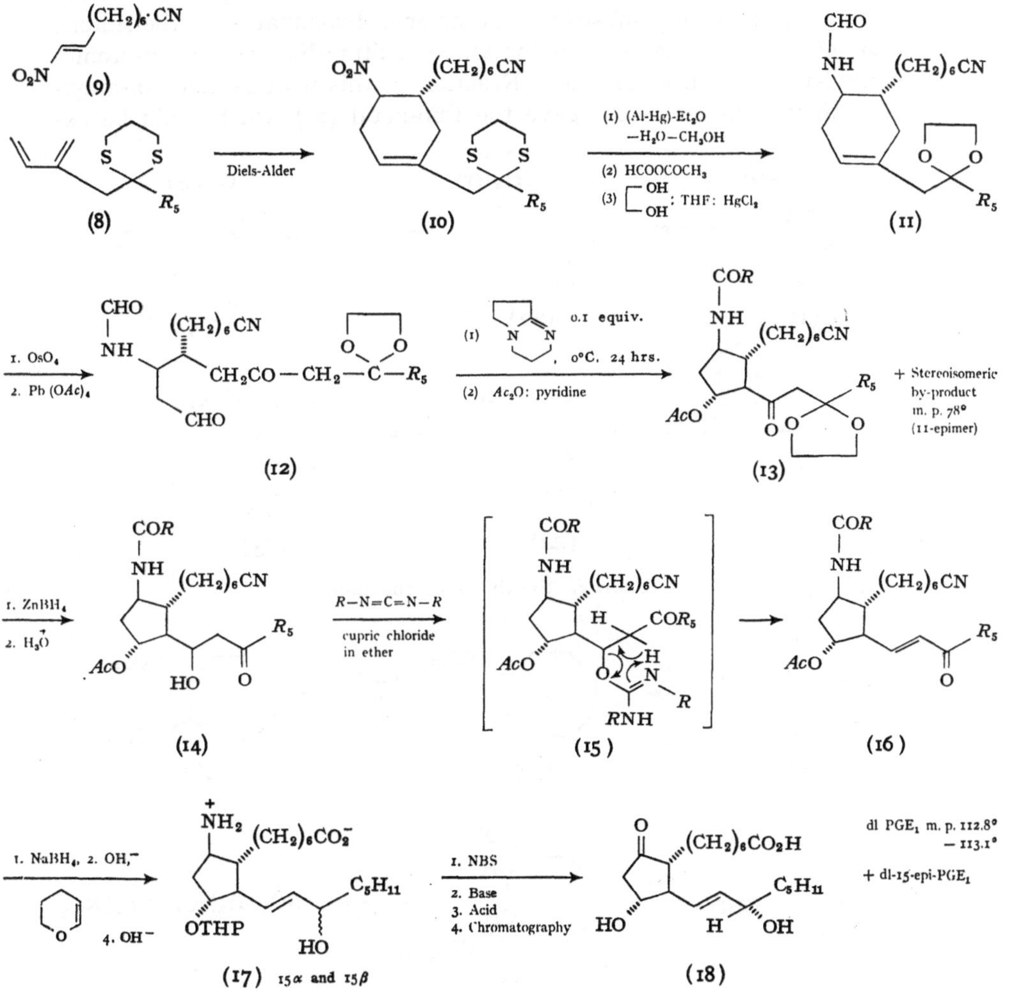

Chart 1. Corey's Initial Synthesis of PGE$_1$ (17)

analytical and biological properties, together with dl-15-*epi* PGE$_1$. The dl-PGE$_1$ was converted to dl-PGA$_1$ and also to dl-PGF$_1\alpha$ and dl-PGF$_1\beta$. A modification of the earlier method (90% acetic acid: 60°) (*53*) for the PGE$_1$ to PGA$_1$ conversion was used involving 0.5 N hydrochloric acid in 1:1 water-tetrahydrofuran (60 hr., 25°); this appears preferable since the earlier route did give some acetylation as a side reaction.

Various features of this elegantly conceived synthesis are worthy of special comment. Some other uses may be found for some of the general ideas. The preparation of the diene component (8) used in the Diels-Alder reaction illustrates an interesting solution to the problem of obtaining a functionally substituted butadiene. Bromination of the sulfone (19) followed by thermal elimination of sulfur dioxide gave 2-bromo-methyl-1,3-butadiene (21) (*40*). Reaction of this with 2-lithio-2-*n*-amyl-1,3-dithiane (*18*) at − 25° gave the thio-ketal (22) which could be ex-

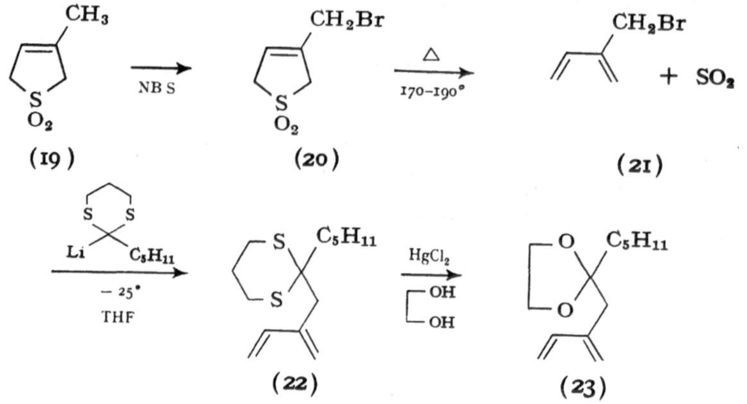

Chart 2. Synthesis of the diene

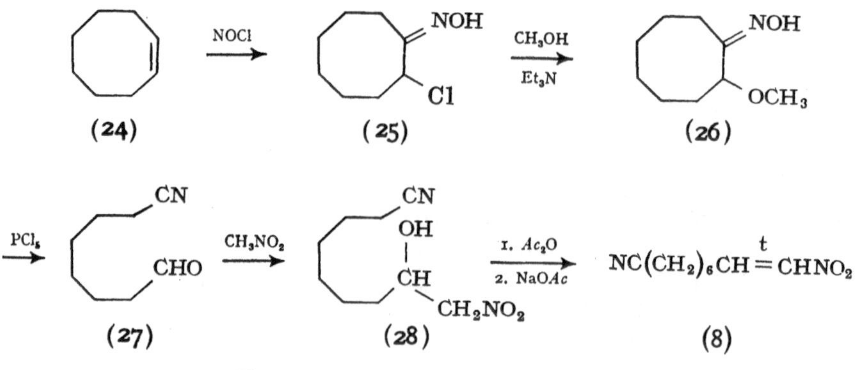

Chart 3. Synthesis of the dieneophile

changed with ethylene glycol in the presence of mercuric chloride to the dioxolane (*23*). The dienophile (*9*) was synthesized from 7-cyanoheptanal (*27*) obtained from the nitrosyl chloride adduct (*25*) of cyclooctene (*49*). Addition of nitromethane under basic conditions gave the nitroalcohol (*28*) which was acetylated and converted to the nitroolefin (*8*) using sodium bicarbonate in ethyl acetate. The two key strategic concepts in the synthesis were the use of a primary amine as the potential 9-ketone, and the formation of the five membered ring by an aldol cyclization, giving a product containing the functional equivalent of the C_{11} to C_{15} enediol unit, namely the mono-ketal of a 1,4-diketone. The need to convert a primary amine to a ketone also led to the development of a new general method for this step in addition to the Ruschig procedure (*57*) (used in the synthesis) or the procedure using argentic picolinate (*5*). This new method (*16*) is illustrated in the case of cyclopentylamine which when treated with 3,5-di-*t*-butyl-1,2-benzoquinone (*30*) at 0° for 20 minutes in methanol gave the adduct (*31*) which on mild acidic hydrolysis at pH 4 produced cyclopentanone (*32*) in 93% overall yield.

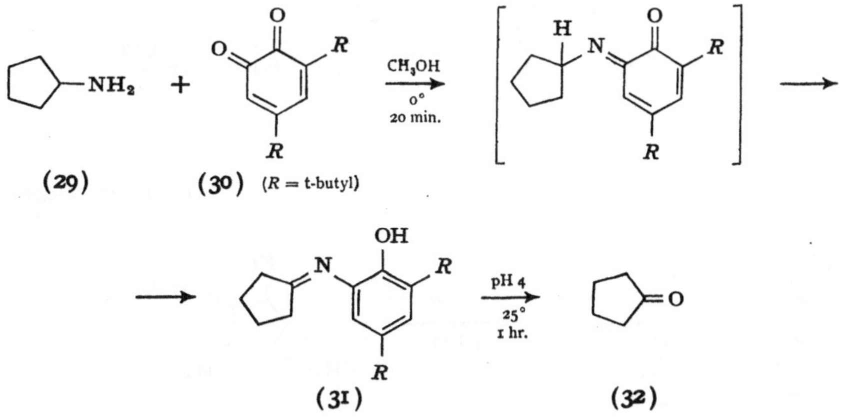

Chart 4. Conversion of Amines to Ketones

A second type of reagent typified by mesityl glyoxal was also developed for this "transamination" reaction type (*16*).

B. Corey's Second Synthesis of dl-PGE₁

This second total synthesis reported by the Harvard group (*19*) again employed the concept of the amine precursor of the 5-ring carbonyl, and also used an aldol condensation at a late stage to form the cyclopentane ring. A major difference was in the use of the unsaturated ketone (*33*) directly as the functional precursor of the 11,15-enediol system (see Chart 5).

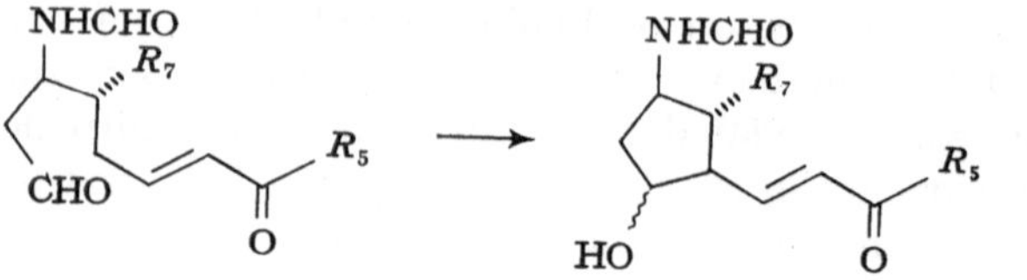

(33)   (34)

Chart 5

A Michael addition of 3-nitropropanal dimethyl acetal (35) to 9-cyano-2-nonenal (36) (prepared by a Wittig reaction of formylmethylene-triphenylphosphorane and 7-cyanoheptanal) in the presence of base gave the adduct (37) which was converted to the unsaturated ketone (38) by

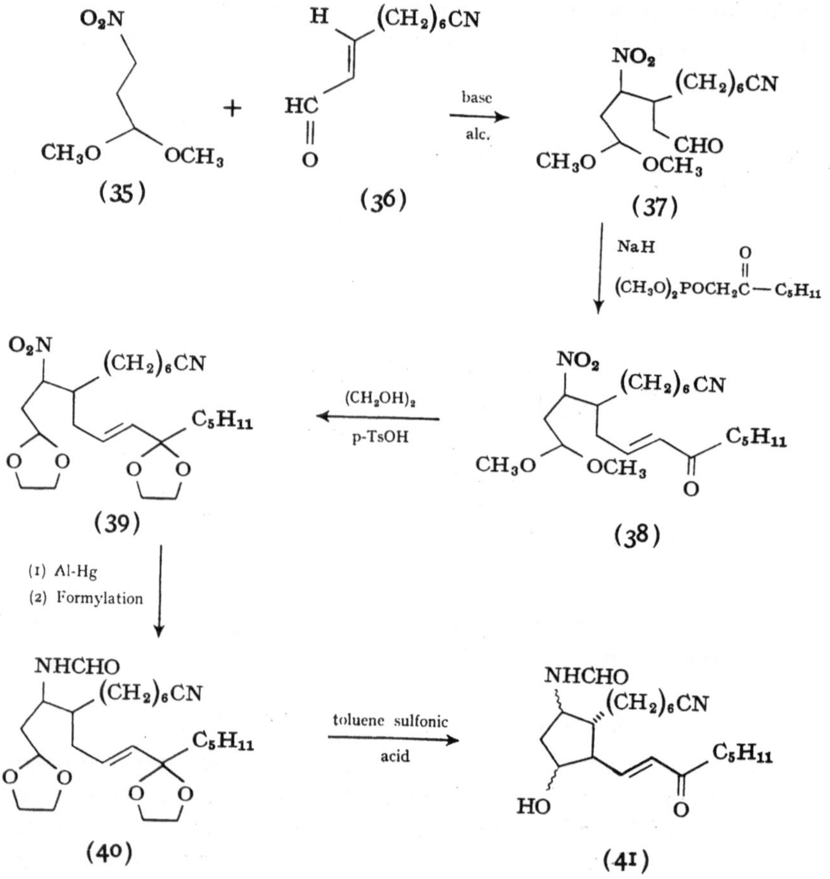

(35)   (36)   (37)

$$\text{NaH}$$
$$\overset{O}{\overset{\|}{(CH_3O)_2POCH_2C-C_5H_{11}}}$$

(39)   (38)

(1) Al-Hg
(2) Formylation

(40)   (41)

the action of the sodio-derivative of dimethyl 2-oxoheptylphosphonate. Reaction with ethylene glycol and acid gave the bis-ketal (39), which could be reduced to the corresponding amine and formylated to produce the protected amine (40). Cyclization of this ketal (40) with p-toluene-sulfonic acid in acetone gave both 11-epimers of the cyclopentanol (41).

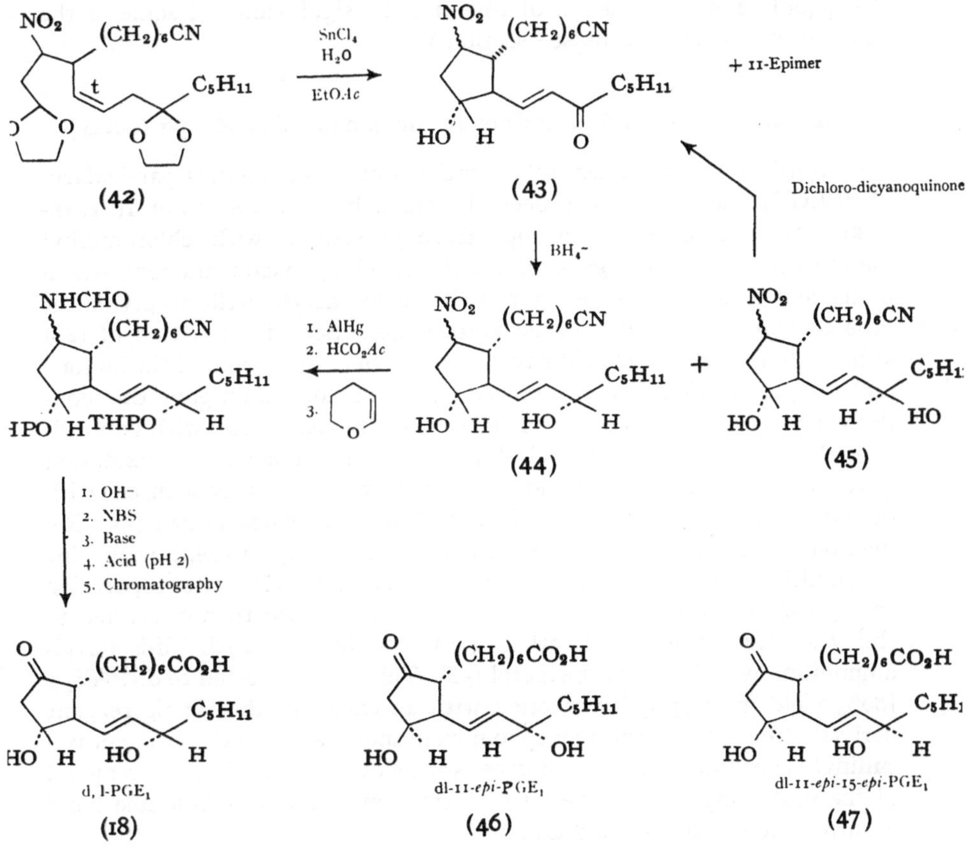

Chart 6. Second Synthesis of Prostaglandin E_1 (COREY et al.)

A variation of the ratio of C_{11}-normal and C_{11}-epi products was found depending on the conditions employed for the cyclization. The four epimers (at 9 and 11) could be separated by chromatography and these were individually transformed to the dl-PGE isomers by the sequence described above in the first synthesis. This also led to the preparation of dl-11-epi-prostaglandin E_1 (46) and dl-11-epi-15-epi-prostaglandin E_1 (47). An alternative was to cyclize the nitroketal (42) employing stannic chloride in acetone, when the product (43) was obtained essentially free of the 11β-epimer (20). An advantage of the use of the nitro group was the more facile chromatographic separation of C_{11} and C_{15} epimers, and also the ready oxidation of the 15-epimer with dichlorodicyanoquinone back to the 15-ketone (43) to enable a recycle. This route has now been adapted to allow a resolution of the natural and enantiomeric forms of PGE_1 (20). For a discussion of this and the significance of some of the epimeric dl-prostaglandins see section VII.

C. A Stereocontrolled Synthesis of Prostaglandins E_2 and $F_2\alpha$ (Corey)

A third, stereocontrolled, total synthesis of prostaglandins particularly of dl-PGE_2 and $PGF_2\alpha$ has been described by Corey's group at Harvard (21). Reaction of cyclopentadienyl sodium with chloromethyl methyl ether at $-55°$ gave methoxymethyl-1,3-cyclopentadiene which underwent a Diels-Alder reaction with 2-chloroacrylonitrile to give after base treatment the anti-bicyclic ketone (48). Selective reaction of (48) with m-chloroperbenzoic acid gave a greater than 95% yield of the lactone (49). Saponification of (49) followed by iodolactonization gave the iodo-derivative (51). Acetylation and removal of the iodo group with tributyl-tinhydride gave (52). Ether cleavage with BBr_3 followed by oxidation gave the unstable acetoxy-aldehyde (53), which was condensed with dimethyl-2-oxoheptylphosphonate to give the trans-enone lactone (55). Reduction of the 15-ketone (prostaglandin numbering system) with zinc borohydride gave a mixture of 15-hydroxy epimers (56). After separation the 15α-epimer was hydrolyzed (K_2CO_3; 1 equiv.) and then converted to the bis-tetrahydropyranyl ether (57). Reduction with di-isobutyl-aluminium hydride led to the lactol (58). This was converted to dl-$PGF_2\alpha$ [80% yield from (56)] by reaction with an excess of the Wittig reagent derived from 5-triphenylphosphoniopentanoic acid and sodiomethyl-sulfinyl carbanide in DMSO to give (59) followed by mild acidic removal of the protecting groups. Selective formation of the cis-5,6-double bond occurred under these conditions.

A minor variation of the procedure was used to prepare dl-PGE_2. Oxidation of the 11,15-bistetrahydropyranyl ether (59) and acid hydrolysis gave dl-PGE_2. Studies on the resolution at an early stage [especially via the hydroxy acid obtained from the lactone (49)] are planned. The important strategic advantages of this approach are the elegant build up of an intermediate (53) with 4 of the 5 asymmetric centers correctly assembled for $PGF_2\alpha$. Addition of the two side chains at a late stage also should allow the synthesis of the primary prostaglandins and a variety of analogs from one precursor.

References, pp. 339—342

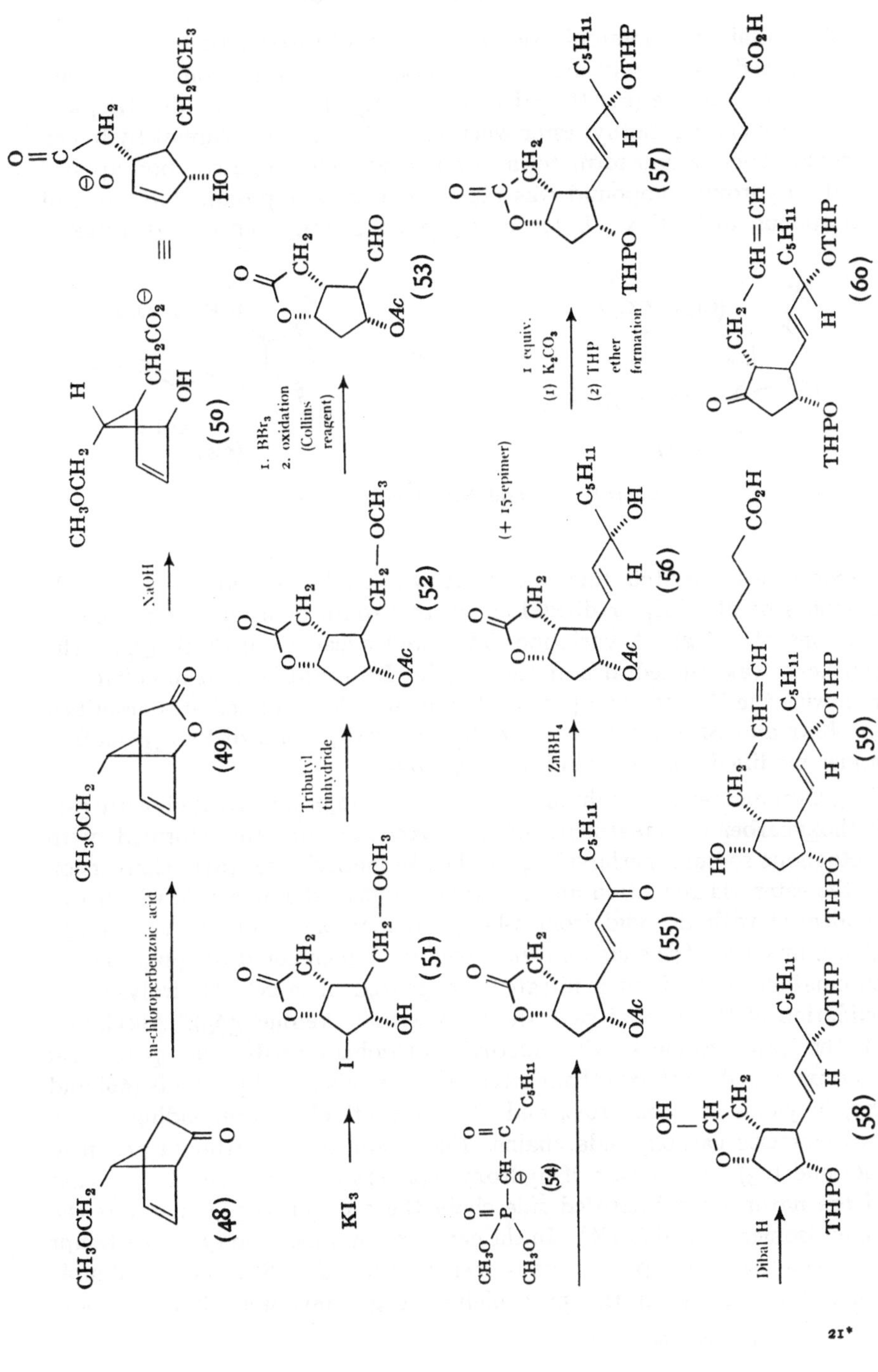

Chart 7. Synthesis of PGE₂ and PGE₂α (COREY et al.) (21)

D. Bicyclohexane Route to Prostaglandins

A preliminary communication by JUST and SIMONOVITCH (McGILL U.)
(35) claimed that a mixture of undisclosed composition containing compounds of structure (61) (R=H and also CH₃) gave impure dl-PGE₁ and
the corresponding methyl ester when treated with a mixture of hydrogen
peroxide and sodium formate in formic acid. The same reaction applied
to the hydroxy compound was reported to give approximately 30% of
amorphous dl-PGF₁α. Later a group at SMITH KLINE and FRENCH

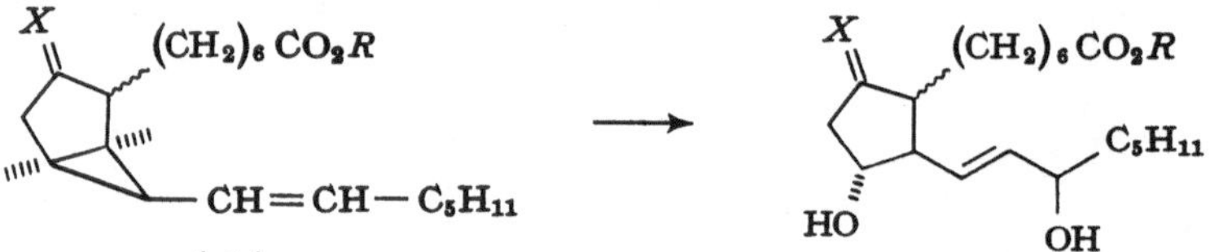

Chart 8. General Bicyclohexane Route

Laboratories repeated this work but was unable to obtain detectable
amounts of dl-PGE₁ or dl-PGF₁α by this route or using various modifications of it (32). A variation did however lead to dl-PGB₁ (32). The
synthesis was studied also at the Upjohn Laboratories and in collaboration with the McGILL group it was found that the intermediates described
by JUST and SIMONOVITCH do produce dl-PGF₁α and dl-PGF₁β methyl
under modified reaction conditions (36, 63).

3-Cyclopentenol tetrahydropyranyl ether (63) was treated with car-
bethoxycarbene. Treatment of the endo : exo mixture formed with
methanolic sodium methoxide gave the exo methyl ester (64). Conversion
of the ester via reduction and re-oxidation gave the aldehyde which was
condensed with the ylid from n-hexyltriphenylphosphonium bromide to
give a mixture of the cis and trans olefines, which could be separated by
chromatography on silver nitrate impregnated silica gel. Hydrolysis and
oxidation of the alcohol gave the cis and trans ketones (65). Alkylation
of the pure isomers with methyl ω-iodoheptanoate and potassium
t-butoxide in dimethoxyethane gave two monoalkylated products (66) and
(73) in yields of about 10% and 25% respectively, correponding to the
two epimeric carboxy side chains. Reduction of the carbonyl group in
(66) and (73) gave a pair of hydroxyl epimers in each case. In the case
of the natural α-substituted side chain the 9α-hydroxy epimer was the
minor isomer formed (10%). In the earlier communication (35) no attempt
had been made to separate the isomer mixtures at C-8, C-9 or the double
bond isomers. When the pure olefins [e. g. (68)] were treated under

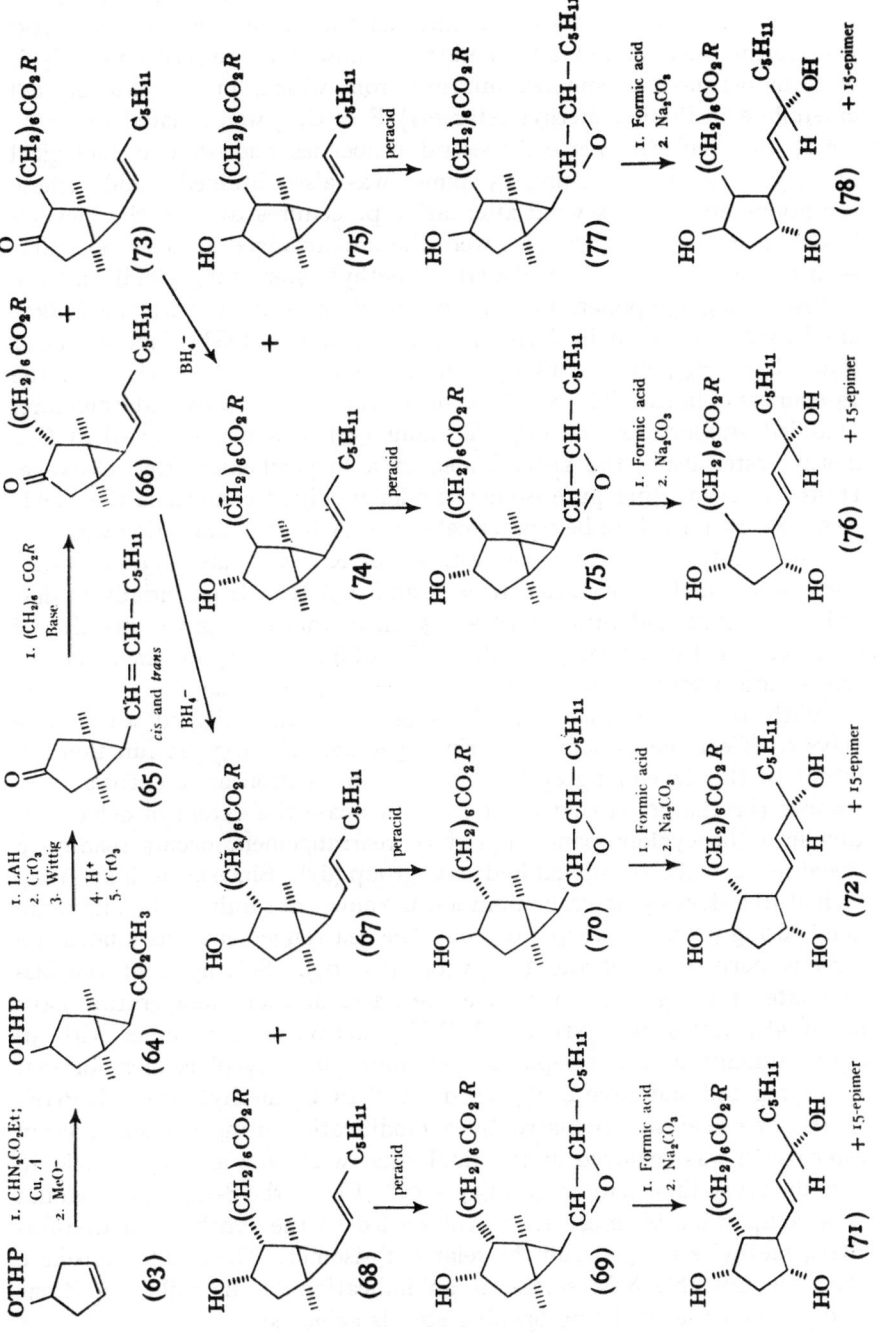

Chart 9. Bicyclohexane Synthesis of prostaglandins — *exo* route via Epoxides (*35, 36*)

especially mild conditions the corresponding epoxides (69) could be isolated. Treatment of these epoxides (as a mixture of isomers at 14,15) with formic acid at room temperature followed by selective hydrolysis of formates gave a complex mixture from which, in the case of (69) crystalline dl-PGF$_1\alpha$ methyl ester.(71) R = CH$_3$ was isolated in 2–3% yield; this had the correct spectral properties and showed biological activity. The corresponding 15-epimer was also obtained. Under these conditions and using several alternative procedures to open the bicyclo-hexane ring the main products were the unrearranged glycols. A similar sequence using (67) led to dl-PGF$_1\beta$ methyl ester (10% yield) and the corresponding 15-epimer. Comparable reactions starting with the β-alky-lated carboxy-side chain derivatives led to dl-8-*iso*-PGF$_1\alpha$ methyl ester [(76) R = CH$_3$], dl-8-*iso*-PGF$_1\beta$ methyl ester [(78) R = CH$_3$] and the 15-epimers of both. This expansion of the earlier work provided, therefore, qualified support for the original claim (35), a synthesis of dl-PGF$_1\alpha$ methyl ester using the general idea of a bicyclohexane ring cleavage. However, even using pure isomers under modified conditions the yields were too low for this to be a practical route. A further limitation was that oxidative solvolysis of the ketoesters (66) R = CH$_3$ and (73) R = CH$_3$ and also of the ketoacids (66) R = H and (73) R = H essentially as des-cribed by JUST and SIMONOVITCH (35) in neither case gave unequivocal evidence for the presence of dl-PGE$_1$, dl-8-iso-PGE$_1$ or their methyl esters, and if present the yield must have been less than 1%.

With these limitations the Upjohn group investigated alternative ways to effect the cyclopropane ring opening. The oxygen function ad-jacent to the developing cyclopropyl carbinyl cation apparently acts to stablize the charge on that carbon and decrease the extent of delocaliza-tion into the cyclopropane ring, since rearrangement occurs readily in the absence of the additional hydroxyl group (65). Since a tosyloxy group vicinal to a developing carbonium ion is known to result in the minimum neighboring group participation (66), the rearrangement was studied on various derivatives of the 1,2-glycols (63, 64). Solvolysis of the *bis*-mesylate of the glycol (79) in acetone-water at room temperature gave about 5% yields of crystalline dl-PGE$_1$ methyl ester together with an equal amount of the 15-epimer. All four pure glycol isomers of (79) were prepared and gave 4–8% yields of dl-PGE$_1$ methyl ester. dl-PGE$_1$ as the free acid was prepared by a modification using a trichloroethyl ester which was removed at the final stage with zinc and acetic acid to give the crystalline acid, m. p. 113.5–115°. Use of the β-alkylated ketones in a comparable bis-mesylate solvolysis led to the synthesis of dl-8-*iso*-PGE$_1$ methyl ester (84) and the related 15-isomer. The ability to isolate the very unstable 8-*iso*-epimer is an indication of the mild conditions under which the final ring opening step is achieved.

References, pp. 339—342

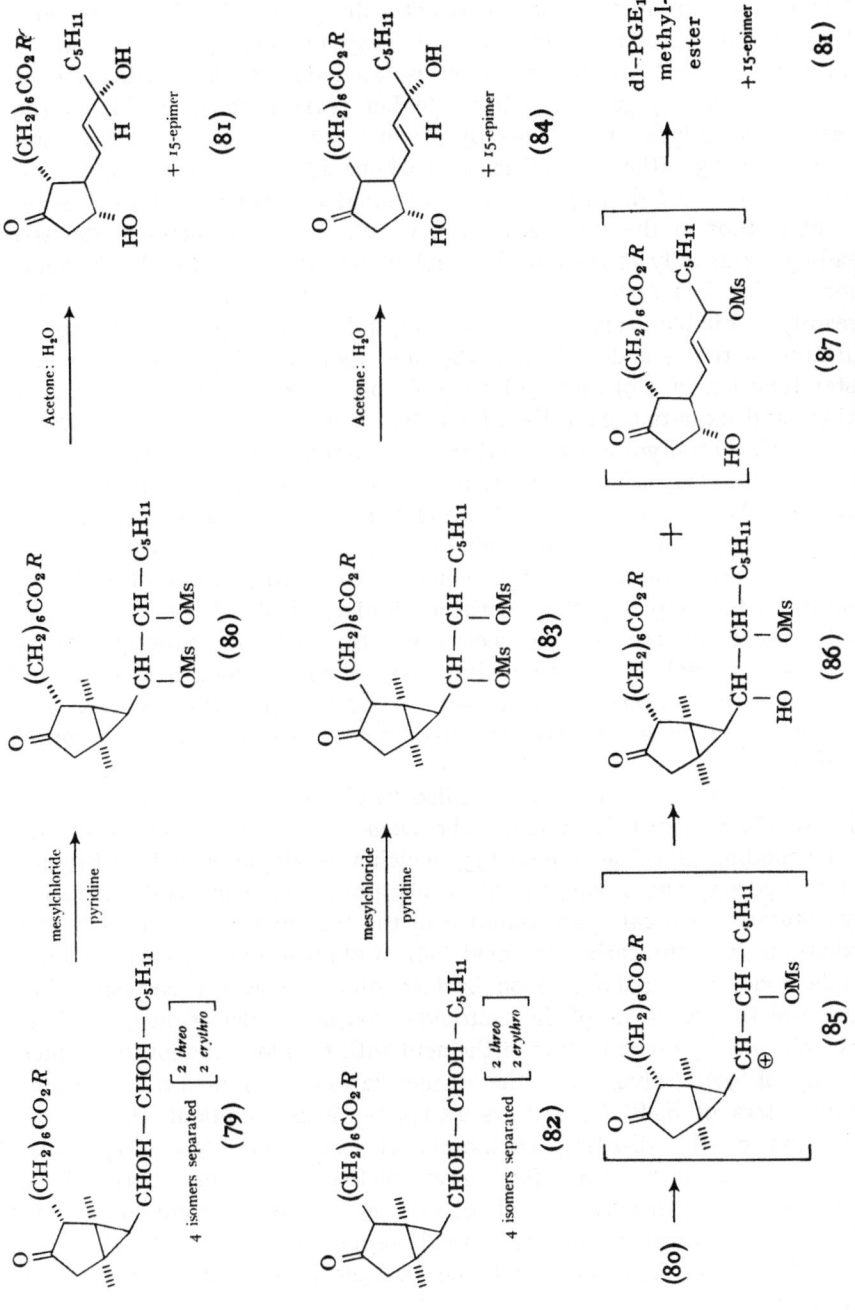

Chart 10. Bicyclohexane route-*exo* route *via* bis-mesylates (Upjohn) (*63, 64*)

The use of this route now made the approach general, but the yield of rearranged product (10–20%) was still disappointingly low. The only other reaction products were monohydroxy monomesylates (86) which could be recycled. A minor isomer isolated during the purification of the 4 glycols (79) gave 2–3 times higher rearrangement yields in the mesylate solvolysis step. This by-product was identified as the *endo*-isomer at C-13, either formed in the carbene addition step and not completely isomerized during the base treatment or isomerized (*exo → endo*) in later steps in the synthesis. A new synthesis was therefore devised leading specifically to the *endo*-bicyclohexane derivatives (4). Hydroboration of bicyclo[3,1,0]-hex-2-ene-6-*endo*-carbocyclic acid methyl ester (88) (readily available from norbornadiene), followed by oxidation gave a mixture of the 2- and 3-alcohols (89) and (90). LiAlH$_4$ reduction of the ester function of (89) and (90) alcohols protected as tetrahydropyranyl ethers and oxidation gave the aldehyde. A Wittig reaction and conversion of the tetrahydropyranyl ethers to a ketone by hydrolysis and oxidation gave (91) after separation of the corresponding 2-ketone by-product. Alkylation of (91) with ω-iodoheptanoate and base gave a good yield of the mono-α-alkylated isomer (92). Hydroxylation with osmium tetroxide gave the glycol (93) which gave a 19% yield of dl-PGE$_1$ methyl ester (94) together with a similar yield of the 15-epimer. Only minor differences were seen using the different pure glycols as starting materials, suggesting that the solvolysis does not occur by a completely concerted mechanism. Interestingly, this synthesis is stereoselective in giving the 11α-hydroxy isomer and the 13,14-*trans* double bond.

This general route (4) was modified to effect the first total synthesis of dl-PGE$_2$ and dl-PGF$_2$α (62). The *endo*-olefin was converted to the corresponding glycol acetonide (95) which was alkylated with 1-bromo-7-tetrahydropyranyloxyhept-2-yne using potassium t-butoxide in tetrahydrofuran. Acid-catalyzed removal of the tetrahydropyranyl ether and oxidation gave the carboxylic acid (96). Catalytic hydrogenation of the triple bond using palladium on barium sulfate gave the corresponding 5,6-*cis*-olefin. Removal of the acetonide grouping under acidic conditions was followed by esterification of the acid with trichloroethanol to produce the glycol (97). Solvolysis of the *bis*-mesylates of (97) gave the trichloroethyl esters of dl-PGE$_2$ and its 15-epimer each in about 15% yield. Reaction of the dl-PGE$_2$ trichloroethyl ester (98) [R = CH$_2$—CCl$_3$] with zinc in acetic acid (67) gave dl-PGE$_2$. Sodium borohydride reduction gave dl-PGF$_2$α and dl-PGF$_2$β. All three compounds had the correct physical and analytical properties compared with the natural isomers, and had half the biological activity of the parent "hormones".

References, pp. 339—342

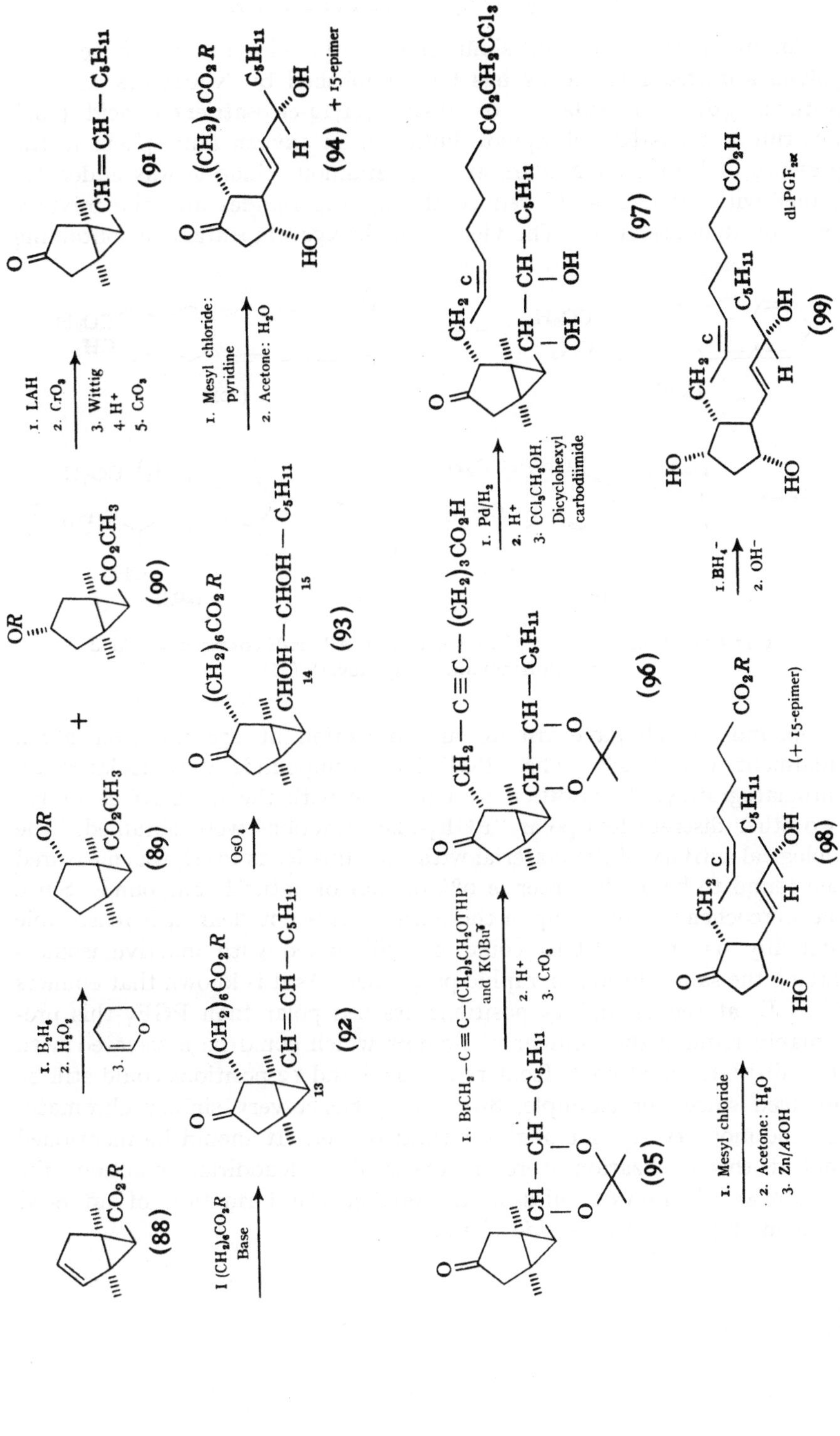

Chart 10 a. Bicyclohexane route-*endo* series (Upjohn) (*4, 62*)

E. Non-Enzymic Cyclization of Fatty Acids

An investigation of the simulation of the biosynthetic route from polyunsaturated fatty acids has been published by NUGTEREN and co-workers (46). Autoxidation of all-cis-8,11,14-eicosatrienoic acid (100) was run in a tris-HCl-phosphate buffer in an oxygen atmosphere in the presence of hemin for 2 hours at 20°. Stannous chloride was added together with citric acid to reduce the hydroperoxides and the mixture kept for 16 hours at 0°. The yield of hydroxycyclopentanone-containing

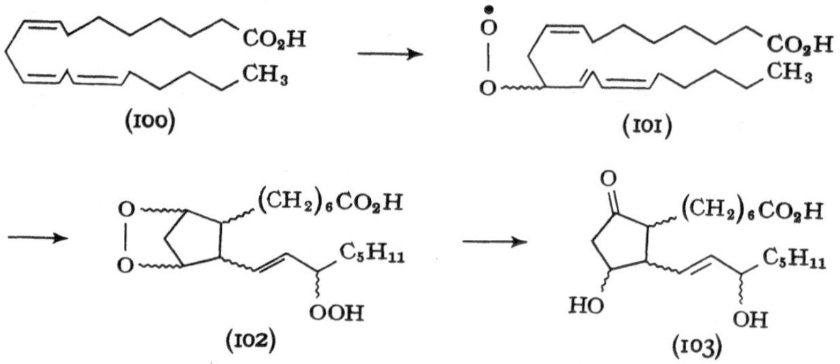

Chart 11. Non-Enzymatic Conversion of All-cis-Eicosatrienoic Acid to Prostaglandins (Unilever) (46)

compounds which gave rise to an absorption at 278 nm after alkali treatment was 1–2%. The "PGE-like" compounds were isolated by chromatography. In addition to a fraction with the same Rf as PGE_1, two other discrete less polar "PGE_1-like" fractions were obtained. The biological activity of the material with the same Rf as PGE_1 was measured and found to be of the order of 6% of that of natural compound. Since the stereochemical make-up of these fractions is not clear, it is reasonable that any natural dl-PGE_1 could be still mixed with inactive isomers having the same chromatographic properties. As it is known that epimers of PGE_1 at the 11 and 15 positions are less polar than PGE_1 this presumably reduces the number of isomers which could be associated with the "dl-PGE_1-like" spot. Isomers at the 8 and 12 positions could still be involved since, for example, 8-iso-PGE_1 has a very similar chromatographic mobility to that of the natural isomer. It should be mentioned that if this cyclization were a concerted cycloaddition reaction, the WOODWARD-HOFMANN rules would predict the formation of an 8-iso compound (i. e. with cis side chains).

IV. Synthetic Routes to Structurally Simplified Prostaglandins

A. Synthesis of dl-13,14-Dihydro-PGE₁ Ethyl Ester

In 1966 BEAL, BABCOCK and LINCOLN described the first total synthesis of a naturally occurring prostaglandin,dl-13,14-dihydro-PGE₁ ethyl ester, which as the free acid is a metabolite of PGE₁, retaining the biological activity of the parent compound (10, 11). The formyl derivative of the 3-ethoxy-2-cyclopentenone (104) was condensed with the ylid from ethyl 6-bromosorbate. Catalytic hydrogenation followed by re-

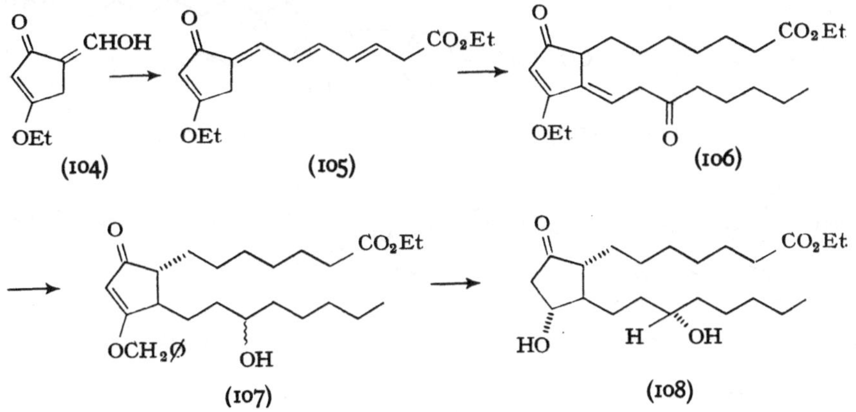

Chart 12. Synthesis of dl-13,14-Dihydro-PGE₁ ethyl ester (BEAL *et al.*) (10, 11)

formylation and a second Wittig reaction introduced the second side chain to give (106). Catalytic hydrogenation of the 12,13 double bond, and acid-catalyzed ester interchange gave the benzyl enol ether (107). The 15-ketone was reduced with lithium tri-t-butoxyaluminium hydride to the corresponding alcohol. Catalytic hydrogenation both cleaved the benzyl ether to the enolic-β-diketone and further reduced the enol double bond to give an 11% yield of material (108) with the same chromatographic properties as authentic dihydro-PGE₁ ethyl ester. A radioisotope dilution method was used to demonstrate the presence of the natural isomer in the synthetic material. Although this route allowed the very facile assembly of the correctly substituted skeleton [i. e. (106)] it suffered from a limitation in the facility with which this could be converted under stereochemical control to the natural isomers.

B. Synthesis of dl-15-Dehydro-PGE₁

MIYANO and DORN have described a facile route to dl-15-dehydro-PGE₁ (42). Condensation of 3-keto-undecan-1,11-dioic acid (109) with styryl glyoxal (110) gave a high yield of the unsaturated ketone (111).

Treatment of (111) with dilute alkali gave (112) in 40–50% yield. Hydroxylation and cleavage of (112) led to the aldehyde (113), which was reduced with zinc in aqueous acetic acid to the 8,12-dihydro derivative (114). Wittig condensation with n-hexanoylmethylene triphenylphosphorane gave a mixture with (115) as the major component together with the corresponding 11-epimer. This route gives good promise as a general route once the problem of selective reduction at C-15 is solved.

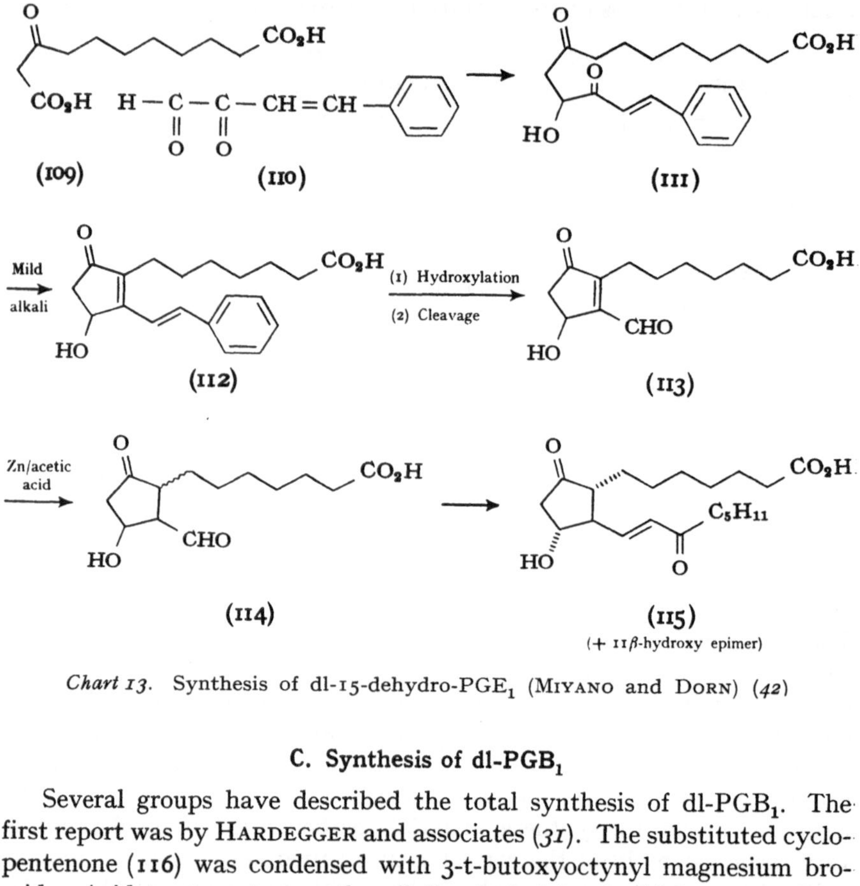

Chart 13. Synthesis of dl-15-dehydro-PGE$_1$ (MIYANO and DORN) (42)

C. Synthesis of dl-PGB$_1$

Several groups have described the total synthesis of dl-PGB$_1$. The first report was by HARDEGGER and associates (31). The substituted cyclopentenone (116) was condensed with 3-t-butoxyoctynyl magnesium bromide. Acid treatment gave the allylic alcohol (118) which was oxidized to the cyclopentenone (119). Catalytic reduction of the triple bond followed by acidcatalyzed cleavage of the t-butyl ether and base-hydrolysis gave racemic-PGB$_1$ (120). A very similar route was published by KLOK, PABON and VAN DORP (38). The main difference was the use of the tetrahydropyranyl ether protecting group rather than a t-butyl ether. Three other groups also published very similar routes to dl-PGB$_1$

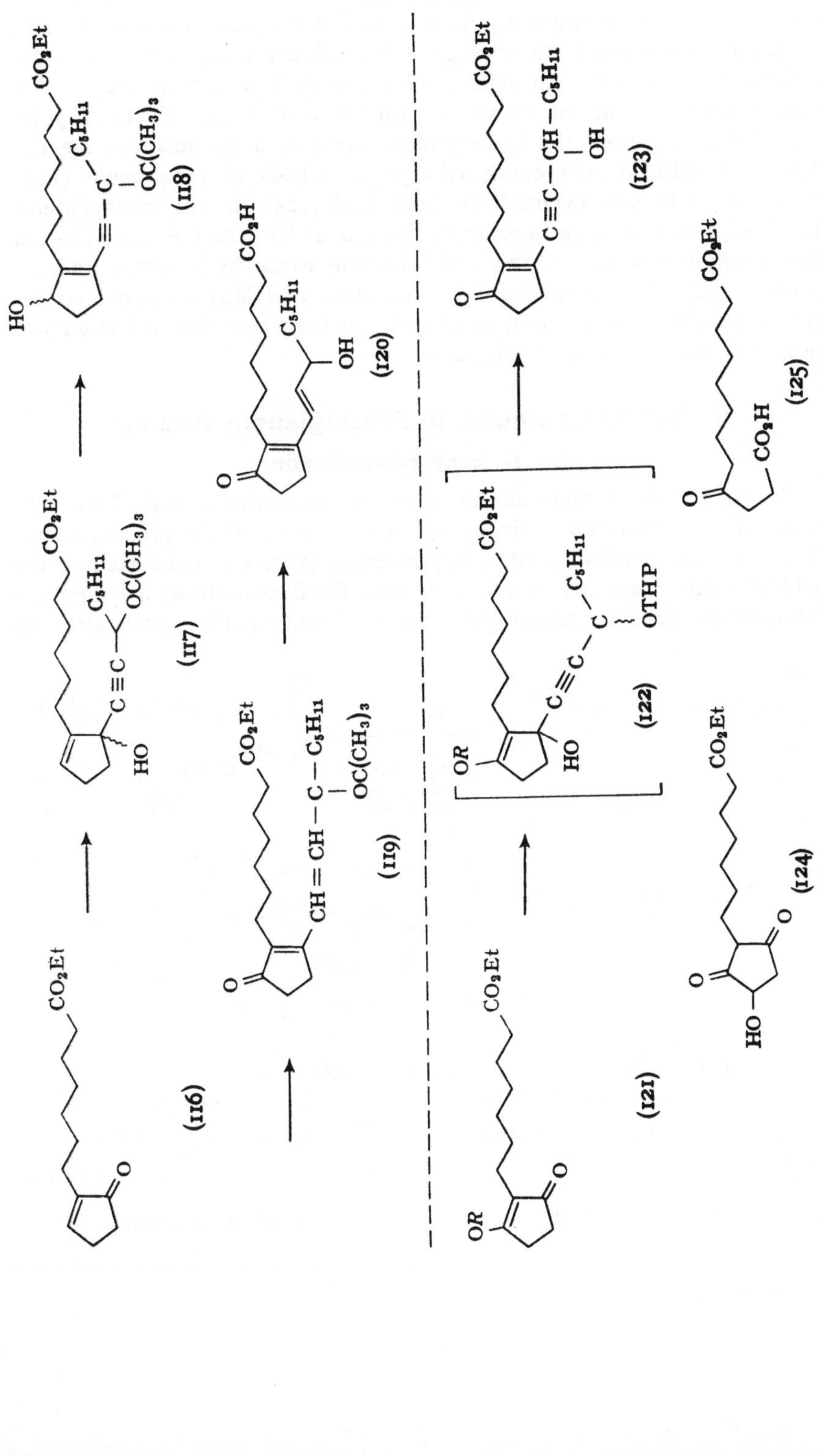

Chart 14. Synthesis of dl-PGB$_1$ (1. HARDEGGER *et al.* (31); 2. KLOK *et al.* (33))

(*14, 37, 68*). In each case the starting material here was the 2-substituted enol ether of cyclopentane 1,3-dione (**121**). A Grignard reaction with an acetylenic magnesium bromide gave the adduct (**122**), which was not isolated but treated with acid to give the cyclopentenone (**123**). This was converted as in the above two routes to dl-PGB₁. COLLINS, JUNG and PAPPO (*14*) used the hydroxydione (**124**) as a precursor of the 1,3-diketone; sulfuric acid catalyzed hydrogenolysis of (**124**) gave (**121**). YURA and IDE (*68*) cyclized the keto acid (**125**) to the cyclopentene-1,3-dione using AlCl₃ and propionyl chloride at 80°. As mentioned earlier the attempted repeat of the bicyclohexane route by HOLDEN and co-workers (*32*) also led to dl-PGB₁. KATSUBE and MATSUI also used the synthetic approach based on (**121**) (*37*). In their case the enol ether was prepared from ethyl 9-oxo-decanote.

V. Synthetic Routes to Prostaglandin Analogs

A. 11-Deoxy-Prostaglandins

BAGLI and co-workers at the Ayerst Laboratories described the first synthesis of a pharmacologically active derivative of the prostaglandins (*6, 8, 9*). The 2-substituted cyclopentenone (**126**) was converted to the acid chloride (**127**) *via* nitrile addition. Condensation with 1-heptyne followed by base treatment gave the enol ether (**128**), which after re-

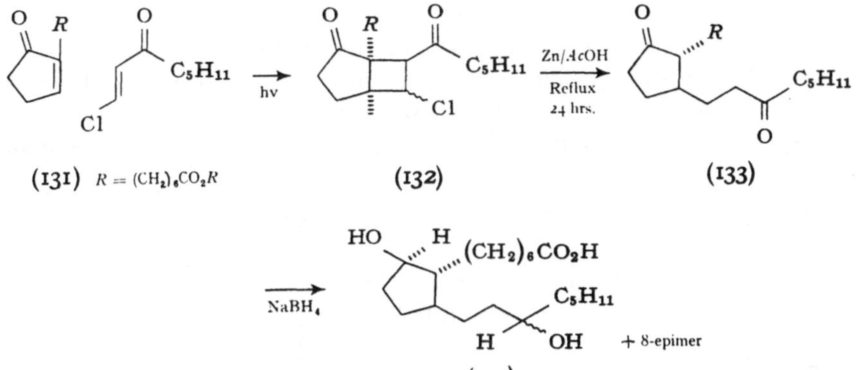

(131) R = (CH₂)₆CO₂R   (132)   (133)

(134)

Chart 15. 11-Deoxyprostaglandins (BAGLI and BOGRI) (6, 8, 9)
Photochemical route to 11-deoxyprostaglandins (7)

duction with sodium borohydride and acid treatment gave the unsaturated ketone (129). Reduction gave a mixture of stereoisomers at C-15. The 11-deoxyprostaglandins showed biological activity in lowering blood pressure, in stimulating smooth muscle contraction and in inhibiting gastric secretion. A careful study established the stereochemistry of the carbon atoms at C-8 and C-12 (6). More recently a novel route to the prostanoic acid structure, again 11-deoxy derivatives, was described by the same group (7). Photochemical addition of the chlorovinyl ketone (131) to the cyclopentenone (126) gave the adduct (132) in 35% yield. Treatment of the photoadduct (132) with zinc in refluxing acetic acid for 24 hrs. gave the ring-opened diketone (133). Sodium borohydride reduction gave the 9,15-diol (134).

B. 7-Oxa-Prostaglandins

FRIED and co-workers have described a synthesis of dl-7-oxa-PGF₁α and related substances which is stereoselective except for the introduction of the hydroxyl at C-15 (27, 28). All-cis-1,2-epoxycyclopentane-3,4-diol was converted to the dibenzyl ether (135). A novel method involving the use of diethyl-octynylalane opened the oxide to the trans-2-octynylcyclopentanol derivative (136). O-Alkylation with t-butyl 6-iodohexanoate gave (137) [R = t-butyl]. Removal of the t-butyl group gave the acetylenic acid [(137) R = H] which could be reduced with lithium in methylamine to the trans olefin, or hydrogenated catalytically using Pd/BaSO₄ to the cis olefin. Oxidation of (138) with selenium dioxide gave dl-7-oxa-PGF₁α (139) and its 15-epimer. More recently this synthesis has been extended to the corresponding 9,11-di and 9,11,15-trideoxy compounds and their ring homologs (29). These have exhibited activity as prostaglandin antagonists on smooth muscle in vitro. A noteworthy aspect of this approach was the development of the diethylalkynyl alane ($Et_2Al-C\equiv C-R$) for the opening of epoxides to form β-hydroxyacetylenes, to overcome the low yields using alkali acetylides, particularly for more highly substituted epoxides. In the dl-7-oxa-$PGF_1\alpha$ synthesis the alane

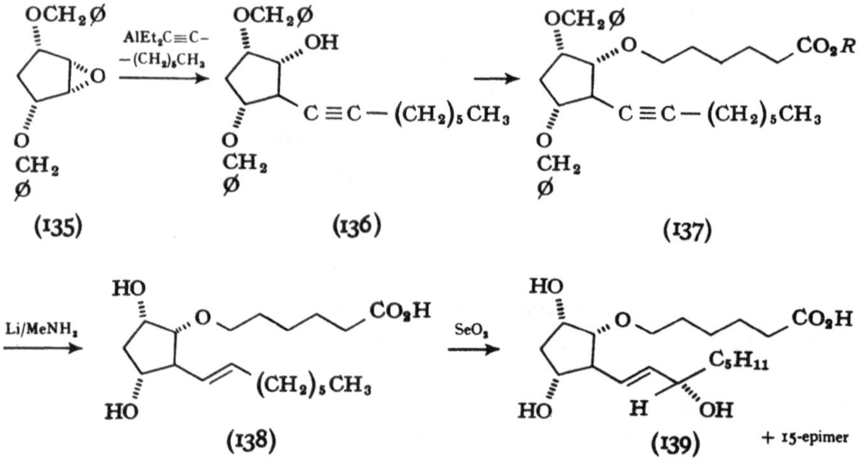

Chart 16. 7-oxa-Prostaglandins (FRIED et al.) (27, 28)

was generated *in situ*; more recently an improved procedure for the preparation of the alanes was described which permits extension to more complicated acetylenes (e. g. $Et_2Al-C\equiv C-CH-C_5H_{11}$) (28).

$$\overset{\displaystyle |}{\text{OTHP}}$$

VI. Miscellaneous Synthetic Approaches

BEAL and co-workers described a synthesis of prostanoic acids in which the side chains attached to the 5-membered ring were *cis* (i. e. 8-*iso*-prostanoic acids) (11). This was based on a Diels-Alder reaction of cyclopentene-3,5-dione and butadiene to give (140). Reduction of the β-diketone and benzylidene formation followed by addition of the alkyl-side chain via the corresponding 6-ring ketone gave (141) which after hydroxylation and cleavage gave the ketoaldehyde (142). A Wittig reaction followed by hydrogenation completed the formation of the carboxy side-chain, and the reduction of the 15-ketone gave dl-13,14-dihydro-8-*iso*-$PGF_1\alpha$ methyl ester [(144) R = CH_3] as a mixture of epimers at C-15. A variation permitted extension to the PGE series.

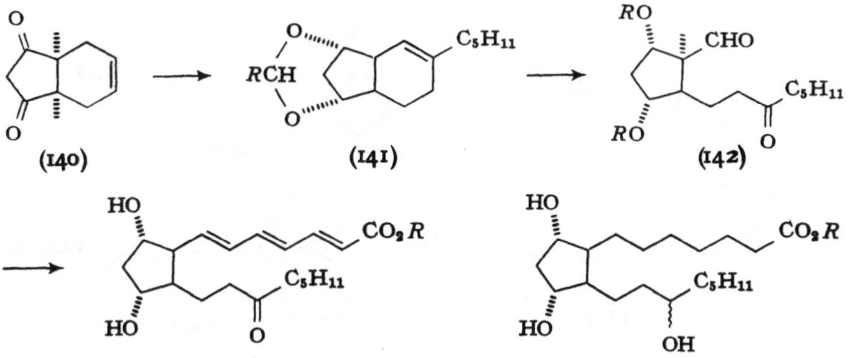

O

(140)

RCH

(141)

C₅H₁₁

RO CHO

RO

(142)

O

C₅H₁₁

HO

CO₂R

C₅H₁₁

HO O

(143)

HO

CO₂R

C₅H₁₁

HO

OH

(144)

Chart 17. Diels-Alder Approach to 8-iso-Prostanoic Acids

An unusual approach to the ring system was based on the nitrite cleavage of the bicycloketone (145) (61). Although several key intermediates in the synthesis of (145) were obtained this route has not yet apparently led to the successful preparation of the required compound.

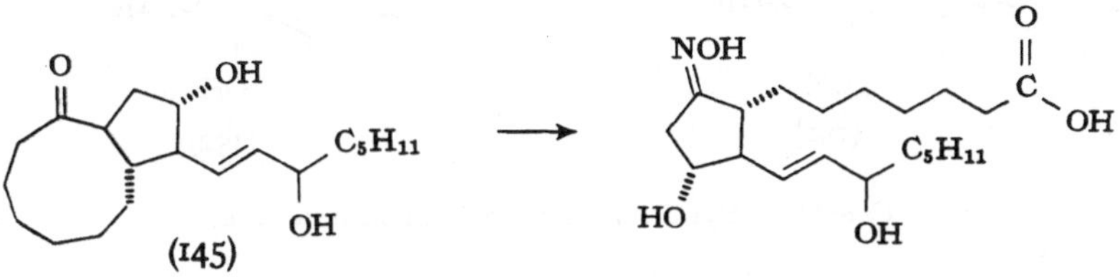

O OH

C₅H₁₁

OH

(145)

NOH

O
‖
C
OH

HO

C₅H₁₁

OH

Chart 18

MORIN, SPRY, HAUSER and MUELLER described a synthetic approach based on the conversion of an aromatic ring to the cyclopentane structure. Among several less successful alternatives the main route is shown in Chart 19. Birch reduction of the substituted anisole (146) available from a byproduct of penicillin-G fermentation, was followed by esterification and ketal formation leading to the olefin (147). Ozonolysis gave the dialdehyde which could be cyclized with 3-azabicyclo[3,2,2]nonane to give (149) together with an equal amount of the alternative unsaturated aldehyde. Wittig condensation gave the dienone (150) which after reduction with sodium borohydride and ketal cleavage gave in low yield a mixture of (151) and (152). A problem was that acid-catalyzed ketal removal caused extensive dehydration of the C-15 hydroxyl.

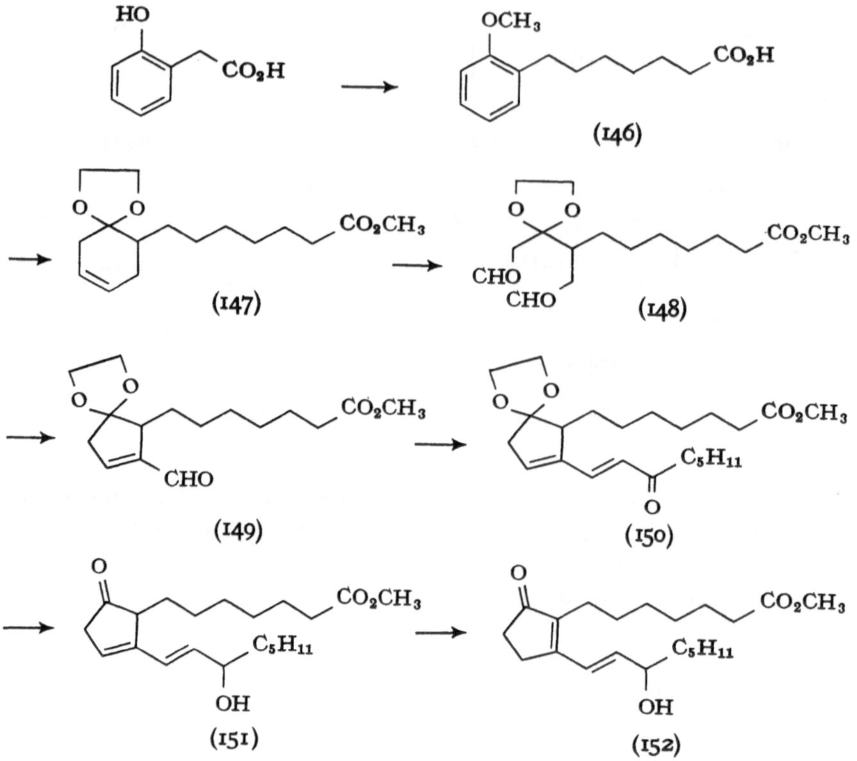

HO

CO₂H → OCH₃ CO₂H

(146)

→ (147) → CHO CHO (148)

→ (149) → (150)

→ (151) → (152)

Chart 19. Synthetic approach of MORIN *et al.*

Some preliminary experiments on an alternative approach to a substituted cyclopentanone structure have been described by ORR and JOHNSON (50). 1,4-cyclohexanedione was transformed to the triol (153) which after conversion to the tosylate (154) and base treatment gave the substituted 5-membered ring (155).

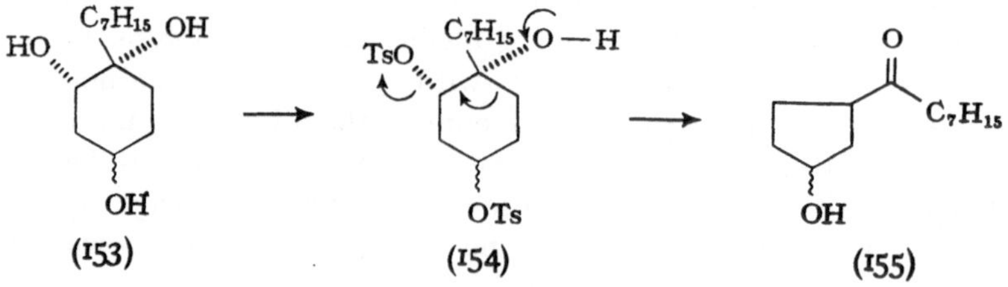

HO C₇H₁₅ OH OH (153) → TsO C₇H₁₅ O—H OTs (154) → O C₇H₁₅ OH (155)

Chart 20

Recently MIYANO has described a novel approach to 15-dehydro-PGB$_1$ and PGE-237 in which the potential 13,14-double bond is "protected" as a bicyclo derivative (41).

VII. Resolution of Racemic Prostaglandins

So far only one report has appeared of a resolution of a synthetic dl-prostaglandin (20). The racemic amine (17) 15α-epimer obtained as an intermediate in the Harvard synthesis was treated with (—)-α-bromo-camphor-π-sulfonic acid and gave a crystalline salt $[\alpha]_{578}$ — 59.6°. Recrystallization gave a single diastereomeric salt, m. p. 157–159° $[\alpha]_{578}$ — 59.6°. The levoamine derived from this salt was converted to prostaglandin E$_1$ as described earlier for the dl-compound and gave material identical with the natural product. By a similar process the racemic amine treated with (+)-α-bromocamphor-π-sulfonic acid gave the enantiomer of prostaglandin E$_1$ $[\alpha]_{578}$ ca. + 57°. The biological activity of the enantiomer of PGE$_1$[ent-PGE$_1$] had one thousandth the activity of the natural isomer in 3-smooth muscle assays and as a vasodepressor. As first pointed out by RAMWELL, SHAW, COREY and ANDERSEN (56) this inactivity is not true of the enantiomers of certain epimeric prostaglandins. For example dl-15-epi-PGA$_1$ is more active than natural 15-epi-PGA$_1$ suggesting pharmacologic activity for the enantiomeric isomer. Again a similar pattern is seen with 11-epimers and dl-11-epi-15-epi-PGE$_1$ has marked bioactivity thought to be associated with the unnatural antipode. Further progress in establishing the activity of these "mirror image" or ent-prostaglandin isomers will depend on a general method of resolution or a total synthesis where a resolution is accomplished at an early stage.

References

1. ANDERSEN, N. H.: Dehydration of Prostaglandins: Study by Spectroscopic Method. J. Lipid Res. **10**, 320 (1969).

2. — Preparative Thin-Layer and Column Chromatography of Prostaglandins. J. Lipid Res. **10**, 316 (1969).

3. AXEN, U.: Synthetic Approaches to Prostaglandins. In "Annual Reports in Medicinal Chemistry 1967", Ed. C. K. Cain, pp. 290. New York: Academic Press. 1968.

4. AXEN, U., F. H. LINCOLN and J. L. THOMPSON: A Total Synthesis of Prostaglandin E$_1$ and Related Substances Via Endo-Bicyclohexane Intermediates. Chem. Commun. 303 (March, 1969).

5. BACON, R. G. R. and W. J. W. HANNA: Metal Ions and Complexes in Organic Reactions. Part V Oxidations of Primary and Secondary Amines With Argentic Picolinate. J. Chem. Soc. 4962 (1965).

6. BAGLI, J. F. and T. BOGRI: Prostaglandins II — An Improved Synthesis and Structural Proof of (±)-11-deoxyprostaglandin F$_1\beta$. Tetrahedron Lett. No. **1**, 5 (1967).

7. Bagli, J. F. and T. Bogri: Prostaglandins III — ± 11-deoxy-13,14-Dihydro-prostaglandin $F_1\alpha$ and $F_1\beta$ — A Novel Synthesis of Prostanoic Acids. Tetrahedron Lett. No. 21, 1639 (1969).

8. Bagli, J. F., T. Bogri, R. Deghenghi and K. Wiesner: Prostaglandins I — Total Synthesis of 9β, 15 — Dihydroxyprost-13-Enoic Acid. Tetrahedron Lett. No. 5, 465 (1966).

9. Bogri, T., J. F. Bagli and R. Deghenghi: An Improved Synthesis of the Prostanoic Acids. Nobel Symposium 2, Prostaglandins, Ed. S. Bergstrom and B. Samuelsson, pp. 231. Stockholm: Almqvist and Wiksell. 1967.

10. Beal III P. F., J. C. Babcock and F. H. Lincoln: A Total Synthesis of A Natural Prostaglandin. J. Am. Chem. Soc. 88, 3131 (1966).

11. Beal, P. F., J. C. Babcock and F. H. Lincoln: Synthetic Approaches in the Prostanoic Acid Series. Nobel Symposium 2, Prostaglandins, Ed. S. Bergstrom and B. Samuelsson, pp. 219. Stockholm: Almqvist and Wiksell. 1967.

12. Bergstrom, S.: Prostaglandins: Members of a New Hormonal System. Science N. Y. 157, 382 (1967).

13. Bergstrom, S., L. A. Carlson and J. R. Weeks: The Prostaglandins: A Family of Biologically Active Lipids. Pharmac. Rev. 20, 1 (1968).

14. Collins, P., C. J. Jung and R. Pappo: Prostaglandin Studies. The Total Synthesis of dl-Prostaglandin B_1. Israel J. Chem. 6, 839 (1968).

15. Corey, E. J.: General Methods for the Construction of Complex Molecules Pure Appl. Chem. 14, 19 (1967).

16. Corey, E. J. and K. Achiwa: A New Method for the Oxidation of Primary Amines to Ketones. J. Am. Chem. Soc. 91, 1429 (1969).

17. Corey, E. J., N. H. Andersen, R. M. Carlson, J. Paust, E. Vedejs, I. Vlattas and R. E. K. Winter: Total Synthesis of Prostaglandins, Synthesis of the Pure dl-E_1, -$F_1\alpha$, -$F_1\beta$, -A_1, and -B_1 Hormones. J. Am. Chem. Soc. 90, 3245 (1968).

18. Corey, E. J., and D. Seebach: Carbanions of 1,3-Dithianes. Reagents for C—C Bond Formation by Nucleophilic Displacement and Carbonyl Addition. Angew. Chem. Intern. Ed. Engl. 4, 1075 (1965).

19. Corey, E. J., I. Vlattas, N. H. Andersen and K. Harding: A New Total Synthesis of Prostaglandins of the E_1 and F_1 Series Including 11-Epiprostaglandins. J. Am. Chem. Soc. 90, 3247 (1968). [See Erratum 90, 5947 (1968).]

20. Corey, E. J., I. Vlattas and K. Harding: Total Synthesis of Natural (Levo) and Enantiomeric (Dextro) Forms of Prostaglandin E_1. J. Am. Chem. Soc. 91, 535 (1969).

21. Corey, E. J., N. M. Weinshenker, T. K. Schaaf and W. Huber: Stereocontrolled Synthesis of Prostaglandins $F_2\alpha$ and E_2 (dl). J. Am. Chem. Soc. 91, 5675 (1969).

22. Crundwell, E., M. A. Pinnegar and W. Templeton: Synthesis of Fatty Acids With Smooth Muscle Stimulant Activity. J. Medicinal Chem. 8, 41 (1965).

23. Daniels, E. G. and J. E. Pike: Isolation of Prostaglandins. Prostaglandin Symposium of the Worcester Foundation for Exp. Biol., Ed. P. W. Ramwell and J. E. Shaw, pp. 379. New York: Interscience. 1968.

24. Daniels, E. G., W. C. Krueger, F. P. Kupiecki, J. E. Pike and W. P. Schneider: Isolation and Characterization of a New Prostaglandin Isomer. J. Am. Chem. Soc. 90, 5894 (1968).

25. Dopp, D.: Notiz zur Hydroborierung von Bicyclo[3.1.0]hexen-(2)-carbonsaure-(6-endo)methyl ester. Chem. Ber. 102, 1081 (1969).

26. Dorp, D. A. van: Essentielle Fettsäuren und Prostaglandine. Naturwissenschaften 56, 124 (1969).

27. FRIED, J., S. HEIM, S. J. ETHEREDGE, P. SUNDER-PLASSMAN, T. S. SANTHAN-AKRISHNAN, J. HIMIZU and C. H. LIN: Synthesis of (±)-7-Oxaprostaglandin $F_1\alpha$. Chem. Commun. 1968, 634.

28. FRIED, J., S. HEIM, P. SUNDER-PLASSMAN, S. J. ETHEREDGE, T. S. SANTHAN-AKRISHNAN and J. HIMIZU: Synthesis of 15-Desoxy-7-oxa-Prostaglandin $F_1\alpha$ and Related Substances. Prostaglandin Symposium of the Worcester Foundation for Exp. Biol., Ed. P. W. RAMWELL and J. E. SHAW, pp. 351. New York: Interscience. 1968. See also FRIED, J., C. H. LIN and S. M. FORD, Tetrahedron Lett. No. 18, 1379 (1969).

29. FRIED, J., T. S. SANTHANAKRISHNAN, J. HIMIZU, C. H. LIN, S. H. FORD, B. RUBIN and E. O. GRIGAS: Prostaglandin Antagonists: Synthesis and Smooth Muscle Activity. Nature 223, 208 (1969).

30. HAMBERG, M.: On the Absolute Configurations of 19-Hydroxy-Prostaglandin B_1. Eur. J. Biochim. 6, 147 (1968).

31. HARDEGGER, E., H. P. SCHENK und E. BORGER: Synthese der Dl-Form eines natürlichen Prostaglandins. Helv. Chim. Acta 50, 2501 (1967).

32. HOLDEN, K. G., B. HWANG, K. R. WILLIAMS, J. WEINSTOCK, M. HARMAN and J. A. WEISBACH: Synthetic Studies on Prostaglandins. Tet. Letters, 1968, 1569.

33. HORTON, E. W.: Hypotheses on Physiological Roles of Prostaglandins. Physiol. Rev. 49, 122 (1969).

34. HINMAN, J. W.: The Prostaglandins. Bioscience 17, 779 (1967).

35. JUST, G. and C. SIMONOVITCH: A Prostaglandin Synthesis. Tetrahedron Lett. No. 22, 2093 (1967).

36. JUST, G., C. SIMONOVITCH, F. H. LINCOLN, W. P. SCHNEIDER, U. AXEN, G. B. SPERO and J. E. PIKE: A Synthesis of Prostaglandin $F_1\alpha$ and Related Substances. J. Am. Chem. Soc. 91, 5364 (1969).

37. KATSUBE, J. and M. MATSUI: Synthetic Studies on Cyclopentane Derivatives. Ag. Biol. Chem. 33, 1078 (1969).

38. KLOK, R., H. J. J. PABON and D. A. VAN DORP: Synthesis of dl-Prostaglandin B_1 and Its Reduction Product dl-Prostaglandin E_1-237. Rec. Trav. Chim. Pays-Bas Belg. 87, 813 (1968).

39. KORNBLUM, N., H. O. LARSON, R. K. BLACKWOOD, D. D. MOOBERRY, E. P. OLIVETO and G. E. GRAHAM: A New Method for the Synthesis of Aliphatic Nitro Compounds. J. Am. Chem. Soc. 78, 1497 (1956).

40. KRUG, R. C. and T. F. YEN: Unsaturated Cyclic Sulfones. I. 3-Bromomethyl-2,5-Dihydrothiophene-1,1-Dioxide. J. Org. Chem. 21, 1082 (1956).

41. MIYANO, M.: Synthetic Studies on Prostaglandins II. Tetrahedron Lett. No. 32, 2271 (1969).

42. MIYANO, M. and C. R. DORN: Total Synthesis of 15-Dehydroprostaglandin E_1. Tetrahedron Lett. No. 20, 1615 (1969).

43. MORIN, R. B., D. O. SPRY, K. L. HAUSER and R. A. MUELLER: Approaches to the Chemical Synthesis of the Prostaglandins. Tetrahedron Lett. No. 57, 6023 (1968).

44. NAGATA, W. and Y. HAYASE: Formylolefination of Carbonyl Compounds. Tetrahedron Lett. 1968, 4559.

45. NUGTEREN, D. H., R. K. BEERTHUS and D. A. VAN DORP: Biosynthesis of Prostaglandins. Nobel Symposium 2, Prostaglandins, Ed. S. BERGSTROM and B. SAMUELSSON, p. 45. Stockholm: Almqvist and Wiksell. 1967.

46. NUGTEREN, D. H., H. VONKEMAN and D. A. VAN DORP: Non-Enzymic Conversion of all-cis 8,11,14-Eicosatrienoic Acid Into Prostaglandin E_1. Rec. Trav. Chim. Pays-Bas Belg. 86, 1237 (1967).

47. NUGTEREN, D. H., D. A. VAN DORP, S. BERGSTRÖM, M. HAMBERG and B. SAMUELSSON: Absolute Configuration of the Prostaglandins. Nature, Lond. 212, 38 (1966).

48. Oediger, H., H. J. Kabbe, F. Moller and K. Eiter: 1,5-Diazo-bicyclo[4.3.0]-nonen-(5). Ein neues Reagenz zur Einführung von Doppelbindungen. Chem. Ber. 99, 2012 (1966).

49. Ohno, M., N. Naruse, S. Torimitsu and M. Okamoto: Reaction of 2-Chloro-cycloalkanone Oximes 1. Their Preparations and Conversion to 2-Alkoxy,2-Acycloxy and 2-Alkylthio Cycloalkanone Oximes. Bull. Chem. Soc. Japan, 39, 1119 (1966).

50. Orr, D. E. and F. B. Johnson: Synthesis of 3-(1,1-ethylenedioxyoctyl)-Cyclopentanone. Can. J. Chem. 47, 47 (1969).

51. Pabon, H. J. J., L. van der Wolf and D. A. van Dorp: Preparation of Primary Alcohols Related to Prostaglandins. Rec. Trav. Chim. Pays-Bas 85, 1251 (1966).

52. Pike, J. E., F. H. Lincoln and W. P. Schneider: Prostanoic Acid Chemistry J. Org. Chem. 34, 3552 (1969).

53. Pike, J. E., F. P. Kupiecki and J. R. Weeks: Biological Activity of the Prostaglandins and Related Analogs. Nobel Symposium 2, Prostaglandins, Ed. S. Bergström and B. Samuelsson, 79. Stockholm: Almqvist and Wiksell. 1967.

54. Ramwell, P. W. and E. G. Daniels: Chromatography of the Prostaglandins. In "Lipid Chromatographic Analysis", Ed. G. V. Marinetti, Vol. 2, p. 313. New York: Marcel Dekker. 1969.

55. Ramwell, P. W., J. E. Shaw, G. B. Clarke, M. F. Grostic, D. G. Kaiser and J. E. Pike: Prostaglandins. In "Progress in the Chemistry of Fats and Other Lipids", Ed. R. T. Holman, Vol. 9, p. 231. Oxford: Pergamon Press. 1968.

56. Ramwell, P. W., J. E. Shaw, E. J. Corey and N. Andersen: Biological Activity of Synthetic Prostaglandins. Nature, Lond. 221, 1251 (1969).

57. Ruschig, H., W. Fritsch, J. Schmidt-Thomé and W. Hode: Über die Herstellung von 17α-oxy-20-Ketosteroiden aus 17(20)-en-20-Acetaminosteroiden. Chem. Ber. 88, 883 (1955).

58. Samuelsson, B.: The Prostaglandins. Angew. Chem. Int. Ed. 4, 410 (1965).

59. Samuelsson, B. and G. Ställberg: Structure and Synthesis of a Derivative of Prostaglandin E_1. Acta Chem. Scand. 17, 810 (1963).

60. Saxe, B. D.: Approaches to the Synthesis of Prostaglandin E_1. Dissertation, Dept. of Chem., Columbia Univ., New York, 1966, 81 pp.

61. Saxe, B. D.: Approaches to the Synthesis of Prostaglandin E_1. Dissertation Abstracts 27, 3870-B (1967).

62. Schneider, W. P.: The Synthesis of (±)-Prostaglandins E_2, $F_2\alpha$, and $F_2\beta$. Chem. Commun. 1969, 304.

63. Schneider, W. P., U. Axen, F. H. Lincoln, J. E. Pike and J. L. Thompson: The Total Synthesis of Prostaglandins. J. Am. Chem. Soc. 90, 5895 (1968) [see Erratum 91, 1043 (1969)].

64. — — — — — The Total Synthesis of Prostaglandin E_1 and Related Substances. J. Am. Chem. Soc. 91, 5372 (1969).

65. Wiberg, K. B. and A. J. Ashe: Solvolysis of Bicyclo [2,1,0]pentane-5-methyl and Bicyclo [3,1,0]hexane-6-methyl tosylates. J. Am. Chem. Soc. 90, 63 (1968).

66. Winstein, S., E. Grunwald and L. L. Ingraham: The Role of Neighboring Groups in Replacement Reactions. XII. Rates of Acetolysis of 2-Substituted Cyclohexyl Benzenesulfonates. J. Am. Chem. Soc. 70, 821 (1948).

67. Woodward, R. B., K. Heusler, J. Gosteli, W. Oppolzer, R. Ramage, S. Ranganathan and H. Vorbrueggen: The Total Synthesis of Cephalosphorin C. J. Am. Chem. Soc. 88, 852 (1966).

68. Yura, Y. and J. Ide: A Total Synthesis of dl-Prostaglandin B_1. Chem. pharm. Bull., Tokyo 17, 408 (1969).

(Received, March 6, 1970)

Chemistry of Cephalosporin Antibiotics

By **R. B. Morin** and **B. G. Jackson**, Madison, Wisconsin

With 1 Figure

I. Introduction

The discovery (*39*) and use of penicillin as the first antibiotic ushered in a new era of modern medicine. Despite the discovery and introduction into medical practice of many other antibiotics, this family of anti-bacterial substances (**1**) maintains a dominant position in the treatment of bacteria-caused diseases (*29, 138*). One of the most important ad-vances in this area within the last ten years has been the discovery of the cephalosporin antibiotics and their use in clinical medicine (*1, 146*).

$$\text{(1)}$$

e. g., Natural penicillins	Semi-synthetic penicillins
R = Benzyl, penicillin G	R = m 2,6-Dimethoxyphenyl, methicillin
R = Phenoxymethyl, penicillin V	R = x-Aminobenzyl, ampicillin

Although cephalosporin C (**2**), from which useful antibiotics have been derived, is not a penicillin, it is closely related to penicillin N (**3**) and the two are coproduced in the fermentation of a *Cephalosporium* organism. They have many structural similarities and are likely biosynthesized from the same precursors. In fact, cephalosporin C was discovered as a by-product in the purification of a degradation product of penicillin N. Cephalosporin C would probably not have been detected by procedures typical of the search for new antibiotics due to its low concentration in the broth produced by the initial *Cephalosporium* strain. Abraham and Newton at Oxford University in England have been largely responsible for this addition to the continuing and fascinating field of β-lactam antibiotics.

(3)   (2)

Penicillin N   Cephalosporin C

Early in the work on cephalosporin C, investigators realized the potential clinical importance of chemical modifications of the antibiotic, particularly after semi-synthetic penicillins had been introduced. Agreements made by the National Research and Development Corporation of Great Britain with a number of pharmaceutical companies in several countries led to extensive efforts which have resulted in the introduction of several cephalosporin derivatives into medicine, cephalothin, cephaloridine, cephaloglycin and cephalexin.

Due to its importance and the urgency necessitated by World War II, penicillin has had an unique position in organic chemistry. The penicillin problem was the first on which modern analytical tools were extensively used to gain insight into the structure of the molecule and to follow the course of degradative and synthetic reactions. Because so many chemists were involved in penicillin chemistry in the forties, their extensive use of physical methods fostered the utilization of these on other problems as well and promoted the introduction and refinement of additional physical techniques. Their rapid acceptance by organic chemists also contributed greatly to the development of stereospecific syntheses of complex natural substances. The use of these techniques is fittingly illustrated by the synthesis of cephalosporin C which was the topic of WOODWARD's Nobel lecture (*154*) on stereospecific synthesis of complex natural substances.

The fact that the β-lactam antibiotics are highly substituted and contain several bonds that are relatively labile presents an interesting but severe challenge to those studying the chemistry and synthesis of biologically important substances. They are formally derivatives of a bisdehydro dipeptide. Many other antibiotics, some of which are important therapeutically, have been recognized in more recent years as derivatives of dehydro peptides (*30*). Increased interest in the chemistry of members of this larger group is likely in the next ten years and could provide a better understanding of the chemistry of simpler peptides.

II. Isolation and Structure of Cephalosporin Antibiotics

A. Historical

In 1945, near the beginning of the intensive search for antibiotic-producing microorganisms, BROTZU, in Sardinia, isolated the fungal organism that produces cephalosporin C. BROTZU (*26*) was able to obtain crude extracts with which he treated several patients. However, he realized the difficulties of isolation and purification. In 1948 his results and a culture of the fungus were transmitted by a British health officer to Sir Howard Florey at the Sir William Dunn School of Pathology, University of Oxford. Florey had played an important role in the development of penicillin.

A collaboration with the Medical Research Council's Antibiotic Research Station at Clevedon was established early in the project and was important throughout the isolation and structure work on cephalosporin C. The initial work by the group showed the presence of two types of antibiotics, lipophilic and hydrophilic substances. The solvent-extractable antibiotics which were acids were named cephalosporins P_{1-5} (27, 50). These substances possessed activity against gram positive bacteria (122) but not against gram negative organisms as had been observed by Brotzu with the crude extract. A continuing investigation at Oxford of the nonextractable antibiotic, originally named cephalosporin N, was led by Dr. E. P. Abraham and his colleague, Dr. G. G. F. Newton. In 1953 (7) these chemists reported degradative evidence suggesting that the substance was a new type of penicillin. Subsequently cephalosporin N was shown (9) to be identical with synnematin B and was renamed (52) penicillin N in order to avoid confusion. In the course of work with impure samples of penicillin N, the presence of an additional antibiotic, ultimately named cephalosporin C, was observed (111).

B. Cephalosporin P

The first group of antibiotics isolated from fermentation broth of the species of *Cephalosporium* was a series of very closely related acids, cephalosporin P_1, P_2, P_3, P_4 and P_5 (27, 50). Only cephalosporin P_1 has been obtained in sufficient quantity and subsequent degradative evidence relates to that antibiotic. The resemblance of cephalosporin P to the antibiotic helvolic acid which had been isolated (33) earlier by Oxford workers was noted (27, 28) at the beginning of the effort on cephalosporin. The two antibiotics were both weak carboxylic acids of similar molecular weight and empirical formula and were cross-resistant in antibacterial testing. Further characterization (28) of cephalosporin P indicated an empirical formula of $C_{32}H_{48}O_8$ and the presence in the molecule of two hydroxyl groups, two acetyl functions and two double bonds, one of which was conjugated with the carboxyl function (UV λ max 211–218 nm, ε 10,000). The acetyl groups could be selectively hydrolyzed by base. Acidification of the bisdesacetyl product afforded a lactone. The conclusion that the compound was tetracyclic on the basis of empirical formula and degrees of unsaturation suggested a steroid nucleus.

Helvolic acid appeared to differ from cephalosporin P by having two keto groups in place of the hydroxyl functions and having an additional double bond conjugated with one of the ketones (UV λ max 232 nm, ε 16,150).

After a lapse of five years, a different group of workers at Oxford University reported (15) additional chemical degradations and postulated structure (4) for cephalosporin P_1. The nature of the sidechain was established as follows. Ozonolysis of cephalosporin P methyl ester furnished acetone. Ozonolysis of dihydro cephalosporin P methyl ester yielded a degradation product which on treatment with acid lost acetic acid and afforded a *cisoid* unsaturated ketone, to which partial structure (5) was assigned. The presence of an α-glycol function in monodesacetyl cephalosporin P_1 was shown by its ready cleavage with lead tetraacetate, the formation of a dioxolane derivative and its oxidation with CrO_3 to a triketone containing a non-enolic α-diketone group. The relative position of the three keto functions to each other was shown by preparation of the dienol diacetate (6), λ max 287 nm, ε 15,400. The lactone prepared from bisdesacetylcephalosporin P had an IR band at 1776 cm^{-1}, indicative of a γ lactone, thus providing evidence for an acetoxyl substituent γ to the carboxyl function in cephalosporin P.

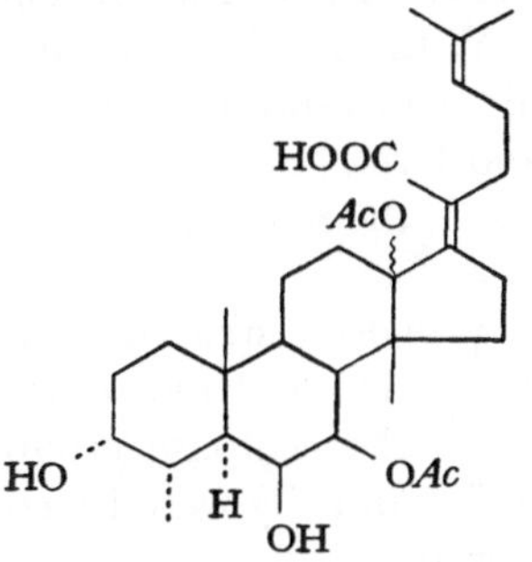

(4) 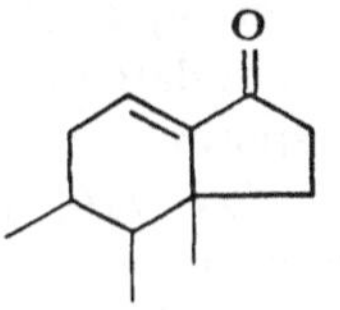   (5) 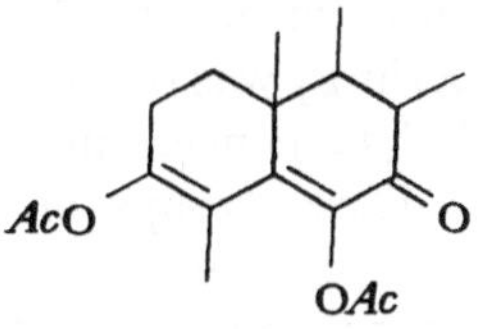   (6)

In 1962 GODTFREDSEN from Leo Pharmaceutical Laboratories in Copenhagen reported (62) the isolation of fusidic acid from *Fusidium coccineum* and shortly thereafter postulated (63) structure (7) for this antibiotic. The steroidal carbon skeleton was established by dehydrogenation. The structure was similar to that proposed for cephalosporin P_1 but differed significantly in having a methyl group at C-8.

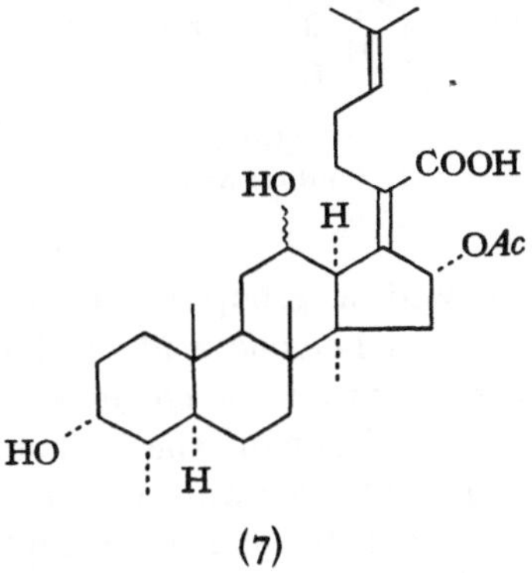

(7) 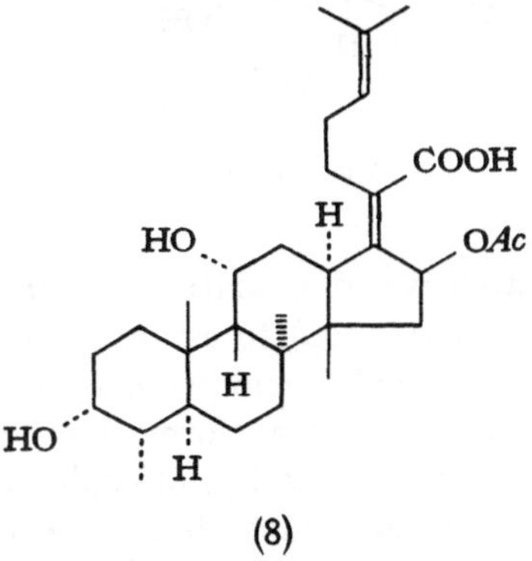   (8)

Fusidic acid

Mass spectral examination of several cephalosporin P derivatives established (98, 72) that the molecular formula for cephalosporin P was $C_{33}H_{50}O_8$, a —CH_2— unit larger than the previously accepted one. NMR evidence (102) gave support to the presence of three tertiary methyl groups and indicated that the two acetoxyl groups were both attached to a carbon bearing a hydrogen.

By a series of elegant chemical transformations which was coupled with careful deductive reasoning, workers in Copenhagen and Zurich modified (64) the structure of fusidic acid to (8). The major structural change involved the assignment of one of the hydroxyl groups to C-11 instead of the C-12. This structure has been confirmed by X-ray analysis (45). The molecule has an unusual *trans-syn-trans* A/B/C ring fusion which forces ring B into a boat conformation. With this type of ring fusion and its methyl substitution pattern fusidic acid represents a biosynthetic intermediate between squalene and lanosterol. Since cephalosporin P_1 was obviously closely related to fusidic acid, the reinterpretation of data concerning cephalosporin P was in order.

Three groups of workers (73, 116, 38) have proposed structure (9) for cephalosporin P_1 on the basis of existing information and new data, particularly from NMR spectra. The assignment of the acetoxyl group to C-6 instead of the C-7 and the configuration at C-6 and C-7 were based on the coupling constants of the two methine protons.

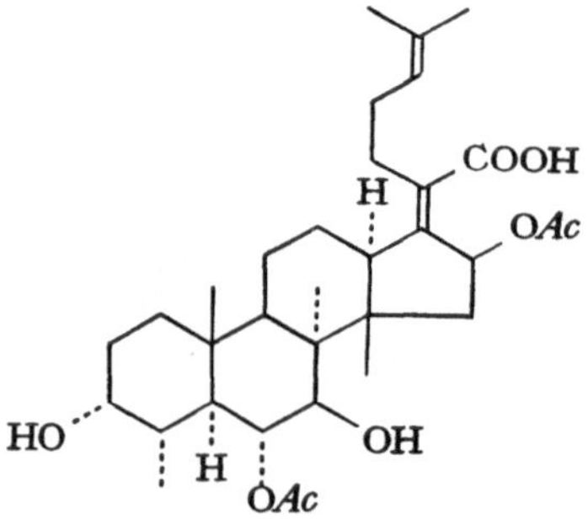

(9)
Cephalosporin P₁

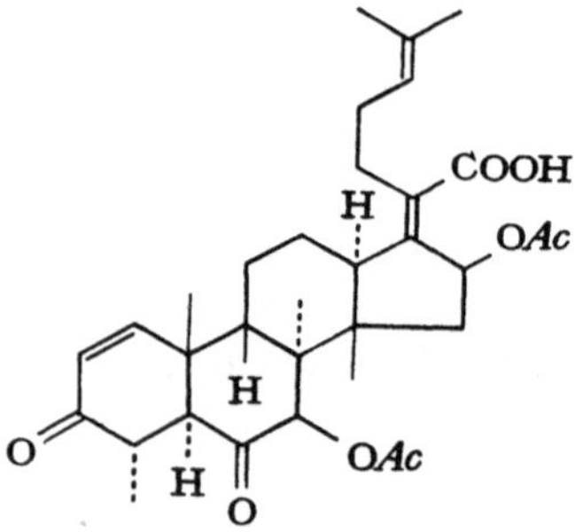

(10)
Helvolic Acid

Extensive degradative experiments allowed a group of Japanese workers (108, 109) to establish structure (10) for helvolic acid. Helvolic acid and fusidic acid have been correlated by means of chemical and microbiological transformations (109, 149). Although the triketone (at carbons 3, 6, and 7) of cephalosporin P methyl ester was very similar to the corresponding derivative of 1,2-dihydro helvolic acid, complete identity of the two substances was not established.

References, pp. 395—403

Cephalosporin P_1, helvolic acid, and fusidic acid are closely related chemically and have similar antibacterial properties. They possess activity against gram positive bacteria. Cephalosporin P is serum inactivated *in vitro* and is not very effective in protecting infected mice (*122*). Sensitive bacterial strains rapidly become resistant to the action of cephalosporin P by being grown *in vitro* with low levels of the antibiotic. Cephalosporin P has not been used for therapy in humans; however, fusidic acid is being used (*91*) clinically, particularly in combination with other antibiotics.

C. Cephalosporin N (Penicillin N)

Cephalosporin P possessed activity against gram positive organisms whereas BROTZU'S extract posessed activity against gram negative as well as gram positive bacteria. Two groups, one at Oxford and the other at the Antibiotic Research Station at Clevedon, isolated (*8*) a new antibiotic, cephalosporin N, which was a nonorganic solvent extractable, hydrophilic substance with activity against both gram positive and gram negative organisms (P, positive; N, negative). Most likely this was the antibiotic responsible for the antibacterial activity observed by BROTZU.

The antibiotic was very difficult to purify (*2, 110*), being unstable in aqueous solution below pH 4 or above 9 and at 7 in presence of heavy metal ions. Very rapidly evidence was accumulated which suggested that cephalosporin N was a new type of penicillin. It was inactivated by penicillinase and induced the production of the enzyme in a strain of *Bacillus cereus*. Acid hydrolysis produced products identical with or similar to those from benzylpenicillin.

Microanalysis (*110*) of the amorphous barium salt suggested a penicillin nucleus with a sidechain containing five or six carbon atoms and one nitrogen. Cephalosporin N behaved as an acid on ion exchange columns and carried a negative charge at pH 6–7; however, in contrast to penicillin G, a basic group could be titrated in the pH range 8–11. Hydrolysis of the antibiotic at 100° at pH 2 for one hour liberated CO_2 and provided substance (11) which corresponded to a new penilloic acid.

More vigorous acid hydrolysis of cephalosporin N yielded a rare amino acid, D-α-aminoadipic acid (12), which must be the unique feature of cephalosporin N. Treatment of the penilloic acid (11) with mercuric chloride at pH 6 gave the mercaptide of penicillamine (13) from which penicillamine was isolated as the hydrochloride and as the thiazolidine derivative of acetone. Oxidation of the penilloic acid (11) gave the corresponding penicillaminic acid (14). From this reaction, an aldehyde was obtained which was oxidized to D-α-aminoadipyl glycine, the structure of which was verified by synthesis (2). This peptide is easily hydrolyzed in acid to glycine and D-α-aminoadipic acid and also undergoes hydrolysis without added acid to glycine and 6-carboxy-piperidone-2.

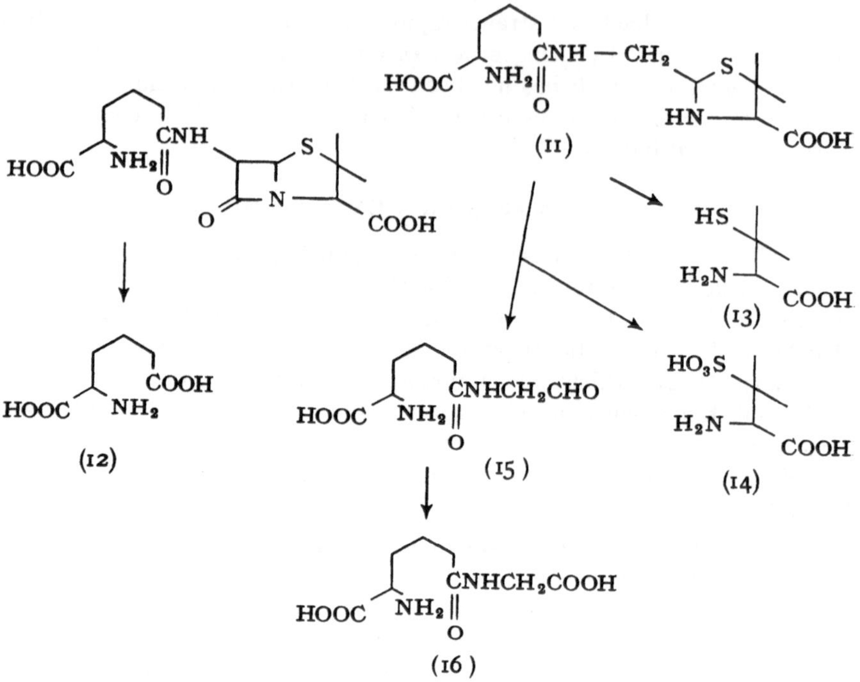

Chart 1. Degradative Reactions of Cephalosporin N

The isolation of penicillamine (13), the aldehyde (15), and CO_2 accounted for all the elements of the proposed penicillin structure. The antibiotic undergoes other reactions typical of a penicillin, such as the formation of the penicilloic acid (17) in alkaline solution (or formation of a hydroxamic acid in the common chemical assay for penicillins) and the formation of the penillic acid (18) in mildly acidic solutions.

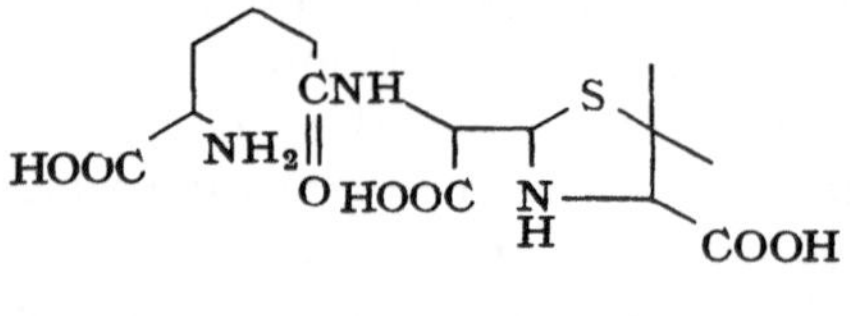

(17)
The penicilloic acid of Pen N

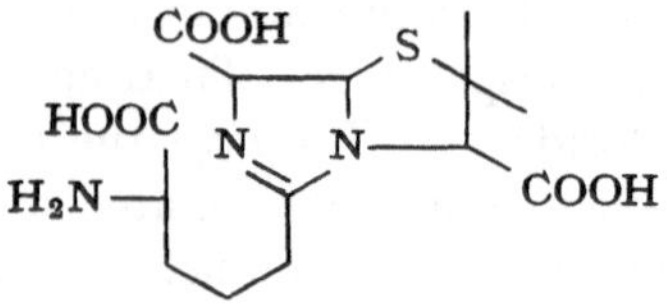

(18)
The penillic acid of Pen N

In the United States, a group of workers from the Michigan Board of Health reported (65) in 1949 that a species of *Cephalosporium* (subsequently named *salmosynnematum*) produced a new water-soluble antibiotic, synnematin. The publication of the chemical and biological properties of synnematin led the English workers to suggest (7) that cephalosporin N and synnematin might well be identical. Further purification (114) yielded two synnematin components one of which, synnematin B, was found to be identical (9) with cephalosporin N.

After the initial isolation and purification of synnematin, several pharmaceutical companies, primarily Abbott Laboratories and Parke Davis & Company, became involved in producing larger quantities and obtaining clinical evaluations. Synnematin possessed 1/100 the gram positive activity of benzylpenicillin but somewhat better activity against certain gram negative organisms, particularly strains of *Salmonella*. The antibiotic was useful in the control of typhoid. Manufacturing problems related to its instability and solubility caused the abandonment of therapeutic use of the substance.

D. Cephalosporin C

During the purification of a sample of the penillic acid (18) from an impure preparation of penicillin N, a fraction was obtained which moved more slowly on an anion-exchange resin column (111) and had a strong UV absorption maximum at 260 nm (ε 9,500). The substance was isolated (112) as a crystalline sodium salt dihydrate, $[\alpha]_D^{20} + 103$, and was shown to be an antibiotic. That acid treatment of purified penicillin N did not produce this substance and that it could be isolated (with greater difficulty) from the broth without acid inactivation indicated that it was a new antibiotic and not a degradation product of penicillin N. The substance, which was named cephalosporin C, could be detected by its UV chromophore in crude samples of penicillin N.

The similarities of penicillin N and cephalosporin C were immediately apparent from the difficulties of separating the two. Indeed the method which originally led to the discovery of cephalosporin C, from acid deactivated broth was routinely used for subsequent isolation of cephalosporin C in larger quantity for structural work. Amberlite IR-4B resin was used with the volatile buffers, ammonium acetate or pyridine acetate. The two antibiotics were separable by paper chromatography utilizing buffered papers and methanol as eluent, although they show the same paper electrophoretic mobility at pH 7. Cephalosporin C by titration possessed two acid groups, pKa < 2.6, and 3.1 and a basic group, pKa 9.8, with an equivalent weight 480 ± 15. Thus, like penicillin N,˙ cephalosporin C behaved as a mono amino-dicarboxylic acid. The compound gave a positive ninhydrin test. Hydrolysis (3) of cephalosporin C, like that of penicillin N, liberated one mole of carbon dioxide and D-α-aminoadipic acid. Further treatment of cephalosporin C with dinitrofluorobenzene followed by acid hydrolysis yielded the DNP derivative of D-α-aminoadipic acid. The infrared spectrum of the new antibiotic possessed a strong band at 1780 cm^{-1} characteristic of the strained β-lactam of penicillins.

Penicillin N and cephalosporin C differed, however, in a number of significant aspects. The micro analytical data for cephalosporin C sodium salt dihydrate best fitted the formula, $C_{16}H_{20}O_8N_3SNa \cdot 2H_2O$ (mol. wt. 473), indicating that the free acid differed from penicillin N by two carbon and two oxygen atoms. Cephalosporin C had a band in the IR at 1735 cm^{-1}, suggestive of an ester or lactone. Relative to penicillin N, cephalosporin C was stable to .lN HCl at room temperature. At pH 11 cephalosporin C was rapidly inactivated, producing two new titratable acid groups and loss of the UV chromophore. Penicillins are similarly inactivated by forming a penicilloic acid which, however, only yields one new acidic group. Cephalosporin C was insensitive (112) or much less sensitive than penicillin N to various preparations of penicillinase. This difference heightened the interest in cephalosporin C. However, cephalosporin C was destroyed by certain lactamase preparations and the loss of antibacterial activity was associated with the loss of the UV chromophore.

Cephalosporin C on acid hydrolysis (3) with lN HCl, 105° for eight hours, liberated two equivalents of NH_3 (or volatile base) whereas, penicillins provided one. In the acid hydrolysis no penicillamine (13) could be detected. However, cephalosporin C closely resembled penicillins because hydrogenolysis with Raney nickel, followed by acid hydrolysis, provided alanine and valine, an observation which indicated the presence of the same carbon skeleton in the two molecules. Significantly, however, the valine from hydrolysis of cephalosporin C was extensively racemized in contrast to the D-valine isolated from benzylpenicillin.

On the basis of this evidence, the suggestion was made, in 1956, that cephalosporin C was a β-lactam antibiotic containing an α-aminoadipyl sidechain linked at C-6 which differed from penicillin N in the thiazolidine ring system and perhaps contained an oxygen substituent (OAc) at C-3.

Additional degradation experiments which were reported in 1958 (6) and 1960 (82) further substantiated the relation between penicillin N and cephalosporin C. Mild acid hydrolysis followed by oxidation with Ag_2O gave α-aminoadipyl glycine (16) which had been obtained earlier from penicillin N. By warming cephalosporin C in aqueous solution at pH 7 or in aqueous pyridine at 37° for three days the English workers isolated a new degradation product, the thiazole (19). The isolation of (16) and (19) is in agreement with partial structure (20). At the time this work was reported degradations of penicillins had not yet resulted in the isolation of a thiazole; however, more recently several reaction sequences have produced thiazoles from penicillins (93, 104).

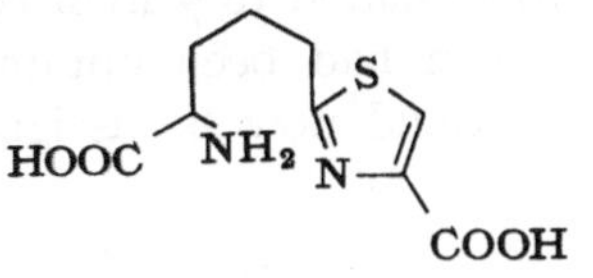

(19)

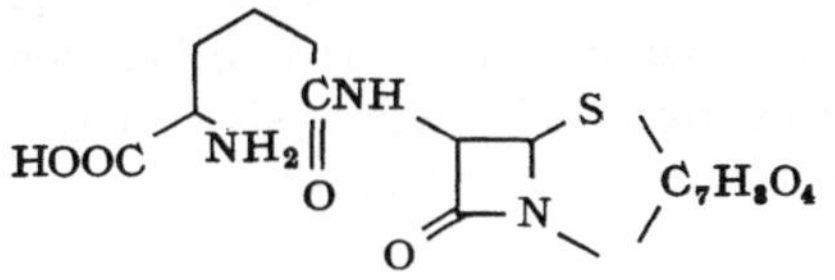

(20)

The isolation (6) of α-keto-isovaleric acid from hydrogenolysis of cephalosporin C with Raney nickel, followed by mild hydrolysis, supported the idea that the antibiotic was a C-3 oxygen substitute d penicillin. However, a fused β-lactam-thiazolidine structure containing a valine carbon skeleton could not conceivably explain the UV chromophore of cephalosporin C (4).

The unknown fragment of (20), $C_7H_8O_4$, must account for the formation of valine on hydrogenolysis and must contain the two "extra" carbons. This two-carbon fragment was shown (4) to be an acetyl group when acetic acid was identified as a product of acid hydrolysis.

A KUHN-ROTH determination indicated the presence of one C-methyl group in cephalosporin C. This result could be easily accounted for by the acetoxyl function and the earlier assumption of a *gem*-dimethyl group in the molecule based on its presumed relationship to the penicillins was questioned. The NMR spectrum of cephalosporin C definitely ruled out a *gem*-dimethyl group.

Structure (3) which was then considered gained rapid acceptance among the workers. This formulation readily explained the results of hydrolysis which produced acetic acid and a neutral species (4, 6), the lactone cephalosporin C_C (21). This substance, which had antibacterial properties possessed a UV chromophore similar to cephalosporin C and had an IR band at 1755 cm⁻¹, characteristic of a γ lactone.

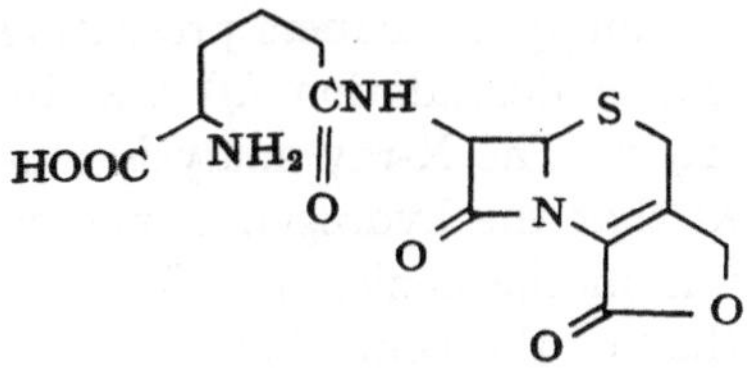

(21)
Cephalosporin C_C

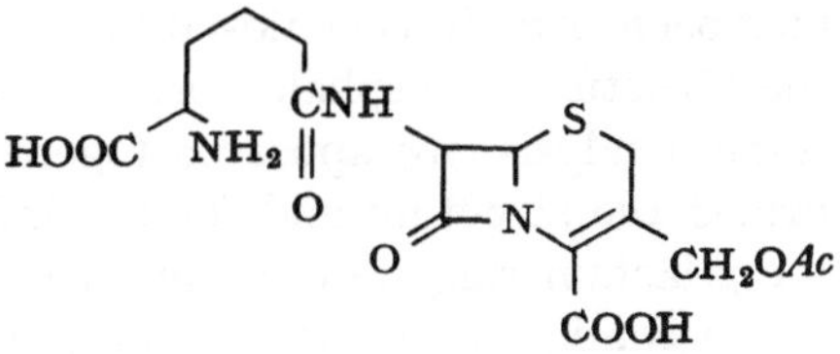

(3)
Cephalosporin C

Treatment of cephalosporin C or C_C with Raney nickel at room temperature, followed by hydrolysis resulted in isolation of a lactone (22) which was identical with a synthetic sample. The more proximate lactone (23), isolated by solvent extraction of the Raney nickel-treated cephalosporin C_C solution, was easily hydrolyzed to (22), thus accounting for the genesis of (22). Compound (23) was hydrogenated to γ hydroxy-valine lactone (24). Two other compounds which had been obtained earlier from acid hydrolysis of cephalosporin C could now be assigned structures (25) and (26).

(22)   (23)   (24)

(25)   (26)

These lactones provide firm evidence for the presence of the dihydro-thiazine ring in cephalosporin C. That the double bond must be in the 3,4 and not the 2,3 position was indicated by the isolation of the dinitro-phenylosazone of hydroxyacetone on ozonolysis and Raney nickel treatment of cephalosporin C. The UV maximum spectrum of cephalosporin C is more consistent with the 3,4 position although it lies at a higher wave length than would be expected (see Section IV). Other properties of cephalosporin C such as hydrolysis of the aminoadipyl sidechain (96) and its reactions with nucleophiles (71) which will be discussed in later sections, further supported expression (3).

During the time the dihydrothiazine structure was first considered, Hodgkin and Maslen detected the presence of a sulfur-containing six-membered ring in cephalosporin C by X-ray analysis. Papers presenting the structure of cephalosporin C by chemical degradation (4) and by X-ray analysis (78) appeared together in 1961. The X-ray analysis con-firmed the structure and further indicated that the hydrogen atoms on the β-lactam ring were cis to each other as in the penicillins. The ab-solute configuration also conforms with that of the penicillins.

Further chemical transformations and syntheses of cephalosporin C confirm the structural findings in every detail.

III. Chemical Transformations of Cephalosporin C

A. Nomenclature

The name cephalosporanic acid has been suggested (4) for the structure (27) analogous to the name penicillanic acid (28) and has been extensively used to describe the many acyl derivatives of the cephalosporin C nucleus. The term cephalosporadesic acid (29) has also been employed for the desacetyl derivatives.

(29)   (27)   (28)

With the isolation of cephalosporin derivatives having a double bond between C-2 and C-3 this nomenclature became cumbersome. The use of the simpler term cepham (30) was suggested (105) in collaboration with the English workers, again in analogy to penam (32) (129) which until very recently has not been extensively used. Compound (31) then becomes Δ^3-cephem. The stereochemistry noted in the structure is implied in the name. The systematic name for cepham is 1-aza-5-thia-6 R-bicyclo-[4,2,0]octan-8-one.

(30)   (31)   (32)

B. Sidechain Amido Functions

1. Reactions of the 7,α Aminoadipamoyl Group

The importance of this function was demonstrated by LODER, NEWTON and ABRAHAM (96), who subjected cephalosporin C to mild acid hydrolysis and obtained low yields of 7-aminocephalosporanic acid (7-ACA, (33)). Reacylation of 7-ACA with phenylacetyl chloride gave a derivative whose antibacterial activity was approximately 100-fold greater than cephalosporin C.

Enzymatic methods which were useful (123) in obtaining 6-APA from natural penicillins were unsuccessful when applied to sidechain cleavage of cephalosporin C. Amidases have been found, however, which are capable of removing relatively non-polar sidechains from cephalo-

23*

sporanates (*43, 89*). Since α-aminoadipic acid is synthesized *de novo* by *Cephalosporium*, it has not been possible to obtain 7-ACA by non-pre-cursed fermentation as was the case with microbiological fermentation of 6-APA by *Penicillium* (*20*).

The search for practical chemical methods which would permit re-moval of the α-aminoadipic sidechain from cephalosporin C led to a process which exploited the sidechain amino group in an intramole-cularly assisted cleavage (*105*). Treatment of cephalosporin C with nitrosyl chloride in formic acid followed by removal of excess nitrosating agent and quenching with water or methanol gave 7-ACA in yields of 35–50%. Three other adipyl cephalosporanates (34, (35), (36) and 7-α-chlorocephalosporanic acid (37) were found to be side products of the reaction (*106*).

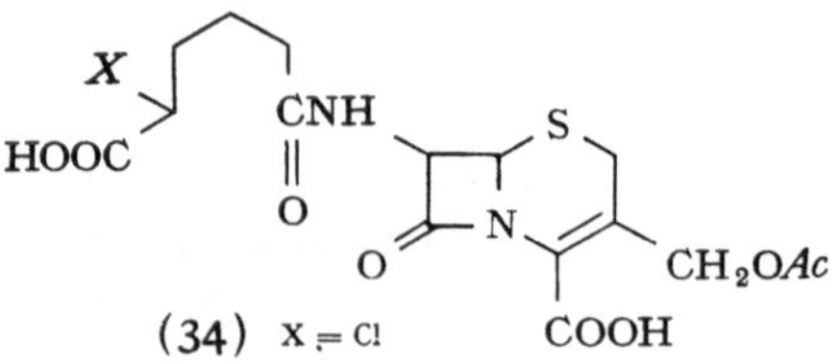

(34) X = Cl
(35) X = OCHO
(36) X = OH

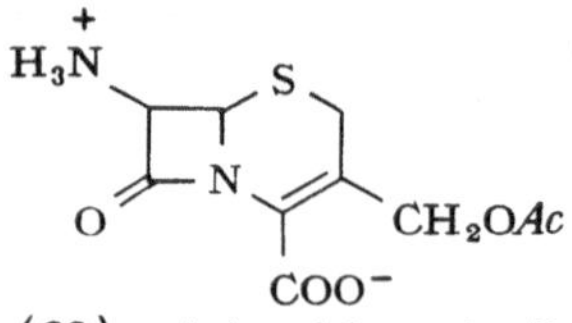

(33) 7-Aminocephalosporanic acid 7-ACA

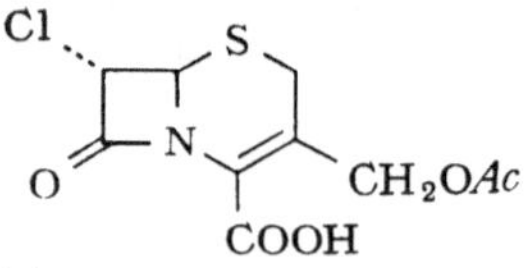

(37) 7-α-Chlorocephalosporanic acid

The proposed mechanism for the formation of (33) involved dia-zotization of the amino function on the sidechain to give (38) which then reacts intramolecularly to produce the iminolactone (39). Solvolysis of the iminolactone then gives (33). Alternately the diazonium intermediate (38) could be substituted intermolecularly by chloride, formate or water

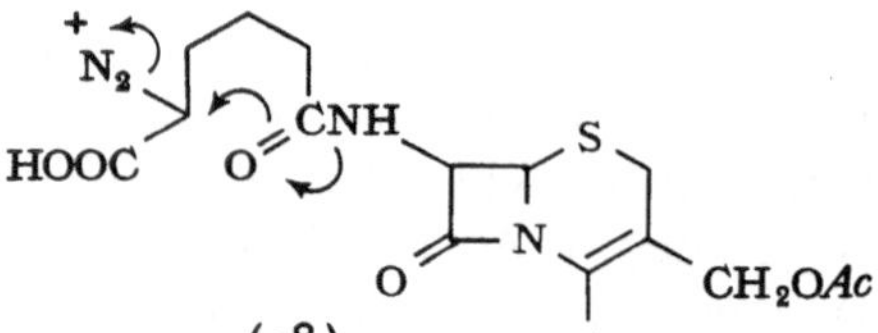

(38)

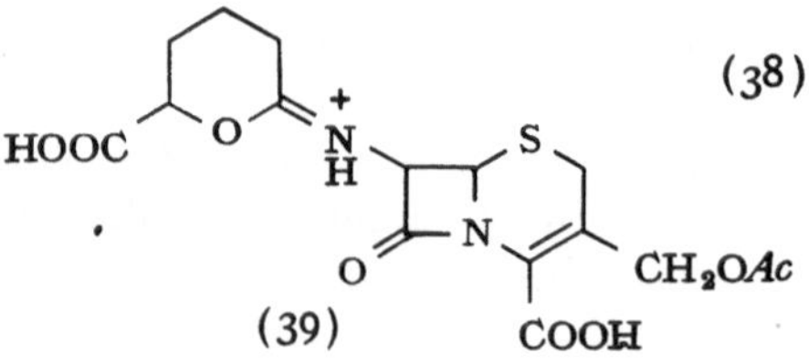

(39)

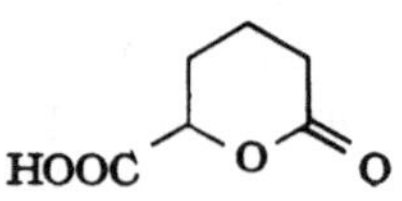

(40)

to produce (34), (35), or (36) respectively. Substance (37) was presumably formed by interaction of nitrosyl chloride with the small amount of (33) formed in the reaction mixture prior to removal of nitrosating agent. The same chloride was produced upon reaction of pure (33) with nitrosyl chloride. Isolation of the lactone (40) (as its methyl ester) lent support to the mechanism.

A more general approach to sidechain cleavage of cephalosporins was developed by the CIBA group (59). A suitably protected cephalosporin C derivative, on treatment with phosphorus pentachloride, formed the iminochloride (41). Methanolysis of the chloride afforded the iminoether (42) which upon subsequent hydrolysis and removal of carboxyl-blocking groups gave (33) in excellent yield (80–85%). Since this method does not depend upon the fortuitous placement of the sidechain amino group it should be generally applicable to all cephalosporins irrespective of the nature of the sidechain.

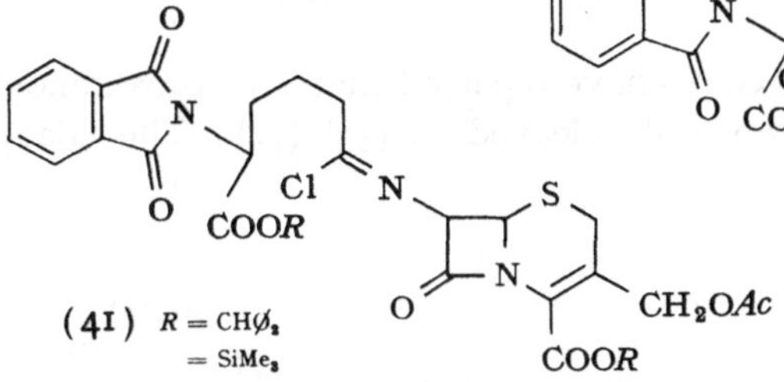

(41) R = CH∅₂
    = SiMe₃

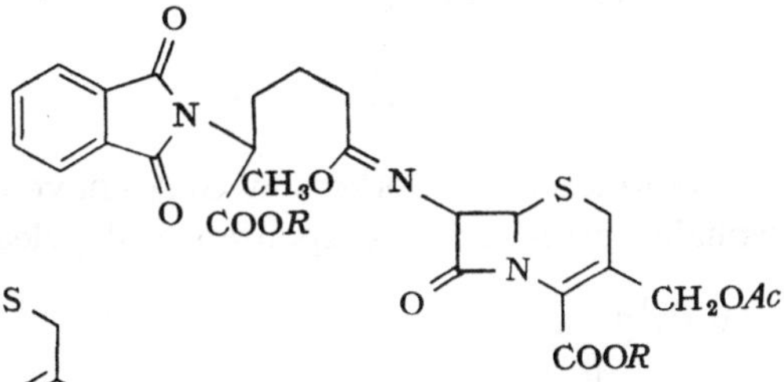

(42) R = CH∅₂
    = SiMe₃

An interesting variation of intramolecular cleavage was also reported (59) by the CIBA scientists. Treatment of cephalosporin C dibenzyl ester (43) with pyridine/acetic acid in methylene chloride at room temperature gave a 50% yield of (44) together with the piperidone (45). The probable

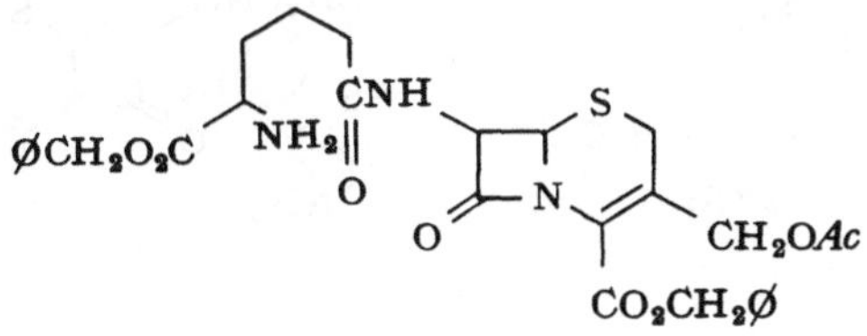

(43)

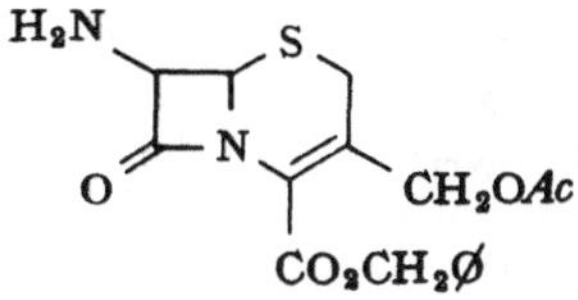

(44)

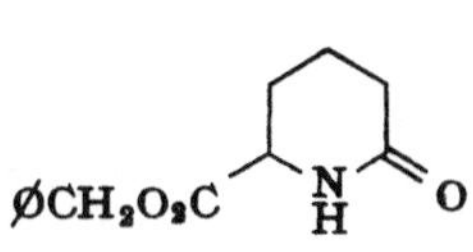

(45)

mechanistic pathway involves the interaction shown in the following partial structures.

(43) 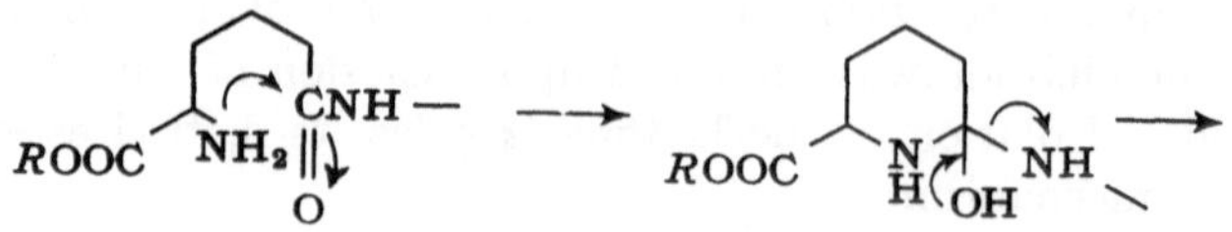 ... → ... → (44) + (45)

The above intramolecularly assisted scissions are formally analogous to observations made by ABRAHAM and NEWTON, reported in 1954 (2). They found that the dipeptide (16) was cleaved to glycine and the piperidone (46) upon heating in water. The first reported acid hydrolysis of cephalosporin C to give (33) could likewise be portrayed as occurring in an intramolecular fashion.

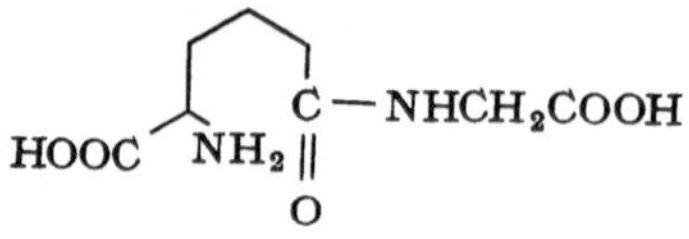

(16) 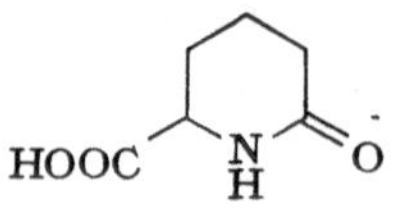   (46)

'COCKER and co-workers at Glaxo have reported that ω-chloroalkanoyl cephalosporins (47) are spontaneously cleaved to (33) (41). The Glaxo

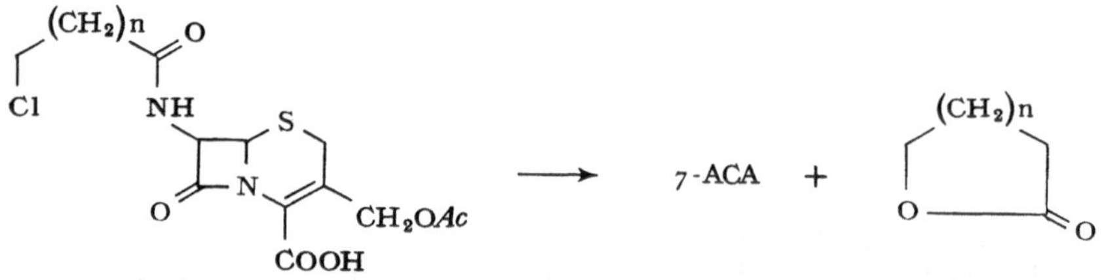

(47)   7-ACA   +

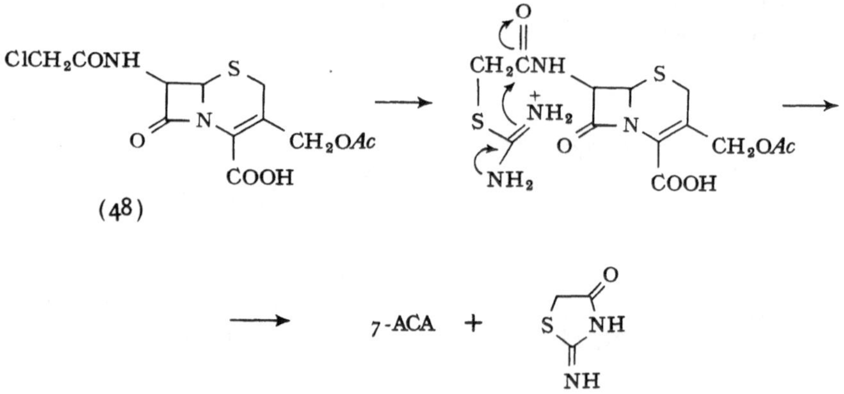

(48)

→   7-ACA   +

Chart 2. Cleavages of Chloroalkanoyl Cephalosporins

group also found that chloroacetamidocephalosporanic acid (48), upon treatment with thiourea, is converted to (33). Both types of transformations have been observed with simpler amides (*135, 32*).

Acid hydrolysis of cephalosporin C under more drastic conditions led to isolation and characterization of cephalosporidine (49) (*11, 22*). This transformation was rationalized by assuming hydrolysis and decarboxylation to give the aldehyde (50) which then undergoes cyclization. Synthesis of (50) and its cyclization to (49) lends support to the postulated intermediacy of the aldehyde.

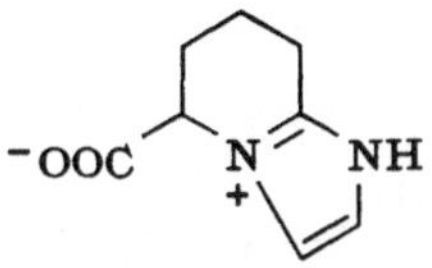

(49) 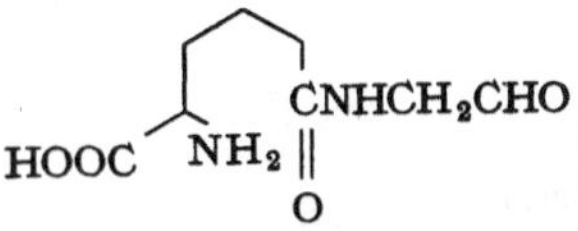 (50)

Another example of apparent intramolecular reaction of the amide function is compound (19). This substance was formed as a minor product when cephalosporin C was heated with aqueous pyridine (*82*). Its probable mode of genesis is shown in the following sequence.

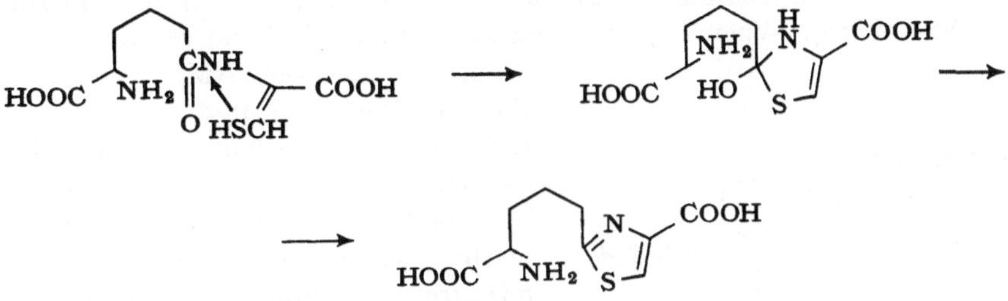

(19)

2. Reactions of the 7-Amino-Group

The most thoroughly studied reaction of this function is its acylation to give 7-acylamido derivatives which possess enhanced antibiotic activity compared with that of cephalosporin C (see Section VII). Acylation has been effected utilizing a variety of methods *viz.*, Schotten-Baumann conditions (*36, 145*), dicyclohexylcarbodiimide (*36, 145*), and mixed anhydride techniques (*132*). When milder acylating agents are used, water competes with the weakly nucleophilic amino function and considerable hydrolysis of the acyl moiety has been observed (*80*).

Two other reactions of the amino group have been reported. One of these has been mentioned in the preceding section, *i. e.*, the reaction of (33) with nitrosyl chloride which gave the 7-α-chloro derivative (37). The inverted configuration at C-7 was based principally on NMR evidence. As for the second reaction, the penicillin analog of 7-ACA, 6-aminopenicillanic

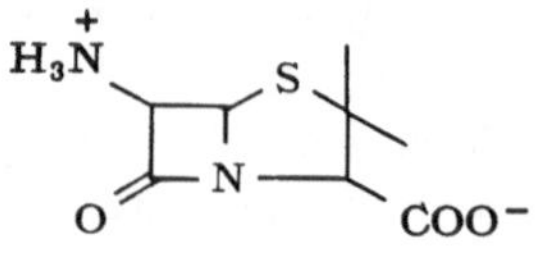

(51)

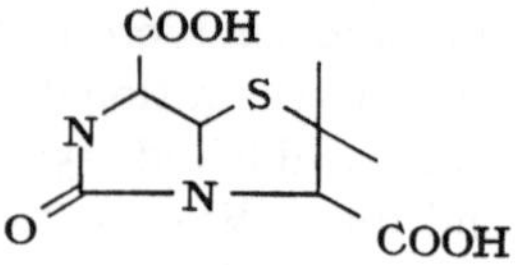

(52)

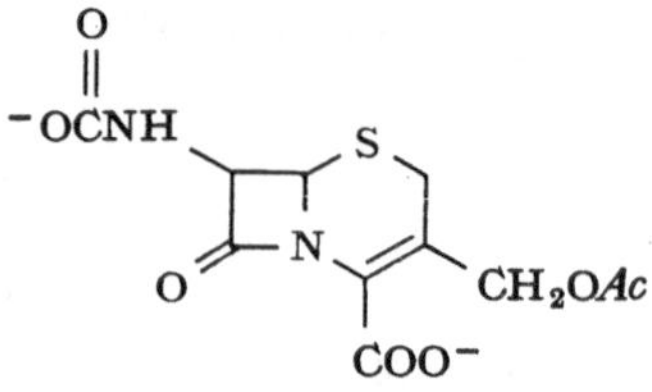

(54)

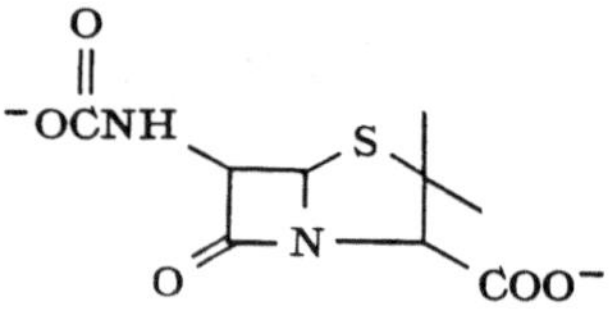

(53)

acid (6-APA) (51) undergoes rearrangement to 8-hydroxypenillic acid (52) in aqueous sodium bicarbonate solution (16, 21, 84), presumably by way of (53). When (33) was treated similarly (13), no rearrangement occurred; however, NMR spectra of the solutions indicated formation of the analogous carbamate (54). The existence of (54) could be substantiated by isolation of the methylcarbonate methyl ester (55) when reaction mixtures containing (54) were treated with methyl iodide and diazomethane.

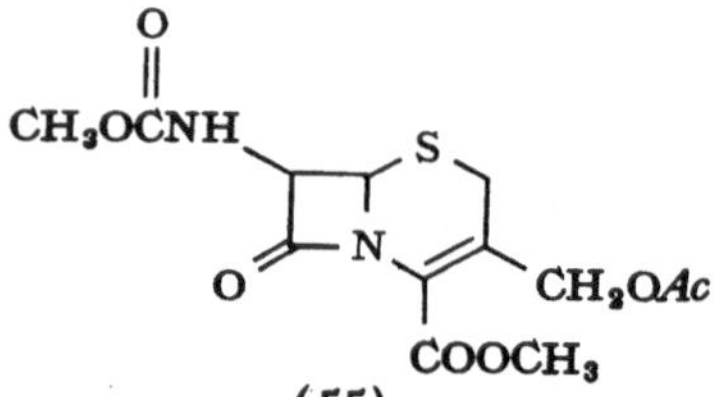

(55)

C. Reactions of the β-Lactam Ring System

Several recent publications (40, 85, 86, 117, 153) have reported base-catalyzed epimerizations of penicillins at C-6 to give 6-epipenicillins (56). The analogous reaction has been reported with cephalosporin sulfoxides (126). Treatment of cephalothin sulfoxide with triethylamine in refluxing chloroform gave two decarboxylated products which were shown to be C-7 epimers, (57 and (58).

A similar observation was made with the ester (59). Triethylamine in dimethyl sulfoxide was found to cause isomerization to (60). Cephalothin, in contrast to its sulfoxide, was not isomerized under conditions which caused epimerization of the sulfoxides.

The structural assignment of the C-7 epimers was based on the known (*101*, *49*) diminution of the H-6, H-7 coupling constant from about 5 Hz in the *cis* compounds to about 2 Hz in the *trans* epimers. Deuterium exchange at C-7 (but not C-6) was also observed.

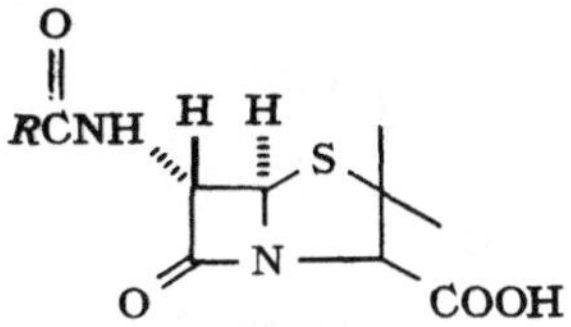

(56) 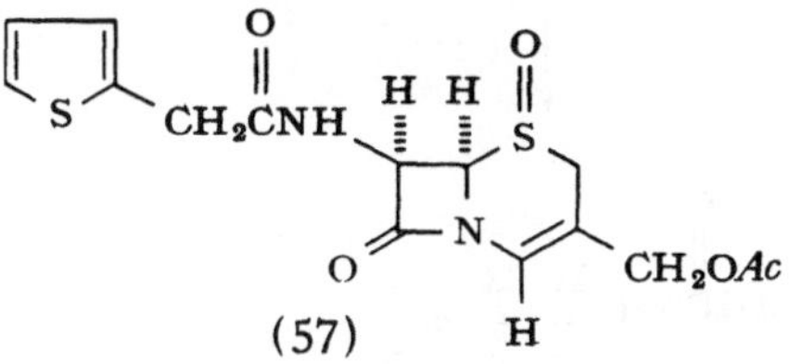   (57)

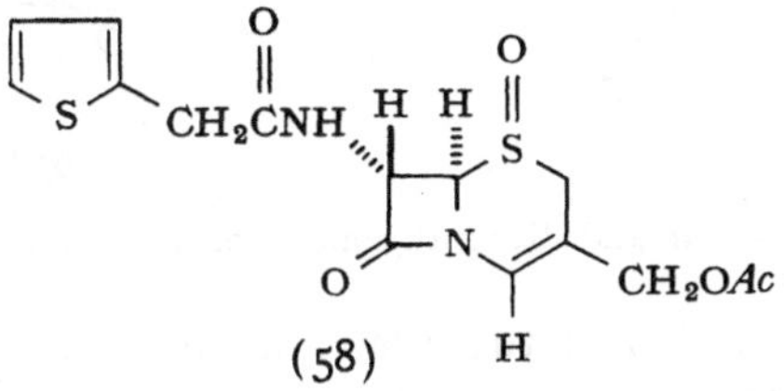

(58)

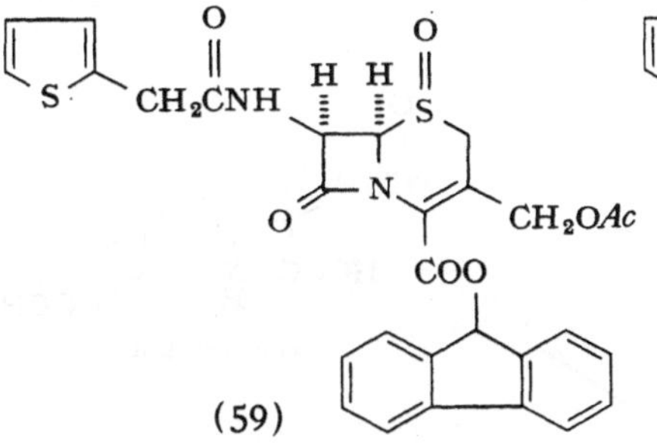

(59) 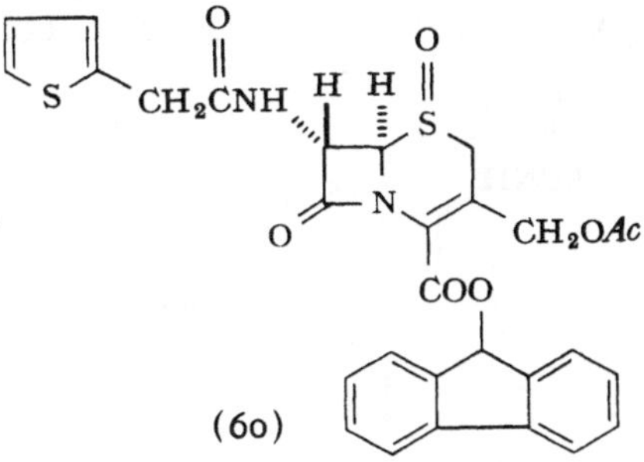   (60)

GUTOWSKI (*69*) has observed epimerization of penicillin sulfoxide under non-basic conditions. The sulfoxide of 6-β-phenoxymethyl penicillin trichloroethyl ester, upon treatment with N,O-bistrimethylsilylacetamide, was transformed into an equilibrium mixture of C-6 epimers (61), (62). Relative quantities of (61) and (62) at equilibrium were approximately 4 : 1 with the 6-β-hydrogen epimer predominating. Rearrangement of the 6-β-hydrogen isomer using conditions similar to those described for the 6-α epimer (*107*) gave the cephem (63). Oxidation of (63) afforded a mixture of sulfoxides epimeric at position — 1, neither of which underwent C-7 isomerization to the sulfoxide of (64) upon treatment with N,O-bistrimethylsilylacetamide.

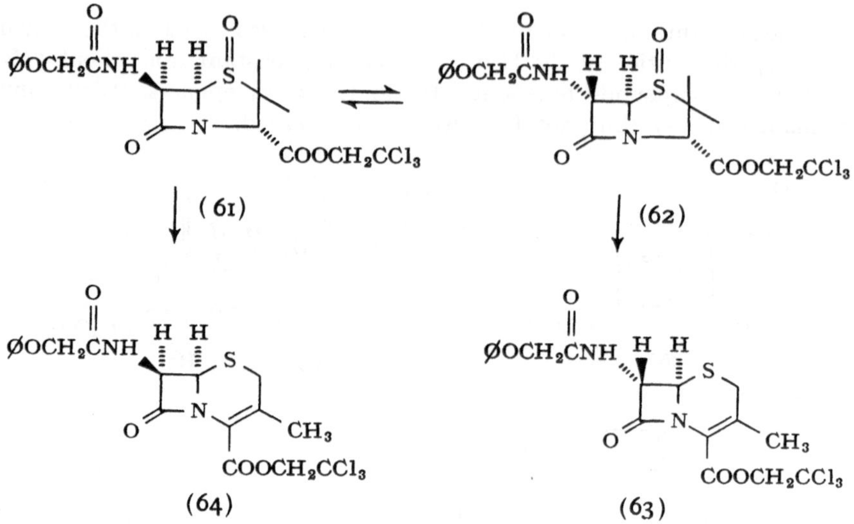

Chart 3. Epimerization and Rearrangement of Penicillin Sulfoxides

The best known reaction of the β-lactam function in the penicillins is its facile cleavage to give a variety of products (39). The following scheme portrays several of these. No parallel reactions for the cephalosporanic acids have been described.

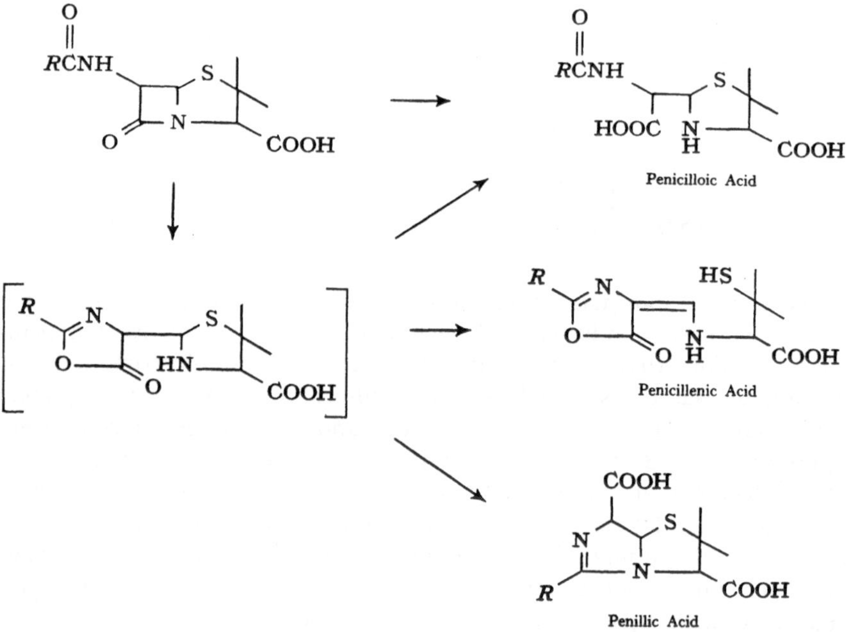

Chart 4. Mild Degradative Reactions of Penicillins

The azetidinone ring of cephalosporins is somewhat more stable to chemical cleavage than that of penicillins. For example, alcohols cleave penicillins to esters of penicilloic acids (*39*); however, cephalosporins are sufficiently stable (*4*) toward alcohols to permit use of methanol as a recrystallization solvent (*81*). Cephalosporins are also more stable toward acids.

The reactivity of the β-lactam has been correlated with the frequency of its infrared absorption band (*107*). In general, the higher the frequency of absorption, the more sensitive the β-lactam toward cleavage by nucleophiles. Penicillin and cephalosporin lactams have IR absorption bands at 1790–1795 cm^{-1} and are, for example, completely cleaved by hydroxylamine. This is the basis for a chemical assay. On the other hand, dihydrocephalosporins (65), whose β-lactam absorption occurs at 1776 cm^{-1}, are resistant to the action of hydroxylamine. Cephalosporins in which the double bond has been isomerized to the 2–3 position (66) (IR max 1784 cm^{-1}) are partially cleaved with hydroxylamine.

(65)   (66)

The sensitivity of penicillin lactams which contrast with that of simple azetidinones has been rationalized (*39*). The fused thiazolidine ring system prevents coplanarity of the —N—C=O and adjacent atoms so that normal amide resonance cannot occur. The lactam carbonyl therefore is more electrophilic and sensitive to nucleophilic attack. The extent of non-coplanarity of the β-lactam is reflected in the distance of the nitrogen from the plane of the three carbons bonded to the nitrogen. SWEET and DAHL (*140*, *141*) have measured this distance for several cephalosporin derivatives. In cephaloridine (67) and cephaloglycin (68) the distance is 0.22–0.24 Å. With the Δ²-cephalosporin (69) the value is only 0.065 Å. In benzylpenicillin the distance was earlier reported to be 0.40 Å (*142*).

SWEET has also correlated these parameters with the length of the carbonyl-nitrogen bond. Electron pair delocalization from nitrogen would result in a shorter bond length due to contribution from structure (70). The measured bond length in the Δ²-isomer (69) was found to be substantially shorter than in penicillins or Δ³-cephalosporins.

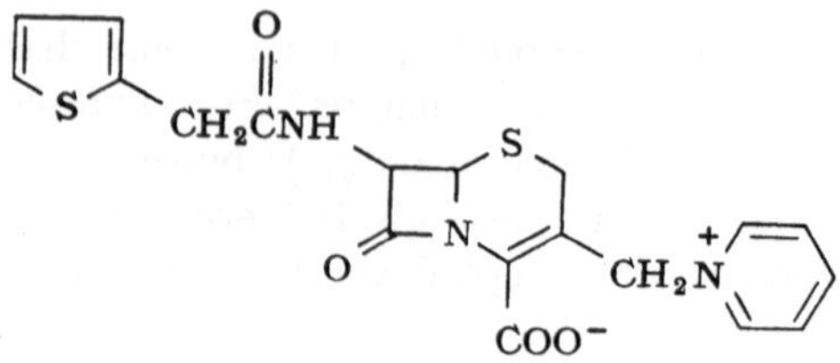

(67)

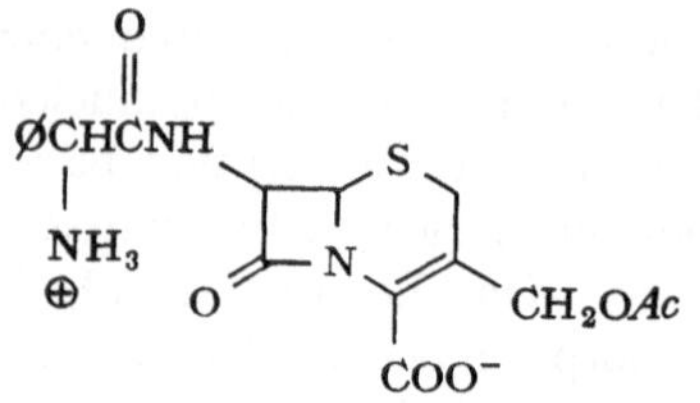

(68)

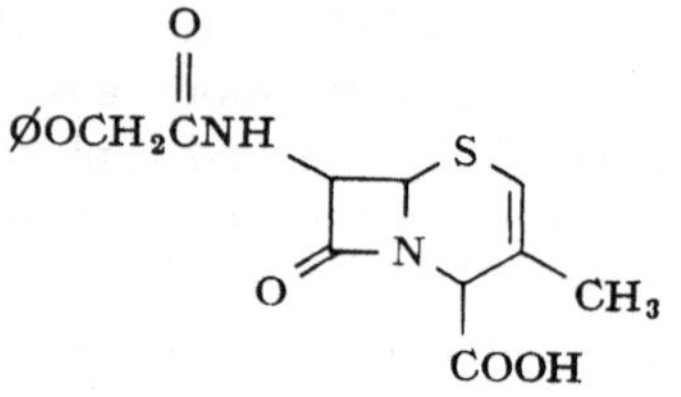

(69)

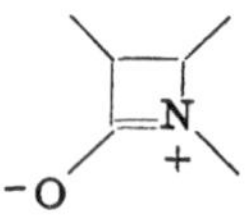

(70)

Scission of penicillin and cephaloporin s azetidinones by enzymes is also known. Several species of bacteria can be induced to produce β-lactamases which are described as penicillinases or cephalosporinases depending upon which type of substrate is sensitive to them. Penicillinases cleave penicillins to penicilloic acids; however, the cleavage of cephalosporins by cephalosporinases is more complex.

SABATH, JAGO and ABRAHAM reported (125) that the β-lactamase from *Pseudomonas pyocyanaea* cleaved not only the β-lactam of cephalosporins but also caused expulsion of the acetate function at C-3'. Similarly, cephaloridine (67) was reported to lose pyridine upon treatment with the enzyme. These degradations were accompanied by loss of the characteristic UV maximum at 260 nm. When cephalosporin C lactone (21) was used as a substrate the UV maximum did not disappear but shifted from 257 to 265 nm. ABRAHAM and SABATH (10) found that β-lactamases from other species of bacteria gave results similar to those reported above.

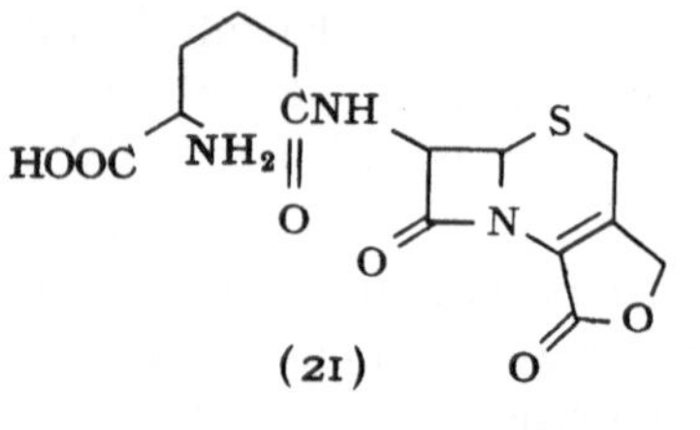

(21)

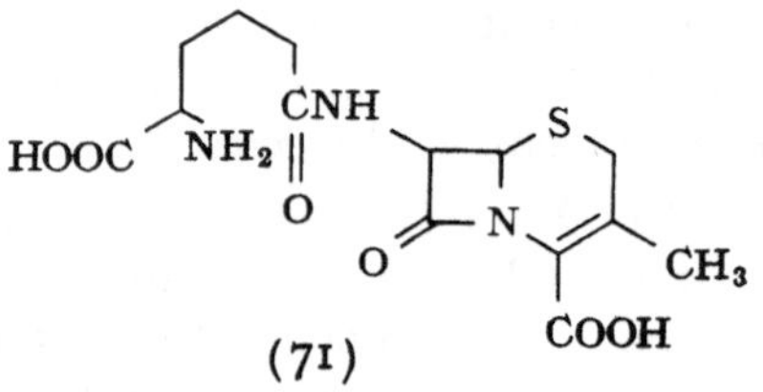

(71)

Correlation of enzymatic and chemical cleavage of cephalosporins using spectral and chromatographic techniques was subsequently reported

References, pp. 395—403

by NEWTON, ABRAHAM and KUWABARA (*113*). Treatment of cephalosporin C with ammonia, ε-aminocaproic acid or β-lactamases gave similar results, specifically rapid loss of the 260 nm chromophore and formation of a new chromophore having λ max 227–230 nm. On the other hand desacetoxycephalosporin C (*71*) was found to be much more stable to the degradation conditions. The 260 nm chromophore disappeared much more slowly and no new 228 nm chromophore was produced.

The results of the preceding degradations were rationalized in the following fashion: rupture of the lactam is accompanied by simultaneous loss of acetate as portrayed in the scheme below. This transformation could account for acetate formation and loss of the 260 nm chromophore.

A compound (*73*) analogous to (*72*) had been postulated as an intermediate in the conversion of (*74*) to (*75*) (*56, 57*). Compound (*75*) was produced on treatment of (*74*) with sodium benzyloxide. Confirmation of the structure of (*75*) was achieved by synthesis.

(72)

(74)   (75)

Additional support for the existence of an intermediate analogous to (*72*) was reported very recently (*74*). Treatment of cephalosporanic acids with ammonia and subsequent ozonolysis produced formaldehyde, which provided evidence for the presence of a $=CH_2$. The course of the reaction of cephalosporanates with deuterated ammonia has been studied with NMR (*75*). These studies indicated that cleavage of the β-lactam was accompanied by simultaneous expulsion of acetate and formation of a diene analogous to (*72*).

A product resulting from cleavage of the β-lactam with aniline has been reported (*25*). Structure (*76*) was proposed for this substance.

(73)   (76)

D. Reactions of the Dihydrothiazine Ring System

In this section the chemistry of the six-membered ring system found in cephalosporin C derivatives will be discussed, more specifically those reactions concerning this ring but not involving a fragmentation of the molecule. Many of the unique reactions of cephalosporins are concerned with this ring system. This system which contains the elements of an enamine, allylic sulfide, allylic acetate and α, β-unsaturated acid presents a more difficult synthetic challenge than the penicillin molecule.

1. Reactions of the Double Bond

a) Formation of Δ^2-Isomers. In the structural work on cephalosporin C the question was raised whether the double bond could be at the 2,3-position. However, the data clearly indicated the contrary.

Subsequent investigations with cephalosporanates provided compounds in which the double bond was in the 2,3-position. Acylation (42) of (33) with acid anhydrides and base under anhydrous conditions provided a mixture of Δ^2- and Δ^3-cephalosporanates. Treatment of various 7-acyl cephalosporanates with base caused equilibration of the double bond between the two positions. The equilibration is more rapidly established in esters or in mixed anhydrides, although the acids themselves will slowly isomerize. With $R' = OAc$, $R^2 =$ thiopheneacetyl (42) or phenoxyacetyl (42, 107) and $R^3 =$ methyl, the equilibrium was 7 : 3 in favor of the Δ^2-isomer (78). With $R' = H$ the ratio at equilibrium was inverted with the Δ^3-isomer predominating (107). The presence of a larger substituent at C-3 most likely favors the Δ^2-isomer in order to minimize the steric repulsion of two cis substituents on the double bond.

(77)   (78)

Although the isomerization of a Δ^3- to a Δ^2-isomer introduces a new asymmetric center, workers in several laboratories report that only a single stable isomer is produced. In the synthesis of (78) ($R' = OH$ or aldehyde attached to C-3) the unstable isomer was produced; however, it epimerizes readily to the normal one (76).

The configuration at C-4 in the desacetoxy series was shown to be the same as at C-3 in the penicillin series. This was accomplished (147) by catalytic reduction of the double bond in the Δ^2-cephalosporanate (69) and subsequent treatment with Raney nickel. The methyl ester of the product was identical with the desthiopenicillin (79) obtained from penicillin V (80). This assignment has been confirmed (140) by an X-ray analysis of (69). A detailed NMR study (46) of (69) and derived sulfoxides has also led to the same conclusion.

(79)

(80)

(69)

The Δ^2-acid (69) in the desacetoxy series can be obtained (107) in excellent yield by mild alkaline hydrolysis of the methyl ester of the corresponding Δ^3-isomer because β,γ-unsaturated esters hydrolyze at a faster rate than the α,β-isomers. The formation of the Δ^2-compounds is often an undesired side reaction, as for example during the original isolation of these derivatives. In a study (35) directed at making esters and amides of the C-4 carboxyl group of cephalothin for biological evaluation, formation of some Δ^2-isomers usually occurred. Isomerization of the double bond to the 2,3-position has been observed (13) on silica chromatography.

However, since an equilibrium can be established between the Δ^2- and Δ^3-isomers and since the β-lactam in the Δ^2-series has a greater stability than in the normal cephalosporanates (42, 107), Δ^2-cephalosporanates have proved to be useful synthetic intermediates. For example, the partial synthesis of cephalosporin C derivatives from penicillin (150) makes use of the Δ^2-compounds for the purpose of functionalizing the methyl group in the desacetoxyl series. The methyl group can be substituted with bromine utilizing N-bromosuccinimide whereas experience with the Δ^3-isomer had been unsuccessful. Also, the hydroxyl group of desacetyl Δ^2-derivatives can be re-esterified (*92, 155*) in contrast to that of Δ^3-compounds (*145*). Thus reactions which are difficult to effect on the Δ^3-compounds can sometimes be accomplished with the double bond isomer. The desired Δ^3-cephalosporanate can then be obtained chromatographically from an equilibrium mixture.

The yields of the Δ^2- to Δ^3-conversion have been improved and the chromatography avoided by an oxidation-isomerization procedure (*87*). Workers (*42*) at Glaxo found that oxidation of the Δ^2-sulfide acids with periodate gave in 30% yield the Δ^3-sulfoxide acids along with neutral material arising from a decarboxylation. The Lilly workers (*46*) found that Δ^2-cephalosporanate esters could be oxidized with a percarboxylic acid in an aprotic solvent to a mixture of Δ^2-sulfoxides which readily isomerized to the Δ^3-compounds in hydroxylic solvents. Reduction of the sulfoxides with acetyl chloride-sodium dithionite completed the synthetic sequence. This Δ^2-, Δ^3-oxidation-isomerization-reduction scheme has allowed the preparation of a variety of acyl substituents on the hydroxymethylene group at C-3 in cephalosporins (*92*).

The acyl groups on the 7-amino function in the compounds used in these isomerizations were the thiopheneacetyl and to a lesser extent phenoxyacetyl and phenylacetyl.

b) Other Reactions of the Double Bond. Several examples of additions to the double bond have been reported. When the acid (**81**) was allowed to stand in a solution of diazomethane, the pyrazoline (**82**) was formed (*13*). Spiro compounds such as (**83**) were obtained when the acetate function was displaced by certain bidentate nucleophiles (*58*). The structure of one of these (**83**), the product from displacement with N-ethyl thiourea, was established by X-ray analysis (*79*). The configuration of this compound at C-4 is opposite to that of the Δ^2-cephalosporins.

Compound (**84**) which arose from a pencillin rearrangement (*107*) is formally an addition product of acetic acid to the double bond. Such additions to the double bond have not been effected. The stereochemistry of (**84**) at C-4 is likely the same as in the Δ^2-cephems and not as in (**83**).

References, pp. 395—403

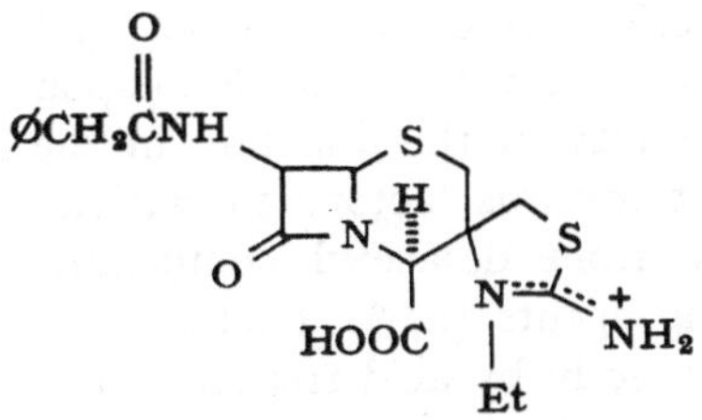

(83)

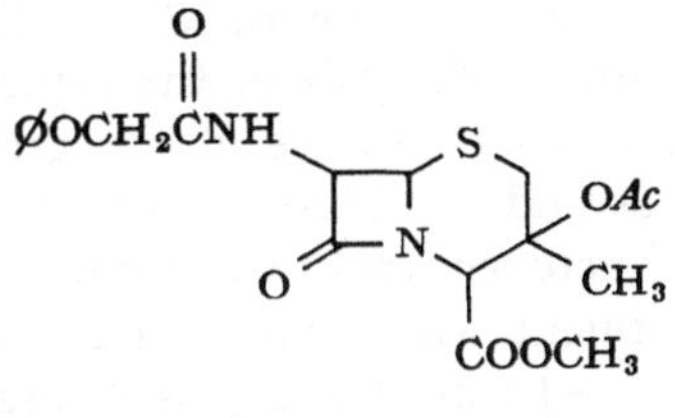

(84)

Catalytic reduction of the double bond in the Δ^3-cephems with Pd or Pt results in dihydro derivatives (107); however, with compounds containing the 3'-acetoxyl group, extensive hydrogenolysis occurred (107, 137). The Δ^2-isomer (R = H) has also been reduced to the dihydro compound (81, 147).

2. Reactions of the C-3 Substituents

a) Formation of Desacetyl Derivatives. The discovery of an acetyl group in cephalosporin C and the acid hydrolysis to the lactone, cephalosporin C_C, were important in the chemical elucidation of the structure of the antibiotic (4).

Although mild alkaline hydrolysis produced desacetylcephalosporin C, further hydrolytic reactions occurred and the expected product (85) could not be isolated. Desacetylcephalosporin C [(85), R = (α-amino-adipyl)] was obtained (83) by the action of citrus acetylesterase on cephalosporin C. This enzymatic hydrolysis has been applied (37) to a variety of cephalosporanates in order to obtain cephalospordesates for biological evaluation and as substrates for further elaboration (145).

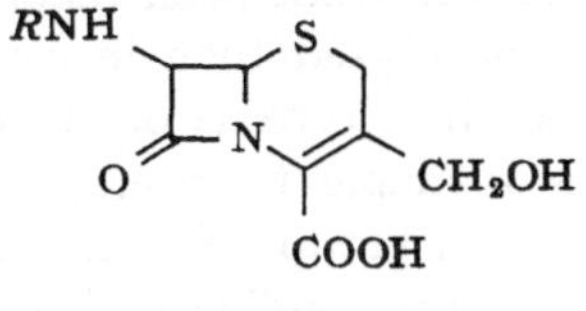

(85)

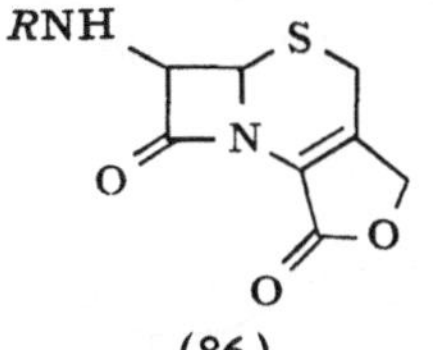

(86)

Desacetyl compounds are formed rapidly in animals and are the major metabolites of cephalosporanates. This degradation has been studied extensively with cephalothin (85) (R = 2-thiopheneacetyl) (94). The same process occurs with cephaloglycin (152). Desacetylcephaloglycin (85), (R = D-phenylglycyl) resisted many attempts at isolation but was finally obtained (92a) in crystalline form by selective acid hydrolysis or enzymatic hydrolysis followed by cellulose chromatography.

Desacetyl compounds of the Δ^2-series could be obtained (92) by alkaline hydrolysis of the cephalosporanate esters or the Δ^2-compounds, presumably because of the greater stability of the β-lactam in the Δ^2-series and the stability of the product towards lactone formation.

Lactones (86) are formed readily from desacetyl compounds on treatment with acid or many acylating agents (145). Lactones can be prepared from the cephalosporanates directly by acid treatment as with cephalosporin C$_C$ and 7-ACA lactone (4, 96).

Since cephalosporanates are metabolized to the desacetyl compounds, and these normally possess less antibacterial activity (37), particularly against gram negative organisms, there has been considerable effort to introduce active, metabolically stable acyl groups at C-3. This effort was hampered by the fact that the hydroxyl group tended to form lactones under the influence of many common acylating agents. Attempts to prepare such compounds by protecting the carboxyl group as an ester which could be removed after acylation have had limited success. The desacetyl ester still has a propensity to lactonize, and often double bond isomerization occurs. Aryl acid chlorides, which are less reactive toward hydrolysis than aliphatic acid chlorides, acylate the hydroxyl group in basic solution (145), and hence a series of such compounds have been made and evaluated (145). A more general procedure (92) to obtain a variety of acyl functions has been described recently which utilizes the Δ^2-desacetyl compounds. Acylation was followed by the oxidation-isomerization-reduction method described in Section III., D., I., a. Acetylation of a Δ^2-compound, followed by double bond isomerization was utilized (155) in the total synthesis of cephalosporins.

The hydroxyl group of desacetylcephalothin was oxidized (34) to the aldehyde (87). The methyl group of the ester could not be selectively hydrolysed. The acid corresponding to (87) has not been reported. The β-lactam of (87) should be more easily hydrolyzed than in cephalothin since the electrons on nitrogen are less available to interact with the carbonyl group. The trichloroethyl ester corresponding to the Δ^2-isomer of (87) is known from synthetic work in Basel (155).

(87)

b) *Nucleophilic Displacement of the Acetate.* The most interesting reaction of the C-3 acetoxy methylene group in cephalosporanates is one which results in alkyl-oxygen cleavage rather than acyl-oxygen scission. During the course of the structure elucidation of cephalosporin C, the English workers used pyridine acetate buffers for ion exchange chromatography.

They discovered that the use of these buffers with cephalosporin C produced a new antibiotic (6, 70) that possessed no net charge at pH 7. This compound, named cephalosporin C_A, was obtained by reacting cephalosporin C with pyridine-water at 37°, followed by chromatography over Dowex IX8. The product had a greater antibacterial activity than cephalosporin C. On the basis of analytical data and knowledge of the structure of cephalosporin C, structure (88) was suggested (71) for cephalosporin C_A. It represents the displacement of the acetate by pyridine. Very quickly a family of new antibiotics was produced (71) by the action of various heterocyclic tertiary amines on cephalosporin C. Many of these showed antibacterial properties superior to that of the parent compound.

(88)

(89)

During the study of fermentation media, DEMAIN et al. at Merck (52) found that addition of sodium thiosulfate increased the potency of the broth. This observation was quickly traced to a new cephalosporin recognized as the Bunte salt (89) resulting from displacement of acetate by thiosulfate.

After the cephalosporin nucleus (33) and other cephalosporanates became readily available, considerable effort in several industrial laboratories was devoted to acetate displacement reactions and a host of derivatives, resulting from the use of a variety of nucleophiles was described. Among the resulting new cephalosporins that have been reported in the scientific literature are those derived from various pyridines (71, 133), other aromatic heterocycles (71, 88, 133), xanthates (148), dithiocarbamates (148), thiobenzoates (41), thiols (41) and anilines (25). Although these have been prepared with a variety of 7-acylamido groups, cephalothin was the most frequently utilized substrate. The products (e. g.), (83) formed by the use of bidentate nucleophiles were discussed in Section III, D., 1., b. Formation of carbon-carbon bonds has been observed (41) with indole and resorcinol as nucleophiles. No displacement with oxygen compounds has been observed. With alkoxides (56) and many primary amines (25) (also as side product with aniline) the other strongly

24*

electrophilic site (*i. e.*, the β-lactam carbonyl) in the molecule becomes the locus of attack. In these cases the β-lactam is broken with simultaneous, internal expulsion of the acetate. Double bond migration or subsequent addition will provide compounds (75) and (76).

The pyridine displacement product of cephalothin has been extensively studied (*134*) and has been introduced in clinical practice as a parenteral broad spectrum antibiotic under the generic name, cephaloridine (67).

(67)   (90)

The formation of cephaloridine from cephalothin has been studied in some detail (*134*, *143*). The displacement proceeds by an S_N1 mechanism (*143*). The evidence indicates a rate-determining ionization to the ion (90) which can then add the nucleophile. The nitrogen of the β-lactam in cephalosporanates has more amine character than expected for normal lactams due to the steric requirements of the β-lactam-dihydrothiazine ring system. The reaction requires a high dielectric constant, protic solvent, normally water. Formation of cephaloridine proceeds in aqueous pyridine and not in pyridine itself. Similar results were found with other nucleophiles in nonaqueous solvents. This is in accord with the postulate of an initial ionization step which requires a polar solvent. Formamide has been the other solvent that has been used successfully. In pyridine-water the maximum yield as calculated from the rate of formation and the rate of decomposition of the product is 54%. The realized yield was 30%. The difference was ascribed to side reactions of the intermediates. With nucleophiles such as thiobenzoate which form an insoluble product on displacement, yields up to 85% have been realized indicating high "yields" of the intermediate ion. The Lilly workers (*134*) have reported that with the addition of a large quantity of non-common ion salts such as potassium iodide or potassium thiocyanate yields of 75% could be realized. The role of thiocyanate in this yield improvement is obscure. The reaction is conducted in nearly a saturated salt solution as the added potassium thiocyanate is several times the weight of the solvent. Isolation of the product is facilitated by the fact that acidification of the reaction mixture produces an insoluble cephaloridine hydrothiocyanate salt.

Displacement reactions have not been reported which utilize esters and lactones as reactants. With the former, double bond equilibration and hydrolysis occur and with the latter, the β-lactam function is more susceptible to hydrolysis. With Δ^3-cephalosporin sulfoxides and Δ^2-cephalosporins initial ionization occurs, but the rate of reaction is much reduced. In the case of the sulfoxide, extensive decarboxylation occurs. The desacetyl compounds do not undergo displacement. The evidence indicates that an ionized carboxyl group facilitates the reaction. Electrostatic stabilization of the cationic intermediate by a neighboring carboxylic anion has been postulated (*143*).

   c) Hydrogenolysis of the Acetate. The acetate function is also subject to displacement by hydrogenolysis (*107, 137*). ABRAHAM and NEWTON (*3*) first demonstrated this by the isolation of valine after Raney nickel desulfurization. These workers also reported the Pd-C catalysed hydrogenation of cephalosporin C under conditions such that one molar equivalent of hydrogen was absorbed. The crude product, which was not further purified, still possessed a UV chromophore at 260 nm although it was reduced in intensity. The reduction of 7-phenoxyacetamido-cephalosporanate (91) ($R = \text{ØOCH}_2\text{CO}$) afforded (92) ($R = \text{ØOCH}_2\text{CO}-$) which had been obtained

(91)   (92)

from penicillin V sulfoxide (*107*). A group at Smith, Kline and French (*137*) which also studied the hydrogenolysis of cephalosporin C obtained desacetoxy cephalosporin C (92) [$R = (\alpha \text{ aminoadipyl})$]. When 7-ACA (33) became available these workers were able to produce 7-ADCA, [(92) $R = \text{H}$)] in relatively good yield by reduction of 7-ACA utilizing Pd-BaSO$_4$ as catalyst. Acylation of the amine group of (92) ($R = \text{H}$) provided a variety of desacetoxyl cephalosporanates for biological evaluation. One of these, made in the Lilly Laboratories, cephalexin (92)

NH$_2$
|
($R = \text{ØCHCO}-$), has been extensively studied (*151*) and is undergoing clinical evaluation as an orally effective broad spectrum antibiotic.

3. Reactions of the Carboxyl Group

   Esters and amides of cephalosporanates have been made (*35*) for biological testing. These were made by activating the carboxyl group by the mixed anhydride or carbodiimide method. The difficulty resulting from isomerization of the double bond to give mixtures of $\varDelta^2$- and $\varDelta^3$-isomers has been discussed earlier. No general procedure for avoiding this isomerization has been reported; however, in certain situations no double bond migration was observed. The diazo compounds, diazomethane (37), phenyldiazomethane (59), diphenyldiazomethane (59, 107), diazofluorene (126) and ethyl diazoacetate (35) have been reported to give the respective esters.

Various esters have been employed to protect the carboxyl group in transformations of cephalosporin-related materials. Among those utilized have been benzyl, and the substituted benzyl esters, p-methoxybenzyl (35) and benzhydryl (59, 107). The latter two can be cleaved by mild acid treatment. The trichloroethyl ester, which can be reductively removed by zinc and acetic acid, was particularly useful in the total synthesis of cephalosporin C (155). The trimethylsilyl ester is readily cleaved in aqueous solution and has found application (59) in the hydrolysis of the aminoadipyl sidechain from cephalosporin C via the iminoester. The t-butyl ester of 7-ACA and acid conditions for ester scission have been reported (136).

The lactones have chemical, physical and biological properties that differ from those of the esters. The lactones of simple acyl cephalosporanates have very limited solubility in most organic solvents; cephalothin lactone has appreciable solubility in only polar aprotic solvents. The β-lactam is more chemically labile and is more subject to hydrolysis than the lactone ring. As a consequence, the lactone has not been opened in good yield without disruption of the lactam. The lactones possess nearly the same *in vitro* antibacterial activity as the parent acids but are inactivated by serum (37). The lactones more resemble penicillins in their deactivation by penicillinase. Esters of cephalosporanates in general do not possess inherent antibacterial activity, as the activity observed most likely is due to hydrolysis to the parent cephalosporanic acid.

The free carboxyl group of the sulfoxides in both $\varDelta^2$- and $\varDelta^3$-isomers is labile to decarboxylation. Treatment (42) of sulfoxides (93) (R = ØCH$_2$CO or 2-thiopheneacetyl) with pyridine at 37° gave the neutral products (94) (R = ØCH$_2$CO or 2-thiopheneacetyl). Workers (126) at the Lederle Laboratories recently have obtained (94) (R = 2-thiophene-

(93)　　　　　　　　　　　　　　　　(94)

acetyl) by heating (93) with tertiary alkyl amines. Their product (94) was a mixture of epimers at C-7, the ratio of which depended upon reaction conditions. Mention has already been made of the periodate oxidation of (95) ($R = \text{ØCH}_2\text{CO}$) to (94) ($R = \text{ØCH}_2\text{CO}$), an oxidative decarboxylation. Compound (96), the origin of which is most likely a decarboxylation of cephaloridine, has been reported in the literature (100). In the rearrangement of penicillin sulfoxide free acid only decarboxylated cephem products could be isolated (107).

(95) (96)

4. Formation of Sulfoxides

The formation of sulfoxide of cephalosporin C was first reported (4) by ABRAHAM and NEWTON. This derivative has very low antibacterial activity and received very little attention until more recently when its utility as an intermediate for further transformation and for characterization was recognized. Periodate, which has found use as a mild selective oxidizing agent for sulfides to sulfoxides will produce the 1-oxide from cephalosporanic acids (42). The Δ^2-series was inert to this reagent. On the other hand both Δ^2- and Δ^3-cephem esters (most of the effort has been in the desacetoxy series) were oxidized by peracids to mixtures of two oxides; the (β) sulfoxide is the predominant isomer in both the Δ^2- and Δ^3-series (46). Like penicillin V sulfoxide, these are the sterically less favored isomers from the standpoint of internal steric congestion and the direction of oxidant approach. However, in the (β) isomers, the intramolecular hydrogen bonding of the oxide to sidechain NH is sterically possible and may account for their preferred formation (46a).

The Δ^2-oxides readily isomerize in hydroxylic solvents to the Δ^3-compounds (46). The ready decarboxylation and C-7 epimerization of cephalothin sulfoxide were discussed in Section III, C. In addition to exchange at C-7, the two protons at C-2 are exchanged for deuterium under basic conditions in the presence of D_2O, but no exchange at C-6 was found (126). This relates to the lack of formation of C-5 substituted products in penicillin sulfoxide rearrangements (107). The steric difficulty of introducting a $\Delta^{1,6}$-double bond ($\Delta^{1,5}$- in penicillin) may account for this. Homoenolization has been invoked to explain the facility of enol formation at C-7 (69).

Sulfoxides have been used to control double bond isomerization. The success of this procedure is dependent upon the ease of reduction of the sulfoxide to the parent sulfide. This has been achieved by a combination of chemical reducing agents and a sulfoxide activating substance such as acetyl chloride (*87, 150*). The stabilization of the β-lactam function in the sulfoxides provides another reason for their utility in modification of β-lactam antibiotics. Since the sulfoxide group is asymmetric and displays a strong magnetic anisotropic effect, the sulfoxides are also useful for NMR studies intended to gain insight into the stereochemistry of other centers (*46, 46a*).

E. Conformation of Cephalosporanates

The two limiting conformations for Δ^3- and Δ^2-cephem derivatives are given in Fig. 1, formulae (97)–(100). X-ray analyses of cephalosporin C

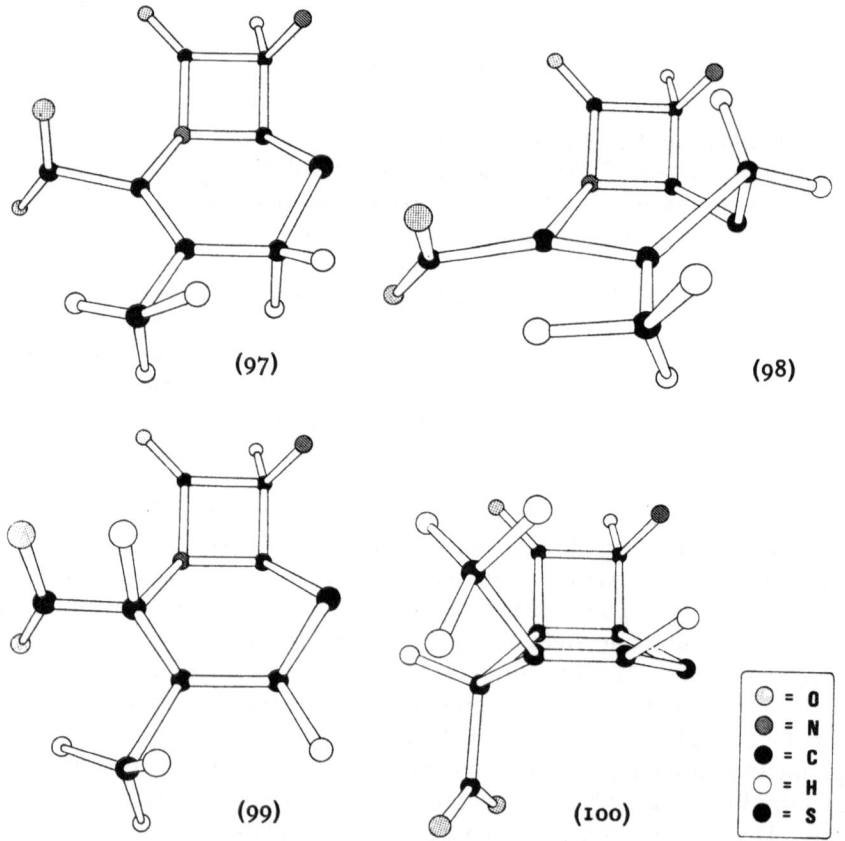

(97)

(98)

(99)

(100)

○	= O
◐	= N
●	= C
○	= H
●	= S

Fig. 1. Limiting conformations of Δ^3- and Δ^2-cephem derivatives

(78), cephalosporin C_C (54), cephaloridine (141) and 7-phenoxyacetamido-3-methyl-ceph-2-em-4-carboxylic acid (140) have been reported. These have shown that, in the solid state, Δ^3-isomers adopt conformation (97) and Δ^2-isomers conformation (99). The preferred conformations are those that possess a less acute angle between the general plane of the six-membered ring and that of the four-membered ring. Penicillin possesses a more acute angle between the ring systems. Detailed NMR analyses (46) of Δ^2- and Δ^3-7-phenoxyacetamido-3-methyl-cephem-4-carboxylic acid esters and their sulfoxides have supported the assumption that these conformations are also preferred in solution. X-ray analysis of the Δ^2-compound indicated one very important difference from the conformation given in (99). The nitrogen of the β-lactam is nearly planar rather than pyramidal (140). This latter fact indicates the molecule is even more flat than shown in (99).

IV. Physicochemical Properties of Cephalosporanates

In the introduction, mention was made of the influence which work on the structure of penicillin exerted on the development of new physico-chemical tools in organic chemistry. The use of such tools was highly significant for the rapid development of cephalosporin chemistry.

Several X-ray structure analyses of cephalosporin derivatives have been reported in addition to the one which confirmed the structure of cephalosporin C (78). Cephalosporin C_C (54), cephaloridine (141) and cephaloglycine (140) are other Δ^3-cephalosporanates that have been examined. Work on the structure of a Δ^2-cephalosporanate has recently been completed (140). Extensive use was made of single crystal X-ray structural techniques in the total synthesis of cephalosporin C. The work done by GOUGOUTAS (155) provided confirmation of the stereochemistry of several important intermediates.

Infrared absorption spectra of cephalosporins have been extensively used and have been discused in detail by GREEN, PAGE and STANIFORTH (67). The most distinguishing band in the infrared is that of the β-lactam carbonyl, 1780–1790 cm^{-1}, which is also found in IR spectra of penicillins. This has been discussed in considerable detail in the penicillin monograph (39) and in early reports on infrared spectroscopy. Workers frequently use this characteristic absorption band to determine the presence or absence of a fused ring β-lactam in reactions of cephalosporin C derivatives. The relationship of the frequency of this absorption to β-lactam chemical reactivity and biological activity has been studied (42, 107, 140).

The UV absorption maximum at 260 nm first called the attention of ABRAHAM and NEWTON to the existence of this class of β-lactam antibiotics. The absorption was inexplicable on the basis of a penam structure (4). However, the maximum was at a considerably longer wave length than that expected of the enamide of a pyruvic acid derivative. The nitrogen atom, due to some steric inhibition of amide resonance, has more amine character than most lactams. Although this likely has some bearing upon the shift to longer wave lengths, it does not account fully for the 260 nm absorption since compound (23) has a UV maximum at

(23)   (101)   (102)

248 nm (18). The cephem compounds lacking a carboxyl group at C-4 have maxima at only slightly shorter wave lengths (105). Resonance structures such as (101) have been invoked to explain the UV absorption curves of cephalosporins (66, 97). This explanation takes into account the greater amine character of the amide nitrogen and the known fact that sulfur will exert a considerable bathochromic effect on a chromophore, even though separated by one saturated carbon atom. The UV absorption maximum of (102) (λ max 285 nm) substantiates this argument (17). The fact that the position in the absorption is not at a longer wave length indicates that the nitrogen in cephalosporins cannot be characterized as an amine.

(103)

The Δ^2-cephalosporanates usually do not exhibit a distinct UV chromophore but often exhibit a shoulder at approximately 240 nm superimposed on the chromophore of the sidechain (42).

The most useful tool in cephalosporin C chemistry has been NMR spectroscopy. In cephalosporins and penicillins the carbons are unsaturated or are highly substituted with heteroatoms; thus the protons are usually widely separated in chemical shift and have simple coupling patterns. Numerous papers have been published which have made extensive use of NMR data and the reader is referred to two (67, 68) in which NMR spectra of cephalosporanates have been explicitly discussed. Solvent induced chemical shifts, the anisotropy of the sulfoxide bond and nuclear Overhauser effects have been utilized recently in chemical studies of cephalosporin C derivatives (46, 46a, 47).

The high resolution mass spectrum of a cephalosporanate (103) has been recorded and discussed (*121*). A similar mass spectral analysis of the corresponding penicillin, penicillin V methyl ester, has been published by BIEMANN (*120*).

V. Syntheses of Cephalosporins

A. Total Synthesis

Efforts directed at total synthesis of cephalosporins were begun in several academic and industrial laboratories soon after the structure had been elucidated. Many of the early attempts followed the SHEEHAN (*128*) penicillin synthesis scheme (Chart 5 below) in which the nitrogen-sulfur heterocycle was constructed by condensation of penicillamine with an appropriately substituted aldehyde. *A priori*, synthesis of

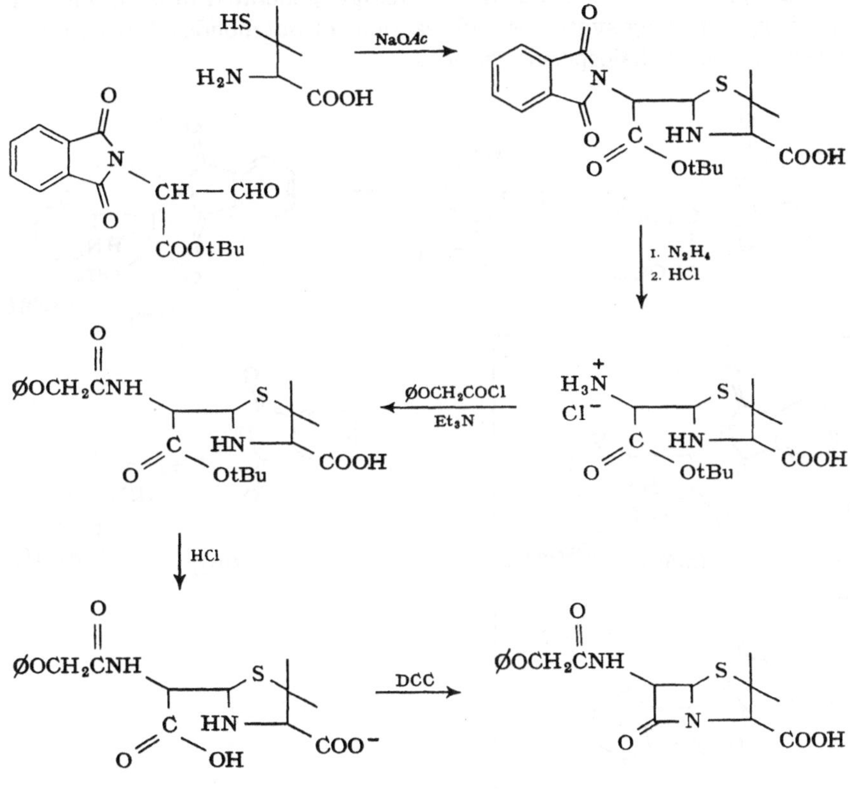

Chart 5. SHEEHAN's Penicillin Synthesis

"cephalosporamine" (104) should allow analogous elaboration of the cephalosporin ring system. Synthesis of such a "cephalosporamine", however, has not yet been achieved.

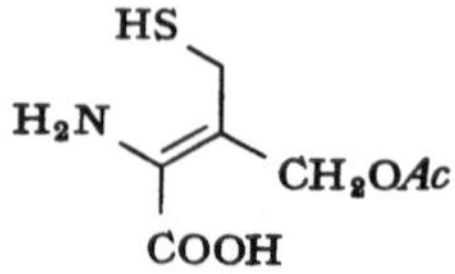

SHEEHAN (130) has reported the synthesis of the saturated cepham ring system, using D,L-homocysteine in place of D-penicillamine. Cleavage of the *t*-butyl ester (105 → 106) was accompanied by decarboxylation to the extent that (107) was the major isolable product. Such extensive decarboxylation was not observed in the penicillin series. Treatment of crude solutions of (106) with dicyclohexylcarbodiimide gave complex mixtures which apparently contained β-lactam substances, but pure (108) was not isolated.

Several early attempts at cephalosporin synthesis utilized substituted butenolides as intermediates which would give cephalosporin lactones as ultimate products. In general, these attempts generated much interesting chemistry regarding syntheses and reactions of butenolides, but with two exceptions no cephalosporin synthesis.

(104)
"Cephalosporamine"

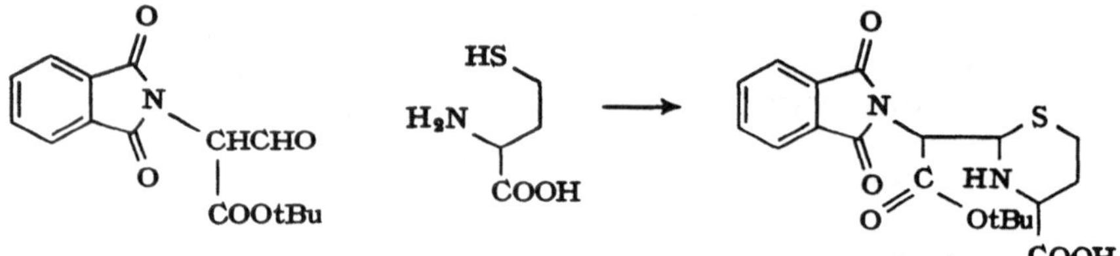

(105)

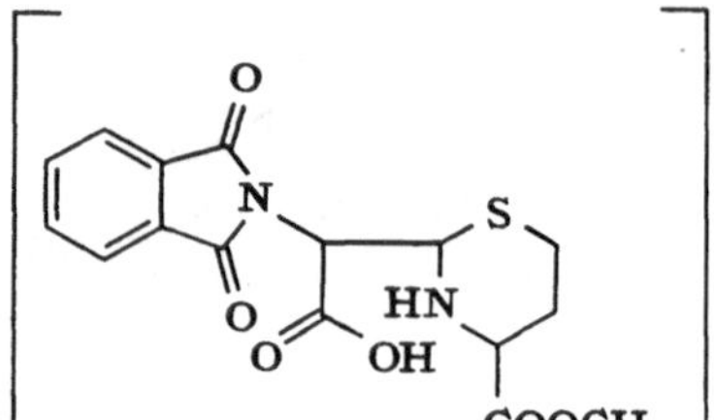

(106)

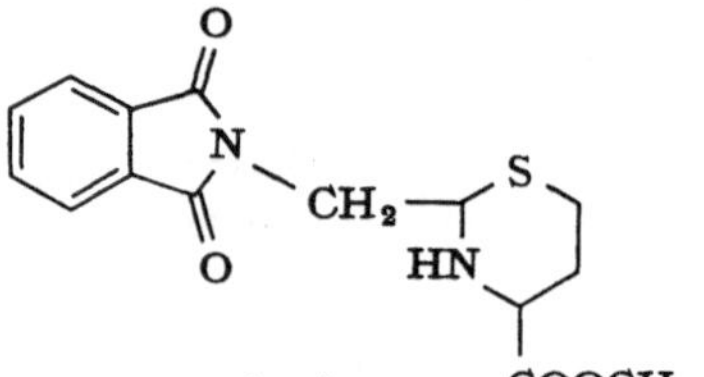

(107)

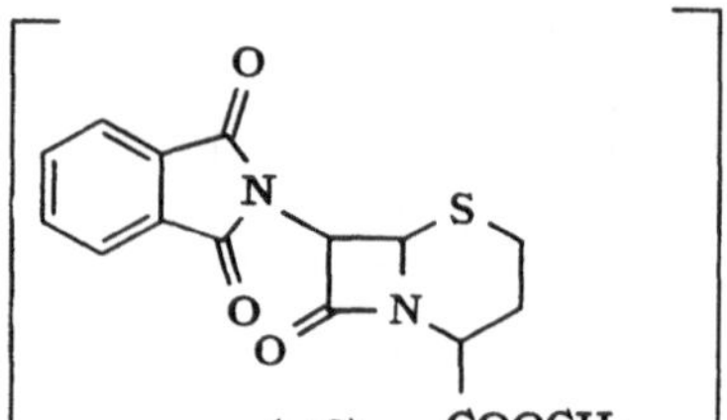

(108)

Investigators at Squibb (*55, 61*) concentrated on the synthesis of "cephalosporamine" lactone (**109**) and the thiol (**110**). These efforts were frustrated by the ready self-condensation of the thiols to give the corresponding sulfides (**111**), (**112**). Compound (**112**) was a degradation product of cephalosporin C which had been isolated and characterized by ABRAHAM and NEWTON (*4*). The trityl derivative (**113**) was stable and cyclization to (**114**) (mixture of diastereomers) was effected.

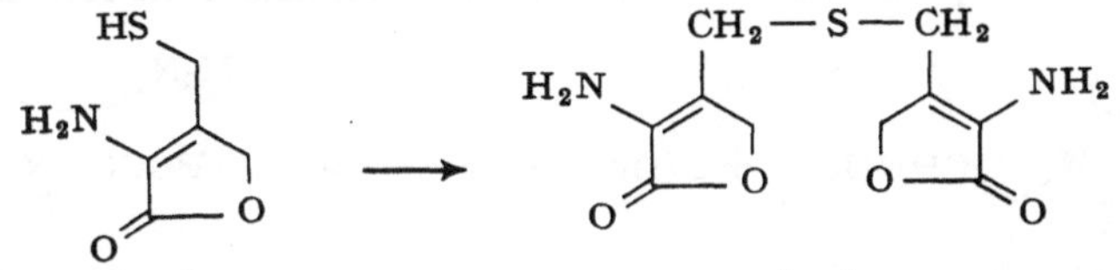

(**109**)   (**111**)

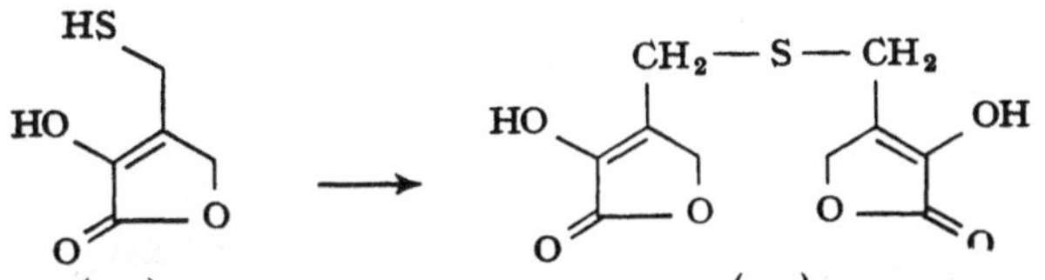

(**110**)   (**112**)

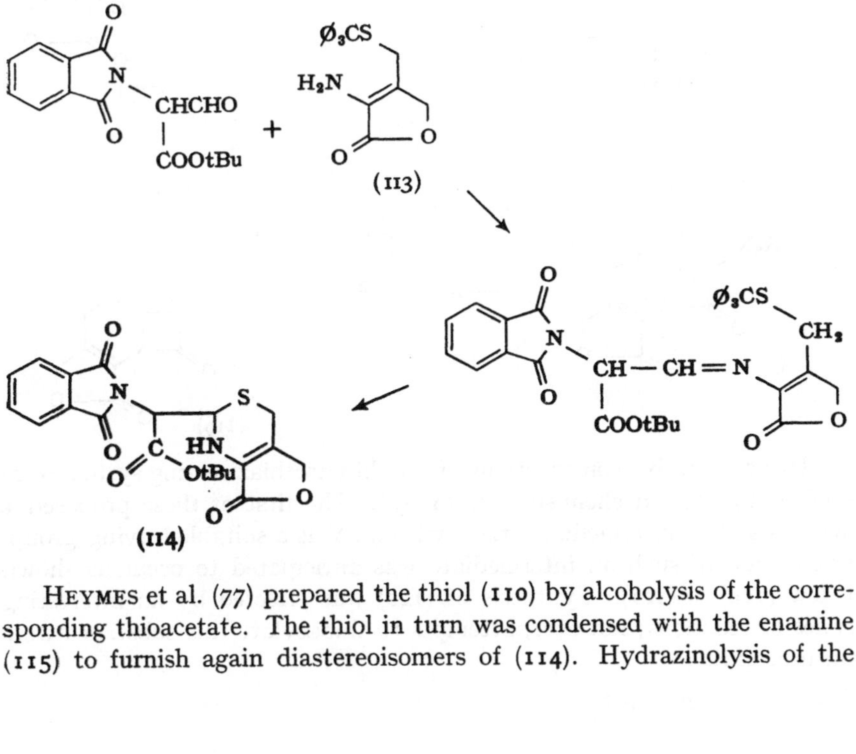

(**113**)

(**114**)

HEYMES et al. (*77*) prepared the thiol (**110**) by alcoholysis of the corresponding thioacetate. The thiol in turn was condensed with the enamine (**115**) to furnish again diastereoisomers of (**114**). Hydrazinolysis of the phthaloyl protecting group from (114) was followed by acid catalyzed cleavage of the *t*-butyl ester. Ester cleavage was accompanied by epimerization at C-7 and a single diastereomer of (116) was obtained. Tritylation of the primary amine and cyclization of the tritylated derivative (117) with dicyclohexylacarbodiimide gave the β-lactam (118). Conversion to racemic desacetylcephalothin lactone (119) was effected by detritylation and reacylation. Spectral comparison of racemic (119) with the lactone prepared from cephalothin established the total synthesis.

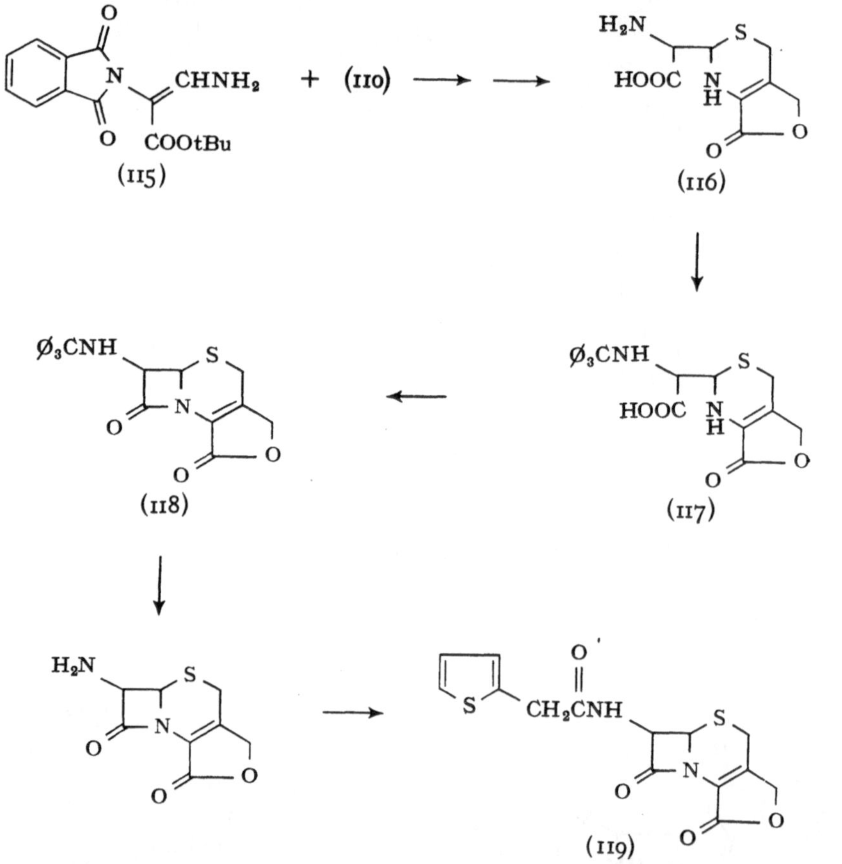

Two alternative constructions of the dihydrothiazine ring system were studied by Oxford chemists (*17, 18, 57*). The first of these proposed a synthesis of an intermediate (120), wherein Y is a suitable leaving group. Cyclization of such an intermediate was anticipated to occur as shown below (120) → (121). Synthesis of (122) was effected by an interesting series of reactions; however, attempts to thioacylate this substance were unsuccessful.

References, pp. 395—403

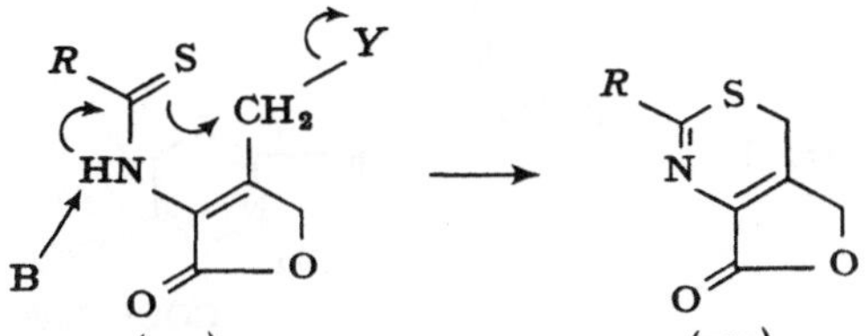

(120)   (121) 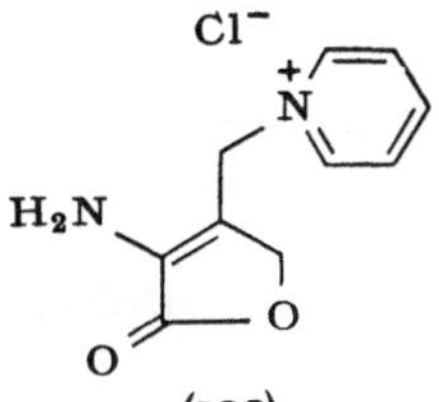   (122)

A second method studied was successful. In a model series, thio-acetamide was condensed with the keto ester (123) to give (124). Reduction with aluminium amalgam and dehydration with hydrochloric acid furnished the desired ring system (102).

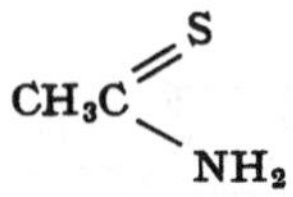

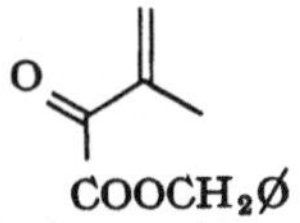

(123) 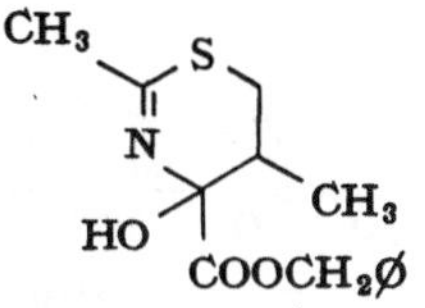   (124)

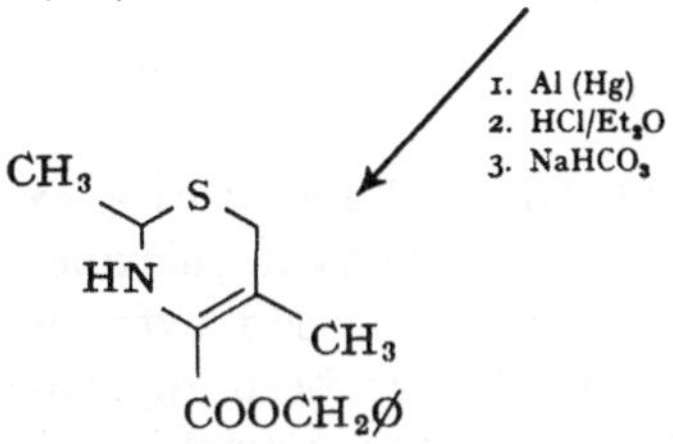

1. Al (Hg)
2. HCl/Et₂O
3. NaHCO₃

(102)

When this process was applied using a thioamide (125) capable of further synthetic exploitation, cyclization occurred in a different direction producing (126).

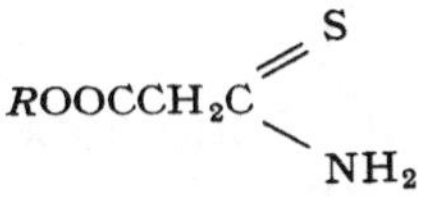

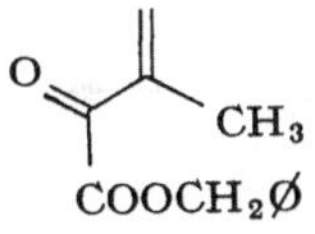

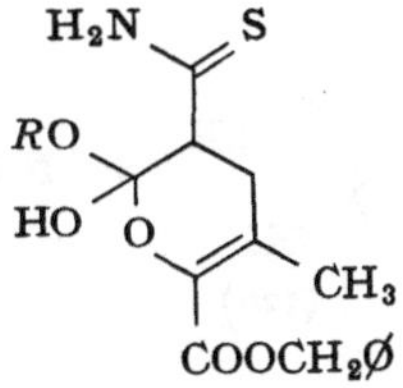

(125)   (123)   (126)

Compound (127) was prepared to circumvent the improper cyclization encountered when (125) and (123) are condensed. Interaction of (127) with (123) did proceed to a thiazine (128). The ring system in (128) was the same as that found in the degradation product (75).

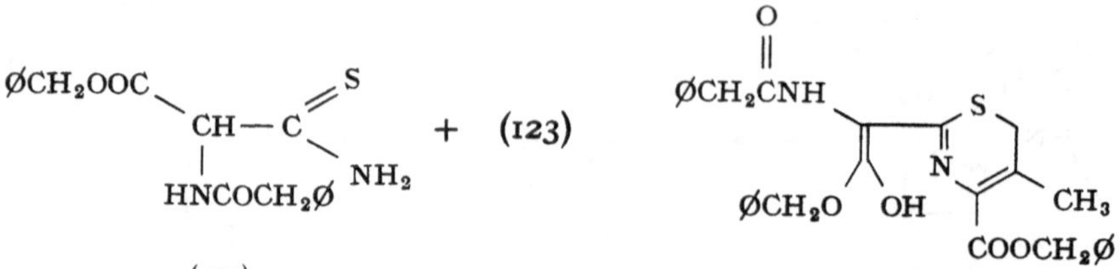

(127)   + (123)   (128)

The desired cyclization of (125) and (123) did occur under acid conditions which also caused dehydration giving the dihydrothiazine (129). Hydrogenolysis of the benzyl esters to give appropriate β-lactam precursors failed.

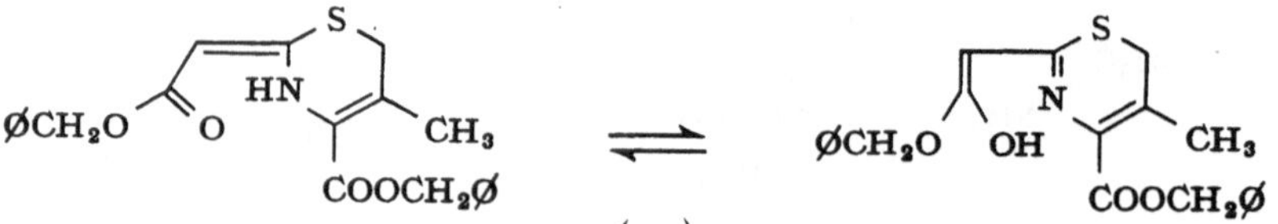

(129)

Investigators at Glaxo Laboratories explored various methods of preparing butenolides and dihydrothiazines similar to what has been described previously (66, 97). No extension of their intermediates to cephalosporin compounds has been reported.

Meyers and Greene (103) reported a synthesis of dihydrothiazines from β-mercaptoethanols and nitriles. The thioalcohol (130) and cyanoacetic ester condensed under acid conditions to give (131). Reduction of (131) with sodium borohydride followed by acid hydrolysis produced the amino acid (132). Ring closure to the β-lactam (133) failed under a variety of conditions.

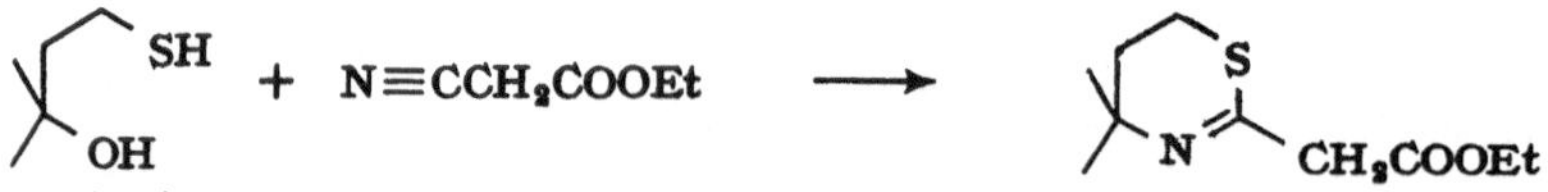

(130)   + N≡CCH₂COOEt   (131)

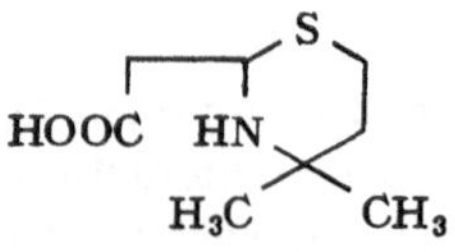

(132) 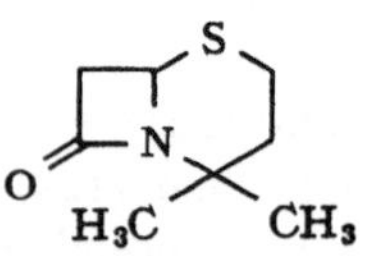   (133)

A novel elaboration of the dihydrothiazine ring system was reported by STORK and CHEUNG (*139*). The azlactone (*134*) was treated successively with hydrogen sulfide and hydrochloric acid in water/acetic acid giving the bicyclic lactone (*135*). Reduction of (*135*) with aluminium amalgam gave the carbinol (*136*) which was then converted to mesylate (*137*). The mesylate underwent ring expansion to (*138*) upon treatment with sodium acetate in dioxane. This ring expansion bears a formal resemblance to the penicillin sulfoxide rearrangement reported earlier.

Similar azlactones were used by REINHOUDT, TAN and BEYERMAN (*119*) to prepare butenolides potentially useful for cephalosporin synthesis. The general method is shown on p. 386. No extension of this work to the synthesis of cephalosporins has been published.

BOSE and co-workers (*24*) have reported the synthesis of analogs of the cepham ring system utilizing condensations of azido acid chlorides with imines. The general sequence is shown on p. 386. This novel lactam synthesis has been successfully applied to give a total synthesis of methyl 6-α-phenoxyacetamido-penicillanate (*139*) (*23*).

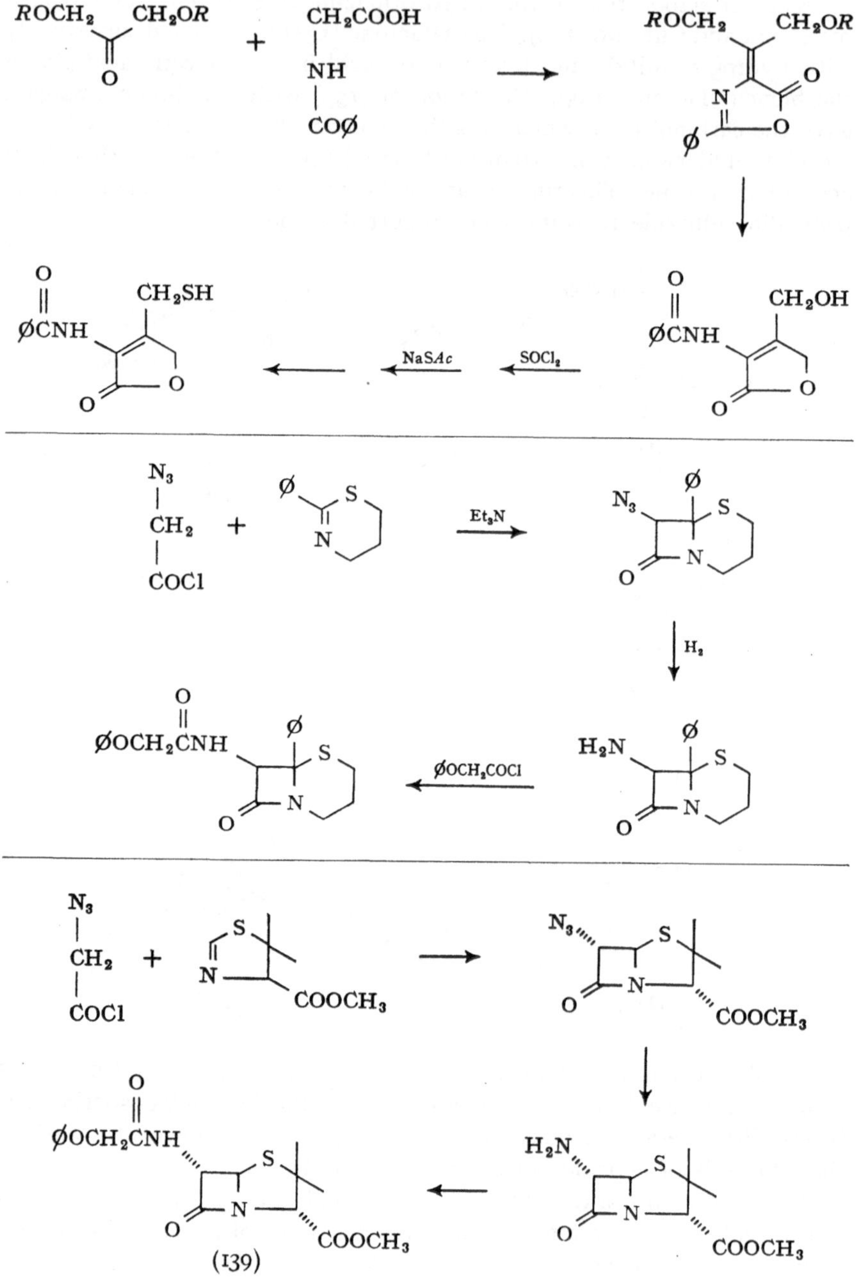

Chart 6. Total Synthesis of Methyl-6-α-Phenoxyacetamidopenicillanate

References, pp. 395—403

Other analogs of the cepham ring system have been reported from several laboratories. These have been reviewed recently in detail (99).

An elegant stereospecific total synthesis of cephalothin and cephalosporin C has been reported by WOODWARD and co-workers (155). A later report by HEUSLER (76) gave additional details of the synthesis. In this unique synthesis, the β-lactam portion was synthesized first. L-(+)-Cysteine was converted to the acetonide (140) and then acylated with t-butyloxycarbonyl chloride to give the carbamate (141). Interestingly, the methyl ester of (140) could not be acylated directly, but acylation proceeded via intramolecular acyl transfer from the t-butyloxycarbonyl mixed anhydride of (140). Esterification of (141) with diazomethane followed by treatment of the resulting methyl ester with dimethyl azodicarboxylate produced the triester (142). Conversion to the carbinol (143) was accomplished by oxidation with lead (IV) acetate and subsequent treatment of the intermediate azo compound with methanolic sodium acetate. The carbinol was first converted to its mesylate and then to the corresponding azide (144). Reduction of the azide function to give (145) was effected by treatment with aluminium amalgam in methanol. Treatment of the aminoester with triisobutyl aluminium in toluene afforded the β-lactam (146). Stereochemistry of the key intermediates (145) and (146) was confirmed by X-ray analysis. Addition of (146) to the unsaturated dialdehyde (147) gave (148) which cyclized to (149) upon treatment

(146)   (147)   (148)

with trifluoroacetic acid. Acylation of (149) with thiophene-2-acety chloride followed by reduction of the aldehyde with diborane yielded the alcohol (150). Acetylation of the alcohol produced the acetate (151).

(149)   (150)

Equilibration of (151) with pyridine partially isomerized the double bond giving a mixture of (151) and (152), from which pure (152) was isolated by chromatography. Cleavage of the trichloroethyl ester with zinc gave synthetic cephalothin identical with that from natural sources.

(151)   (152)

Condensation of (149) with (153) afforded the acid (154). Following esterification with trichloroethanol, the analogous sequence of reduction, acetylation, double bond isomerization, isomer separation and zinc reduction gave synthetic cephalosporin C.

References, pp. 395—403

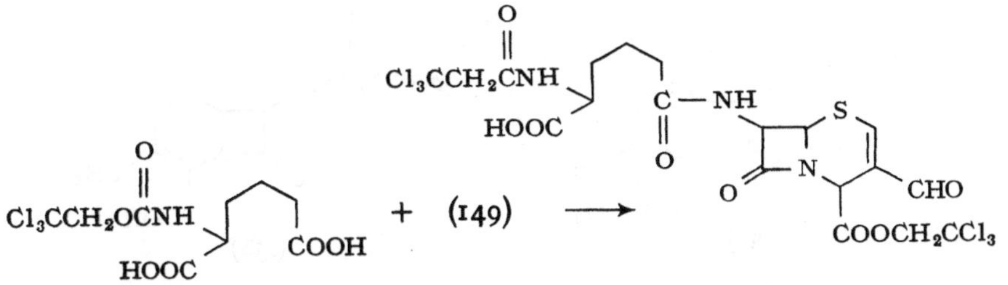

$$
\text{Cl}_3\text{CCH}_2\overset{\overset{\text{O}}{\|}}{\text{OCNH}} \underset{\text{HOOC}}{\diagup\diagdown\diagdown}\text{COOH} \quad + \quad (149) \quad \longrightarrow
$$

(153)

(154)

B. Partial Synthesis

The obvious similarity between the cephalosporins and the penicillins suggests that interconversions between the two series of antibiotics should be feasible. Scientists at the Lilly Research Laboratories reported, in 1963 (*105a*), the transformation of penicillin V into desacetoxycephalosporanates. Full details of this and related conversions were published in 1969 (*107*).

Penicillin V sulfoxide methyl ester (155), when heated with acetic anhydride, is converted to the acetates (156) and (84).

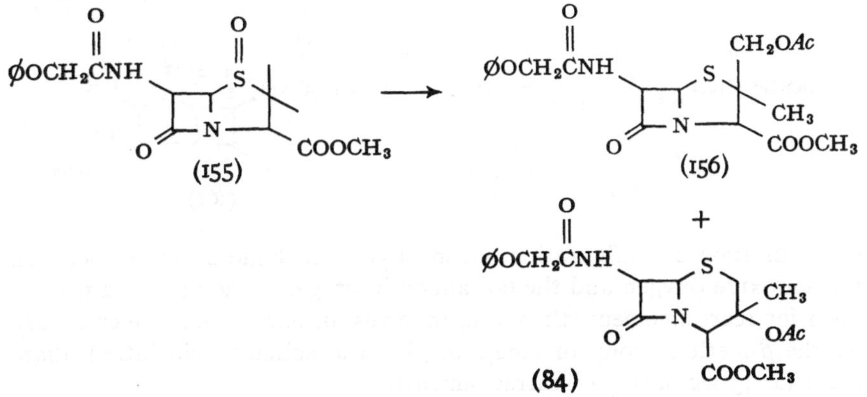

(155)

$\longrightarrow$

(156)

+

(84)

When heated with a trace of acid, the sulfoxide (155) rearranges to a desacetoxy cephalosporin (157). Hydrogenolysis of 7-phenoxyacetamido-cephalosporanic acid methyl ester also produced (157), thus correlating the two families of antibiotics.

Without carboxyl protection, attempts to prepare the acid (158) from penicillin V sulfoxide were essentially unsuccessful, the principal product being the decarboxylated ceph-3-em (159).

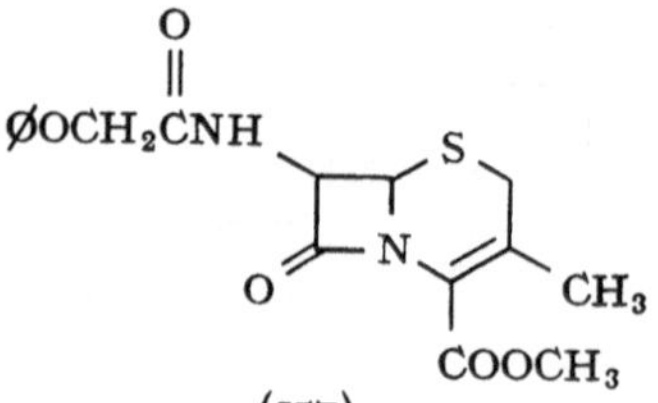

(157)

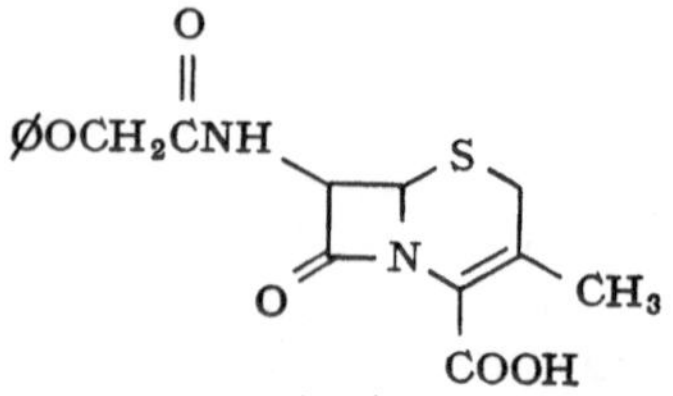

(158)

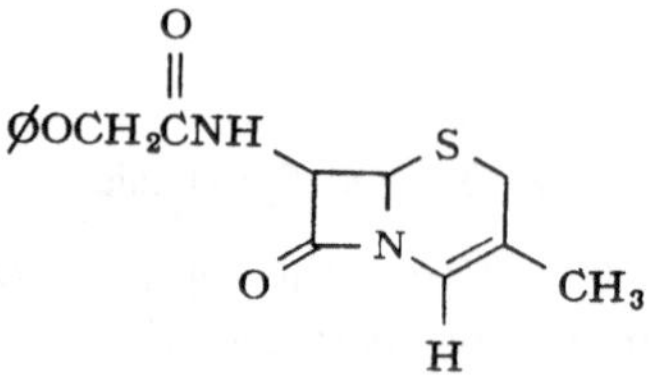

(159)

The fact that only one isomer of (156) and (84) was found in the reaction mixture suggests that the rearrangement was stereospecific. Based on NMR arguments, the stereochemistry of (156) was assigned as shown in (160) (*107*). Detailed NMR and X-ray crystallographic studies have proved (*46a*) that the sulfoxide oxygen in (161) has the β configura-

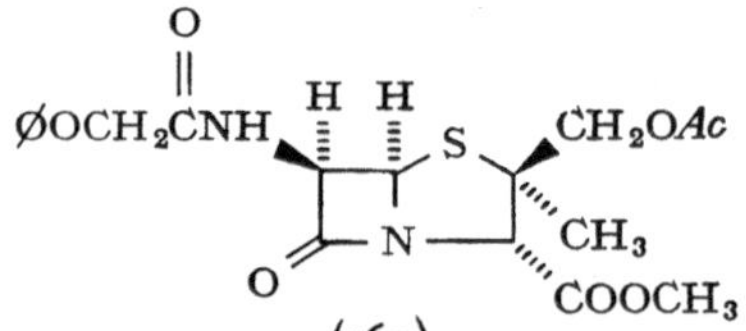

(160)

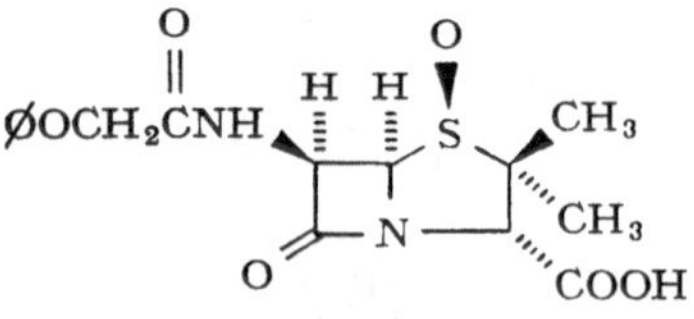

(161)

tion. In such a configuration, strong hydrogen bonding exists between the sulfoxide oxygen and the 6-β-amido hydrogen. The proposed mechanism for rearrangement, therefore, involves an elimination preferentially on the β-methyl group of (162), to give the sulfenic acid intermediate (163) (a 1,5 sigmatropic rearrangement).

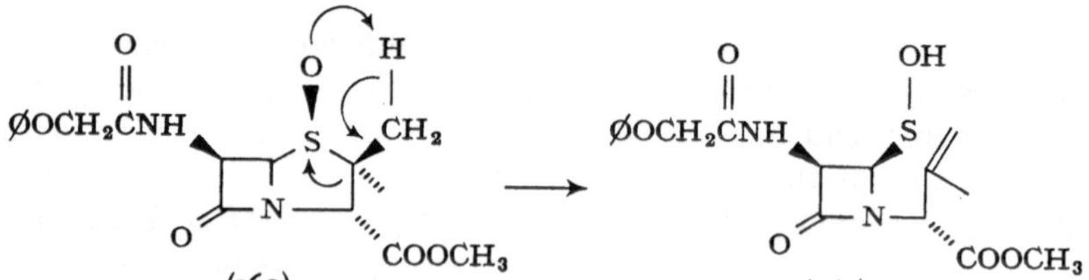

(162)

(163)

Addition of the sulfenic acid function and of acetate to the double bond would thus give the two acetates (84) and (156). The formation of (157) was rationalized as involving an addition of activated sulfenic acid and subsequent elimination of hydrogen ion. Compound (159) likely results from a concerted decarboxylation-addition reaction. The stereochemistry assigned to (160) was further substantiated by the observation that its sulfoxide did not undergo rearrangement to (103). Apparently, the presence of a methyl group *cis* to the sulfoxide oxygen (β) is a prerequisite for rearrangement.

$$\text{ØOCH}_2\text{CNH}$$

(103)

Formation of only a single sulfoxide upon oxidation of penicillins has been described and was rationalized on the basis that the 6-β-amido hydrogen exerts control by hydrogen bonding with the incipient sulfoxide oxygen or the oxidant (46a). Oxidation of penicillins not possessing such an amide hydrogen has been studied, (19), (47) and has been found to give rise to α-oriented sulfoxides. Use of iodobenzene dichloride in aqueous pyridine as oxidant has produced both sulfoxides in approximately equal amounts (19). Although known methods for inverting sulfoxide configurations were unsuccessful when applied to the β-sulfoxide, inversion has been accomplished by photochemical means (12). No descriptions of attempted rearrangements of α-sulfoxides have been published.

In addition to the β-lactam-containing products described above, three non-β-lactam products (164)–(166) were isolated from sulfoxide rearrangement mixtures (107). Compound (167) has been obtained from rearrangement of penicillin V sulfoxide and has also been found as a degradation product of penicillin V (93). The small quantities isolated from sulfoxide rearrangement may arise from the presence of some unoxidized penicillin or by disproportionation.

Conversion of the ester (157) to a cephalosporin was reported recently by WEBBER et al. (150). As reported in a previous section, basic hydrolysis of (157) gave the Δ^2-acid (69). The acid was converted to its corresponding p-methoxybenzyl ester and brominated to give (168). Treatment of the crude bromide with potassium acetate afforded an equilibrium mixture of Δ^2- and Δ^3-esters (169). Oxidation of the mixture of esters with m-chloroperbenzoic acid gave a single sulfoxide, (170). Following reduction of the sulfoxide with sodium dithionite-acetyl chloride, the ester group was cleaved with trifluoroacetic acid giving the cephalosporin (171), identical with that prepared from cephalosporin C.

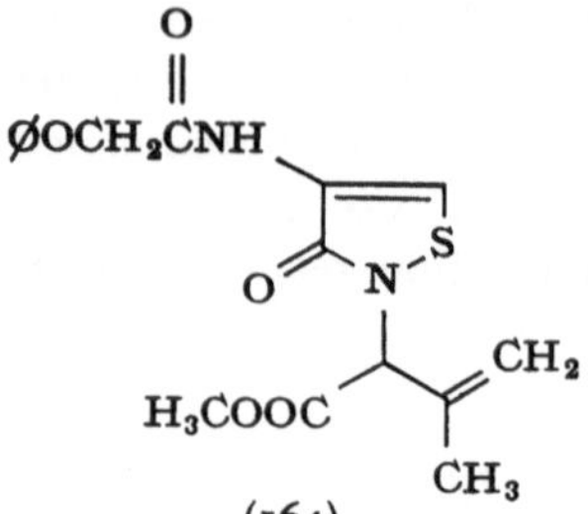

(164)

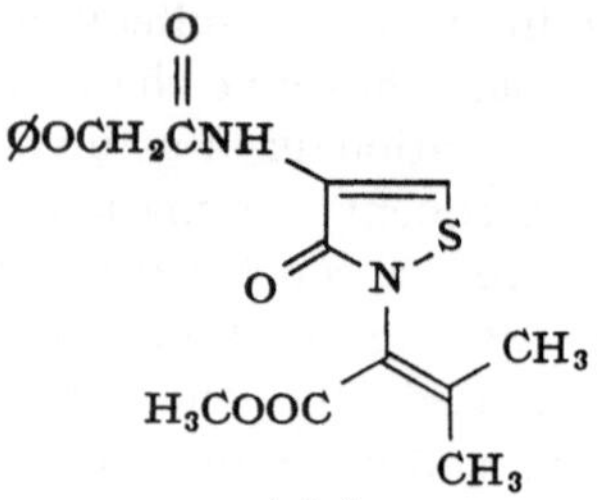

(165)

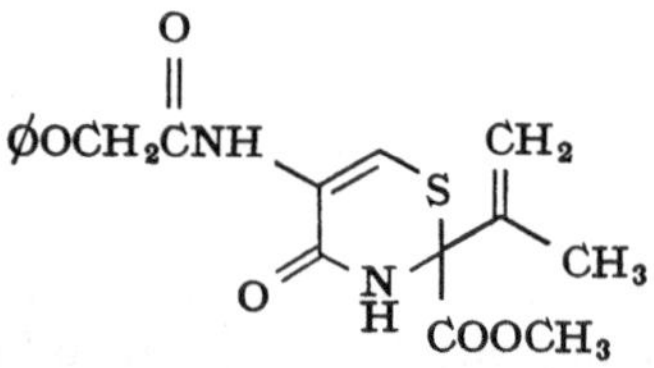

(166)

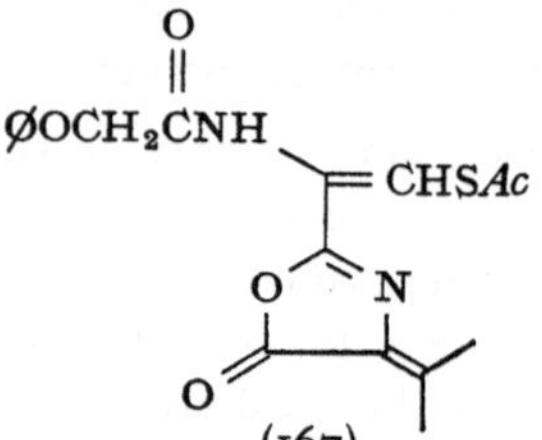

(167)

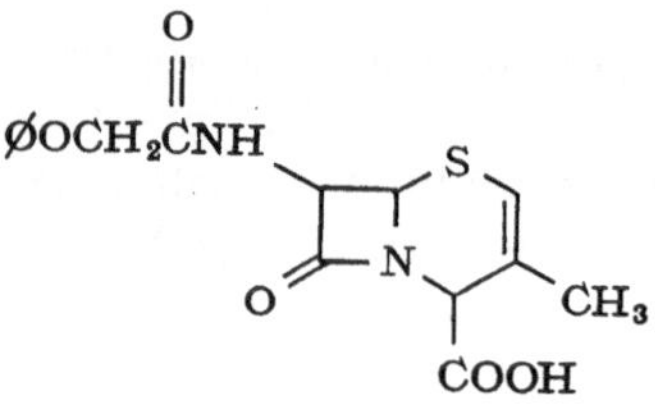

(69)

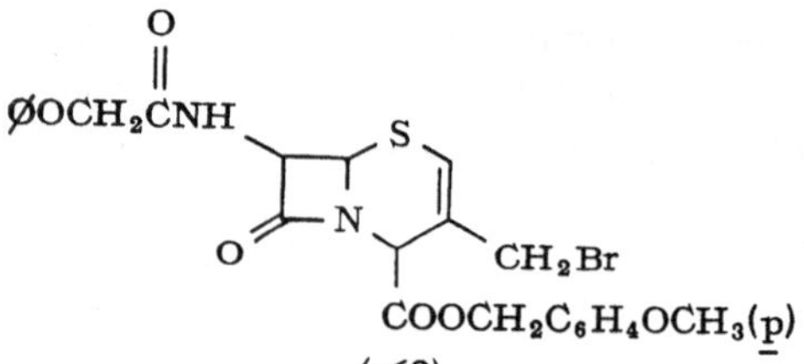

(168)

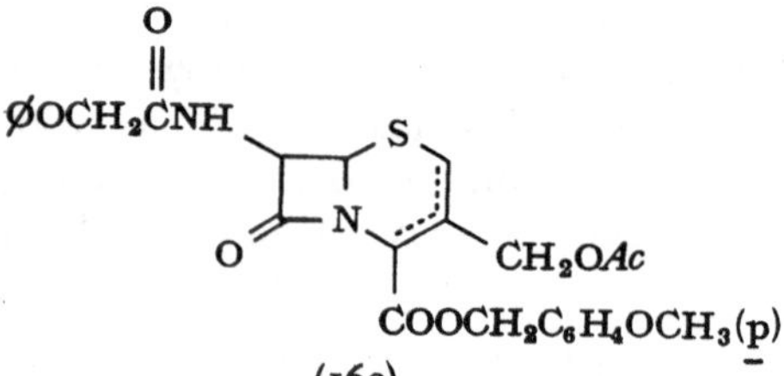

(169)

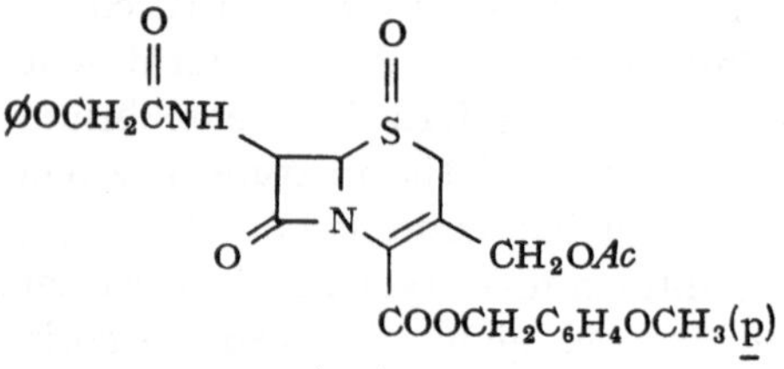

(170)

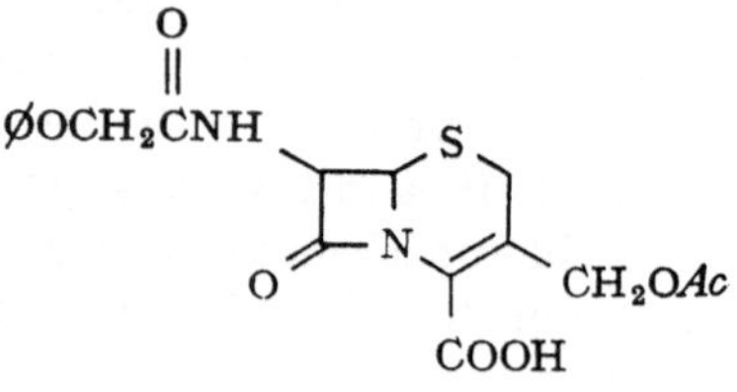

(171)

In a study related to the penicillin sulfoxide rearrangement the sulfoxide ester (**172**) was treated with triethyl phosphite and gave excellent yields of the thiazoline (**173**) (*48*).

Compound (**173**) probably arises from elimination to a sulfenic acid related to (**163**), followed by a reduction and cyclization.

(172)   (173)

VI. Biosynthesis

Since penicillin N and cephalosporin C have many structural similarities and are coproduced by the same organism, the biosynthesis of the two antibiotics is undoubtedly closely related. L-α-aminoadipic acid, L-cysteine and L-valine are incorporated into the sidechain, the β-lactam ring and the sulfur-nitrogen ring system of cephalosporin C (*144*) respectively. Similarly, penicillins are synthesized by *Penicillium chrysogenum* from a sidechain acid, in this case a mono-substituted acetic acid, cysteine and valine (*5, 51*).

(174)

The tripeptide δ (α-aminoadipyl) cysteinyl valine (**174**) has been isolated from *Penicillium chrysogenum* and has been suggested (*14*) as a precursor of normal penicillins. The discovery (*44, 60*) in this organism of isopenicillin N, which differs from penicillin N by having an L-α-aminoadipyl residue, gave credence to this postulate. The final step in this postulation of the biosynthesis of penicillin is an acyl transfer reaction or the production of 6-APA if precursor acid is not added. The fact that α-aminoadipic acid stimulated the production of benzyl penicillin and blocked the inhibitory effect of added L-lysine gave further support to this scheme (*131*).

The tripeptide (**174**) also appears to be formed in *Cephalosporium sp.*, and is rapidly labelled upon addition of 1-^{14}C-valine. The tripeptide is found in small quantities in the intracellular fluid and undergoes rapid turnover. The intermediacy of this substance in the biosynthesis of β-lactam antibiotics in both *Penicillium chrysogenum* and *Cephalosporium sp.* has been widely accepted. However, this has not yet been experimentally verified. The existence of permeability barriers to uptake of peptides and the presence in the extracellular fluid of amidases which rapidly hydrolyse peptides have prevented the testing of this hypothesis (*95*). Until cell-free antibiotic-producing systems are found, precursor-product relationships will be very difficult to establish.

Methionine has been found (*53, 90, 115*) to stimulate the production of β-lactam antibiotics in *Cephalosporium sp.* The sulfur but not carbon atoms of the amino acid are incorporated into the antibiotics (*31*). Cysteine and homocysteine do not provide the same stimulatory effect. The yield of benzyl pencillin in *Penicillium chrysogenum* is not increased by the addition of methionine.

Cephalosporium sp. differ further from *Penicillium chrysogenum* in that no sidechain amidases or acyl transferases are known to be produced. No 7-ACA has been reported found in the fermentation. This important difference has necessitated chemical manipulation of cephalosporin C to obtain clinically useful antibiotics.

VII. Biological Activity of Cephalosporanates

Several reports and reviews of the antibacterial activities of cephalosporins have been published (*1, 37, 127, 146*). Some general comments on activity follow, but for details the reader is referred to the cited references.

As is the case with penicillins, presence of an amide function at C-7 is essential for useful antibiotic action. In general, the cephalosporins are less active than the corresponding penicillins when measured against the standard *Staphyloccus aureus* organism; however, the cephalosporins are often more effective against penicillin-resistant *Staphylococci*. In addition, increased activity toward gram negative organisms is observed with certain sidechains. Cephalothin (**175**), the first cephalosporin to be marketed, is active against both resistant *Staphylococci* and gram negative organisms.

Hydrolysis of the acetate moiety at C-3' reduces the biological activity by a factor of approximately $1/2$. Replacement of the acetate with various nucleophiles has given antibiotics with enhanced activity; for example, in some test systems cephaloridine (67) is some 2–3 times as potent as cephalothin (134).

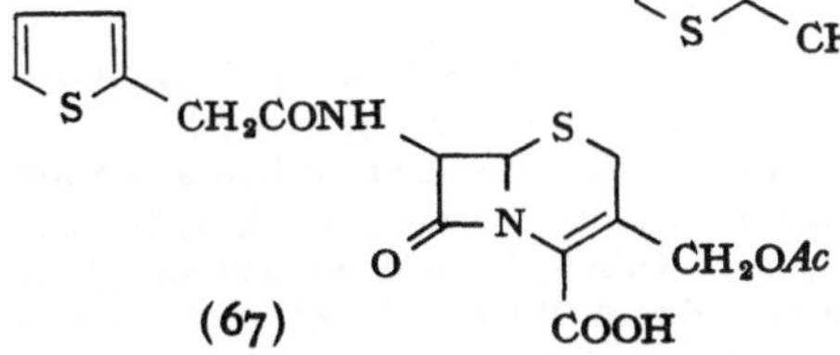

(67) 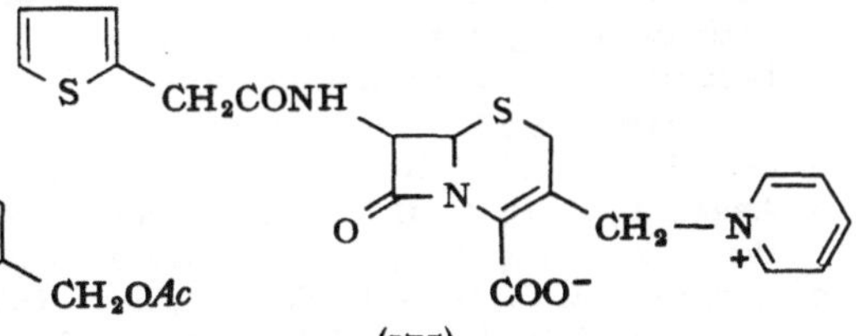 (175)

Neither cephalothin nor cephaloridine is particularly effective when administered orally. Presence of an α-amino group on the C-7 amide does give compounds which are orally absorbed. Cephaloglycin (68) (132) is an example of such aminoacyl cephalosporins.

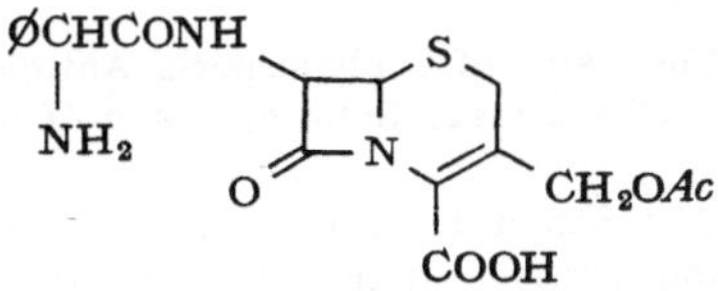

(68) 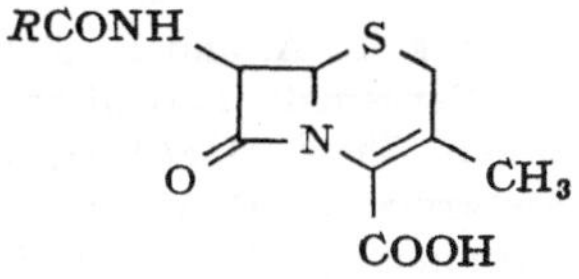 (176)

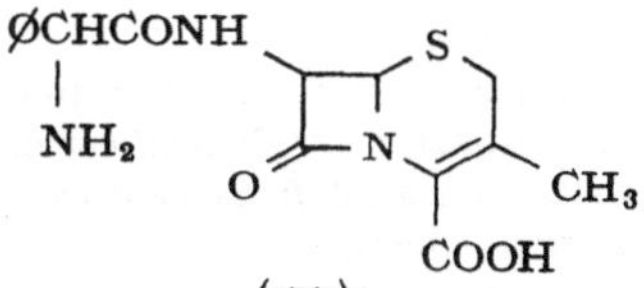

(177)

As a class, the desacetoxy cephalosporins (176) (124, 137) are considerably less active than their acetylated counterparts. Cephalexin (177), however, is absorbed from the gastro-intestinal tract to such a high degree that its activity relative to cephaloglycin is compensated. In fact, the effective dose of cephalexin is often less than that of cephaloglycin (151).

References

1. ABRAHAM, E. P.: The Cephalosporin C Group. Quart. Rev. (Chem. Soc., London) 21, 231 (1967).
2. ABRAHAM, E. P. and G. G. F. NEWTON: Synthesis of D-δ-Amino-δ-carboxylvalerylglycine (A Degradation Product of Cephalosporin N) and of DL-δ-Amino-δ-carboxyvaleramide. Biochem. J. 58, 266 (1954).

3. Abraham, E. P. and G. G. F. Newton: Experiments on the Degradation of Cephalosporin C. Biochem. J. 62, 658 (1956).
4. — — The structure of Cephalosporin C. Biochem. J. 79, 377 (1961).
5. — — Penicillins and Cephalosporins. In: Antibiotics II Biosynthesis, p. 1–16. Gottlieb, D. and P. D. Shaw, Ed. New York: Springer. 1967.
6. — — Structure and Function of Some Sulfur-containing Peptides. CIBA Foundation Symposium on Amino Acids and Peptides with Antimetabolic Activity, p. 205–225. Wolstenholme, G. E. W. and C. M. O'Connor, Ed. Boston: Little, Brown and Co. 1968.
7. Abraham, E. P., G. G. F. Newton, K. Crawford, H. S. Burton, C. W. Hale: Cephalosporin N: A New Type of Penicillin. Nature 171, 343 (1953).
8. Abraham, E. P., G. G. F. Newton and C. W. Hale: Purification and Some Properties of Cephalosporin N, a New Penicillin. Biochem. J. 58, 94 (1954).
9. Abraham, E. P., G. G. F. Newton, B. H. Olson, D. M. Schullrmans, J. R. Schenck, M. P. Hargie, M. W. Fisher and S. A. Fusari: Identity of Cephalosporin N and Synnematin B. Nature 176, 551 (1955).
10. Abraham, E. P. and L. D. Sabath: Enzymatic Behavior of the Cephalosporins. Enzymologia 29, 223 (1965)
11. Abraham, E. P. and P. W. Trown: Structure and Synthesis of Cephaloridine, a Degradation Product of Cephalosporin C. Biochem. J. 86, 271 (1963).
12. Archer, R. A. and P. V. Demarco: Photochemical Preparation and Conformational Analysis by Proton Magnetic Resonance of Penicillin (R)-Sulfoxides. J. Amer. Chem. Soc. 91, 1530 (1969).
13. Archer, R. A. and B. S. Kitchell: Chemistry of Cephalosporin Antibiotics. VI. Carbamate Formation in Aqueous Bicarbonate Solutions of 7-ACA. J. Organ. Chem. (USA) 31, 3409 (1966).
14. Arnstein, H. R. V. and D. Morris: The Structure of a Peptide, Containing α-Aminoadipic Acid, Cystine and Valine, Present in the Myceuum of Penicillium chrysogenum. Biochem. J. 76, 357 (1960).
15. Baird, B. M., T. G. Halsall, E. R. H. Jones and G. Lowe: Cephalosporin P₁. Proc. Chem. Soc. 257 (1961).
16. Ballio, A., E. B. Chain, F. Dentice di Accadia, M. Mauri, K. Rauer, M. J. Schlesinger and S. Schlesinger: Identification of a Compound Related to 6-Aminopenicillanic Acid Isolated from Culture Media of Penicillium chrysogenum. Nature 191, 909 (1961).
17. Barrett, G. C., S. H. Eggers, T. R. Emerson and G. Lowe: Studies Related to Cephalosporin C. Part II. A Synthetical Route to 5,6-Dihydro-4 H-1,3-thiazines and 3,6-Dihydro-2 H-1,3-thiazines. J. Chem. Soc. (London) 788 (1964).
18. Barrett, G. C., V. V. Kane and G. Lowe: Studies Related to Cephalosporin C. Part I. 3-Hydroxy- and 3-Amino-furan-2(5 H)-ones. J. Chem. Soc. (London) 783 (1964).
19. Barton, D. H. R., F. Comer and P. G. Sammes: Stereoisomerism of Penicillin Sulfoxides. J. Amer. Chem. Soc. 91, 1529 (1969).
20. Batchelor, F. R., F. P. Doyle, J. H. C. Nayler and G. N. Rolinson: Synthesis of Penicillin: 6-Aminopenicillanic Acid in Penicillin Fermentations. Nature 183, 257 (1959).
21. Batchelor, F. R., D. Gazzard and J. H. C. Nayler: Action of Carbon Dioxide on 6-Aminopenicillanic Acid. Nature 191, 910 (1961).
22. Bishop, E. O. and R. E. Richards: A Nuclear-Magnetic-Resonance Study of Cephalosporidine and Related Compounds. Biochem. J. 86, 277 (1963).

23. BOSE, A. K., G. SPIEGELMAN and M. S. MANHAS: Studies on Lactams. X. Total Synthesis of 5,6-*trans*-Penicillin V Methyl Ester. J. Amer. Chem. Soc. 90, 4506 (1968).

24. BOSE, A. K., V. SUDARSANAM, B. ANJANEYULU and M. S. MANHAS: Studies on Lactams-XI. Synthesis of Some Cepham Derivatives. Tetrahedron 25, 1191 (1969).

25. BRADSHAW, J., S. EARDLEY and A. G. LONG: Cephalosporanic Acids. Part VI. Action of Primary and Secondary Aromatic Amines on Cephalosporanic Acids. J. Chem. Soc. (London) 801 (1968).

26. BROTZU, G.: Lav. Ist. Igiene, Cagliari (1948).

27. BURTON, H. S. and E. P. ABRAHAM: Isolation of Antibiotics from a Species of *Cephalosporium*. Cephalosporins P_1, P_2, P_3, P_4 and P_5. Biochem. J. 50, 168 (1951).

28. BURTON, H. S., E. P. ABRAHAM and H. M. E. CARDWELL: Cephalosporin P_1 and Helvolic Acid. Biochem. J. 62, 171 (1956).

29. BUTLER, K.: Penicillins. Kirk-Othmer Encyclopedia of Chemical Technology, Second Edition 14, p. 652—707. A. Standen, Exec. Ed. New York: Interscience Publishers. 1967.

30. BYCROFT, B. W.: Structural Relationships in Microbial Peptides. Nature 224, 595 (1969).

31. CALTRIDER, P. G. and H. F. NISS: Role of Methionine in Cephalosporin Synthesis. Appl. Microbiol. 14, 746 (1966).

32. CAPON, B.: Neighboring Group Participation. Quar. Rev. (Chem. Soc., London). 18, 45 (1964).

33. CHAIN, E., H. W. FLOREY, M. A. JENNINGS and T. I. WILLIAMS: Helvolic Acid, an Antibiotic Produced by *Aspergillus Fumigatus, Mut. Helvola* Yuill. Brit. J. Exp. Path. 24, 108 (1943).

34. CHAMBERLIN, J. W. and J. B. CAMPBELL: Chemistry of Cephalosporin Antibiotics. X. Synthesis of Methyl 3-Formyl-7-(thiophene-2-acetamido) 3-cepham-4-carboxylate, a New Cephalosporin Derivative. J. Med. Chem. 10, 966 (1967).

35. CHAUVETTE, R. R. and E. H. FLYNN: Chemistry of Cephalosporin Antibiotics. V. Amides and Esters of Cephalothin. J. Med. Chem. 9, 741 (1966).

36. CHAUVETTE, R. R., E. H. FLYNN, B. G. JACKSON, E. R. LAVAGNINO, R. B. MORIN, R. A. MUELLER, R. P. PIOCH, R. W. ROESKE, C. W. RYAN, J. L. SPENCER and E. VAN HEYNINGEN: Chemistry of Cephalosporin Antibiotics. II. Preparation of a New Class of Antibiotics and the Relation of Structure to Activity. J. Amer. Chem. Soc. 84, 3401 (1962).

37. — — — — — — — — — — Structure-Activity Relationships Among 7-Acylaminocephalosporanic Acids. Antimicrobial Agents and Chemotherapy 687 (1962).

38. CHOU, T. S., E. J. EISENBRAUN and R. T. RAPALA: The Chemistry of Cephalosporin P_1. Tetrahedron Letters 409 (1967); The Chemistry of Steroid Acids from *Cephalosporium acremonium*. Tetrahedron 25, 3341 (1969).

39. CLARKE, H. T., J. R. JOHNSON and R. ROBINSON, Ed.: The Chemistry of Penicillin. Princeton, New Jersey: Princeton University Press. 1949.

40. CLAYTON, J. P., J. H. C. NAYLER, R. SOUTHGATE and E. R. STOVE: Penicillanic Acids: Requirements for Epimerization at C-6. Chem. Commun. 129 (1969).

41. COCKER, J. D., B. R. CROWLEY, J. S. G. COX, S. EARDLEY, G. I. GREGORY, J. K. LAZENBY, A. G. LONG, J. C. P. SLY and G. A. SOMERFIELD: Cephalosporanic Acids. Part II. Displacement of the Acetoxyl Group by Nucleophiles. J. Chem. Soc. (London) 5015 (1965).

398   R. B. Morin and B. G. Jackson:

42. Cocker, J. D., S. Eardley, G. I. Gregory, M. E. Hall and A. G. Long: Cephalosporanic Acids. Part IV. 7-Acylamidoceph-2-em-4-carboxylic Acids. J. Chem. Soc. (London) 1142 (1966).
43. Cole, M.: Properties of the Penicillin Deacylase Enzyme of *Escherichia coli.* Nature **203**, 519 (1964).
44. Cole, M. and F. R. Batchelor: Aminoadipylpenicillin in Penicillin Fermentations. Nature **198**, 383 (1963).
45. Cooper, A.: The Crystal Structure of Fusidic Acid Methyl Ester 3-*p*-Bromobenzoate. Tetrahedron **22**, 1379 (1966).
46. Cooper, R. D. G., P. V. Demarco, C. F. Murphy and L. A. Spangle: Chemistry of Cephalosporin Antibiotics. Part XVI. Configurational and Conformational Analysis of Deacetoxy-Δ^2- and -Δ^3-Cephalosporins and their Corresponding Sulphoxide Isomers by Nuclear Magnetic Resonance. J. Chem. Soc. (C) (London) 340 (1970).
46a. Cooper, R. D. G., P. V. Demarco, J. C. Cheng and N. D. Jones: Structural Studies on Penicillin Derivatives. I. The Configuration of Phenoxymethyl Penicillin Sulfoxide. J. Amer. Chem. Soc. **91**, 1408 (1969).
47. Cooper, R. F. G., P. V. Demarco and D. O. Spry: Structural Studies on Penicillin Derivatives. II. The Configuration of Phthalimidopenicillin and Epiphthalimidopenicillin Sulfoxides. J. Amer. Chem. Soc. **91**, 1528 (1969).
48. Cooper, R. D. G. and F. José: Structural Studies on Penicillin Derivatives. IV. A Novel Rearrangement of Penicillin Sulfoxides. J. Amer. Chem. Soc. **92**, 2575 (1970).
49. Corey, E. J. and A. M. Felix: A New Synthetic Approach to the Penicillins. J. Amer. Chem. Soc. **87**, 2518 (1965).
50. Crawford, K., N. G. Heatley, P. F. Boyd, C. W. Hale, B. K. Kelly, G. A. Miller and N. Smith: Antibiotic Production by a Species of *Cephalosporium.* J. Gen. Microbiol. **6**, 47 (1952).
51. Demain, A. L.: Biosynthesis of Penicillins and Cephalosporins. In Biosynthesis of Antibiotics, Vol. I, p. 29–94. Snell, J. F., Ed. New York: Academic Press. 1966.
52. Demain, A. L. and J. F. Newkirk: Biosynthesis of Cephalosporin C. Appl. Microbiol. **10**, 321 (1962).
53. Demain, A. L., J. F. Newkirk and D. Hendlin: Effect of Methionine, Norleucine, and Lysine Derivatives on Cephalosporin C Formation in Chemically Defined Media. J. Bacteriol. **85**, 339 (1963).
54. Diamond, R. D.: Ph. D. Thesis, Oxford University (1963).
55. Dolfini, J. E., J. Schwartz and F. Weisenborn: Synthesis of Dihydrothiazines Related to Deacetylcephalosporin Lactones. An Alternate Total Synthesis of Deacetylcephalosporin Lactones. J. Organ. Chem. (USA) **34**, 1582 (1969).
56. Eggers, S. H., T. R. Emerson, V. V. Kane and G. Lowe: The Synthesis of a New Fragmentation Product of a Cephalosporanic Acid Derivative. Proc. Chem. Soc. (London) 248 (1963).
57. Eggers, S. H., V. V. Kane and G. Lowe: Studies Related to Cephalosporin C. Part III. A Synthetical Route to 6H-1,3-Thiazines and the Synthesis of a New Fragmentation Product of a Cephalosporanic Acid Derivative. J. Chem. Soc. (London) 1262 (1965).
58. Fazakerley, H., D. A. Gilbert, G. T. Gregory, J. K. Lazenby and A. G. Long: Cephalosporanic Acids. Part V. The Action of Bidentate Nucleophiles on Cephalosporanic Acids. J. Chem. Soc. (London) 1959 (1967).

59. FECHTIG, B., H. PETER, H. BICKEL and E. VISCHER: Modifikationen von Antibiotika 2. Mitteilung über die Darstellung von 7-Aminocephalosporansäure. Helv. Chim. Acta 51, 1108 (1968).

60. FLYNN, E. H., M. H. McCORMICK, M. C. STAMPER, H. DE VALERIA and C. W. GODZESKI: A New Natural Penicillin from Penicillium chrysogenum. J. Amer. Chem. Soc. 84, 4594 (1962).

61. GALANTAY, E., H. ENGEL, A. SZABO and J. FRIED: Synthetic Studies Related to Cephalosporin C Sulfur-Containing α-Tetronic Acids and α-Aminobutenolides. J. Organ. Chem. (USA) 29, 3560 (1964).

62. GODTFREDSEN, W. O., S. JAHNSEN, H. LORCH, K. ROHOLT and L. TYBRING: Fusidic Acid: A New Antibiotic. Nature 193, 987 (1962).

63. GODTFREDSEN, W. O. and S. VANGEDAL: The Structure of Fusidic Acid. Tetrahedron 18, 1029 (1962).

64. GODTFREDSEN, W. O., W. VON DAEHNE, S. VANGEDAL, A. MARQUET, D. ARIGONI and A. MELERA: The Stereochemistry of Fusidic Acid. Tetrahedron 21, 3505 (1965).

65. GOTTSHALL, R. Y., J. M. ROBERTS, L. M. PORTWOOD and J. C. JENNINGS: Synnematin, an Antibiotic Produced by Tilachlodium. Proc. Soc. Exptl. Biol. Med. 76, 307 (1951).

66. GREEN, D. M., A. G. LONG, P. J. MAY and A. F. TURNER: α-Tetronic Acids and the 3,6-Dihydro-2 H-1,3-thiazine Ring. J. Chem. Soc. (London) 766 (1964).

67. GREEN, G. F. H., J. E. PAGE and S. E. STANIFORTH: Cephalosporanic Acids. Part I. Infrared Absorption and Proton Magnetic Resonance Spectra of Cephalosporin and Penicillin Analogues. J. Chem. Soc. (London) 1595 (1965).

68. — — — Proton Magnetic Resonance Spectra of Penicillin and Cephalosporin Derivatives. Chem. Commun. 597 (1966).

69. GUTOWSKI, G. E.: 6-Epi Penicillins and 7-Epi Cephalosporins. Tetrahedron Letters. 1779 (1970).

70. HALE, C. W., E. P. ABRAHAM and G. G. F. NEWTON: Properties of Some Derivatives of Cephalosporin C and Heterocyclic Tertiary Bases. International Congress of Biochemistry, 4th, Vienna, 1958. Abstracts of Communications.

71. HALE, C. W., G. G. F. NEWTON and E. P. ABRAHAM: Derivatives of Cephalosporin C Formed with Certain Heterocyclic Teriary Bases. Biochem. J. 79, 403 (1961).

72. HALSALL, T. G., E. R. H. JONES and G. LOWE: The Molecular Formula of Cephalosporin P_1. Proc. Chem. Soc. (London) 16 (1963).

73. HALSALL, T. G., E. R. H. JONES, G. LOWE and C. E. NEWALL: Cephalosporin P_1. Chem. Commun. 685 (1966).

74. HAMILTON-MILLER, J. M. T., G. G. F. NEWTON and E. P. ABRAHAM: Products of Aminolysis and Enzymic Hydrolysis of the Cephalosporins. Biochem. J. 116, 371 (1970).

75. HAMILTON-MILLER, J. M. T., E. RICHARDS and E. P. ABRAHAM: Changes in Proton-Magnetic-Resonance Spectra During Aminolysis and Enzymic Hydrolysis of Cephalosporins. Biochem. J. 116, 385 (1970).

76. HEUSLER, K.: Advances in the Total Synthesis of β-Lactam Antibiotics. In Topics in Pharmaceutical Sciences Vol. I. PERLMANN, D., Ed., p. 33–51. New York: Interscience Publishers. 1968.

77. HEYMÈS, R., G. AMIARD and G. NOMINÉ: Sur une Synthèse de la D,L-Déacétyl-céphalothine-lactone. C. R. Acad. Sc. Paris Série C 263, 170 (1966).

78. HODGKIN, D. C. and E. N. MASLEN: The X-ray Analysis of the Structure of Cephalosporin C. Biochem. J. 79, 393 (1961).

79. Hunt, D. and D. Rogers: Unpublished results as quoted in Reference 58, Footnote 15b.

80. Jackson, B. G.: In Press.

81. Jackson, B. G.: Personal observation.

82. Jeffery, J. D'A., E. P. Abraham and G. G. F. Newton: Further Degradation Products of Cephalosporin C Isolation and Synthesis of 2(4-amino-4-carboxy-butyl) Thiazole-4-carboxylic Acid. Biochem. J. **75**, 216 (1960).

83. — — — Deacetylcephalosporin C. Biochem. J. **81**, 591 (1961).

84. Johnson, D. A. and G. A. Hardcastle, Jr.: Reaction of 6-Aminopenicillanic Acid with Carbon Dioxide. J. Amer. Chem. Soc. **83**, 3534 (1961).

85. Johnson, D. A. and D. Mania: Epi-6-aminopenicillanic Acid and Epipenicillin G. Tetrahedron Letters 267 (1969).

86. Johnson, D. A., D. Mania, C. A. Panetta and H. H. Silvestri: Epihetacillin. Tetrahedron Letters 1903 (1968).

87. Kaiser, G. V., R. D. G. Cooper, R. E. Koehler, C. F. Murphy, J. A. Webber, I. G. Wright and E. M. Van Heyningen: Chemistry of Cephalosporin Antibiotics. XIX. Transformation of Δ^2-Cepham to Δ^3-Cephem by Oxidation-Reduction at Sulfur. J. Organ. Chem. (USA). **35**, 2430 (1970).

88. Kariyone, K., H. Harada, M. Kurita and T. Takano: Cefazolin, a New Semisynthetic Cephalosporin Antibiotic. I. Synthesis and Chemical Properties of Cefazolin. J. Antibiotics **23**, 131 (1970).

89. Kaufmann, W. and L. Bauer: Variety of Substrates for a Bacterial Benzyl Penicillin-splitting Enzyme. Nature **203**, 520 (1964).

90. Kavanagh, F., D. Tunin and G. Wild: D-Methionine and the Biosynthesis of Cephalosporin N, Arch. Biochem. Biophys. **77**, 268 (1958).

91. Korzybski, T., Z. Kowszyk-Gindifer and L. Kuryxowicz: Antibiotics Origen, Nature and Properties, Volume II, p. 1360–1362. Translated by E. Paryski, Pergamon Press. Warszawa: PWN-Polish Scientific Publishers. 1967.

92. Kukolja, S.: Chemistry of Cephalosporin Antiobiotics. XX. Synthesis and Biological Properties of 3-Acyloxymethyl-7-[2-(thienyl)acetamido]-3-Cephem-4-carboxylic Acid and Related Derivatives. J. Med. Chem. In Press.

92a. — Chemistry of Cephalosporin Antibiotics. XI. Preparation and Properties of Desacetylcephaloglycin and Its Lactone. J. Med. Chem. **11**, 1067 (1968).

93. Kukolja, S., R. D. G. Cooper and R. B. Morin: Structural Studies on Penicillin Derivatives. Part III. Rearrangement and Fragmentation of Penicillin V. Tetrahedron Letters 3381 (1969).

94. Lee, C. C., E. B. Herr, Jr. and R. C. Anderson: Pharmacological and Toxicological Studies on Cephalothin. Clin. Med. **70**, 1123 (1963).

95. Loder, P. B., E. P. Abraham and G. G. F. Newton: Behavior of α-Amino-adipylcysteine and Glutamylcysteine in the Presence of Intact and Disrupted Mycelium of a *Cephalosporium sp.* Biochem. J. **112**, 389 (1969).

96. Loder, P. B., G. G. F. Newton and E. P. Abraham: The Cephalosporin C Nucleus (7-Aminocephalosporanic Acid) and Some of its Derivatives. Biochem. J. **79**, 408 (1961).

97. Long, A. G. and A. F. Turner: Derivatives of α-Tetronic Acid. Tetrahedron Letters 421 (1963).

98. Lynch, J. E., J. M. Wilson, H. Budzikiewicz and C. Djerassi: Determination of Mass Spectra of Non-Volatile Substances. Experientia **19**, 211 (1963).

99. Manhas, M. S. and A. K. Bose: Synthesis of Penicillin, Cephalosporin C and Analogs. New York: Marcel Dekker. 1969.

100. MARTIN, J. L. and W. H. C. SHAW: in Proc. of Soc. Anal. Chem. Conf., Nottingham, 1965. Cambridge, England: W. Heffer & Sons, Ltd. 1965.

101. McMILLAN, I. and R. J. STOODLEY: A Novel Rearrangement of Methyl 6-Chloro Penicillinate. Tetrahedron Letters 1205 (1966).

102. MELERA, A.: Zur Konstitution von Helvolinsäure und Cephalosporin P_1. Experientia 19, 565 (1963).

103. MEYERS, A. I. and J. M. GREENE: Thiazine Derivatives. III. The Synthesis of Some 2-Substituted 5,6-Dihydro-1,3(4H)-thiazines and Tetrahydro-1,3-thiazines Related to Cephams. J. Organ. Chem. (USA) 31, 556 (1966).

104. MORIN, R. B.: Unpublished research.

105. MORIN, R. B., B. G. JACKSON, E. H. FLYNN and R. W. ROESKE: Chemistry of Cephalosporin Antibiotics I. 7-Aminocephalosporanic Acid from Cephalosporin C. J. Amer. Chem. Soc. 84, 3400 (1962).

105a. MORIN, R. B., B. G. JACKSON, R. A. MUELLER, E. R. LAVAGINO, W. B. SCANLON and S. L. ANDREWS: Chemistry of Cephalosporin Antibiotics. III. Chemical Correlation of Penicillin and Cephalosporin Antibiotics. J. Amer. Chem. Soc. 85, 1896 (1963).

106. MORIN, R. B., B. G. JACKSON, E. H. FLYNN, R. W. ROESKE and S. L. ANDREWS: Chemistry of Cephalosporin Antibiotics. XIV. The Reaction of Cephalosporin C with Nitrosyl Chloride. J. Amer. Chem. Soc. 91, 1396 (1969).

107. MORIN, R. B., B. G. JACKSON, R. A. MUELLER, E. R. LAVAGNINO, W. B. SCANLON and S. L. ANDREWS: Chemistry of Cephalosporin Antibiotics. XV. Transformation of Penicillin Sulfoxide. A Synthesis of Cephalosporin Compounds. J. Amer. Chem. Soc. 91, 1401 (1969).

108. OKUDA, S., S. I. IWASAKI, M. I. SAIR, Y. MACHIDA, A. INOUE and K. TSUDA: Stereochemistry of Helvolic Acid. Tetrahedron Letters 2295 (1967).

109. OKUDA, S., Y. SATO, T. HATTORI and M. WAKABAYASHI: Isolation and Structural Elucidation of 3-Oxo-16B-acetoxyfusida-1,17(20) [16,21-cis],24-trien-21-oic-acid. Tetrahedron Letters 4847 (1968).

110. NEWTON, G. G. F. and E. P. ABRAHAM: Degradation, Structure and Some Derivatives of Cephalosporin N. Biochem. J. 58, 103 (1954).

111. — — Cephalosporin C, a New Antibiotic Containing Sulfur and D-α-Amino-adipic Acid. Nature 175, 548 (1955).

112. — — Isolation of Cephalosporin C, a Penicillin-like Antibiotic Containing D-α-Aminoadipic Acid. Biochem. J. 62, 651 (1956).

113. NEWTON, G. G. F., E. P. ABRAHAM and S. KUWABARA: Preliminary Observations on the Formation and Breakdown of "Cephalosporoic Acids". Antimicrobial Agents and Chemotherapy, p. 449—455. G. L. HOBBY, Ed. American Society for Microbiology. Ann Arbor, Michigan: 1968.

114. OLSON, B. H., J. C. JENNINGS and A. J. JUNER: Separation of Synnematin into Components A and B by Paper Chromatography. Science 117, 761 (1953).

115. OTT, J. L., C. W. GODZESKI, D. PAVEY, J. D. FARRAN and D. R. HORTON: Biosynthesis of Cephalosporin C. I. Factors Affecting the Fermentation. Appl. Microbiol. 10, 515 (1962).

116. OXLEY, P.: Cephalosporin P_1 and Helvolic Acid. Chem. Commun. 729 (1966).

117. RAMSAY, B. G. and R. J. STOODLEY: Studies Related to Penicillins. Part III. 6β-Phthalimidohomopenicillanic Acid. J. Chem. Soc. (London) C 1319 (1969).

118. RAPAIA, R. T.: Personal communication.

119. REINHOUDT, D. N., H. S. TAN and H. C. BEYERMAN: Synthesis of Some Derivatives of Cephalosporamine. On the Synthesis of Cephalosporin Antibiotics. II. Rec. trav. Chim. Pays-Bas 87, 1153 (1968).

120. RICHTER, W. and K. BIEMANN: Hochauflösungs-Massenspektren von Penicillin-Derivaten. Monatsh. Chem. **95**, 766 (1964).

121. — — Hochauflösungs-Massenspektren von Cephalosporin-Derivaten. Monatsh. Chem. **96**, 484 (1965).

122. RITCHIE, A. C., N. SMITH and H. W. FLOREY: Some Biological Properties of Cephalosporin P₁. Brit. J. Pharmacol. **6**, 430 (1951).

123. ROLINSON, G. N., F. R. BATCHELOR, D. BUTTERWORTH, J. CAMERON-WOOD, M. COLE, G. C. EUSTACE, M. V. HART, M. RICHARDS and E. B. CHAIN: Formation of 6-Aminopenicillanic Acid from Penicillin by Enzymatic Hydrolysis. Nature **187**, 236 (1960).

124. RYAN, C. W., R. L. SIMON and E. M. VAN HEYNINGEN: Chemistry of Cephalosporin Antibiotics. XIII. Desacetoxycephalosporins. The Synthesis of Cephalexin and Some Analogs. J. Med. Chem. **12**, 310 (1969).

125. SABATH, L. D., M. JAGO and E. P. ABRAHAM: Cephalosporinase and Penicillinase Activities of a β-Lactamase from *Pseudomonas pyocyanea*. Biochem. J. **96**, 739 (1965).

126. SASSIVER, M. L. and R. G. SHEPHERD: Epimerization of Some Cephalosporin Sulfoxides. Tetrahedron Letters 3993 (1969).

127. SCHUMACHER, G. E.: Cephalosporin Antibiotics. J. Amer. Pharm. Assoc. **6**, 430 (1966).

128. SHEEHAN, J. C. and K. R. HENERY-LOGAN: The Total Synthesis of Penicillin V. J. Amer. Chem. Soc. **81**, 3089 (1959).

129. SHEEHAN, J. G., K. R. HENERY-LOGAN and D. A. JOHNSON: The Synthesis of Substituted Penicillins and Simpler Structural Analogs. VII. The Cyclization of a Penicilloate Derivative to Methyl Phthalimidopenicillanate. J. Amer. Chem. Soc. **75**, 3292 (1953).

130. SHEEHAN, J. C. and J. A. SCHNEIDER: A Synthetic Approach to Cephams. J. Organ. Chem. (USA) **31**, 1635 (1966).

131. SOMERSON, N. L., A. L. DEMAIN and T. D. NUNHEIMER: Reversal of Lysine Inhibition of Penicillin Production by α-Aminoadipic or Adipic Acid. Arch. Biochem. Biophys. **93**, 238 (1961).

132. SPENCER, J. L., E. H. FLYNN, R. W. ROESKE, F. Y. SIU and R. R. CHAUVETTE: Chemistry of Cephalosporin Antibiotics. VII. Synthesis of Cephaloglycin and Some Homologs. J. Med. Chem. **9**, 746 (1966).

133. SPENCER, J. L., F. Y. SIU, E. H. FLYNN, B. G. JACKSON, M. V. SIGAL, H. M. HIGGINS, R. R. CHAUVETTE, S. L. ANDREWS and D. E. BLOCH: Chemistry of Cephalosporin Antibiotics. VIII. Synthesis and Structure-Activity Relationships of Cephaloridine Analogues. Antimicrobial Agents and Chemotherapy 573 (1966).

134. SPENCER, J. L., F. Y. SIU, B. G. JACKSON, H. M. HIGGINS and E. H. FLYNN: Chemistry of Cephalosporin Antibiotics. IX. Synthesis of Cephaloridine. J. Organ. Chem. (USA) **32**, 500 (1967).

135. SPEZIALE, A. J. and P. C. HAMM: Preparation of 2-Substituted Acetamides. J. Amer. Chem. Soc. **78**, 5580 (1956).

136. STEDMAN, R. J.: t-Butyl Ester of 7-Aminocephalosporanic Acid. J. Med. Chem. **9**, 444 (1966).

137. STEDMAN, R. J., K. SWERED and J. R. E. HOOVER: 7-Aminodesacetoxycephalosporanic Acid and Its Derivatives. J. Med. Chem. **7**, 117 (1964).

138. STEWART, G. T.: The Penicillin Group of Drugs. New York: Elsevier Publishing Co. 1965.

139. STORK, G. and H. T. CHEUNG: Total Synthesis of the Cephalosporin Antibiotics. I. The Dihydrothiazine System of Cephalosporin Cc. J. Amer. Chem. Soc. **87**, 3783 (1965).

140. SWEET, R. M.: Ph. D. Thesis, University of Wisconsin (1970).

141. SWEET, R. M. and L. F. DAHL: The Structure of Cephaloridine Hydrochloride Monohydrate. Biochem. Biophys. Res. Comm. **34,** 14 (1969).

142. — — In Press.

143. TAYLOR, A. B.: Cephalosporanic Acids. Part III. Reactions with Pyridine — Kinetics and Mechanism. J. Chem. Soc. (London) 7020 (1965).

144. TROWN, P. W., B. SMITH and E. P. ABRAHAM: Biosynthesis of Cephalosporin C from Amino Acids. Biochem. J. **86,** 284 (1963).

145. VAN HEYNINGEN, E.: Chemistry of Cephalosporin Antibiotics. III. Acylation of Cephalosporadesates. J. Med. Chem. **8,** 22 (1965).

146. — Cephalosporins. Advan. in Drug Res. **4,** 1—70. N. J. HARPER and A. B. SIMMONDS, Ed. London: Academic Press. 1967.

147. VAN HEYNINGEN, E. and L. K. AHERN: Chemistry of Cephalosporin Antibiotics. XII. Configuration of the Carboxyl Group in Δ^2-Cephalosporins. J. Med. Chem. **11,** 933 (1968).

148. VAN HEYNINGEN, E. and C. N. BROWN: Chemistry of Cephalosporin Antibiotics. IV. Acetoxyl Replacements with Xanthates and Dithiocarbamates. J. Med. Chem. **8,** 174 (1965).

149. VON DAEHNE, W., H. LORCH and W. O. GODTFREDSEN: Microbiological Transformations of Fusidane-type Antibiotics. A Correlation Between Fusidic Acid and Helvolic Acid. Tetrahedron Letters 4843 (1968).

150. WEBBER, J. A., E. M. VAN HEYNINGEN and R. T. VASILEFF: Chemistry of Cephalosporin Antibiotics. XVII. Functionalization of Deacetoxycephalosporin. The Conversion of Penicillin into Cephalosporin. J. Amer. Chem. Soc. **91,** 5674 (1969).

151. WICK, W. E.: Cephalexin, a New Orally Absorbed Cephalosporin Antibiotic. Appl. Microbiol. **15,** 765 (1967).

152. — Unpublished results.

153. WOLFE, S. and W. S. LEE: A Ready C-6 Epimerization of the Penicillin Nucleus. Chem. Commun. 242 (1968).

154. WOODWARD, R. B.: Recent Advances in the Chemistry of Natural Products. Science **153,** 487 (1966).

155. WOODWARD, R. B., K. HEUSLER, J. GOSTELI, P. NAEGELI, W. OPPOLZER, R. RAMAGE, S. RANGANATHAN and H. VORBRÜGGEN: The Total Synthesis of Cephalosporin C. J. Amer. Chem. Soc. **88,** 852 (1966).

(Received, April 24, 1970)

Oligosaccharide der Frauenmilch

Von **H. Wiegandt** und **H. Egge**, Marburg/Lahn

Inhaltsübersicht

I. Einleitung

Die Fortschritte der Untersuchungen über die Oligosaccharide der Frauenmilch lassen sich in drei Phasen einteilen*. Die erste Phase ist verbunden mit den Namen G. Esbach (*31*) und G. Deniges (*25*). Esbach beobachtete als erster eine im Vergleich zum Lactoserum der Kuhmilch geringere Rechtsdrehung des Lactoserums der Frauenmilch. Die Drehung des Lactoserums der Frauenmilch kann bezogen auf das Reduktionsäquivalent $+ 37°$ bis $+ 47°$ betragen, gegenüber einem theoretischen Wert von $[\alpha]_D^{20} = + 52,3°$ für Lactose. Wenig später zog Deniges aus diesen Befunden den Schluß, daß neben der von ihm aus Frauenmilch isolierten Lactose auch andere zusätzliche Bestandteile vorhanden sein müssen, die für das ungewöhnliche optische Verhalten verantwortlich sind. Die zweite Phase der Untersuchungen der Oligosaccharide der Frauenmilch stand im Zeichen der Arbeiten von M. Polonovsky und A. Lespangnol, die von J. Montreuil (*109*) ausführlich referiert wurden. Mit Hilfe der Fällung und Kristallisation sowie der Bestimmung des sogenannten „Kupferindex ϱ", d. h., des Verhältnisses der erhältlichen Mengen Cu_2O vor und nach Hydrolyse (Kupferindex ϱ für Lactose

---

* Vergleiche hierzu die Übersichten (*68—74, 85, 105—108*).

Literaturverzeichnis: SS. 422—428

= 1,40), gelang den Autoren die Charakterisierung einer stark links drehenden *Gynolactose*, sowie einer schwach rechts drehenden ($[\alpha]_D$ = + 25°) *Allolactose* genannten Fraktion. Die für die Gynolactose erhaltenen Kupferindices von 2,5 bis 4,2 deuteten schon damals auf das Vorliegen mehrerer, auch höherer Oligosaccharide hin.

Der kristallisierten *Allolactose* wurde von Polonovsky und Lespagnol (*120* bis *123*) die Struktur einer O-β-D-Galaktopyranosyl-(1 → 6)-D-glucose zugeteilt, da sie auch in ihren Derivaten mit den Helferich (*50*) publizierten Daten für die chemisch synthetisierte Verbindung gut übereinstimmte.

Die *Allolactose* konnte jedoch später weder von Kuhn und Mitarb. (*72*) noch von Montreuil und Mitarb. (*109*) in frischer Frauenmilch nachgewiesen werden. Eine Klärung dieser Diskrepanz geben Befunde von Wallenfels (*143*) und Aronson (*3*). Sie erhielten durch Einwirken von β-Galaktosidase aus *E. coli*, Schimmelpilzen, *Helix pomatia* oder Kälberdarm auf Lactose oder Glucose ein Disaccharid, welches sie Lactobiose benannten. Die Identität dieser Lactobiose mit der Allolactose von Helferich wurde von Kuhn und Mitarb. (*77*) sicher bewiesen. Die Leichtigkeit, mit der die Allolactose durch verschiedene Enzymsysteme gebildet werden kann, läßt den Schluß zu, daß bei Polonovsky und Lespagnol die Allolactose durch eine bakterielle Infektion der untersuchten Milchproben entstanden ist.

Die dritte Phase der Untersuchung der Oligosaccharide der Frauenmilch bringt die Aufklärung der chemischen Strukturen dieser Zucker durch die Arbeiten der Laboratorien Richard Kuhns und Jean Montreuils. Das Interesse an der näheren Erforschung der Milchzucker wurde wesentlich angeregt durch die Beobachtung biologischer Wirkungen dieser Kohlenhydrate als Bifidusfaktoren, Blutgruppensubstanzen und durch ihre Verwandtschaft zu den Influenzavirusrezeptoren.

a) Bifidusfaktoren

1899 entdeckte Tissier (*138*) in Paris im Stuhl von brusternährten Säuglingen Mikroorganismen, die er *Lactobacillus bifidus* oder *Bacterium bifidum* nannte. Im Darmtrakt brusternährter Kinder findet sich nach übereinstimmenden Befunden mehrerer Arbeitsgruppen (*45*, *112—114*) eine praktisch 98% reine Lactobazillus bifidus-Flora. Schon Tissier fand, daß bei mit Kuhmilch ernährten Flaschenkindern dieses Bakterium nicht vorkommt. Die erhöhte Resistenz von Brustkindern, insbesondere gegenüber Infektionen des Magen-Darm-Traktes, die sich auch heute noch in hochzivilisierten Ländern nachweisen läßt (*45*, *46*, *114*), wird von vielen Pädiatern auf die Anwesenheit einer Lactobazillus bifidus-Flora im Darmtrakt dieser Säuglinge zurückgeführt. Bereits vor den Arbeiten von Polonovsky und Lespagnol hatte Schönfeld (*133*) gezeigt, daß sich der für die Bifidusflora wesentliche Stoff in der proteinfreien Magermilch der Frauenmilch befand. Der enge Zusammenhang zwischen den Befunden von Tissier, Schönfeld und Polonovsky und Lespagnol wurde erst 1950 durch die Untersuchungen von György (*115*, *3*, *135*, *141*) offenbar. Es gelang der Arbeitsgruppe um György auch, eine Mutante des *L. bifidus* zu isolieren, die nur in Gegenwart von Muttermilch wuchs, den *Lactobacillus bifidus* var. Penn. (*42*, *44*). Diese Mutante besitzt, wie später gezeigt wurde, einen absoluten Bedarf für β-D-N-Acetylglucosaminide (*45*, *69*, *96*). Die Gynolactose erwies sich als ein Wuchsstoff für *L. bacillus bifidus* var. Penn.

In der Literatur werden neben diesen Oligosacchariden auch noch andere Bifidusfaktoren angeführt, wie z. B. die Lactulose, (O-β-D-Galaktopyranosyl-($1 \rightarrow 4$)-D-fructose) (*119*, *1*). Bis heute scheint es jedoch keineswegs eindeutig geklärt zu sein, wie die Bifidusflora die Gesundheit des Säuglings beeinflußt.

b) Blutgruppensubstanzen

Auf den Zusammenhang der Oligosaccharide der Frauenmilch mit den Blutgruppensubstanzen wird in Abschnitt V ausführlicher eingegangen.

c) „Rezeptoren" für das Influenzavirus

Influenzaviren wie auch andere Myxoviren haben die Eigenschaft, gewaschene Erythrozyten in der Kälte zu agglutinieren. Die auf die Agglutination folgende Elution wies auf einen enzymatischen Prozeß hin (*2*, *20*). In der Tat besitzen Myxoviren ein Enzym, die *Neuraminatglykohydrolase* (EC 3.2.1.18), das bestimmte α-ketosidisch gebundene Sialinsäurereste abzuspalten vermag. Ähnliche Neuraminidasen wurden auch in Kulturfiltraten einiger Bakterien gefunden (*36*). Die Virusneuraminidase und die an der Erythrozytenoberfläche gebundenen Sialinsäurereste spielen bei der Hämagglutination eine wesentliche, jedoch nicht die alleinige Rolle, da auch noch andere Faktoren von Bedeutung sind (*28*). Eine Reihe der Oligosaccharide der Frauenmilch enthalten Sialinsäurereste in verschiedenen Stellungen, sind also „Rezeptoren" für Influenzaviren. Diese niedermolekularen Verbindungen erlauben das Studium der Substratspezifitäten von Neuraminidasen unterschiedlicher Herkunft. (Vgl. Abschnitt II, 2 d.)

II. Isolierung und Analyse der Oligosaccharide

1. Isolierung

Polonovsky und Lespagnol hatten durch fraktionierte Fällungen mit Aethanol, Methanol und Aceton kristallisierte *Allolactose* erhalten (*109*). Die Hauptmenge an Protein und Lactose läßt sich an Stelle des von A. Gauhe (*34*) ursprünglich eingesetzten Bariumhydroxyds und Zinksulfats durch 68%iges Äthanol ausfällen (*64*), wobei die höheren Oligosaccharide in der Mutterlauge verbleiben. Mittels Dialyse können 40—75% der niedermolekularen bifidusaktiven Bestandteile der entrahmten Milch in der Außenlösung erhalten werden (*43*). Zur Anreicherung der höhermolekularen Oligosaccharide wurde von Montreuil (*37*, *109*) die fraktionierte Ammonsulfatfällung herangezogen. Er erhielt dabei Oligosaccharide mit Molgewichten von 1444 bis 2020 (*37*).

Die weitere Reinigung und Abtrennung der einzelnen Oligosaccharide erfolgt vorwiegend durch chromatographische Methoden.

Literaturverzeichnis: SS. 422—428

a) Chromatographie an Kohle

Durch Chromatographie an Kohle-Celite-Säulen nach dem von WHISTLER und DURSO (148, 149) angegebenen Prinzip konnten KUHN und Mitarb. (90) Lacto-N-Tetraose in kristallisierter Form darstellen. Daneben wurden auch fucose- und sialinsäurehaltige Oligosaccharide angereichert. Kohle-Zellulose-Säulen (55) und Stearinsäure-behandelte Kohle (51) wurden ebenfalls zur Trennung von Oligosacchariden empfohlen. Bei der Elution der Oligosacchride mit Wasser-Aethanolgemischen hat sich die Anwendung eines kontinuierlichen Gradienten als überlegen erwiesen (6). Bei nicht säuregewaschener Kohle muß den Elutionsmitteln Säure zugesetzt werden (137).

b) Chromatographie an Sephadex

In letzter Zeit sind vielfach an die Stelle der etwas mühsam herzustellenden Kohle-Celite-Säulen Trennungen an Sephadex getreten (32, 33, 64, 100, 101, 116).

c) Chromatographie an Ionenaustauschern

Die Chromatographie an Ionenaustauschern ist vorwiegend für die Reinigung der Sialo-Oligosaccharide herangezogen worden (52, 54, 83), aber auch die Trennung neutraler Oligosaccharide auf Austauscherharzen mit borathaltigen Elutionsmitteln ist beschrieben worden (57).

d) Hochspannungselektrophorese

Neutrale Oligosaccharide als Boratkomplexe sowie saure Zucker wurden in der Hochspannungselektrophorese auf Papier getrennt (19, 37).

2. Analytische Methoden

Allgemeine Übersichten über die Methoden der Analyse von Oligosacchariden, komplexen Polysacchariden, Glykoproteiden und Mucopolysacchariden sind mehrfach erschienen (6, 7, 23, 130, 131, 149).

a) Papierchromatographie

Die Papierchromatographie ist die für die Trennung und Identifizierung der Oligosaccharide und ihrer Spaltprodukte am meisten verwendete Methode. Die Tabelle 1 gibt eine Übersicht über die für die Oligosaccharide gebräuchlichsten Lösungsmittelsysteme.

Tabelle 1. *Lösungsmittelsysteme für die Papierchromatographie*

Lösungsmittel		
1. Äthylacetat/Pyridin/Wasser/Eisessig	=	5: 5 : 3 : 1
2. Äthylacetat/Pyridin/Wasser/Eisessig	=	5: 5 : 4 : 1
3. Äthylacetat/Pyridin/Wasser/Eisessig	=	5: 5 : 5 : 1
4. Äthylacetat/Pyridin/Wasser/Eisessig	=	5: 5 : 3,5 : 1
5. Äthylacetat/Wasser/Eisessig	=	7: 2 : 2
6. n-Butanol/n-Propanol/0,1 N HCl	=	1: 2 : 1
7. n-Butanol/Eisessig/Wasser	=	4: 1 : 5 (Oberphase)
8. n-Butanol/Pyridin/Wasser	=	6: 4 : 3
9. Äthylacetat/Pyridin/Wasser	=	10: 7 : 3
10. n-Buthylacetat/Eisessig/Wasser	=	3: 2 : 1
11. Propanol/Wasser	=	4: 1
12. Phenol/Ameisensäure/2-Propanol/Wasser	=	80: 10 : 5 : 100 (Unterphase)
13. Pyridin/Äthylacetat/Wasser	=	1: 3,6 : 1,15
14. Phenol/Wasser/konz. Ammoniak	=	150: 40 : 1
15. Äthylacetat/Eisessig/Wasser	=	3: 1 : 3 (Oberphase)
16. Äthylacetat/Pyridin/Wasser	=	2: 1 : 2 (Oberphase)
17. Äthylacetat/Eisessig/Wasser	=	9: 2 : 2·
18. Butanol, Wasser gesättigt		

Zum Nachweis der Zucker auf Chromatogrammen dienen eine Reihe von Sprühreagenzien, die auf Grund bestimmter Farbreaktionen auch eine versuchsweise Zuordnung von Bindungstypen erlauben. Diese Reaktionen können zusammen mit anderen Daten, wie z. B. R_F-Werten wertvolle Hinweise zur Identifikation der Zucker liefern. So erhält man mit Triphenyltetrazoliumchlorid bei Vorliegen einer Bindung an C_2 des reduzierenden Zuckers keine oder nur eine sehr schwache Färbung (*142*). Mit Anilin-Diphenylamin-Phosphorsäure-Reagenz entsteht eine blaue oder purpurne Färbung, soweit das Oligosaccharid am reduzierenden Ende eine Glucose besitzt, die in 4-Stellung substituiert ist. Bei Verknüpfung in 6-Stellung erhält man schiefergraue bis grüne Farbtöne (*134*). Die Reaktion nach Morgan-Elson mit *p*-Dimethylaminobenzaldehyd fällt negativ aus, wenn das Hexosamin an C_4 substituiert ist (*87, 92, 145*).

b) Oxydation mit Perjodat

Die Oxydation mit Perjodat hat als Mikromethode zur Strukturanalyse komplexer Oligosaccharide große Erfolge gezeigt. Mehrere ausführliche Übersichten über die Reaktionen und deren Ergebnisse sind erschienen (*10, 11, 30, 91, 130*). Neben der Bestimmung des Perjodatverbrauches und der Menge an freigesetzter Ameisensäure oder Formaldehyd hat die Methode bei der Identifizierung von Verknüpfungsstellen der Fucose oder der Sialinsäure wertvolle Dienste geleistet.

c) Permethylierung

Die Permethylierung von Oligo- und Polysacchariden als klassische Methode (*48, 49, 125, 149*) zur Strukturaufklärung von Oligo- und Polysacchariden hat lange unter dem Mangel gelitten, große Mengen Substanz zu benötigen. Seit den Arbeiten von KUHN, PERILA und BISHOP, WALLENFELS und HAKOMORI (*4, 9, 47, 75, 84, 93, 118, 144*) ist es jedoch möglich, im mg-Bereich durch Kombination der Permethylierung mit der gaschromatographischen Analyse der permethylierten Produkte Aufschluß über die Strukturen komplexer Oligosaccharide zu gewinnen.

d) Enzymatischer Abbau

In jüngerer Zeit wurden mehrere gereinigte Enzyme (*35, 97, 98, 117, 131*) erfolgreich bei der Analyse komplexer Oligosaccharidstrukturen eingesetzt. So konnte mittels dreier aus *Diplokokkus pneumoniae* isolierter Hydrolasen, einer Neuraminidase (E. C. 3.2.1.18), β-Galaktosidase (E. C. 3.2.1.23) und β-N-Acetylglucosaminidase (E. C. 3.2.1.30), nachgewiesen werden, daß im α_1-Glykoprotein aus Humanplasma etwa in 7—8% der Zuckerketten die terminale Sequenz N-Acetylneuraminyl-β-D-galaktosyl-2-acetamido-2-deoxyglucose vorliegt (*53*). Analoge Untersuchungen wurden mit einer N-Acetylgalaktosaminidase aus Helix pomatia an Blutgruppen-A-aktiven Substanzen aus Ovarialcysten und Schweinemagen-Mucin durchgeführt (*140*).

Die Neuraminidase des Newcastle disease Virus spaltet die (α, 2 → 3) ketosidische Bindung in der 3'-Sialyllactose etwa ebenso leicht wie die (α, 2 → 8)-Bindung der Disialyllactose (*29*). Die (α, 2 → 6)-Bindung der Sialinsäure an Lactose wird praktisch nicht angegriffen (*27*). Interessant ist, daß die Neuraminidase des Virus der klassischen Geflügelpest, das ebenfalls zur Gruppe der Myxoviren gehört, zwar die (α, 2 → 3)-, nicht aber (α, 2 → 8)-Bindungen spaltet (*29*).

e) Gaschromatographie

Eine Übersicht über die gaschromatographische Trennung flüchtiger Zuckerderivate gibt BISHOP (*9*). Über gaschromatographische Trennungen von methylierten Spaltprodukten von Oligo- und Polysacchariden nach Permethylierung und Methanolyse wurde mehrfach berichtet (*4, 63, 84, 144*). Einen wesentlichen Fortschritt brachten Silylierungsreagenzien wie Hexamethyldisilazan, Bis-Trimethylsilylacetamid und Bis-Trimethylsilyltrifluoracetamid, die eine schnelle und quantitative Silylierung erlauben. Sie ermöglicht gaschromatographische Trennung von Monosacchariden, Di- und Trisacchariden. Eine Erweiterung dieser Methode über Tetrasaccharide hinaus dürfte jedoch wegen der hohen Siedepunkte und der thermischen Instabilität dieser Verbindung nur schwer möglich sein.

f) Massenspektrometrie

In einer breit angelegten Studie untersuchten Sweeley und Mitarb. (24) den Einfluß verschiedener Substituenten auf die Massenspektren von O-Trimethylsilylaethern der Glucose, Galaktose und des N-Acetylgalaktosamins. Die aus dieser Untersuchung gewonnenen Erkenntnisse wurden erfolgreich zur massenspektrometrischen Analyse einiger in ihrer Struktur schon bekannte Glykosphingolipide angewendet (132). Wesentlich bei dieser Methode ist, daß auch Verbindungen mit Molgewichten von weit über 1500 die für die Struktur entscheidenden Fragmente bei Massenzahlen unter $m/e = 850$ liefern.

Wie am Beispiel der Cerebroside (128) und der Sulfatide (61) gezeigt, wird die Massenspektrometrie kombiniert mit Gaschromatographie sicherlich auch bei der Analyse von Gemischen der Di- und Trisaccharide mit Erfolg angewendet werden. Durch Verwendung stabiler Isotope lassen diese Methoden auch Stoffwechseluntersuchungen am Menschen zu.

g) Automatische Bestimmungsmethoden

Automatische Bestimmungsmethoden, basierend auf konventionellen Farbreaktionen, sind mehrfach beschrieben worden: Glucosamin und Galaktosamin können über die Ninhydrinreaktion an Aminosäureanalysatoren bestimmt werden (12, 13, 26). Auch für sialinsäurehaltige Zucker (62) und andere komplexe Kohlenhydrate (56) sind automatische Bestimmungen angegeben worden.

III. Beschreibung der Zucker

Die einzigen Milcharten, die hinsichtlich ihres Gehaltes an Oligosacchariden genauer untersucht sind, stammen vom Rind und vom Menschen. Beide enthalten hauptsächlich Lactose. Frauenmilch unterscheidet sich aber deutlich durch das Vorkommen höherer reduzierender Zucker der „Gynolactose", die in Kuhmilch nicht oder nur in relativ geringerer Konzentration gefunden werden.

Die Zusammensetzung der Milch über den Zeitraum der gesamten Laktation bleibt nicht konstant. Die Menge der höheren Oligosaccharide, die im Colostrum am größten ist, sinkt im Laufe der Laktation ab, während diejenige der Lactose ansteigt (124, 108, 121). Beginnend etwa mit der zweiten Hälfte der Schwangerschaft findet auch ein bedeutender Anstieg der Konzentration von reduzierenden Oligosacchariden im Urin statt (22). Die während der Laktation im Urin ausgeschiedene Zuckermenge beträgt etwa 1 g pro Liter, im Vergleich zu 4—13 g pro Liter in der Milch. Sowohl die im Schwangerenharn vorkommenden Oligosaccharide, als auch die von normalen Personen im Urin ausgeschiedenen Zucker erwiesen sich als weitgehend identisch mit den in der Milch aufgefundenen (22, 54).

Literaturverzeichnis: SS. 422—428

Als Grundbausteine der freien, reduzierenden Oligosaccharide konnten die folgenden Zucker gefunden werden: D-Glucose, D-Galaktose, L-Fucose, D-2-Desoxy-2-acetamido-glucose (N-Acetylglucosamin), D-2-Desoxy-2-acetamido-galaktose (N-Acetylgalaktosamin), und Sialinsäuren, d. h. N-Acetylneuraminsäure und N-Glycolylneuraminsäure. In Glycoproteinen der Milch kommen hierzu noch Mannose und D-2-Desoxy-2-acetamido-mannose (N-Acetylmannosamin) vor.

Aus der großen Mannigfaltigkeit der Verknüpfungsmöglichkeiten dieser einzelnen Monosaccharidbausteine untereinander sind offenbar ganz bestimmte bevorzugt.

Ein Vergleich der chemischen Strukturen läßt folgende Aufbauprinzipien erkennen:

1) *Glucose* steht immer nur terminal am reduzierenden Ende der freien Oligosaccharide der Milch. Das entspricht der Stellung der Glucose am mit Ceramid verbundenen Ende der Oligosaccharide der Glykosphingolipide.

2) Die meisten Oligosaccharide sind durch lineare Verknüpfung der Grundbausteine aufgebaut. Verzweigungen treten durch Anheften von Fucose oder Sialinsäure auf.

3) Die weit überwiegende Anzahl der in der Milch bzw. im Urin aufgefundenen freien, reduzierenden Oligosaccharide leiten sich von der *Lactose* her ab. Diese kann selbst schon mit Fucose- oder Sialinsäureresten bestückt sein, oder das Disaccharid wird weiter unter gleichzeitigem Einbau von Fucose- oder Sialinsäureverzweigungen zu einer linearen Zuckerkette mit 4, 5 oder mehr Gliedern verlängert.

Folgende glycosidische Verknüpfungsarten wurden bei den Milchzuckern gefunden:

O-β-D-Galaktopyranosyl-(1 → 4)-D-glucose,
O-β-D-Galaktopyranosyl-(1 → 3)-D-N-acetylglucosamin,
O-β-D-Galaktopyranosyl-(1 → 4)-D-N-acetylglucosamin,
O-β-D-Galaktopyranosyl-(1 → 3)-D-N-acetylgalaktosamin*,
O-α-D-Galaktopyranosyl-(1 → 3)-D-Galaktose**,
O-α-D-N-Acetylgalaktosaminopyranosyl-(1 → 3)-D-galaktose***,
O-β-D-N-Acetylglucosaminopyranosyl-(1 → 3)-D-galaktose,
O-β-D-N-Acetylglucosaminopyranosyl-(1 → 4)-D-galaktose,
O-β-D-N-Acetylglucosaminopyranosyl-(1 → 6)-D-galaktose,
O-α-L-Fucopyranosyl-(1 → 3)-D-glucose,

---

* Kommt vor in freien Zuckern des Urins sowie in Glykosphingolipiden bzw. Glykoproteinen.

** Terminale Gruppierung der Blutgruppensubstanzen *B*. Vorkommen in Milch wahrscheinlich.

*** Endgruppe der Blutgruppensubstanzen *A*. Vorkommen in Milch wahrscheinlich.

O-α-L-Fucopyranosyl-(1 → 2)-D-galaktose,
O-α-L-Fucopyranosyl-(1 → 3)-D-N-acetylglucosamin,
O-α-L-Fucopyranosyl-(1 → 4)-D-N-acetylglucosamin,
O-α-N-Acetylneuraminyl-(2 → 3)-D-galaktose,
O-α-N-Acetylneuraminyl-(2 → 6)-D-galaktose,
O-α-N-Acetylneuraminyl-(2 → 6)-D-N-acetylglucosamin,
O-α-N-Acetylneuraminyl-(2 → 8)-N-Acetylneuraminsäure.

Manche Zucker können auch Sulfat- (*17, 127*) bzw. Phosphatreste (*21*) tragen.

Die im folgenden aufgeführten freien, reduzierenden Oligosaccharide konnten bisher aus Milch isoliert und ihre chemische Struktur aufgeklärt werden.

   1) O-α-N-Acetylneuraminyl-(2 → 3)-D-galaktose

Gal

3

↑

2 α NANA

Isolierung aus Kuhcolostrum: (*91*).
Die durch Hydrolyse und Perjodatoxydation bewiesene Struktur war auch für das Kohlenhydrat des Hirngangliosids G_{Gal} 1 NANA* gefunden worden.

   2) O-α-N-Acetylneuraminyl-(2 → 6)-O-β-D-galaktopyranosyl-(1 → 4)-D-2-desoxy-2-acetamido-glucose, 6'-N-Acetylneuraminyllactosamin

Gal β 1 → 4 GlcNAc
6

↑

2 α NANA

Das Kohlenhydrat konnte sowohl aus Kuhcolostrum, wie auch aus Urin isoliert werden (*91, 54*).

   3) [N-Acetylneuraminyl-]-O-α-N-acetylneuraminyl-(2 → 3)-O-β-D-galaktopyranosyl-(1 → 3)-D-2-desoxy-2-acetamido-galaktose

```
         ┌ Gal β 1 → 3 GalNAc
         │ 3
NANA-    │ ↑
         └ 2 α NANA
```

Das Oligosaccharid, das zwei Moleküle Sialinsäure enthält, wurde bisher nur aus Urin (*54*) isoliert. Die Stellung eines Sialinsäurerestes ist noch unbekannt.

Die drei beschriebenen Sialo-Oligosaccharide 1—3 sind bislang die einzigen in Milch bzw. Urin aufgefundenen, die nicht die Grundstruktur der Lactose in ihrem Aufbau zeigen (vgl. Biosynthese der Lactose und des Lactosamins).

---

* Vgl. zur Nomenklatur der Ganglioside (*150*).

Literaturverzeichnis: SS. 422—428

Fucose und Sialinsäureabkömmlinge der Lactose

4) O-β-D-Galaktopyranosyl-[O-α-L-fucopyranosyl-(1 → 3)-]-D-glucose, 3-Fucosido-lactose (*123, 107*).

Gal β 1 → 4 Glc
3
↑
1 α Fuc

5) O-α-L-Fucopyranosyl-(1 → 2)-O-β-D-galaktopyranosyl-(1 → 4)-D-glucose, 2'-Fucosido-lactose

Gal β 1 → 4 Glc
2
↑
1 α Fuc $[\alpha]_D^{20}$: — 57°*

Der Zucker kommt sowohl in der Frauenmilch, als auch im Schwangerenharn (*80, 22*) vor. In der Milch beträgt die Menge 10% der höheren Oligosaccharide (ohne Lactose).

6) O-α-L-Fucopyranosyl-(1 → 2)-O-β-D-galaktopyranosyl-(1 → 4)-[O-α-L-fucopyranosyl-(1 → 3)-]D-glucose, 3,2'-Difuco-lactose, Lacto-difuco-tetraose

Gal β 1 → 4 Glc
2   3
↑   ↑
1 α Fuc   1 α Fuc $[\alpha]_D^{20}$: — 106°

Auch dieser Zucker wurde aus der Frauenmilch (*86*), als auch aus dem Urin Schwangerer (*22*) isoliert.

Die Substanz zeigte in Hemmversuchen schwache Le[b]-Blutgruppenaktivität.

7) O-α-N-Acetylneuraminyl-(2 → 3)-O-β-D-galaktopyranosyl-(1 → 4)-D-glucose, 3'-N-Acetylneuraminyl-lactose

Gal β 1 → 4 Glc
3
↑
2 α NANA

Die Verbindung wurde ursprünglich aus Kuhcolostrum isoliert (*83*), kommt aber auch in der Frauenmilch (*139*) sowie im Urin (*54*) vor. Das Sialo-oligosaccharid entspricht dem Kohlenhydratteil des weit verbreiteten Gangliosids G_{Lact} 1 NANA (*150*).

8) O-α-N-Glycolylneuraminyl-(2 → 3)-O-β-D-galaktopyranosyl-(1 → 4)-D-glucose, 3'-N-Glycolylneuraminyl-lactose

---

* Die angegebenen Werte beziehen sich immer auf wäßrige Lösungen.

Gal β 1 → 4 Glc
3
↑
2 α NGNA

Kommt in geringerer Menge im Kuhcolostrum vor (*91*).

Das Hauptgangliosid der Organe des reticuloendothelialen Systems bei Pferd und Rind ist G_{Lact} 1 NGNA. Es enthält als Zuckerteil die 3'-N-Glycolylneuraminyl-lactose.

9) O-α-N-Acetylneuraminyl-(2 → 6)-β-D-galaktopyranosyl-(1 → 4)-D-glucose, 6'-N-Acetylneuraminyl-lactose

Gal β 1 → 4 Glc
6   .
↑
2 α NANA

Dieses Oligosaccharid wurde sowohl aus Frauenmilch, wie aus Kuhmilch isoliert (*74, 91*). Während in der Frauenmilch nur wenig 3'-Acetylneuraminyl-lactose neben dem 6'-Isomeren vorkommt, so ist das Verhältnis der beiden Monosialo-lactosen im Kuhcolostrum umgekehrt.

10) O-α-N-Acetylneuraminyl-(2 → 8)-O-α-N-acetylneuraminyl-(2 → 3)-O-β-D-galaktopyranosyl-(1 → 4)-D-glucose, 3'-Di-N-acetylneuraminyl-lactose.

Gal β 1 → 4 Glc
3
↑
2 α NANA 8 ← 2 α NANA

Die Disialo-lactose kommt im Kuhcolostrum vor (*91*). Sie bildet den Kohlenhydratanteil des Gangliosids G_{Lact} 2 NANA das aus Rinder bzw. Menschenhirn (*150*) und Rindernieren (*126*) extrahiert werden konnte. Die auch in einer Reihe von anderen Hirngangliosiden vorkommende Di-Sialo-Gruppierung NANA α 2 → 8 NANA- scheint in besonderem Maße zur Lactonbildung zu neigen. In saurem Milieu (pH 2) liegt der endständige der beiden Sialinsäurereste nahezu vollständig als Lacton vor (A. Gauhe unveröff., H. Wiegandt unveröff.).

11) O-α-D-N-Acetylneuraminyl-(2 → 3)-O-β-D-galaktopyranosyl-6-sulfat-(1 → 4)-D-glucose

HO_3S → 6 Gal β 1 → 4 Glc
3
↑
2 α NANA

Isolierung: (*17*) aus Rattenmammaextrakten.
Strukturaufklärung: (*127*).

12) O-β-D-Galaktopyranosyl-3-phosphat-(1 → 4)-D-glucose, Lactose-3'-phosphat

Literaturverzeichnis: SS. 422—428

Gal β 1 → 4 Glc

3

↑

PO₃H₂

Isolierung und Strukturaufklärung: (*21*).

Alle weiteren, höheren Milchzucker lassen sich von zwei stickstoff-haltigen Tetraosen, der *Lacto-N-tetraose* (Typ I) und der *Lacto-N-neo-tetraose* (Typ II) her ableiten (vgl. zur immunol. Bedeutung (*99*)).

Oligosaccharide des Typs I

13) O-β-D-Galaktopyranosyl-(1 → 3)-O-β-D-2-desoxy-2-acetamido-glucopyranosyl-(1 → 3)-O-β-D-galaktopyranosyl-(1 → 4)-D-glucose, Lacto-N-tetraose.

Gal β 1 → 3 GlcNAc β 1 → 3 Gal β 1 → 4 Glc   [α]D: + 25,5°

Diese Tetraose macht etwa 15% der höheren Oligosaccharide (ohne Lactose) der Frauenmilch aus (zirka 500 mg pro Liter) (*78, 90*). Sie konnte auch aus Schwangerenurin in kristalliner Form erhalten werden (*22*).

14) O-α-L-Fucopyranosyl-(1 → 2)-O-β-D-galaktopyranosyl-(1 → 3)-O-β-D-2-desoxy-2-acetamid-oglucopyranosyl-(1 → 3)-O-β-D-galaktopyranosyl-(1 → 4)-D-glucose, Lacto-N-fucopentaose I.

Gal β 1 → 3 GlcNAc β 1 → 3 Gal β 1 → 4 Glc

2

↑

1 α Fuc   [α]$_D^{23}$: − 11,0 → − 16,3 (Mutarotation der α-Form), Fp. 216°

Die Menge dieser Pentaose beträgt etwa 8% der höheren Oligosaccharide (ohne Lactose) der Frauenmilch (*79*).

15) O-β-D-Galaktopyranosyl-(1 → 3)-[O-α-L-fucopyranosyl-(1 → 4)-]O-β-D-2-desoxy-2-acetamido-glucopyranosyl-(1 → 3)-O-β-D-galaktopyranosyl-(1 → 4)-D-glucose, Lacto-N-fucopentaose II

Gal β 1 → 3 GlcNAc β 1 → 3 Gal β 1 → 4 Glc

4

↑

1 α Fuc

Das aus Frauenmilch (4% der höheren Oligosaccharide) (*81*) oder Schwangerenurin (*22*) isolierte Kohlenhydrat besitzt im immunologischen Test Leª-Aktivität.

16) O-α-L-Fucopyranosyl-(1 → 2)-O-β-D-galaktopyranosyl-(1 → 3)-[O-L-fuco-pyranosyl-(1 → 4)-]O-β-D-2-desoxy-2-acetamido-glucopyranosyl-(1 → 3)-O-β-D-galaktopyranosyl-(1 → 4)-D-glucose, Lacto-N-difucohexaose I

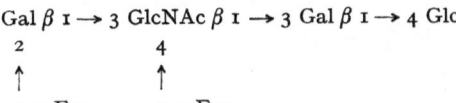

Gal β 1 → 3 GlcNAc β 1 → 3 Gal β 1 → 4 Glc

2   4

↑   ↑

1 α Fuc   1 α Fuc

Dieses Oligosaccharid der Frauenmilch (*82*) kommt wahrscheinlich auch im Schwangerenurin vor (*22*). Es ist immunologisch Le^b-positiv.

17) O-β-Galaktopyranosyl-(1 → 3)-[O-α-L-fucopyranosyl-(1 → 4)-] O-β-D-2-desoxy-2-acetamido-glucopyranosyl-(1 → 3)-O-β-D-galaktopyranosyl-(1 → 4)-[O-α-L-fucopyranosyl-(1 → 3)-]D-glucose, Lacto-N-difuco-hexaose II

Gal β 1 → 3 GlcNAc β 1 → 3 Gal β 1 → 4 Glc
     4   3
     ↑   ↑
     1 α Fuc   1 α Fuc

Das Kohlenhydrat besitzt, wie auch die Lacto-N-fucopentaose II, Blutgruppenaktivität Le^a. Es wurde aus Frauenmilch (*81*) in kristalliner Form erhalten.

18) O-α-N-Acetylneuraminyl-(2 → 3)-O-β-D-galaktopyranosyl-(1 → → 3)-O-β-D-2-desoxy-2-acetamido-glucopyranosyl-(1 → 3)-O-β-D-galaktopyranosyl-(1 → 4)-D-glucose, „Pentasaccharid a"

Gal β 1 → 3 GlcNAc β 1 → 3 Gal β 1 → 4 Glc
3
↑
2 α NANA

Isolierung aus Frauenmilch: (*89*).
Strukturaufklärung: (*91*).

19) O-β-D-Galaktopyranosyl-(1 → 3)-[O-α-N-acetylneuraminyl-(2 → → 6)-]O-β-D-2-desoxy-2-acetamido-glucopyranosyl-(1 → 3)-O-β-D-galaktopyranosyl-(1 → 4)-D-glucose, „Pentasaccharid b".

Gal β 1 — 3 GlcNAc β 1 — 3 Gal β 1 — 4 Glc
       6
       |
     2 α NANA $[\alpha]_D^{24}$: + 14,5 — 15,0°

Isolierung aus Frauenmilch: (*89*).
Konstitutionsermittlung: (91).

20) O-α-N-Acetylneuraminyl-(2 → 3)-O-β-D-galaktopyranosyl-[O-α-N-acetylneuraminyl-(2 → 6)-]O-β-D-2-desoxy-2-acetamido-glucopyranosyl-(1 → 3)-O-β-D-galaktopyranosyl-(1 → 4)-D-glucose, Di-N-Acetylneuraminyl-lacto-N-tetraose

Gal β 1 → 3 GlcNAc β 1 → 3 Gal β 1 → 4 Glc
    3   6
    ↑   ↑
   2 α NANA 2 α NANA $[\alpha]_D^{20}$: — 2,7°

Isolierung aus Frauenmilch: (*37*).
Chemische Konstitution: (*38*).

Literaturverzeichnis: SS. 422—428

Oligosaccharide des Typs II

21) O-β-D-Galaktopyranosyl-(1 → 4)-O-β-2-desoxy-2-acetamido-glucopyranosyl-(1 → 3)-O-β-D-galaktopyranosyl-(1 → 4)-D-glucose, Lacto-N-neotetraose.

Gal β 1 → 4 GlcNAc β 1 → 3 Gal β 1 → 4 Glc

$$[\alpha]_D^{22}: + 33 → + 27° \text{ Mutarotation}$$

Neben der Lacto-N-tetraose ist die Lacto-N-neotetraose in der Frauenmilch nur in geringer Menge enthalten. Sie ergibt keinen Farbstoff mit Dimethylamino-benzaldehyd in der Reaktion nach MORGAN-ELSON (vgl. Abschnitt II, 2a).
Strukturaufklärung: (88).
Die Lacto-N-neotetraose zeigt immunologisch eine Spezifität nach Art der Pneumokokken Typ XIV-polysaccharide.

22) O-β-D-galaktopyranosyl-(1 → 4)-[O-α-L-fucopyranosyl-(1 → 3)-] O-β-D-2-desoxy-2-acetamido-glucopyranosyl-(1 → 3)-O-β-D-galaktopyra-nosyl-(1 → 4)-D-glucose, Lacto-N-fucopentaose III

Gal β 1 → 4 GlcNAc β 1 → 3 Gal β 1 → 4 Glc

3
↑
1 α Fuc $[\alpha]_D^{22}: + 7,35;$ F. p. 275—277° (zers.)

Darstellung aus Frauenmilch und Strukturaufklärung: (64). Die Verbindung inhibiert die Le^a-anti-Le^a-Hämagglutination nicht.

23) O-α-N-Acetylneuraminyl-(2 → 6)-O-β-D-galaktopyranosyl-(1 → → 4)-O-β-D-2-desoxy-2-acetamido-glucopyranosyl-(1 → 3)-O-β-D-galakto-pyranosyl-(1 → 4)-D-glucose, ,,Pentasaccharid c"

Gal β 1 → 4 GlcNAc β 1 → 3 Gal β 1 → 4 Glc

6
↑
2 NANA $[\alpha]_D^{24}: + 13°$

Isolierung: (89).
Strukturaufklärung: (91).
Eine Reihe weiterer Mono- und Disialoderivate der Lacto-N-neotetraose wurden als Kohlenhydratkomponenten extraneuraler Ganglioside isoliert (151).

24) O-α-N-Glycolylneuraminyl-(2 → 3)-O-β-D-galaktopyranosyl-(1 → → 4)-O-β-D-2-desoxy-2-acetamido-glucopyranosyl-(1 → 3) -O-β-D-galakto-pyranosyl-(1 → 4)-D-glucose, Des-sphingosino-gangliosid* G$_{LNnT}$ 1 NGNA

Gal β 1 → 4 GlcNAc β 1 → 3 Gal β 1 → 4 Glc

3
↑
2 α NGNA

---

* Des-sphingosino-gangliosid bezeichnet das durch Abspaltung von Fettsäure und Sphingosin aus Gangliosid erhaltene saure Oligosaccharid.

Darstellung aus dem Gangliosid G_{LNnT} I NGNA der Erythrozyten, Milz und Nieren von Rindern: (152).

25) O-β-D-Galaktopyranosyl-(1 → 4)-O-β-D-2-desoxy-2-acetamido-glucopyranosyl-(1 → 6)-[O-β-D-galaktopyranosyl-(1 → 3)-]O-β-D-galakto-pyranosyl-(1 → 3)-O-β-D-2-desoxy-2-acetamido-glucopyranosyl-(1 → 3)-O-β-D-galaktopyranosyl-(1 → 4)-D-glucose.

Gal β 1 → 4 GlcNAc β 1 → 6
$\qquad\qquad\qquad\qquad\qquad$ Gal β 1 → 3 GlcNAc β 1 → 3 Gal β 1 → 4 Glc
$\qquad$ Gal β 1 → 3

Der Zucker wurde aus Frauenmilch isoliert von A. Kobata, B. F. Torain und V. Ginsburg, Fed. Proc. 29 (1970) 410, Abstr. 929. Die Autoren berichten außerdem über das Vorkommen von fünf Abkömmlingen dieser Heptaose (2 mono-fuco-, 1 difuco-1 monosialo-. und 1 monosialo-monofuco-Derivat).

Grimmonprez und Montreuil (37) berichteten über die Isolierung von weiteren höheren, reduzierenden Oiligosacchariden aus Frauenmilch. Diese Zucker wurden entsprechend ihrer elektrophoretischen Wanderungs-geschwindigkeit mit Ziffer 1—5 bezeichnet. Sie besitzen folgende mole-kulare Zusammensetzung:

Oligosaccharid Nr.	Glc	Gal	GlcNAc	Fuc	NANA
1	1	4	3	2	0
2	1	4	3	2	1
3	1	3	2	1	1
4	1	3	2	1	1
5	1	2	1	1	1

Die Aufklärung ihrer genauen chemischen Struktur ist noch nicht abgeschlossen.

IV. Vergleich der freien Kohlenhydrate der Milch und des Urins mit den konjugierten Oligosacchariden der Glykosphingolipide

Ein Vergleich des Aufbaus der Kohlenhydratkomponenten der Glyko-sphingolipide mit den in Milch und Urin sezernierten Zuckern zeigt gewisse auffallende Ähnlichkeiten. Mit Ausnahme des nur in sehr geringer Menge im Hirn vorkommenden Gangliosids G_{Gal} I NANA, dessen Zuckerrest der 3-N-Acetyl-neuraminyl-galaktose des Kuhcolostrums entspricht, beginnt bei den höheren Glykosphingolipiden die Reihenfolge der Mono-saccharidbausteine vom am Ceramid gebundenen Ende her stets mit

Literaturverzeichnis: SS. 422—428

Glucose, die in 4-Stellung β-glycosidisch mit Galaktose verbunden ist. Von hier aus verzweigt sich der Syntheseweg und es können verschiedene Kettenarten, von denen bisher vier bekannt sind, gebildet werden:

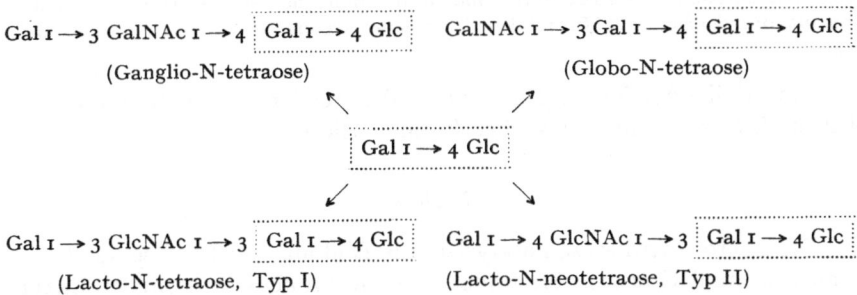

Während in der Reihe der *Globo*saccharide bisher weder Fucose noch Sialinsäurederivate aufgefunden wurden, existieren die *Ganglio*saccharide mit Sialinsäureresten bestückt. Man hat bei ihnen noch keine Fucoside entdeckt. Die Zucker der Lacto-N-tetraosereihen vom Typ I und Typ II dagegen können sowohl Fucose als auch Sialinsäurereste tragen*. Einen Überblick über die Lokalisation dieser Oligosaccharidtypen gibt folgende Aufstellung:

	Ganglio-N-tetraose	Globo-N-tetraose	Lacto-N-tetraosen Typ I und Typ II
Hirn	+*	—	—
Viscerale Organe/ Erythrocyten	+*	+*	+*
Milch, Urin	—	—	+

* Als Bestandteile von Glykosphingolipiden.

In der Milch überwiegen quantitativ die Saccharide der Lacto-N-tetraose (Typ I) jene der Lacto-N-neo-tetraose (Typ II). Bei den Gangliosiden der Reihen G_{LNT} bzw. G_{LNnT} der visceralen Organe ist das Verhältnis umgekehrt*. Hier werden in größerer Menge diejenigen gefunden, die sich von der Lacto-N-neotetraose (Typ II) ableiten (*94*, *151*).

V. Biosynthese und Blutgruppenmerkmale

Es gilt heute als gesichert, daß der Aufbau komplexer Oligosaccharide der freien, reduzierenden, wie der konjugierten, durch schrittweises An-

---

* L-Fucose-haltige Sphingoglykolipide aus menschlichem Tumorgewebe mit Le[b]- und Le[a]-Aktivität sind von S. I. HAKOMORI (Biochim. Biophys. Acta **202**, 227 (1970)) beschrieben worden.

knüpfen von nucleotid-aktivierten Monosaccharidbausteinen an die Acceptormoleküle erfolgt*.

Die an Protein oder Ceramid gebundenen komplexen Oligosaccharidketten an den Zelloberflächen werden durch membrangebundene Glykosyltransferasen aufgebaut, die vielleicht als Multiglykosyltransferasesysteme (S. Roseman) im Konzert agieren.

Eine Reihe der für die Biosynthese der Zucker der Milch verantwortlichen Enzyme kann in löslicher Form vorliegen.

Lactose

Die Lactosesynthetase katalysiert die Bildung von Lactose aus UDP-Galaktose und Glucose. Sie findet sich sowohl als mikrosomales Enzym in Milchdrüsen laktierender Tiere (146) als auch in löslicher Form in der Milch (5). Das Enzym katalysiert ebenfalls die Übertragung von Galaktose auf N-Acetylglucosamin unter Entstehung von N-Acetyllactosamin (5, 146).

Die Lactosesynthetase ließ sich in zwei Proteinkomponenten (A und B) auftrennen. Die Komponenten A und B besitzen für sich alleine keine Lactosesynthetaseaktivität. Ein wieder hergestellter Komplex aus Protein A und B besitzt volle Enzymwirksamkeit (15). Das Protein B konnte mit α-Lactalbumin identifiziert werden (16). Die Proteine von Mensch, Rind, Schaf und Ziege sind immunologisch unterschiedlich, für die Herstellung der Lactosesynthetaseaktivität aber vertauschbar (136). Das Protein A für sich alleine katalysiert die Übertragung von Galaktose auf N-Acetylglucosamin. Durch Protein B, das selbst keine Enzymwirksamkeit besitzt, wird die Substratspezifität von Protein A für N-Acetyl-glucosamin in eine solche für Glucose umgewandelt. Man hat deshalb Protein B einen „specifier" genannt (14). Die Menge an Protein A in Milchdrüsen der Maus steigt während der Schwangerschaft stetig an; dagegen bleibt der Spiegel an Protein B konstant und relativ niedrig und wird erst zu Beginn der Laktation wesentlich erhöht. Während der Laktation ist das Verhältnis der Mengen von Protein A und B ähnlich dem, wie es in mit Insulin bzw. Hydrocortison vorbehandelten Milchdrüsen und Organkulturen nach Stimulierung mit Prolaktin gefunden wird (141).

Die enzymatische Verknüpfung von L-Fucose (39) bzw. Sialinsäure (59) mit Lactose wurde mit partikelgebundenen Enzymen aus Rattenmamma in vitro aufgezeigt. Sie erfolgt offensichtlich an einem auch für die Lactosebildung verantwortlichen Multiglycosyltransferasesystem.

Die Bildungsrate von Neuraminyllactose aus UDP-Galaktose plus Glucose war höher als die aus Lactose in einem in vitro System mit Subzellulärpartikeln aus Rattenmilchdrüsen. Es konnte im ersten Falle ein Partikel-Lactose-Komplex isoliert werden (8).

---

* Im Ziegencolostrum wurden durch UDP-aktivierte Oligosaccharide gefunden (58). Ein Aufbau höherer Zucker durch Verknüpfung kleinerer Oligosaccharide aber ist noch nicht beobachtet worden.

Literaturverzeichnis: SS. 422—428

Die in der Milch oder im Urin vorkommenden Oligosaccharide können in individuellen Proben entsprechend der Blutgruppe des Spenders unterschiedlich sein. Am deutlichsten wurden die Zusammenhänge für die L-Fucose-haltigen Zucker der Milch aufgeklärt (*101—104, 111*).

80% der Bevölkerung (Sekretoren) sezernieren lösliche A, B bzw. O (*18*) blutgruppenaktive Substanzen entsprechend ihrer speziellen Blutgruppenzugehörigkeit (*110, 147*). Etwa 20% der Bevölkerung (Nonsekretoren) scheiden statt dessen nur Le^a-positive Kohlenhydrate aus (*60*). E. F. GROLLMAN und V. GINSBURG verglichen den Sekretorstatus und das Vorkommen von 2'-Fucosyllactose in der Milch. Sie konnten zeigen, daß die oben beschriebene Sekretionsfähigkeit von der Gruppierung

$$\text{Fuc } \alpha \text{ I} \rightarrow \text{2 Gal} -$$

und somit von der Anwesenheit einer GDP-L-Fucose: D-Galaktose α-2-Fucosyl-transferase abhängt (*40*).

Die Le^a-Spezifität der von Nonsekretoren ausgeschiedenen Substanzen beruht auf folgender Kohlenhydratengruppierung:

$$\text{Gal } \beta \text{ I} \rightarrow \text{3 GlcNAc} -$$
$$4$$
$$\uparrow$$
$$\text{I } \alpha \text{ Fuc}$$

Die Le^b-Aktivität bedingende Kohlenhydratstruktur unterscheidet sich durch die Hinzufügung einer weiteren Molekül L-Fuc in α-I $\rightarrow$ 2 glykosidischer Bindung an die Galaktose:

$$\text{Gal } \beta \text{ I} \rightarrow \text{3 GlcNAc} -$$
$$2 \qquad\qquad 4$$
$$\uparrow \qquad\qquad \uparrow$$
$$\text{I } \alpha \text{ Fuc} \qquad \text{I } \alpha \text{ Fuc}$$

Dementsprechend wurden Oligosaccharide mit einer O-α-L-Fuc-(I $\rightarrow$ 2)-O-β-D-gal-gruppierung nur in der Milch von Le (b+) und nicht bei Le (a+)-Spendern gefunden (*66*). „Non-sekretoren" können nicht Le (b+)-Blutgruppe besitzen (*129*).

Die Milch von Frauen mit Blutgruppe Le (a+) oder (b+) enthält eine GDP-L-Fucose: N-Acetyl-β-D-glucosaminylsaccharid α-4-L-Fucosyltransferase, die bei Le (a—, b—)-Personen fehlt.

Der Le (a—, b—)-Anteil der Bevölkerung beträgt etwa 5%. Bei diesem enthält die Milch weder Lacto-N-fucopentaose II, noch die Lacto-N-difucohexaosen I und II (*64*). Wegen der schwierigen Abtrennbarkeit von der Lacto-N-fucopentaose II konnte daher die Lacto-N-fucopentaose III erstmals aus Milch von Le (a—, b—)-Spendern rein isoliert werden (*64*). Dieser Zucker gehört zu den *Lacto-N-neotetraosesacchariden* (Typ II) und besitzt keine bekannte serologische Aktivität (*99*). Bei etwa 1% der

Le (a—, b—)-Spender fehlt offenbar auch die Lacto-N-fucopentaose III
in der Milch. Es ist noch nicht geklärt, ob diese Personen nicht in der Lage
sind, die Kette der Lacto-N-neotetraose zu bilden oder ob ihnen eine
GDP-L-Fucose: N-Acetyl-β-glucosaminylsaccharid α-3-L-Fucosyltrans-
ferase fehlt.

Die Antigendeterminante, die verantwortlich ist für die Blutgruppen
A — Spezifität sezernierter Substanzen, enthält N-Acetylgalaktosamin in α-glycosidischer Verknüpfung an Galaktose, die einen Fucose-
rest trägt:

$$\text{Gal NAc } \alpha \ 1 \longrightarrow 3 \ \text{Gal}—$$
$$2$$
$$\uparrow$$
$$1 \ \alpha \ \text{Fuc}$$

Von V. Ginsburg und Mitarb. (65) wurde erwiesen, daß nur in der
Milch von Frauen der Blutgruppe A bzw. AB eine bestimmte N-Acetyl-
galaktosamintransferase vorkommt. Das Enzym überträgt N-Acetylgalaktosamin nur auf ein Galaktosemolekül, das schon in 2-Stellung
einen α-L-Fucopyranosyl-Rest trägt. „Non-Sekretoren" können daher
trotz möglicher Anwesenheit der N-Acetyl-gakatosamintransferase keine
lösliche Blutgruppensubstanz A synthetisieren (67).

Literaturverzeichnis

1. Adachi, S. und S. Patton: Presence and significance of lactulose in milk
    products. J. Dairy Sci. **44**, 1375 (1961).
2. Anderson, S. G.: Hemagglutination by animal viruses. In F. M. Burnett:
    The Viruses III, 22 Acad. Press (1959).
3. Aronson, M.: Transglycosidations during lactose hydrolysis. Arch. Biochem.
    Biophys. **39**, 370 (1952).
4. Aspinall, G. O.: Gas-liquid partition chromatography of methylated and
    partially methylated methyl-glycosides. J. Chem. Soc. (London) 1676 (1963).
5. Babad, H. und W. Z. Hassid: Soluble UDP-D-galactose: D-glucose β-4-
    galactosyltransferase from bovine milk. J. Biol. Chem. **241**, 2672 (1966).
6. Bailey, R. W.: Oligosaccharides. Pergamon Press. 1965.
7. Bailey, R. W. und J. B. Pridham: Adv. Carbohydrate Chem. **17**, 121 (1962).
8. Barra, H. S., Cumar und R. Caputto: The synthesis of neuramin-lactose by
    preparations of rat mammary gland and its relation to the synthesis of lactose.
    J. Biol. Chem. **244**, 6233 (1969).
9. Bishop, C. T.: Separation of carbohydrate derivatives by gas-liquid chromato-
    graphy. Meth. of biochem. Anal. **10**, 1 (1962) Ed. D. Glick.
10. Bobbit, J. M.: Periodate oxidations of carbohydrates. Adv. Carbohydrate
    Chem. **11**, 1—41 (1956).
11. Breck, W. G., R. D. Corbett und G. W. Hay: A micro analysis for periodate
    oxidation. Chem. Commun. **1967**, 604.
12. Brendel, K., N. O. Roszel, R. W. Wheat und E. A. Davidson: Ion-exchange
    separation and automated assay of some hexosamins. Anal. Biochem. **18**,
    147 (1967).

13. BRENDEL, K., R. S. STEELE, R. W. WHEAT und E. A. DAVIDSON: Ion-exchange separation and automated assay of complex mixtures of amino acids and some hexosamins.

14. BREW, K., T. C. VANAMAN und R. L. HILL: The role of α-lactalbumin and the A protein in lactose-synthetase: A unique mechanism for the control of a biological reaction. Proc. Nat. Acad. Sci. USA **59**, 491 (1968).

15. BRODBECK, U. und K. E. EBNER: Resolution of a soluble lactose synthetase into two protein components and solubilisation of microsomallactose-synthetase. J. Biol. Chem. **241**, 762 (1966).

16. BRODBECK, U., W. L. DENTON, N. TANAHASHI und K. E. EBNER: Isolation and identification of the B protein of lactose synthetase as α-lactalbumin. J. Biol. Chem. **242**, 1391 (1967).

17. CARUBELLI, R., L. C. RYAN, R. E. TRUCCO und R. CAPUTTO: Neuraminlactose-sulfate, a new compound isolated from the mammary glands of rats. J. Biol. Chem. **236**, 2381 (1961).

18. CEPPELINI, R.: On the genetics of secretor and Lewis characters: a family study. Proc. 5th Intern. Congr. of Blood Transfusion, p. 207. Paris 1954.

19. CLOTTEN, R. und A. CLOTTEN: Hochspannungselektrophorese, S. 463. Stuttgart: G. Thieme. 1962.

20. COHEN, A.: Mechanisms of cell infection. In WILSON SMITH ed.: Mechanisms of virus infection, p. 153. Academic Press. (1963).

21. CUMAR, F. A., P. A. FERCHMIN und R. CAPUTTO: Isolation and identification of a lactose phosphate from cow colostrum. Biochem. Biophys. Res. Commun. **20**, 60 (1965).

22. DATE, J. W.: Udskillelsen af nogle Kuhlhydrater i urinen under gravidität og diegivning. Holsterbro: Ducky Grafik. 1965.

23. DAVIDSON, E. A.: Analysis of sugars found in mucoploysaccharides. Meth. Enzymol. **8**, 52 (1966).

24. DE JONGH, D. C., J. D. HRIBOR, I. D. RADFORD, S. HANESSIAN, M. BIEBER, G. DAWSON und C. C. SWEELEY: Analysis of trimethylsilyl derivatives of carbohydrates by gaschromatography and mass spectrometry. J. Amer. Chem. Soc. **91**, 1728 (1969).

25. DENIGES, G. et BONANS: Zitiert nach MONTREUIL (109). J. Pharm. et Chim. **5**, 363 (1888).

26. DONALD, A. S. R.: The separation of glucosamine and galaktosamine and their glucitols on an amino acid analyser. J. Chromatog. **35**, 106 (1968).

27. DRZENIEK, R.: Differences in splitting capacity of virus and V. cholerae neuraminidases on sialic acid type substrates. Biochem. Biophys. Res. Commun. **26**, 631 (1967).

28. DRZENIEK, R., H. FRANK und R. ROTT: Electron microscopy of purified influenza virus neuraminidase. Virology **36**, 703 (1968).

29. DRZENIEK, R. und A. GAUHE: Differences in substrate specifity of myxovirus neuraminidases. Biochem. Biophys. Res. Commun., in press (1970).

30. DYER, J. R.: Use of periodate oxidation in biochemical analysis. Meth. in biochem. Analysis III (1956) 111. Ed. D. Glick.

31. ESBACH, G.: Zitiert nach MONTREUIL (109). J. Pharm. et Chim. **17**, 5. serie, 533 (1888).

32. FLODIN, P. und U. ASPBERG: Separation of oligosaccharides with gel filtration. Biological structure and function. Vol. 1, (New York) (1961) 345.

33. FLODIN, P., J. D. GREGORY und L. RODIN: Separation of acid oligosaccharides by gel filtration. Anal. Biochem. **8**, 424 (1964).

34. Gauhe, A., P. György, J. R. E. Hoover, R. Kuhn, C. S. Rose, H. W. Ruelius und F. Zilliken: Bifidus factor IV. Preparations obtained from human milk. Arch. Biochem. Biophys. 48, 214 (1954).
35. Ginsburg, V. und E. F. Neufeld: Complex heterosaccharides of animals. Ann. Rev. Biochem. 38, 371 (1969).
36. Gottschalk, A.: Chemistry and biology of sialic acids and related substances. Cambridge Univ. Press 1960.
37. Grimmoprez, L. und J. Montreuil: Étude physico-chimique de six nouveaux oligosides du lait de femme. Bull. soc. chim. biol. 50, 843 (1968).
38. Grimmoprez, L., G. Takerkart und J. Montreuil: Structure d'un hexaose isolé du lait de femme. C. R. Acad. Sci. (Paris) 265, 2124 (1968).
39. Grollman, A. P. und V. Ginsburg: Biosynthesis of the oligosaccharides of milk. Federat. Proc. 22, Nr. 2 (1963).
40. Grollman, E. F. und V. Ginsburg: Correlation between secretor status and the occurence of 2'-fucosyllactose in human milk. Biochem. Biophys. Res. Comm. 28, 50 (1967).
41. György, P.: A hitherto unrecognized biochemical difference between human milk and cow milk. Pediatrics 11, 98 (1953).
42. György, P., R. F. Norris und C. S. Rose: Bifidus factor I. A variant of lactobacillus bifidus requiring a special growth factor. Arch. Biochem. Biophys. 48, 193 (1954).
43. György, P., J. R. E. Hoover, R. Kuhn und C. S. Rose: Bifidus factor III. The rate of dialysis. Arch. Biochem. Biophys. 48, 209 (1954).
44. György, P., R. Kuhn, C. S. Rose und F. Zilliken: Bifidus factor II. Its occurrence in milk from different species and in other natural products. Arch. Biochem. Biophys. 48, 202 (1954).
45. György, P.: Nutrition and intestinal flora in man. 4. Congr. international de la Nutrition, Paris 1957. Ann. Nutrit. et Alim. 11, 189 (1957).
46. — Orientation in infant feeding. Federat. Proc. 20, (1961) Suppl. 7, 169.
47. Hakomori, S. I.: A rapid permethylation of glycolipid and polysaccharide catalysed by methyl sulfinyl carbanion in dimethylsulfoxide. J. Biochemistry (Tokyo) 55, 205 (1964).
48. Haworth, W. N., E. L. Hirst und J. I. Webb: Polysaccharides. Part II. The acetylation and methylation of starch. J. Chem. Soc. (London) 1928, 2681.
49. Haworth, W. N. und H. Machemer: Polysaccharides. Part X. Molecular structure of cellulose. J. Chem. Soc. (London) 1932, 2270.
50. Helferich, B. und G. Sparnberg: Kristallisierte O-β-D-Galaktosido-D-glucose. Ber. deutsch. chem. Gesell. 66, 806 (1933).
51. Hoban, N. und J. M. White: Gradient elution of disaccharides on stearic acid treated charcoal columns. Anal. Chem. 30, 1294 (1958).
52. Horowitz, S. T., S. Roseman und H. J. Blumenthal: Preparation of glucosamine oligosaccharides. I Separations. J. Amer. Chem. Soc. 79, 5046 (1957).
53. Hughes, R. C. und R. W. Jeanloz: The extracellular glycosidases of diplococcus pneumoniae. I. Purification and properties of a neuraminidase and a β-galactosidase. Action on the α_1 acid glycoprotein of human plasma. Biochemistry 3, 1535 (1964).
54. Huttunen, J. K.: Neuraminic acid containing oligosaccharides in human urin. Dissertation, Helsinki 1966. Annales Medicinae Experimentalis et Biologiae Fenniae 44, (1966) Suppl. 12.
55. Jermyn, M. A.: Separation of monomethylated p-nitrophenyl-β-glucosides on a carbon column. Austral. J. Chem. 10, 455 (1957).

56. JOLLEY, R. L. und M. L. FREEMAN: Automated carbohydrate analysis of physiologic fluids. Clin. Chem. 14, 538 (1968).
57. JONES, J. K. N., R. A. WALL und A. O. PICTET: Separation of sugars on ion exchange resins. Chem. and Ind. (London) 1959, 1196.
58. JOURDIAN, G. W., F. SHIMIZU und S. ROSEMAN: Isolation of nucleotide-oligosaccharides containing sialic acid. Federat. Proc. 20, 161 (1961).
59. JOURDIAN, G. W., D. M. CARLSON und S. ROSEMAN: The enzymatic synthesis of sialyl-lactose. Biochem. Biophys. Res. Comm. 10, 352 (1963).
60. KABAT, E. A.: Blood group substances. New York: Academic Press, Inc. 1956.
61. KARLSSON, K. A., B. E. SAMUELSSON und G. O. STEEN: Mass spectrometry of polar complex lipids, analysis of a sulfatide derivative. Biochem. Biophys. Res. Comm. 37, 22 (1969).
62. KENDAL, A. P.: An automated determination of sialic acids. Anal. Biochem. 23, 150 (1968).
63. KIRCHER, ST. W.: Gas-liquid partition chromatography of methylated sugars. Analyt. Chemistry 32, 1103 (1960).
64. KOBATA, A. und V. GINSBURG: Oligosaccharides of human milk, Lacto-N-fucopentaose III. J. Biol. Chem. 244, 5496 (1969).
65. KOBATA, A., E. F. GROLLMAN und V. GINSBURG: An enzymic basis for blood type A in humans. Arch. Biochem. Biophys. 124, 609 (1968).
66. KOBATA, A., V. GINSBURG und M. TSUDA: Oligosaccharides of human milk. Arch. Biochem. Biophys. 130, 509 (1969).
67. KOBATA, A. und V. GINSBURG: UDP-N-acetyl-D-galactosamine: D-galactose α-3-N-acetylgalactosamine-transferase of the gene, that determines blood type A in man. J. Biol. Chem. 245 (1970), in press.
68. KUHN, R.: Über die Oligosaccharide der Frauenmilch. Angew. Chem. 67, 184 (1955).
69. — Aminozucker. Angew. Chem. 69, 23 (1957).
70. — Biological importance of aminosugars. Proc. IVth Int. Congr. Biochem., Wien 1958, Vol. 1, p. 67. Pergamon Press 1959.
71. — Sugars in human milk. Bibliotheka Paediat. Suppl. Ann. Paediat. 66, 205 (1958).
72. — Bifidus factors of human breast milk. Proc. Soc. Biol. Chem. India 17, 18 (1958).
73. — Les oligosaccharides du lait. Bull. Soc. Chim. Biol. 40, 297 (1958).
74. — Biochemie der Rezeptoren und Resistenzfaktoren. Naturwiss. 46, 43 (1959).
75. KUHN, R. und H. H. BAER: Zur Permethylierung von N-Acetyl-galaktosamin-Derivaten. Liebig's Ann. Chem. 611, 236 (1957).
76. KUHN, R., H. H. BAER und A. GAUHE: Fucosido-lactose, das Trisaccharid der Frauenmilch. Chem. Ber. 88, 1135 (1955).
77. — — Chemische und enzymatische Synthese des O-β-D-galaktosido-N-acetyl-D-glucosamins; enzymatische Synthese der Allolactose. Chem. Ber. 88, 1713 (1955).
78. — — — Die Konstitution der Lacto-N-tetraose. Chem. Ber. 89, 504 (1956).
79. — — — Kristallisation und Konstitutionsermittlung der Lacto-N-fucopentaose I. Chem. Ber. 89, 2514 (1956).
80. — — — Kristallisierte Fucosido-lactose. Chem. Ber. 89, 2513 (1956).
81. — — — Die Konstitution der Lacto-N-fucopentaose II. Chem. Ber. 91, 364 (1958).
82. — — — 2-α-L-Fucopyranosyl-D-galaktose und 2-α-L-Fucopyranosyl-D-talose. Liebig's Ann. 611, 242 (1958).

426   H. Wiegandt und H. Egge:

83. Kuhn, R. und R. Brossmer: Über das durch Viren der Influenzagruppe spalt-
bare Trisaccharid der Milch. Chem. Ber. **89**, 2013 (1956).
84. Kuhn, R. und H. Egge: Über Ergebnisse der Permethylierung der Ganglioside G I und G II. Chem. Ber. **91**, 3338 (1963).
85. Kuhn, R., H. Egge, R. Brossmer, A. Gauhe, P. Klesse, W. Lochinger,
E. Röhm, H. Trischmann und D. Tschampel: Über die Ganglioside des Ge-
hirns. Angew. Chem. **72**, 805 (1960).
86. Kuhn, R. und A. Gauhe: Über die Lacto-difuco-tetraose der Frauenmilch.
Ann. **611**, 249 (1958).
87. — — Über ein krist. Le^a aktives Hexasaccharid aus Frauenmilch. Chem.
Ber. **93**, 647 (1960).
88. — — Die Konstitution der Lacto-N-neotetraose. Chem. Ber. **95**, 518 (1962).
89. — — Über drei saure Pentasaccharide aus Frauenmilch. Chem. Ber. **95**,
513 (1962).
90. — — Bestimmung der Bindungsstelle von Sialinsäureresten in Oligosacchariden
mit Hilfe von Perjodat. Chem. Ber. **98**, 395 (1965).
91. Kuhn, R., A. Gauhe und H. H. Baer: Ein stickstoffhaltiges Tetrasaccharid aus
Frauenmilch. Chem. Ber. **86**, 827 (1953).
92. — — — Einfluß von Substituenten auf die Farbreaktion von N-Acetylglucosamin mit p-Dimethyl amino benzaldehyd. Chem. Ber. **87**, 1138 (1954).
93. Kuhn, R., H. Trischmann und J. Löw: Zur Permethylierung von Zuckern und
Glykosiden. Angew. Chem. **67**, 32 (1955).
94. Kuhn, R. und H. Wiegandt: Über ein glucosaminhaltiges Gangliosid. Z.
Naturforsch. **19b**, 80 (1964).
95. — — Weitere Ganglioside aus Menschenhirn. Z. Naturforsch. **19b**, 256
(1964).
96. Lambert, R. und F. Zilliken: Novel growth factor for lacto-bacillus bifidus
var. pennsylvanicus. Arch. Biochem. Biophys. **110**, 544 (1965).
97. Li, Y. T.: Presence of a D-mannosidic linkage in glycoproteins. Liberation of
a D-Mannose from various glycoproteins bei a α-mannosidase isolated from
Jack Bean meal. J. Biol. Chem. **241**, 1010 (1966).
98. Li, Y. T. und S. C. Li: Studies on the glycosidases in Jack Bean meal II.
Separation of various glycosidases by isoelectric focussing. J. Biol. Chem.
243, 3994 (1968).
99. Lloyd, K. O. und E. A. Kabat: Immunochemical studies on blood groups,
XLI, Proposedstructures for the carbohydrate-portions of blood group A, B,
H. Lewis^a and Lewis^b substances. Proc. Nat. Acad. Sci. USA **61**, 1470 (1968).
100. Lundblad, A. und A. J. Berggard: Gel filtration of the low molecular weight
carbohydrate components of normal urine. B. B. A. **57**, 129 (1962).
101. — Excretion of oligosaccharides in the urine of secretors and non-secretors
belonging to different bloodgroups. Biochim. Biophys. Acta **130**, 130 (1966).
102. — Two urinary oligosaccharides characteristic of A and B secretors: isolation
and partial characterisation. Student Service, Uppsala 1966.
103. — Two urinary oligosaccharides characteristic of A$_1$ and B secretors: Isolation
and partial characterisation[1]. Biochim. Biophys. Acta **148**, 151 (1967).
104. — Isolation and characterisation of a urinary oligosaccharide characteristic
of bloodgroup O(H)-secretors. Biochim. Biophys. Acta **165**, 202 (1968).
105. Montreuil, J.: Application de chromatographie en papier à l'étude des glucides
en glycoproteins du liquids biologiques. 3e Coll. Hôp. St.-Jean, Bruges **1955**,
209.
106. — Structure of two trisaccharides isolated from human milk. C. R. Acad.
Sci. **242**, 192 (1956).

107. MONTREUIL, J.: Les glucides du lait de femme. Bull. Soc. Chim. Biol. (Paris) **39**, 395 (1957).

108. MONTREUIL, J. und S. MULLET: Constitution glucidique du lait de femme au cours de la lactation. C. R. Soc. Biol. **153**, 1364 (1959).

109. MONTREUIL, J.: Les glucides du lait. Bull. Soc. Chim. Biol. (Paris) **42**, 1399 (1960).

110. MORGAN, W. T. J.: A contribution to human biochemical genetics; the chemical basis for blood group specificity. Proc. Roy. Soc. (London) Ser. B **151**, 308 (1960).

111. — Chemische Grundlagen der menschlichen Blutgruppenspezifität. Naturwiss. **46**, 181 (1959).

112. MORO, E.: Morphologische und bakteriologische Untersuchungen über die Darmbakterien des Säuglings. I. Die Bakterienflora des normalen Frauenmilchstuhles. Jahrbuch Kinderh. **61**, (5) 687 (1905).

113. — II. Die Verteilung und die Schicksale der normalen Bakterien im Säuglingsdarm. Jahrbuch Kinderh. **61**, 870 (1905).

114. — Morphologische und biologische Untersuchungen über die Darmbakterien des Säuglings. Jahrbuch Kinderh. **62**, 467 (1905).

115. NORRIS, R. F., T. FLANDERS, P. GYÖRGY und R. TOMARELLI: The isolation and cultivation of lactobacillus bifidus: A comparison of branched and unbranched strains. J. Bacteriol. **60**, 681 (1950).

116. ÖHMAN, R. und O. HYGSKATT: The isolation of sialyl-lactosides with the aid of gel filtration. Anal. Biochem. **23**, 391 (1968).

117. PAZUR, J. H. und C. F. BISHOP: Enzymatic hydrolysis of a glucomannan from Jack pine (pinus buchsiana). Can. J. Chem. **39**, 815 (1961).

118. PERILA, O. und C. F. BISHOP: The mechanism of enzymatic synthesis of galactosyl-oligosaccharides. J. Biol. Chem. **208**, 439 (1954).

119. PETUELY, F.: Biochemische Untersuchungen zur Regulation der Dickdarmflora des Säuglings (Über den Bifidusfaktor). Verlag Notring der wissenschaftlichen Verbände 1957.

120. POLONOWSKI, M. und A. LESPAGNOL: Sur deux nouveaux sucres du lait de femme, le gynolactose et l'allolactose. C. R. Acad. Sci. **192**, 1319 (1931).

121. POLONOWSKI, M., A. LESPAGNOL und H. WAREMBOURG: Variations physiologiques de concentration des different glucides du lait de femme. C. R. Soc. Biol. **107**, 303 (1931).

122. — — Nouvelles acquisitions sur les composés glucidiques du lait de femme. Bull. Soc. Chim. Biol. **15**, 320 (1933).

123. POLONOWSKI, M. und J. MONTREUIL: Étude chromatographique des polyosides du lait de femme. C. R. Acad. Sci. (Paris) **238**, 2263 (1954).

124. PREVOST, H.: Dissertation. Pharmacol., Lille 1934.

125. PURDIE, T. und J. C. IRVIN: The alkylation of sugars. J. Chem. Soc. (London) **83**, 1021 (1903).

126. PURO, K.: Carbohydrate components of bovine kidney gangliosides. Dissertation, Helsinki 1969.

127. RYAN, L. C., R. CARUBELLI, R. CAPUTTO und R. E. TRUCCO: Studies on the structure of neuramin-lactose-sulfate. Biochim. Biophys. Acta **101**, 252 (1965).

128. SAMUELSSON, K. und B. SAMUELSSON: Gas-liquid chromatography-mass spectrometry of cerebrosides as trimethylsilyl ether derivatives. Biochem. Biophys. Res. Com. **37**, 15 (1969).

129. SHEN, L., E. F. GROLLMAN and V. GINSBURG: An enzymatic basis for secretor status and bloodgroup substance specificity in humans. Proc. Nat. Acad. Sci. USA **59**, 224 (1968).

130. Smith, F. und R. Montgomery: The chemistry of plant gums and mucilages. New York: Verlag Reinhold. 1959.

131. Spiro, R. G.: Analysis of sugar found in glycoproteins. Methods in Enzymology 8 (1966); Ed. Kolowick.

132. Sweeley, C. C. und G. Dawson: Determination of glycosphingolipidstructures by massspectrometry. Biochem. Biophys. Res. Comm. 37, 6 (1969).

133. Schönfeld, H.: Über die Beziehungen der einzelnen Bestandteile der Frauenmilch zur Bifidus Flora. Jahrbuch Kinderh. 113, 19 (1926).

134. Schwimmer, S. und H. Bevenue: Reagent for differentiation of 1,4 and 1,6-linked glucosaccharides. Science 123, 543 (1956).

135. Tomarelli, R. M., R. F. Norris, P. György, J. B. Hassinen und F. W. Bernhart: Nutrition of variants of lactobacillus bifidus. J. Biol. Chem. 181, 879 (1949).

136. Tanahashi, N., U. Brosbeck and K. E. Evner: Enzymic and immunological activity of various B-Proteins of lactose-synthetase. Biochim. Biophys. Acta 154, 247 (1968).

137. Taylor, P. M. und W. J. Whelan: Fractionating sugars on charcoal. Chem. & Industry 44 (1962).

138. Tissier, H.: Thesis Paris (1900). Recherches sur la flore intestinale des nourissons (état normale et pathologique). G. Carré et C. Naud, Paris 1900.

139. Trucco, R. E. und R. Caputto: Neuraminlactose, a new compound isolated from the mammary glands of rats. J. Biol. Chem. 206, 901 (1954).

140. Tuppy, H. und W. L. Staudenbauer: The action on soluble blood group A substances of an α-N-acetyl-galactosaminidase from helix pomatia. Biochemistry 5, 1742 (1967).

141. Turkington, R. W., K. Brew, T. C. Vanaman und R. L. Hill: The hormonal control of lactose-synthetase in the developing mouse mammary gland. J. Biol. Chem. 234, 3382 (1968).

142. Wallenfels, K.: Nachweis reduzierender Zucker auf Papierchromatogrammen und quantitative Auswertung. Naturwiss. 37, 491 (1950).

143. Wallenfels, K., E. Berut und G. Limberg: Isolierung von Lactobiose, Lactotriose und Galaktobiose aus enzymatischen Hydrolysaten der Lactose. Liebigs Ann. Chem. 584, 113 (1953).

144. Wallenfels, K., G. Bechter, R. Kuhn, H. Trischmann und H. Egge: Permethylierung von oligomeren und polymeren Kohlenhydraten und quantitative Analyse der Spaltungsprodukte. Angew. Chem. 75, 1014 (1963).

145. Watkins, W. M.: Enzymatic synthesis of nitrogen containing disaccharides by α-galactosyltransfer. Nature 181, 117 (1958).

146. Watkins, W. M. und W. Z. Hassid: The synthesis of lactose by particulate enzyme preparations from guineapig and bovine mammary gland. J. Biol. Chem. 237, 1422 (1962).

147. Watkins, W. M.: Blood group-substances. Science 152, 172 (1966).

148. Whistler, R. L. und D. F. Durso: Separation of sugars on Charcoal. J. Am. Chem. Soc. 72, 677 (1950).

149. Whistler, R. L.: Methods in Carbohydrate-chemistry. Academic Press 1965.

150. Wiegandt, H.: Struktur und Funktion der Ganglioside. Angew. Chem. 80, 89 (1968); Angew. Chem. Int. Ed. 7, 87 (1968).

151. Wiegandt, H. und H. W. Bücking: Carbohydrate components of extraneuronal gangliosides. Europ. J. Biochem. in press.

152. Wiegandt, H. und B. Schulze: Spleen gangliosides: The structure of ganglioside G_{LNnT} I NGNA. Z. Naturforsch. 24 b, 945 (1969).

(Eingelaufen am 18. Februar 1970)

Glucagon: Chemistry and Action

By **W. Bromer,** Indianapolis, Indiana

With 4 Figures

I. Introduction

Glucagon is a polypeptide hormone secreted by the α_2-cells of the pancreas in response to a variety of stimuli. The hormone exhibits many biologic effects in several different target organs, but it is best known for its role in the homeostatic control of blood glucose. Many of the actions of glucagon result from interaction with cell membrane-bound adenyl cyclase, thus increasing the intracellular conversion of ATP to adenosine $3'$-$5'$-phosphate (cyclic AMP), the "second messenger".

The aim of this article is to provide for chemists a concise overview of glucagon as an important biochemical molecule. Chemical aspects receive a more thorough coverage than biological aspects; many hundreds of additional biological reports are available and are ably reviewed by UNGER and EISENTRAUT (98), FOA (32), and WEINGES (105). The selection of citations is quite arbitrary, with emphasis on recent reports.

II. Assay

KIMBALL and MURLIN (46) suggested in 1923 that glucagon existed, but lack of a convenient and accurate assay inhibited progress in understanding glucagon for about thirty years. Shortly after the isolation and crystallization of the hormone in 1953 by STAUB, SINN, and BEHRENS (89), a number of improved assay methods were devised. These have provided the basis for recent progress in understanding the functions of the hormone. A more detailed review of assay methods for glucagon is available (7).

1. Physicochemical Methods

The concentration of crystalline glucagon in aqueous solution (pH 1–9) is readily determined from knowledge of the ultraviolet absorption coefficient (cf. III). Several electrophoretic and chromatographic methods (51, 53, 88, 108, 115) have been devised to partially separate glucagon from other proteins; these could be developed as a basis for assay if relatively large amounts of hormone are to be determined.

2. Bioassay

a) Measurement of the Hyperglycemic Action of Glucagon

Methods have been developed using a variety of experimental animals; however, only two methods are employed extensively at present. SMITH and ROBBINS (83) modified the method of STAUB and BEHRENS (87) wherein anesthetized cats are injected at two dose levels in a modified Latin square design. Tarding and coworkers (95) employed cortisone-pretreated rabbits in a twin cross-over design somewhat analogous to the classical insulin assay. Both assays are based on the four-point (high dose-low dose), parallel-line approach, giving confidence intervals (P = = 0.95) in the range $\pm$ 20% when 16–32 animals are used. Glucagon can be measured in concentrations of 10^{-8} M by means of the cat method, but this sensitivity is not sufficient for measurement of the hormone in blood. Both methods are simple and reliable, but are not highly specific and require considerable investment in time and animals. Although a number of other biologic effects of glucagon could also be employed as a basis for a quantitative bioassay, none have been routinely used in such manner.

b) Measurement of the Glycogenolytic Effect of Glucagon

A variety of in-vitro methods is used to provide a quantitative measure of glucagon. Among them, the liver-slice (VUYLSTEKE and DE DUVE (104)), the liver-homogenate (MAKMAN and SUTHERLAND (56)), and the slice-homogenate (UI et al. (97)) methods are used most extensively. These methods are based on the glucagon-induced formation of cyclic AMP which activates phosphorylase, which in turn is responsible for glycogenolysis; measurement of nucleotide, phosphorylase, or phosphate provides a basis for the assay.

These methods are about ten times more sensitive than in-vivo assays, are probably more specific, and are more economical of time, animals, and labor. However, they are more difficult to use and probably provide no marked advantages in terms of precision.

3. Radioimmune Assays

Virtually all radioimmune methods depend on a competition for antibody-binding sites between an unknown quantity of hormone and a known amount of labeled hormone; as the quantity of hormone increases, the amount of labeled hormone bound to antibodies decreases, and vice versa. In the glucagon assay, separation and measurement of free ^{125}I-glucagon from antibody-bound ^{125}I-glucagon give an indirect measure of unlabeled hormone. Separation is obtained by precipitation of the antigen-antibody complex with salts (GRODSKY et al. (38); UNGER et al. (99); PROBST and COLWELL (75)), with alcohol (TAYLOR et al. (96)), or with a second antibody (HAZZARD and coworkers (4)); SHIMA and FOA (82)); others use the preferential adsorption of the free hormone by cellulose (NOVAKA and FOA (67)) or by charcoal (HERBERT et al. (43)) or an electrophoretic separation (ROSSELIN et al. (77)).

Three problems plagued the development and application of the assay: high-titer antiglucagon serum is difficult to prepare (glucagon is weakly immunogenic), ^{125}I- or ^{131}I-glucagon is of limited radiochemical stability, and glucagon in crude extracts or serum is rapidly destroyed by proteolysis. Although many different approaches were taken to enhance the immunogenicity of glucagon, no single method of developing high-titer antisera is widely accepted. The approach of ASSAN et al. (2) and others in coupling glucagon to albumin is a promising one.

TAYLOR and colleagues (96) and PROBST and COLWELL (75) called attention to apparent radiochemical instability of ^{131}I- or ^{125}I-labeled glucagon; PROBST and COLWELL solved the problem by preparing a purified ^{131}I-glucagon antibody complex which provided greater stability for the ^{131}I-glucagon. The problem of the instability of glucagon in blood or in crude tissue extracts was largely overcome (EISENTRAUT et al. (26)) through the use of pancreatic trypsin inhibitor.

If strong antisera are obtained, the radioimmune methods are exquisitely sensitive, permitting the measurement of 10^{-10} M glucagon. Such sensitivity is required for the assay of glucagon in serum. Precision of the method is inherently high since many replicates may be run simultaneously and since biological variation is not a factor. Specificity in the physicochemical sense is quite high, although caution is required in interpreting results on the basis of biologically active glucagon; no guarantee exists that the hormonal sites of action are identical to the antigenic determinants. A case in point involves the glucagon 1–21 fragment which interacts strongly with at least one antiglucagon serum but which has little or no biologic activity (BROMER and STOŘVICK (13)). Aside from the fact that some degradation products of glucagon may give false results with the radioimmune assay, other entities like secretin and "enteroglucagon" (cf. section IV, 1) may also interfere because of similarities to the amino acid sequence of glucagon.

III. Chemical Aspects

1. Isolation, Properties, and Primary Structure

Since STAUB, SINN, and BEHRENS (88, 89) crystallized porcine glucagon in 1953, comparatively little attention has been given to the isolation of the hormone directly from pancreas. Only porcine and bovine hormones have been isolated in highly purified, crystalline form. The relatively low concentration of glucagon ($< 10\mu g$ per g tissue) found in most mammalian pancreas has no doubt been a discouraging factor in the development of new isolation methods.

LOCHNER, ESTERHUIZEN, and UNGER (53) demonstrated that glucagon can be separated from insulin in human pancreatic extracts by electrophoresis on strips of cellulose acetate. The electrophoretic behavior of crude human glucagon was found to be similar to that of the bovine-porcine hormone. The method is not applicable to the isolation of crystalline glucagon; the need remains for a simple, reproducible method for the extraction and purification of glucagon from pancreas.

Crystals of glucagon obtained from mildly alkaline, aqueous solution conform to the isometric system and appear as rhombic dodecahedra (cf. Fig. 1). The hormone crystallizes readily in the absence of metals, but crystalline metal-glucagon complexes, such as zinc glucagon, may readily be prepared. Zinc glucagon is less soluble than glucagon and exhibits a prolonged biologic action (9, 106). The isoelectric point of glucagon in aqueous solution lies near pH 7.0 (115); the hormone is relatively insoluble in water through the pH range 3–9 (cf. Table 1).

Fig. 1. Photomicrograph of crystalline glucagon (ca. 100 X)

A glucagon-like impurity comprises about 10% of the weight of most crystalline glucagon preparations. This minor component exhibits the same amino acid composition and molecular size. It appears to have one more negative charge (at pH 8.5) and at least 50–75% of the biological activity of the major component (8). The exact chemical difference between glucagon and the minor component is not known but probably involves the loss of an amide group. The available crystalline glucagon preparations usually contain a trace (0.01–0.1%) of insulin unless the insulin is destroyed by reduction.

Glucagon readily forms fibrils and gels at concentrations of 0.1% or greater in acidic (pH 1–3) aqueous solution; fibrils form upon rapid warming and cooling of the solution. Freshly prepared fibrils redissolve if the pH is raised to about 10.5.

Table 1. *Some Properties of Crystalline Glucagon*

Property	Value
$E_{1\,cm}^{0.1\%}$ (276 mμ, water, pH 1–8)	2.12 $\pm$ 0.05
Isoelectric point	$\sim$ pH 7
Molecular weight	3485
Zinc content	< 0.01%
Solubility (aqueous) pH 3.5–8.5	50 μg per ml or less
pH 2–3 and 9–10	> 10 mg per ml
Solubility (nonaqueous) Dimethylsulfoxide 90% Formic Acid 90% Chloroethanol 70% Dimethyl formamide	5 mg per ml or greater
Stability (0.2–1 mg per ml) Aqueous solutions, pH 2.5–3	3 months, 4° C
Aqueous solutions, pH 9–10	4 hours, 25° C; 50% after 4 months, 4° C

Table 2. *Amino Acid Composition of Porcine or Bovine Glucagon*

Amino Acid	Number of Residues
Aspartic Acid	4
Threonine	3
Serine	4
Glutamic acid	3
Glycine	1
Alanine	1
Valine	1
Methionine	1
Leucine	2
Tyrosine	2
Phenylalanine	2
Histidine	1
Lysine	1
Arginine	2
Tryptophan	1
Total	29

The amino acid composition of porcine glucagon (*12*) is given in Table 2; bovine glucagon has an identical composition (*8*). Interestingly, the bovine hormone was recently shown (*8*) to have an amino acid sequence identical to porcine glucagon (9) (cf. Fig. 2). All other protein hormones isolated to date from bovine and porcine tissues show some differences in the arrangement of their amino acids. The significance of this similarity in the evolutionary sense is unknown, since the structures of the hormone from other species are still unknown.

2. Synthesis

Efforts from several laboratories to synthesize glucagon were successfully culminated in 1967 by WÜNSCH and his collaborators (112). The high content of functional side chains and the presence of tryptophan 25, methionine 27, and N-terminal histidine created many synthetic problems. An outline of the synthetic approach in Figure 3 provides a hint of the complexity of the total synthesis. Side chain and amino terminal protecting groups had to be chosen carefully to avoid subsequent destruction of tryptophan by acid and to circumvent poisoning of the catalyst by methionine during hydrogenolysis. Thus, the amino termini of most fragments were protected with the 2-nitrophenylsulfenyl group (NPS), although the carbobenzoxy group (Cbz) had been used extensively in building the fragments. The unusual di-adamantyloxycarbonyl (di-AdOC) protection was used for N-terminal histidine. Tertiary butyl ethers (tBu) and esters (OtBu) were employed extensively for side-chain protection of hydroxyl and carboxyl groups, respectively. This approach gave a product with about half the specific biological activity of natural glucagon. Gel filtration on columns of G-50 Sephadex (WÜNSCH, JAEGER, and SCHARF (113)) yielded a fraction that subsequently was crystallized and was found to be identical to natural glucagon by a wide variety of criteria (107, 113).

3. Conformation

The relatively low solubility of glucagon in aqueous solutions in the pH range 4–9 discouraged until recently the application of biophysical methods of conformational analysis. This solubility behavior of glucagon, along with ease of fibril and crystal formation, suggested early that glucagon undergoes strong self-association, perhaps of several varieties. A molecular weight of about 4000 was observed by sedimentation-velocity studies in 4 M guanidine HCl (12), and 3985 was observed by KAY and MARSH (45) using light-scattering methods. KAY and MARSH (45) estimated from optical rotatory dispersion that glucagon in dilute alkali (pH 10) maintained about 10% helical structure. GRATZER, BAILEY, and BEAVEN (35), GRATZER et al. (36), SRERE and BROOKS (86), and EDELHOCH and LIPPOLDT (22) have examined dilute solutions of glucagon ($<$ 1 mg per ml) from pH 2–10 using a wide variety of approaches, e. g., fluorescence, circular dichroism, optical rotatory dispersion, sedimentation, thermal

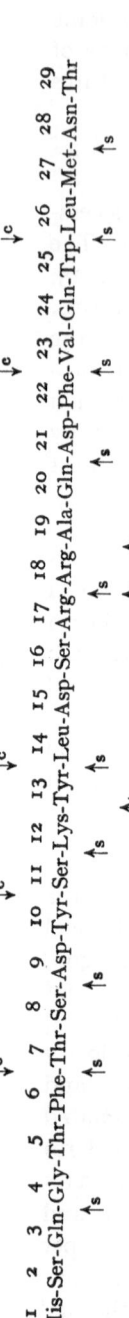

Fig. 2. The amino acid sequence of porcine glucagon. Arrows denote bonds split by enzymes: c chymotrypsin, s subtilisin, t trypsin

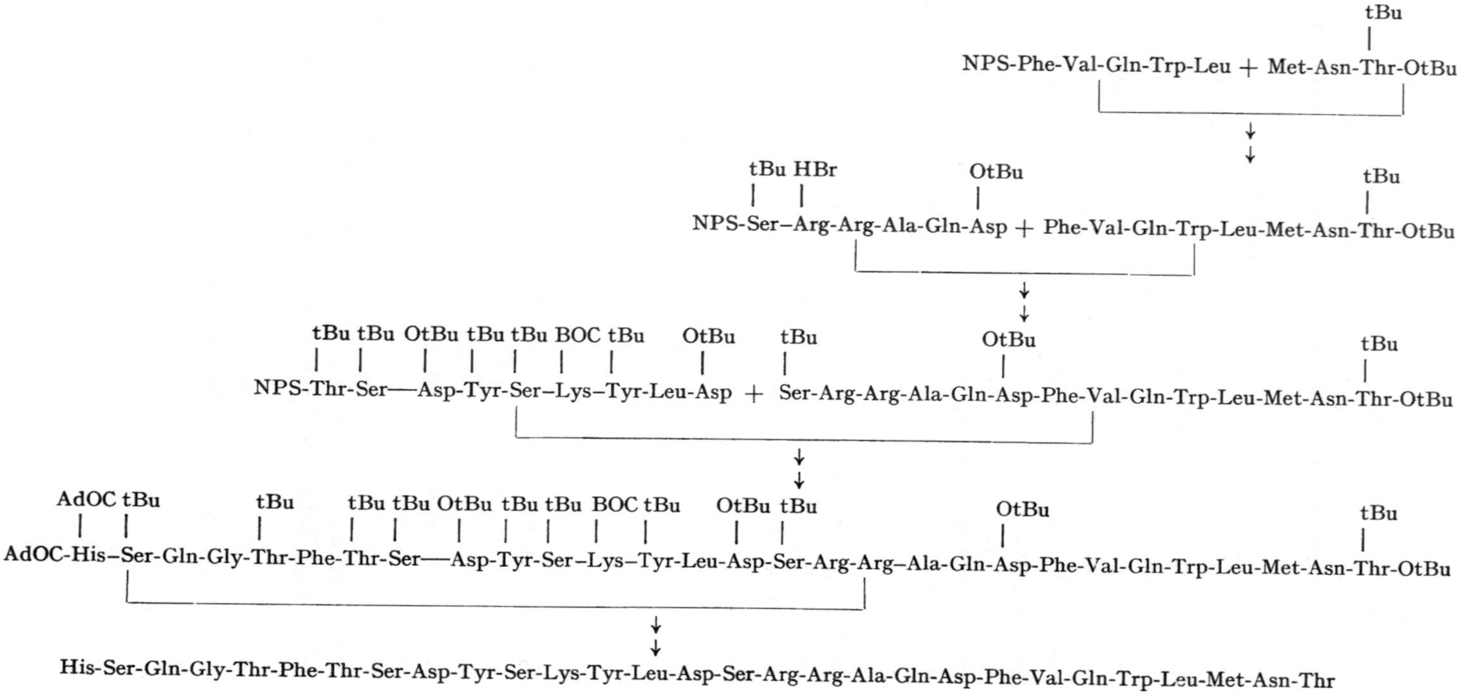

Fig. 3. Summary of the chemical synthesis of glucagon by Wünsch (112). Abbreviations for blocking groups are explained in text

difference spectra, and nuclear magnetic resonance. All of the data are in substantial agreement that glucagon has very little ordered structure in dilute aqueous solutions. However, at high concentrations of hormone (6, 36) or in 2-chloroethanol solutions, the α-helix content appears to increase from about 10% to values greater than 50%.

BLANCHARD and KING (6) had suggested from X-ray crystallographic studies that the aggregation of glucagon during crystallization involved the formation of α-helix. On the basis of X-ray evidence at 8 Å resolution, KING (47) proposed that glucagon in the crystalline state was entirely in α-helical form, giving rise to a cylinder approximating 40 Å long and 6.8 Å in diameter. SWANN and HAMMES (94), using sedimentation equilibrium, gel filtration, and concentration difference spectra, concluded that this presumed coil-helix transition of glucagon involved primarily a dimerization process followed by formation of hexamers. Several workers noted an increased viscosity followed by gel and fibril formation in pH 2 solutions of glucagon; GRATZER and his coworkers (35, 36) studied the birefringent gels and the fibrils by infrared spectroscopy, concluding that these forms consist of glucagon in antiparallel β-structure. Thus, glucagon appears to be a versatile polypeptide capable of dramatic (temperature, pH, or concentration-dependent) conformational changes from random coil in dilute solution to either β-structure in acidic gels or α-helix in crystalline form. In all likelihood, the random coil form predominates under physiologic conditions; however, this does not preclude the possibility of a conformational change in the environment of the hormonal receptor.

4. Relationships of Structure and Function

As with all hormones, studies of "structure and function" with glucagon are limited at the present time to observing the gross biological consequences of chemical modifications. Until receptor sites have molecular identity, the fundamental relationships of hormonal structure to function will remain obscure. Nonetheless, valuable information can sometimes be gained about functional groups required for biologic action.

During the studies of the amino acid sequence of glucagon, BROMER et al. (9) observed that none of the peptides (cf. Fig. 2) resulting from the proteolytic actions of trypsin, chymotrypsin, or subtilisin retained hyperglycemic activity. In addition, eleven amino acids were released through prolonged digestion with carboxypeptidase A, also leading to a loss of hyperglycemic activity. On the basis of kinetic data, BROMER and REMUS (10) reported a tentative correlation of this loss of activity with the removal of phenylalanine in position 22. WÜNSCH and WENDL-BERGER (114) could not confirm this finding with the synthetic tricosapeptide, and subsequent studies (8, 85) confirmed the lack of hyperglycemic activity of the 22-residue fragment. A recent report (62) suggested that removal of about eight C-terminal residues from glucagon results in no loss of lipolytic activity, but this result is in contradiction to data in our laboratory; both laboratories agree that glucagon 1–18 retains no lipolytic activity and that oxidation of Met_{27} to the sulfoxide results in no loss of activity. No lipolytic activity was lost by removal of the C-terminal dipeptide, Asn_{28}-Thr_{29}, and by conversion of Met_{27} to homoserine as obtained with BrCN treatment (62); this derivative also retained partial ability to activate adenyl cyclase (85). Trp_{25} was modified with N-bromosuccinimide, also giving an active product (62), while His_1 removal via the Edman degradation resulted in complete loss of lipolytic action. The latter finding is consistent with enzymatic studies (44, 60) wherein removal of N-terminal His-Ser led to loss of activity. On the basis of these preliminary results, a tentative conclusion can be reached that functional groups in the regions of residues 1 and 18–26 are important for expression of biologic function. Such a hypothesis is on a particularly tenuous basis until such time that evidence for chemical characterization and biologic quantitation are reported.

IV. Site of Formation and Release

1. Pancreas

Available evidence strongly supports the view that glucagon is produced in the α_2-cells of the islets of Langerhans in a wide variety of species; more specifically, glucagon is most likely synthesized via the Golgi apparatus and is stored in the α granules of the α_2-cells. As much as 3% of the dry weight of horse α_2-cells behaves immunologically like porcine pancreatic glucagon (Lundquist and coworkers (55)). The hormone has never been synthesized in a cell-free system, although there is little reason to suspect that its synthesis involves other than the usual nucleic acid-ribosomal system.

Unger and his coworkers (76) have obtained evidence for the existence in pancreatic extracts of a biologically inactive polypeptide ("large glucagon") which cross-reacts with antiglucagon sera but which has a molecular weight about twice that of glucagon. Less than 10% of the total glucagon immunoreactivity in pancreatic extracts has been found in the large molecular weight fraction. The analogy to the biosynthetic precursor of insulin, proinsulin, is excellent up to the point of proteolytic conversion of the "large glucagon" to glucagon. No means has yet been found to effect the conversion, and this may be extremely difficult to demonstrate, since glucagon itself is very readily split by the usual pancreatic proteolytic enzymes. Proinsulin appears to be an important biosynthetic precursor of insulin because it facilitates folding of the

References, pp. 446—452

protein, especially formation of the proper disulfide bonds. Similar reasons for the existence of a proglucagon do not appear to exist. Determining the actual function of "large glucagon" may not be an easy task, if this is dependent upon its isolation from pancreas in a concentration estimated at about 1 μg per g tissue. Additional characterization work is certainly required before a definitive role can be assigned to "large glucagon".

2. Intestine

Although the pancreas is generally recognized as the major site of formation of the biologically effective hormone, glucagon-like substances have been long associated with intestinal tissue. The fact is well established that some cells in the gastrointestinal tract of mammals are very similar, if not identical, to the α-cells of the pancreas. This information, along with the finding (SUTHERLAND and DE DUVE (92); MAKMAN and SUTHERLAND (56)) that extracts of canine duodenum contained a glycogenolytic factor, lent credence to the idea that glucagon may also be produced in the gut. The most extensive studies of this "enteroglucagon" have been made by UNGER and coworkers (24, 69, 102) using the radio-immune assay. Two glucagon-like fractions have been separated by gel filtration, one having a molecular size in the range of pancreatic glucagon and one about twice this size. Some of the properties of the fractions are given in Table 3. Suggestions have also been made that enteroglucagon is secreted in response to large oral doses of glucose; the enteroglucagon thus formed may be a potent stimulus for the release of insulin. Clarification of the role of enteroglucagon (as is the case with "large" glucagon) depends in large measure on its isolation and characterization. Apparently both large glucagon and enteroglucagon have some antigenic determinants in common with pancreatic glucagon.

Table 3. *Some Properties of Enteroglucagon*

Property	High molecular weight fraction	Low molecular weight fraction
Proportion found	~ 50%	~ 50%
Immunologic reactivity	Cross reacts* with antiglucagon sera	
Glycogenolytic action	Inactive	Active
Lipolytic action	Active	Active**
Insulin release	Active	?

* Not in a manner identical to pancreatic glucagon.
** Probably contains secretin, which also has lipolytic action.

3. Release

The development of sensitive immunoassay methods has permitted study of factors which affect the release of pancreatic glucagon into the circulation. Such measurements in peripheral blood are complicated by enteroglucagon which by definition reacts with most antiglucagon sera and which may comprise 90% of the total plasma glucagon-like immuno-reactivity. However, antisera have been prepared recently which react strongly with pancreatic glucagon and relatively poorly with entero-glucagon (24), thus providing one method to distinguish between the two polypeptides. Another method (102) involves triple catheterization so that changes in glucagon concentration can be detected in vena caval, pancreaticoduodenal, and mesenteric venous blood. Still a third method involves the in-vitro study of glucagon release from pancreas or gut preparations. Results from such studies are compiled in Table 4. The data show that pancreatic glucagon release is stimulated by hypogly-cemia, high levels of amino acids, pancreozymin, or starvation. On the other hand, hyperglycemia or sulfonyl urea treatment inhibits glucagon secretion.

Table 4. Factors Influencing the Release of Pancreatic Glucagon

Stimulus	Reference	Effect on release of Glucagon
Hypoglycemia	(15, 69)	Increase
Hyperglycemia	(15, 17, 69)	Depress
Amino acids	(68)	Increase
Pancreazymin	(16, 68)	Increase
Theophylline	(17)	Increase
Oubain	(17)	Increase
Starvation	(1)	Increase
Sulfonyl urea	(80)	Depress

The release of enteroglucagon appears to occur especially after administration of oral glucose; pancreozymin and intraduodenal amino acids apparently have little or no effect on the release of enteroglucagon (16, 68, 102).

V. Glucagon in Blood

1. Concentration

Evidence is accumulating that enteroglucagon may account for at least 90% of the glucagon-like immunoreactivity in peripheral plasma during fasting (24). Many previous studies had demonstrated that glucagon-like immunoreactivity in human serum corresponded to about

1–10 ng per ml. Use of specific antisera has shown that the concentration of pancreatic glucagon in human plasma is only about 0.13 ng per ml. Even these data are relative since porcine (or bovine) glucagon is routinely used as a standard; no highly purified human glucagon has been prepared and tested for completeness of cross-reaction with anti(porcine)glucagon sera. All of the values obtained for glucagon in plasma are in reality porcine pancreatic glucagon equivalents. Clarification of the various forms of glucagon in blood is also needed.

2. Catabolism

Studies with both labeled and unlabeled exogenous hormone indicate that the half-life of glucagon in the circulation is between 5 and 10 minutes (3, 100). A similar result is obtained from measuring the disappearance rate of pancreozymin-stimulated glucagon obtained from the pancreatic vein (101). When ^{131}I-labeled glucagon is injected intravenously, the radioactivity appears to accumulate very rapidly in tissues like liver and kidney, followed by a rapid decline (98). The data are usually interpreted that glucagon is quickly bound to the tissues and is destroyed by proteolytic enzymes, with the resulting peptide fragments being released into the circulation. Most of the known proteolytic enzymes degrade and inactivate glucagon. In particular, two enzymes have been isolated from liver which rapidly inactivate glucagon by releasing the N-terminal dipeptide, His-Ser (44, 60). One of the enzymes has been identified as cathepsin c, a dipeptidyl aminopeptidase I. The precise mode of degradation of endogenous glucagon is unknown; however, all available data suggest that the hormone has a rapid turnover in the circulation.

VI. Glucagon Action

The biologic actions of glucagon may be considered from several viewpoints, e. g., effects on the whole animal, on particular tissues, on metabolic systems, or on specific enzymes. At none of these levels is knowledge complete for glucagon or for any hormone. Furthermore, glucagon exhibits such a variety of effects that the assignment of physiological or functional roles is particularly difficult. Subcutaneous or intravenous administration of glucagon to an experimental animal produces within about thirty minutes all the gross effects listed in Table 5.

1. On Adenyl Cyclase

Although all the biologic actions of glucagon may not be mediated by cyclic AMP, the evidence is strong enough in many cases (74) to warrant brief consideration of the concept (93) as a background for many of the biologic actions of glucagon. Glucagon released from the pancreas following various stimuli quickly reaches target tissues by way of the circulation. At the target cell membrane, the hormone interacts with a receptor, probably the tissue-specific enzyme (or enzymes), adenyl cyclase. The possibility still exists that the hormone interacts with a separate receptor system which, in turn, acts on adenyl cyclase. In any case, such interaction results in activation of the enzyme (conformational change?) which converts ATP to intracellular cyclic AMP; the resulting higher intracellular concentration of the "second messenger" appears to be responsible for intracellular activation (phosphorylation?) of specific control proteins, such as enzymes or histones. Specificity of hormonal action appears to reside both at the adenyl cyclase and cellular function levels. Many other hormones act in an analogous fashion, making the cyclase-cyclic AMP system a focal point of current hormone research. Purification of adenyl cyclase is a major unsolved problem. The enzyme appears to be intimately associated with cell membrane; efforts to solubilize it have usually resulted in loss of hormone responsiveness.

Table 5. *Some Gross Biologic Effects of Glucagon in Animals*

Site	Effect
Blood	Hyperglycemia
	Increased free fatty acids
	Decreased amino acids
	Hypokalemia, hypocalcemia, hypophosphatemia
	Increased insulin
	Increased epinephrine and norepinephrine
Urine	Increased electrolytes
	Increased urea
Tissues	Decreased glycogen in liver, heart
	Increased hepatic blood flow
	Increased renal flow and filtration
	Increased hepatic uptake of amino acids
	Increased coronary rate and contractile force
	Increased hepatic gluconeogenesis
	Decreased gut motility and gastric secretion

2. In Pancreatic β-cells

Numerous workers had noted that injections of glucagon were followed by a rise in the concentration of venous insulin; this result was commonly interpreted as a secondary response to hyperglycemia. However, SAMOLS and coworkers (78) in 1965 showed that glucagon exerts a direct effect on the pancreatic β-cells to release insulin. Although the glucagon-stimulated release of insulin appears to be potentiated by hyperglycemia (79), numerous reports have confirmed that glucose is not necessary (25) and, in fact, the response to glucagon is more rapid (37) than that to glucose.

The proximity of α- and β-cells in pancreatic islets invites the speculation that endogenous glucagon is released from α-cells and acts directly on adjacent β-cells. Evidence exists that insulin release is mediated by cyclic AMP (57, 91), suggesting the possibility that glucagon may act on a membrane-bound cyclase in β-cells.

3. In Liver

The liver is clearly a major target organ for glucagon. In addition to the gross effects listed in Table 5, more specific actions of glucagon in liver have been observed as shown in Table 6. In almost all cases, cyclic AMP mediates these diverse effects of glucagon on hepatic protein, lipid, and particularly carbohydrate metabolism. Glucagon activates membrane-bound adenyl cyclase in liver tissue (5, 58, 74).

Table 6. *Some Actions of Glucagon in Liver*

Effect	Reference
Enhances gluconeogenesis from alanine, lactate, and pyruvate. Increases urea production. Increases glycogenolysis via phosphorylase activation	(*30, 33*)
Activates or induces:	
Triglyceride lipase	(*110*)
Pyruvate carboxylase	(*110*)
Phosphoenol pyruvate carboxykinase	(*23, 33*)
Carbamoyl phosphate synthetase	(*61*)
Argininosuccinate synthetase	(*61*)
Argininosuccinase	(*61*)
Tyrosine α-ketoglutarate transaminase	(*14*)
Phenylalanine pyruvate transaminase	(*14*)
Inhibits:	
Glycogen synthetase	(*21*)

A summary of the actions of glucagon in liver is presented in Fig. 4. The hormone in some fashion activates adenyl cyclase, promoting the synthesis of cyclic AMP; this second messenger apparently acts in a concerted fashion in the liver to make glucose available for energy intracellularly and extracellularly (hyperglycemia). This is accomplished by the actions of cyclic AMP (A) to activate phosphorylase, causing the breakdown of glycogen to glucose (glycogenolysis), (B) to inhibit glycogen synthetase, preventing the resynthesis of glycogen and permitting the accumulation of glucose, and (C) to activate triglyceride lipase and amino acid transaminases (and probably phosphoenol pyruvate carboxykinase), leading to the breakdown of lipids and amino acids, and their conversion to glucose (gluconeogenesis). Physiologic concentrations of glucagon suffice for these dramatic effects on glycogenolysis and gluconeogenesis (84). Some other gross effects of glucagon (cf. Table 5) are also consistent

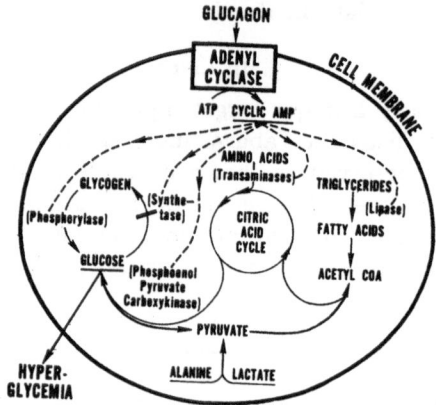

Fig. 4. Metabolic scheme of major actions of glucagon in liver

with this concept; namely, the uptake of amino acids by the liver, their disappearance from the blood, and their conversion to urea which is excreted in the urine.

4. In Adipose Tissue

Glucagon has a marked lipolytic action in adipose tissue and in isolated fat cells (39, 90). Again the hormone appears to act via cyclic AMP, which activates an adipose tissue lipase, resulting in a release of free fatty acids and glycerol (103); ketone bodies (4) have been observed in some laboratories. Insulin appears to act in an opposite manner, presumably by lowering the concentration of cyclic AMP (50).

5. In Heart

FARAH and TUTTLE (31) observed in 1960 that glucagon and epinephrine exhibit similar inotropic effects in isolated heart muscle. The implications of the report went essentially unnoticed for several years, until at this writing the inotropic action of glucagon is under intensive investigation. The mechanism of action of glucagon is not clear.

References, pp. 446—452

Although both the catecholamines and glucagon exert similar glyco-genolytic effects via cyclic AMP in the perfused heart (48), the action of the catecholamines, but not glucagon, is inhibited by β-adrenergic blocking agents (34, 49, 54, 63).

Dichloroisoproteronol blocks the inotropic action but not the stimula-tion of adenyl cyclase provided by glucagon (64). Thus, glucagon stimu-lates heart adenyl cyclase by a different mechanism than the catechola-mines (63); perhaps separate allosteric sites exist on the cyclase, or the enzyme is not the receptor. Additional studies are needed to clarify the role of cyclic AMP in the inotropic action of glucagon.

6. In the Gastrointestinal System

Glucagon is now established as a general depressant of gut activity, inhibiting gut motility and meal-, gastrin-, or histamine-stimulated gastric secretion (42, 52, 66). The mechanism of these dramatic effects is un-known; cyclic AMP has thus far not been implicated. Suggestions have been made that the depressing effects may be mediated by the central nervous system (73) or by hyperglycemia. Glucagon does appear to suppress appetite in normal humans (72). However, the inhibitory effects of glucagon do not require intact vagal innervation (59), and do not correlate well with the transient hyperglycemia produced by the hormone. The role of "enteroglucagon", if any, in the gastrointestinal system is also not known.

7. On Plasma Electrolytes and Renal Function

The diuretic effect of glucagon is well established; an increased ex-cretion of sodium, calcium, potassium, phosphate, bicarbonate, chloride, iodine, uric acid, urea, and water are observed following glucagon administration (19, 28, 29). The hormone probably acts by increasing the glomerular filtration rate and the renal plasma flow (29, 81). The molecular mechanism of action is not known; however, cyclic AMP may mediate the permeability of nephrons (40).

Serum concentrations of phosphate, potassium, and calcium are mildly depressed following glucagon administration (18, 27, 65, 70, 111). The hypokalemia, which follows a very brief hyperkalemia, is probably attributable both to electrolyte excretion and release of insulin. Glucagon apparently initiates an increased turnover of phosphorus (20). The hypocalcemic effect may be attributable both to excretion and to replace-ment of calcium in bone (109). Paloyan and his coworkers (71) have suggested that some of the symptoms of chronic pancreatitis (hyper-thyroidism, hypocalcemia, and hypophosphatemia) may result from an over-production of glucagon.

VII. Physiologic Role

A simple statement of the physiologic role of glucagon is not possible, partly because no common syndromes of glucagon excess or deficiency have been observed or produced. A possible exception is a transient form of diabetes produced in experimental animals through prolonged treatment with huge doses of glucagon. Whether or not this condition bears any resemblance to a physiologic function of glucagon is open to question.

The existence of enteroglucagon is another obstacle to defining a clear physiologic role for the pancreatic hormone; clarification of the relationship of enteroglucagon to pancreatic glucagon is sorely needed. No obvious reasons may be put forward for the apparently high concentration of enteroglucagon in serum. At the present time the only physiologic function tentatively ascribable to enteroglucagon is the release of insulin following an oral glucose load. Still another reason that the role of glucagon defies simple explanation is that the metabolic effects of this hormone are intimately interrelated with those of insulin, growth hormone, the catecholamines, and cyclic AMP. Such considerations are beyond the scope of this brief review, but the reader may gain additional insight from the discussion of Foa (32).

Pancreatic glucagon plays an important role in the mechanisms for maintaining "normal" concentrations of blood glucose and possibly of free fatty acids and amino acids. However, the overall thrust of these hormonal actions involving glycogenolysis, lipolysis, and gluconeogenesis is to help distribute energy (glucose, fatty acids) to the heart, peripheral tissues, and the central nervous system, particularly in the fasting state. Other related functions of glucagon include the release of insulin and the regulation of protein catabolism. In addition, the hormone probably helps in the maintenance of normal cardiac function, in the regulation of the urinary excretion of electrolytes, in the control of renal and hepatic blood flow, and in the regulation of appetite, gastric secretion, and gut motility.

Many of these physiologic functions of glucagon are mediated by cyclic AMP, formed from ATP in response to the hormonal activation of adenyl cyclase. The mechanisms of action of glucagon are well defined in comparison to many of the protein and polypeptide hormones, but the fact remains that almost nothing is known of the fundamental molecular interactions between hormone and receptor.

References

1. Aguilar-Parada, E., A. M. Eisentraut and R. H. Unger: Effects of Starvation on Plasma Pancreatic Glucagon in Normal Man. Diabetes 18, 717 (1969).
2. Assan, R., G. Rosselin, J. Drouet, J. Dolais and G. Tchobroutsky: Glucagon Antibodies. Lancet 11, 590 (1965).

3. BERSON, S. A., R. S. YALOW and B. W. VOLK: In Vivo and In Vitro Metabolism of Insulin-I[131] and Glucagon-I[131] in Normal and Cortisone-Treated Rabbits. J. Lab. Clin. Med. **49**, 331 (1957).

4. BEWSHER, P. D. and J. ASHMORE: Ketogenic and Lipolytic Effects of Glucagon on Liver. Biochem. Biophys. Res. Commun. **24**, 431 (1966).

5. BITENSKY, M. W., V. RUSSELL and W. ROBERTSON: Evidence for Separate Epinephrine and Glucagon Responsive Adenyl Cyclase Systems in Rat Liver. Biochem. Biophys. Res. Commun. **31**, 706 (1968).

6. BLANCHARD, M. H. and M. V. KING: Evidence of Association of Glucagon from Optical Rotatory Dispersion and Concentration-Difference Spectra. Biochem. Biophys. Res. Commun. **25**, 298 (1966).

7. BROMER, W. W.: Glucagon. In: R. I. DORFMAN (Ed.), Methods of Hormone Res., Vol. 2 A. New York: Academic Press. 1969.

8. — Unpublished data (1970).

9. BROMER, W. W. and R. E. CHANCE: Zinc Glucagon Depression of Blood Amino Acids in Rabbits. Diabetes **18**, 748 (1969).

10. BROMER, W. W. and B. J. REMUS: The Action of Carboxypeptidase A on Glucagon. VI Int. Congr. Biochem. Abstracts **II**, 142, New York (1964).

11. BROMER, W. W., L. G. SINN and O. K. BEHRENS: The Amino Acid Sequence of Glucagon. V. Location of Amide Groups, Acid Degradation Studies, and Summary of Sequential Evidence. J. Am. Chem. Soc. **79**, 2807 (1957).

12. BROMER, W. W., A. STAUB, E. R. DILLER, H. L. BIRD, Jr., L. G. SINN and O. K. BEHRENS: The Amino Acid Sequence of Glucagon. I. Amino Acid Composition and Terminal Amino Acid Analysis. J. Am. Chem. Soc. **79**, 2794 (1957).

13. BROMER, W. W. and W. STORVICK: Unpublished data (1970).

14. BROWN, C. B. and M. CIVEN: Control of Rat Liver Aromatic Amino Acid Transaminases by Glucagon and Insulin. Endocrinology **84**, 381 (1969).

15. BUCHANAN, K. D., J. E. VANCE, K. DINSTL and R. H. WILLIAMS: Effect of Blood Glucose on Glucagon Secretion in Anesthetized Dogs. Diabetes **18**, 11 (1969).

16. BUCHANAN, K. D., J. E. VANCE, A. MORGAN and R. H. WILLIAMS: Effect of Pancreozymin on Insulin and Glucagon Levels in Blood and Bile. Am. J. Physiol. **215** (6), 1293 (1968).

17. CHESNEY, T. M. and J. G. SCHOFIELD: Studies on the Secretion of Pancreatic Glucagon. Diabetes **18**, 627 (1969).

18. CRAWFORD, A. L., M. J. HENDERSON, R. D. HAWKINS and R. E. HAIST: The Effect of Glucagon on Blood Sugar and Inorganic Phosphorus Levels in Normothermic and Hypothermic Rats. Physiol. Pharmacol. **43**, 601 (1965).

19. DALLE, X., J. TANGHE and W. GRIJSPEERDT: Effect of Glucagon on Renal Excretion of Electrolytes. Arch. Int. Pharmacodyn. **120**, 505 (1959).

20. DEVENANZI, F., C. D. ALTARES and J. FORERO: Organ Distribution of Radioactive Orthophosphate and Total Phosphorus in the Rat. Diabetes **13**, 609 (1965).

21. DEWULF, H. and H. G. HERS: The Role of Glucose, Glucagon and Glucocorticoids in the Regulation of Liver Glycogen Synthesis. Europ. J. Biochem. **6**, 558 (1968).

22. EDELHOCH, H. and R. L. LIPPOLDT: Studies on Polypeptide Hormones. I. Fluorescence. J. Biol. Chem. **244**, 3876 (1969).

23. EISENSTEIN, A. B. and I. STRACK: Effect of Glucagon on Carbohydrate Synthesis and Enzyme Activity in Rat Liver. Endocrinology **83**, 1337 (1968).

24. EISENTRAUT, A. M., A. OHNEDA, E. PARADA and R. H. UNGER: Immunologic Discrimination between Pancreatic Glucagon and Enteric Glucagon-like Immunoreactivity in Tissues and Plasma. Diabetes **17** (Suppl. 1), 321 (1968).

25. Eisentraut, A. M. and R. H. Unger: Effect Upon Insulin Secretion of Physiologic Doses of Glucagon Administered Via the Portal Vein. Diabetes **16**, 283 (1967).

26. Eisentraut, A. M., N. Whissen and R. H. Unger: Incubation Damage in the Radioimmunoassay for Human Plasma Glucagon and Its Prevention with "Trasylol". Am. J. Med. Sci. **255**, 137 (1968).

27. Ellis, S. and S. B. Beckett: Mechanism of the Potassium Mobilizing Action of Epinephrine and Glucagon. J. Pharmacol. Exp. Therapy **142**, 318 (1963).

28. Elrick, H., E. R. Huffman, C. J. Hlad, N. Whipple and A. Staub: Effects of Glucagon on Renal Function in Man. J. Clin. Endocrinol. **18**, 813 (1958).

29. Elrick, H., N. Whipple, Y. Arai and C. J. Hlad: Further Studies on the Renal Action of Glucagon. J. Clin. Endocrinol. **19**, 1274 (1959).

30. Exton, J. H. and C. R. Park: The Role of Cyclic AMP in the Control of Liver Metabolism. Adv. in Enzyme Regulation **6**, 391 (1968).

31. Farah, A. and R. Tuttle: Studies on the Pharmacology of Glucagon. J. Pharmacol. Exptl. Therap. **129**, 49 (1960).

32. Foa, P. P.: Glucagon. Ergebnisse der Physiol. **60**, 147 (1968).

33. Garcia, A., J. R. Williamson and G. F. Cahill, Jr.: Studies on the Perfused Rat Liver. II. Effect of Glucagon on Gluconeogenesis. Diabetes **15**, 188 (1966).

34. Glick, G., W. W. Parmley, A. S. Wechsler and E. H. Sonnenblick: Glucagon: Its Enhancement of Cardiac Performance in the Cat and Dog and Persistence of its Inotropic Action Despite Beta-Receptor Blockade with Propranolol. Circulation Res. **22**, 789 (1968).

35. Gratzer, W. B., E. Bailey and G. H. Beaven: Conformational States of Glucagon. Biochem. Biophys. Res. Commun. **28**, 914 (1967).

36. Gratzer, W. B., G. H. Beaven, H. W. E. Rattle and E. M. Bradbury: A Conformational Study of Glucagon. Europ. J. Biochem. **3**, 276 (1968).

37. Grodsky, G. M. and L. L. Bennett: Time Sequence in the Release of Insulin. The Effect of Glucose, Glucagon, and Potassium. Proc. 26th Meeting Amer. Diabetes Assoc., p. 18, Chicago, 1966.

38. Grodsky, G. M., T. Hayashida, C. T. Peng and I. I. Geschwind: Production of Glucagon Antibodies and Their Role in Metabolism and Immunoassay of Glucagon. Proc. Soc. Exptl. Biol. Med. **107**, 491 (1961).

39. Hagen, J. H.: Effects of Glucagon on the Metabolism of Adipose Tissue. J. Biol. Chem. **236**, 1023 (1960).

40. Handler, J. S. and J. Orloff: Cysteine Effect on Toad Bladder Response to Vasopressin, Cyclic AMP, and Theophylline. Am. J. Physiol. **206**, 505 (1964).

41. Hazzard, W. R., P. M. Crockford, K. D. Buchanan, J. E. Vance, R. Chen and R. H. Williams: A Double Antibody Immunoassay for Glucagon. Diabetes **17**, 179 (1968).

42. Heimburg, R. L. and G. A. Hallenbeck: Inhibition of Gastric Secretion in Dogs by Glucagon Given Intraportally. Gastroenterology **47**, 531 (1964).

43. Herbert, V., K.-S. Lau, C. W. Gottlieb and S. J. Bleicher: Coated Charcoal Immunoassay of Insulin. J. Clin. Endocrinol. Metab. **25**, 1375 (1965).

44. Kakiuchi, S. and H. H. Tomizawa: Properties of a Glucagon-Degrading Enzyme of Beef Liver. J. Biol. Chem. **239**, 2160 (1964).

45. Kay, C. M. and M. M. Marsh: Some Optical Properties of Fetuin and Glucagon. Biochem. Biophys. Acta **33**, 251 (1959).

46. Kimball, C. P. and J. R. Murlin: Aqueous Extracts of Pancreas. III. Some Precipitation Reactions of Insulin. J. Biol. Chem. **58**, 337 (1923).

47. King, M. V.: A Low-Resolution Structural Model for Cubic Glucagon Based on Packing of Cylinders. J. Mol. Biol. **11**, 549 (1965).

48. KREISBERG, R. A. and J. R. WILLIAMSON: Metabolic Effects of Glucagon in the Perfused Rat Heart. Am. J. Physiol. 207, 721 (1964).

49. LaRAIA, P. J., R. J. CRAIG and W. J. REDDY: Glucagon: Effect on Adenosine 3',5'-Monophosphate in the Rat Heart. Am. J. Physiol. 215, 968 (1968).

50. LEFEBVRE, P. J. and A. S. LUYCKX: Effect of Insulin on Glucagon Enhanced Lipolysis In Vitro. Diabetologica 5, 195 (1969).

51. LIGHT, A. and M. V. SIMPSON: Studies on the Biosynthesis of Insulin. I. The Paper Chromatographic Isolation of ^{14}C-Labeled Insulin from Calf Pancreas Slices. Biochem. Biophys. Acta 20, 251 (1956).

52. LIN, T. M., D. N. BENSLAY and R. H. TUST: Further Study of the Effect of Glucagon on Meal-, Histamine-, Gastrin-, and Sham Feeding-Induced Gastric HCl Secretion. Physiologist 6, 225 (1963).

53. LOCHNER, J. DeV., A. C. ESTERHUIZEN and R. H. UNGER: Separation of Human Insulin, Glucagon and Other Pancreatic Proteins. Diabetes 13, 387 (1964).

54. LUCCHESI, B. R.: Cardiac Actions of Glucagon. Circulation Research 22, 777 (1968).

55. LUNDQUIST, G., S. E. BROLIN, R. H. UNGER and A. M. EISENTRAUT: The Cellular Origin of Pancreatic Glucagon. Personal communication.

56. MAKMAN, M. H. and E. W. SUTHERLAND: Use of Liver Adenyl Cyclase for Assay of Glucagon in Human Gastrointestinal Tract and Pancreas. Endocrinology 75, 127 (1964).

57. MALAISSE, W. J., F. MALAISSE-LAGAE, D. A. MAYHEW, P. H. WRIGHT and J. ASHMORE: Hormonal Regulation of Insulin Secretion. Clin. Res. 15, 325 (1967).

58. MARINETTI, G. V., T. K. RAY and V. TOMASI: Glucagon and Epinephrine Stimulation of Adenyl Cyclase in Isolated Rat Liver Plasma Membranes. Biochem. Biophys. Res. Commun. 36, 185 (1969).

59. MAYO, H. W., Jr. and D. M. EMERSON: Endocrine Influences of the Pancreas on Gastric Secretion. I. The Effects of Glucagon and Alloxan on Heidenhain Pouch Secretion. Surgery 44, 91 (1958).

60. McDONALD, J. K., B. B. ZEITMAN, T. J. REILLY and S. ELLIS: New Observations on the Substrate Specificity of Cathepsin C (Dipeptidyl Aminopeptidase I). J. Biol. Chem. 244, 2693 (1969).

61. McLEAN, P. and F. NOVELLO: Influence of Pancreatic Hormones on Enzymes Concerned with Urea Synthesis in Rat Liver. Biochem. J. 94, 110 (1965).

62. MITCHELL, W. M. and P. W. FELTS: Correlation of Structure and Lipolytic Activity of Glucagon. Abstr. 158th Amer. Chem. Soc. Meeting, September 8, 1969, No. 122.

63. MURAD, F. and M. VAUGHAN: Effect of Glucagon on Rat Heart Adenyl Cyclase. Biochem. Pharmacol. 18, 1053 (1969).

64. NAMM, O. H. and S. E. MAYER: The Role of Cyclic AMP in Myocardial Contractility and the Phosphorylase Activity Pathway. Pharmacologist 10, 145 (1968).

65. NATELSON, S., J. B. PINCUS and G. RAUNAZZISI: Dynamic Control of Calcium, Phosphate, Citrate, and Glucose Levels in Blood Serum. Effect of ACTH, Adrenaline, Noradrenaline, Hydrocortisone, Parathormone, Insulin, and Glucagon. Clin. Chem. 9, 31 (1963).

66. NECHELES, H., J. SPORN and L. WALKER: Effect of Glucagon on Gastrointestinal Motility. Amer. J. Gastroent. 45, 34 (1966).

67. NOVAKA, K. and P. P. FOA: A Simplified Glucagon Immunoassay and Its Use in a Study of Incubated Pancreatic Islets. Proc. Soc. Exp. Biol. Med. 130, 330 (1969).

68. Ohneda, A., E. Aguilar-Parada, A. M. Eisentraut and R. H. Unger: Characterization of Response of Circulating Glucagon to Intraduodenal and Intravenous Administration of Amino Acids. J. Clin. Invest. **47**, 2305 (1968).

69. — — — — Control of Pancreatic Glucagon Secretion by Glucose. Diabetes **18**, 1 (1969).

70. Paloyan, E., D. Paloyan and P. V. Harper: Glucagon-Induced Hypocalcemia. Metabolism **16**, 35 (1967).

71. — — — The Role of Glucagon Hypersecretion in the Relationship of Pancreatitis and Hyperparathyroidism. Surgery **62**, 167 (1967).

72. Penick, S. B. and L. E. Hinkle, Jr.: Depression of Food Intake Induced in Healthy Subjects by Glucagon. New England J. Med. **264**, 893 (1961).

73. Penick, S. B. and G. P. Swith: The Effect of Glucagon on Food Intake and Body Weight in Man. J. Obesity **1**, 1 (1964).

74. Pohl, S. L., L. Birnbaumer and M. Rodbell: Glucagon-Sensitive Adenyl Cyclase in Plasma Membrane of Hepatic Parenchymal Cells. Science **164**, 566 (1969).

75. Probst, G. W. and R. W. Colwell: Glucagon Immunoassay by Tracer Displacement. Biochemistry **5**, 1209 (1966).

76. Rigopoulou, D., I. Valverde, J. Marco, G. Faloona and R. H. Unger: Large Glucagon Immunoreactivity (LGI) in Extracts of Pancreas. J. Biol. Chem. **245**, 496 (1970).

77. Rosselin, G., G. Tchobroutsky, R. Assan, J. Drouet, J. Dolais, M. Freychet and M. Derot: La Methode Radio-immunologique D'etude des Hormones Proteiques et de leurs Anti-corps chez L'homme. Ann. Endocrinol. (Paris) **26**, 449 (1965).

78. Samols, E., G. Marri and V. Marks: Promotion of Insulin Secretion by Glucagon. Lancet **11**, 415 (1965).

79. Samols, E., J. Tyler, G. Marri and V. Marks: Stimulation of Glucagon Secretion by Oral Glucose. Lancet **11**, 1257 (1965).

80. Samols, E., J. M. Tyler and P. Mialhe: Suppression of Pancreatic Glucagon Release by the Hypoglycaemic Suphonylureas. Lancet **1**, 174 (1969).

81. Senato, M. and D. P. Early: Effect of Glucagon on Renal Functions in the Dog. Proc. Soc. Exp. Biol. **102**, 701 (1959).

82. Shima, K. and P. P. Foa: A Double Antibody Assay for Glucagon. Clin. Chim. Acta **22**, 511 (1968).

83. Smith, F. A. and E. B. Robbins: Glucagon. In: U. S. Pharmacopeia XVII, p. 268.

84. Sokal, J. E.: Effect of Glucagon on Gluconeogenesis by the Isolated Perfused Rat Liver. Endocrinology **78**, 538 (1966).

85. Spiegel, A. M. and M. W. Bitensky: Effects of Chemical and Enzymatic Modifications of Glucagon on Its Activation by Adenyl Cyclase. Endocrinology **85**, 638 (1969).

86. Srere, P. A. and G. C. Brooks: The Circular Dichroism of Glucagon Solutions. Arch. Biochem. Biophys. **129**, 708 (1969).

87. Staub, A. and O. K. Behrens: The Glucagon Content of Crystalline Insulin Preparations. J. Clin. Invest. **33**, 1629 (1954).

88. Staub, A., L. Sinn and O. K. Behrens: Purification and Crystallization of Glucagon. J. Biol. Chem. **214**, 619 (1955).

89. — — — Purification and Crystallization of Hyperglycemicglycogenolytic Factor (HGF). Science **117**, 628 (1953).

90. STEINBERG, D., E. SHAFRIR and M. VAUGHAN: Direct Effect of Glucagon on Release of Unesterified Fatty Acids (UFA) from Adipose Tissue. Clin. Res. 7, 250 (1959).

91. SUSSMAN, K. E. and G. D. VAUGHAN: Insulin Release after ACTH, Glucagon, and Cyclic in AMP the Perfused Isolated Rat Pancreas. Diabetes 16, 449 (1967).

92. SUTHERLAND, E. W. and C. DE DUVE: Origin and Distribution of the Hyper-glycemic-Glycogenolytic Factor of the Pancreas. J. Biol. Chem. 175, 663 (1948).

93. SUTHERLAND, E. W. and G. A. ROBISON: The Role of Cyclic-3′,5′-AMP in Response to Catecholamines and Other Hormones. Pharmacol. Rev. 18, 145 (1966).

94. SWANN, J. C. and G. G. HAMMES: Self-Association of Glucagon. Equilibrium Studies. Biochemistry 8, 1 (1969).

95. TARDING, F., P. NIELSEN, B. KEISNER-NIELSEN and AA. V. NIELSEN: Biological Assay of Glucagon in Rabbits. Diabetologica 5, 146 (1969).

96. TAYLOR, K. W., S. L. HOWELL, W. MONTAGUE and J. C. EDWARDS: Immuno-assay of Insulin and Glucagon. Clin. Chim. Acta 22, 71 (1968).

97. UI, M., B. KOBAYOSHI and Y. ITO: A New Assay Method for Hyperglycemic-glycogenolytic Factor. Endocrinol. Japon. 3, 191 (1956).

98. UNGER, R. H. and A. M. EISENTRAUT: Glucagon. In: C. H. GRAY and A. L. BACHARACH (Edits.), Hormones in Blood. New York: Academic Press. 1967.

99. UNGER, R. H., A. M. EISENTRAUT and L. L. MADISON: The Effects of Total Starvation Upon the Levels of Circulating Glucagon and Insulin in Man. J. Clin. Invest. 42, 1031 (1963).

100. UNGER, R. H., M. EISENTRAUT, M. S. McCALL and L. L. MADISON: Glucagon Antibodies and an Immunoassay for Glucagon. J. Clin. Invest. 40, 1280 (1961).

101. UNGER, R. H., H. KETTERER, J. DUPRÉ and A. M. EISENTRAUT: The Effects of Secretin, Pancreozymin, and Gastrin on Insulin and Glucagon Secretion in Anesthetized Dogs. J. Clin. Invest. 46, 630 (1967).

102. UNGER, R. H., A. OHNEDA, I. VALVERDO, A. M. EISENTRAUT and J. EXTON: Characterization of the Responses of Circulating Glucagon-like Immuno-reactivity to Intravenous Administration of Glucose. J. Clin. Invest. 47, 48 (1968).

103. VAUGHAN, M., J. E. BERGER and D. STEINBERG: Hormone-Sensitive Lipase and Monoglyceride Lipase Activities in Adipose Tissue. J. Biol. Chem. 239, 401 (1964).

104. VUYLSTEKE, C. A. and C. DE DUVE: The Assay of Glucagon on Isolated Liver Slices. Arch. Int. Pharmacodyn. III, 437 (1957).

105. WEINGES, K. F.: Glucagon. Monograph in Biochemie und Klinik series. Stutt-gart: G. Thieme. 1968.

106. — Influence of a Prolonged Action Glucagon on Blood Sugars, Inorganic Serum Phosphate, and the Total Amino Acids in Serum. Arch. Exp. Path. Pharmk. 237, 22 (1959).

107. WEINGES, K. F., E. WÜNSCH, G. BIRO, H. KETTL and M. MITZUNO: The Im-munological Reactivity and Biological Activity of Synthetic Glucagon. Dia-betologica 5, 97 (1969).

108. WIESEL, L. L., V. POSITANO, Y. KOLOGLU and G. E. ANDERSON: Chromato-graphic Separation of Glucagon and Insulin from Serum by Resin-Impregnated Paper. Proc. Soc. Exp. Biol. Med. 112, 515 (1963).

109. WILLIAMS, G. A., E. N. BOWSER and W. J. HENDERSON: Mode of Hypocalcemic Action of Glucagon in the Rat. Endocrinology 85, 537 (1969).

110. Williamson, J. R.: Mechanism for the Stimulation In Vivo Hepatic Gluco-neogenesis by Glucagon. Biochem. J. **101**, 11c (1966).

111. Wolfson, S. K., Jr. and S. Ellis: Effects of Glucagon on Plasma Potassium. Proc. Soc. Exp. Biol. Med. **91**, 226 (1956).

112. Wünsch, E.: Die Totalsynthese des Pankreas-Hormons Glucagon. Z. Natur-forschg. **22b**, 1269 (1967).

113. Wünsch, E., E. Jaeger and R. Scharf: Zur Synthese des Glucagons. XIX. Reindarstellung des Synthetischen Glucagons. Chem. Ber. **101**, 3664 (1968).

114. Wünsch, E. and G. Wendlberger: Zur Synthese des Glucagons. XVIII. Dar-stellung der Gesamtsequenz. Chem. Ber. **101**, 3659 (1968).

115. Ziegler, M. and H. G. Lippmann: Quantitative Electrophoretische Trennung von Insulin und Glukagon. Naturwissenschaften **55**, 181 (1968).

(Received, February 16, 1970)

Namenverzeichnis. Author Index

Kursiv gedruckte Seitenzahlen beziehen sich auf Literaturverzeichnisse

Page numbers printed in *italics* refer to References

TRUCCO, R. E. 412, 413, 414, *427, 428.*
TRUSCHEIT, E. 205, 209, 229, 235, *251, 254.*
TSCHESCHE, R. 123, 160, 183, 184, 185, 186, 192, 193, 194, 195, 196, 197, 200, *202, 203.*
TSCHAMPEL, D. *426.*
TSCHIERSCH, B. 85, 96, 100, *108.*
TSUCHYA, T. 270, *308.*
TSUDA, K. 286, *309,* 346, *410.*
TSUGA, H. 32, 56, 67.
TSUNEDA, K. 285, 186, 295, *309.*
TSUYUKI, T. 207, *254.*
TSUZUKI, K. 32, 56, 67.
TUCK, B. 5, *71.*
TUMLINSON, J. H. 240, *254.*
TUNIN, D. 394, *400.*
TUPPY, H. 409, *428.*
TURKINGSTON, R. W. 405, 420, *428.*
TURNER, A. F. 378, 384, *399, 400.*
TUST, R. H. 445, *449.*
TUTTLE, R. 444, *448.*
TYBRING, L. 347, 377, *399.*
TYLER, J. M. 431, 440, 443, *450.*
TYLER, V. E. 129, *160.*

UCHIYAMA, M. 19, 32, *72.*
UBAEV, K. 116, 117, *146, 160.*
UEBEL, E. C. 246, 249, *250.*
UEDA, S. 141, *152.*
UI, M. 431, *451.*
UMEZAWA, H. 126, 129, 130, 137, *153, 156.*
UNDERHILL, E. W. 99, 102, *106, 108.*
UNDERWOOD, W. G. E. 5, 14, 25, 26, 28, 31, 55, 42, *68, 70.*
UNGER, R. H. 430, 431, 432, 438, 439, 440, 441, 443, 446, *447, 449, 450, 451.*
UNRAU, A. M. 135, *152.*
URIBE, E. G. 96, *108.*
UYEO, S. *153.*

VALVERDO, I. 438, 439, 440, *451.*
VANAMAN, T. C. 405, 420, *423, 428.*
VANCE, J. E. 440, 447, *448.*
VAN DER BURG, W. J. 299, *306.*
VANG, K. S. 132, 134, *150.*
VANGEDAL, S. 347, 348, *399.*
VAN HEYNINGEN, E. M. 344, 359, 367, 368, 369, 370, 371, 394, 395, *402, 403.*
VANOFSKY, C. 128, *148.*

VARADY, J. 5, 17, 66, *72.*
VASLEFF, R. T. 367, 376, 391, 394, *403.*
VAUGHAN, M. 444, 445, *449, 451.*
VAUGHAN, P. F. T. *72.*
VAZQUES, D. 126, *160.*
VEDEJS, E. 316, *340.*
VELGOVA, H. 290, *311.*
VENKATARAMAN, K. 2, *72.*
VENKATARAMANI, B. 19, *63.*
VERBIT, L. 30, 34, *72.*
VIHERVAARA, K. 129, 131, *159.*
VINING, L. C. 129, 135, 136, *149, 150.*
VISCHER, E. 357, 374, *399.*
VISWANATHAN, N. 119, *151.*
VLATTAS, I. 316, *340.*
VOLK, B. W. 441, *447.*
VOLLPRECHT, P. 129, 131, *154.*
VON DAEHNE, W. 348, *399, 400.*
VONKEMAN, H. 318, 329, 330, *341.*
VORBRUEGGEN, H. 328, 342, 368, 370, 377, 387, *403.*
VUYLSTEKE, C. Y. 431, *451.*

WAHLBERG, K. 115, 117, 141, *160.*
WAISS, A. C. 61, *72.*
WAKABAYASHI, N. 218, *254,* 348, *401.*
WALDVOGEL, G. 248, *251, 253,* 280, 281, *283, 305, 306, 308, 312.*
WALKER, B. H. 113, *148.*
WALKER, J. 126, 145, *148, 151.*
WALKER, L. 445, *449.*
WALL, E. N. 247, *255.*
WALL, R. A. 407, *425.*
WALLENFELS, K. 405, 408, *428.*
WALLER, G. R. 129, 132, 134, 140, 141, *147, 150, 151, 152, 157, 160.*
WAN, A. S. C. 120, *160.*
WANZLICK, H. W. 45, *72.*
WARBURTON, W. K. 2, *72.*
WAREMBOURG, H. 410, *427.*
WARNHOFF, E. W. 166, 172, 183, 185, 190, 192, 193, 194, 195, 196, 197, 200, *203.*
WASADA, N. 279, *308.*
WASTL, H. 175, 193, *203.*
WATANABE, T. 204, 205, 207, *251.*
WATER, R. M. 204, 216, 217, *254.*
WATKINS, W. M. 408, 420, *428.*
WEBB, J. I. 408, *424.*
WEBBER, J. A. 367, 368, 370, 376, 391, 394, *400, 403.*
WEBER, F. *160.*

Sachverzeichnis. Subject Index